Texts in Computer Science

Series Editors

Orit Hazzan ⓘ, Faculty of Education in Technology and Science, Technion—Israel Institute of Technology, Haifa, Israel

Frank Maurer, Department of Computer Science, University of Calgary, Calgary, Canada

Titles in this series now included in the Thomson Reuters Book Citation Index!

'Texts in Computer Science' (TCS) delivers high-quality instructional content for undergraduates and graduates in all areas of computing and information science, including core theoretical/foundational as well as advanced applied topics. TCS books should be reasonably self-contained and aim to provide students with modern and clear accounts of topics ranging across the computing curriculum. As a result, the books are ideal for semester courses or for individual self-study in cases where people need to expand their knowledge. All texts are authored by established experts in their fields, reviewed internally and by the series editors, and provide numerous examples, problems, and other pedagogical tools; many contain fully worked solutions.

The TCS series is comprised of high-quality, self-contained books that have broad and comprehensive coverage and are generally in hardback format and sometimes contain color. For undergraduate textbooks that are likely to be more brief and modular in their approach, Springer offers the flexibly designed *Undergraduate Topics in Computer Science* series, to which we refer potential authors.

K. Erciyes

Guide to Graph Algorithms

Sequential, Parallel and Distributed

Second Edition

 Springer

K. Erciyes
Computer Engineering Department
Yaşar University
Izmir, Türkiye

ISSN 1868-0941   ISSN 1868-095X (electronic)
Texts in Computer Science
ISBN 978-3-032-05293-3   ISBN 978-3-032-05294-0 (eBook)
https://doi.org/10.1007/978-3-032-05294-0

To the memories of Semra, Seyhun, Şebnem and Hakan

Preface to the Second Edition

It has been about seven years since the publication of this book during which we have witnessed breakthrough developments in computation and algorithms, and their implementations in diverse fields. Two fundamental achievements have been in Artificial Intelligence (AI) and Quantum Computation. Machine Intelligence is at the core of AI and research and development in this field is commonly accompanied by Deep Learning which is implemented by multi-layer neural networks in general. Multi-layer neural networks and their implementations existed for decades but their widely usage in areas like natural language processing, image processing and finance was made possible by the technological developments in computer technology.

Quantum computation on the other hand, has brought a new and interesting view which is radically different than the classical computation although the physical realization of quantum computers is still at an early stage with promising results emerging. The integration of these two fields, namely AI and quantum computation was inevitable and we are now witnessing this collaboration as Quantum Neural Networks as one research field where classical neural network functions such as forward and backward propagation are carried out by quantum computer circuits resulting in much better performances than the classical neural networks.

Graphs are widely used to model real-life data such as computer networks, social networks and biological networks as we have elaborated in the first edition of this book. In the last decades, we have witnessed relatively recent implementations of advanced graph data structures such as knowledge graphs and uncertain graphs which are effectively used in various applications such as natural language processing and social network analysis.

Again, the commonsense expectation is the integration of the usage of classical and advanced graph data structures and AI and this is what exactly happened in a relatively recent time frame. Graph Neural Networks (GNNs) can input graph data, perform some additional neural network functions in this data during the training phase and provide outputs. The GNNs can input knowledge graphs and uncertain graphs in various implementations including natural language processing. A Quantum Graph Neural Network (QGNN) is an integration of a GNN and quantum computing which can be used for computationally hard tasks such as drug discovery and complex network analysis. With the foregoing in mind, the following new chapters and updates have been included in the second edition of the book:

- **Chapter 15: Advanced Graph Structures and Algorithms**: This chapter introduces the knowledge graph and uncertain graph structures and their algorithms. It also includes graph mining methods that are used to discover repeating subgraphs in large graphs.
- **Chapter 16: Graph Machine Intelligence**: A short introduction to machine intelligence and neural networks is provided with description of GNNs; a brief introduction to quantum computation and quantum algorithms is provided, QGNNs and their implementations are outlined.
- Additional text and Python code are provided in Chaps. 6, 12, 13 and 17.

Izmir, Türkiye   K. Erciyes

Preface to the First Edition

Graphs are key data structures for the analysis of various types of networks such as mobile ad hoc networks, the Internet, and complex networks such as social networks and biological networks. Study of graph algorithms is needed to efficiently solve a variety of problems in such networks. This study is commonly centered around three fundamental paradigms: Sequential graph algorithms; parallel graph algorithms and distributed graph algorithms. Sequential algorithms in general assume a single flow of control and such methods are well established. For intractable graph problems that do not have sequential solutions in polynomial time, approximation algorithms which have proven approximation ratios to the optimum solutions can be used. Many times however, the approximation algorithms are not known to date and the only choice is the use of heuristics which are common sense rules that are shown experimentally to work for a wide range of inputs. The algorithm designer is frequently confronted with this task of knowing what to search or not; and what road to follow if the solution does not exist. The first aim of this book is to provide a comprehensive and in-depth analysis of sequential graph algorithms and guide the designer on how to approach a typical hard problem by showing how to inspect an appropriate heuristic which is commonly needed in many cases.

Parallel algorithms are needed to provide speed-up in the running of graph algorithms. Shared memory parallel algorithms synchronize using a common memory and distributed memory parallel algorithms communicate by message-passing only. Distributed graph (or network) algorithms are aware of network topology and can be used for various network related tasks such as routing. *Distributed algorithms* is the common term used for distributed memory and distributed graph algorithms, however, we will call shared memory and distributed memory *parallel graph algorithms* parallel graph algorithms and distributed graph or network algorithms as *distributed algorithms*. Design and analysis of parallel and distributed algorithms as well as sequential algorithms for graphs will be the subject of this book.

A second and a fundamental goal of this book is to unify these three seemingly different methods of graph algorithms where applicable. For example, the minimum spanning tree (MST) problem can be solved by four classical sequential algorithms: Boruvka's, Prim's, Kruskal's, and Reverse-Delete algorithms all with similar complexities. A parallel MST algorithm will attempt to find the MST of a large network on a fewer number of processors solely to obtain a speedup. In a distributed MST algorithm, each processor is a node of the network graph and participates to find the MST of the network. We will describe and compare all three paradigms for this and many other well-known graph problems by looking at the same problem from three different angles, which we believe will help to understand the problem better and form a unifying view.

A third and an important goal of this work will be the conversions between sequential, shared, and distributed memory parallel and distributed algorithms for graphs. This process is not commonly implemented in literature although there are opportunities in many cases. We will exemplify this concept by maximal weighted matching algorithm in graphs. The sequential approximation algorithm for this purpose with 0.5 ratio has a time complexity of $O(mlog(m))$ with m being the number of edges. Preis provided a faster localized sequential algorithm based on the first algorithm with better $O(m)$ complexity. Later, Hoepmann provided a distributed version of Preis' algorithm. More recently, Manne provided sequential form of Hoepman's distributed graph algorithm and parallelized this algorithm. The sequence of methods employed has been sequential $\rightarrow$ sequential $\rightarrow$ distributed $\rightarrow$ sequential $\rightarrow$ parallel for this graph problem. Although this example shows a rather long transformation sequence, sequential $\rightarrow$ parallel and sequential $\rightarrow$ distributed are commonly followed by researchers mostly by common sense. Parallel graph algorithms $\leftrightarrow$ distributed graph algorithms conversion of algorithms is very seldom practiced. Our aim will be to lay down the foundations of these transformations between paradigms to convert an algorithm in one domain to another. This may be difficult for some types of algorithms but graph algorithms are a good premise.

As more advanced technologies are developed, we are confronted with the analysis of big data of complex networks which have tens of thousands of nodes and hundreds of thousands of edges. We also provide a part on algorithms for big data analysis of complex networks such as the Internet, social networks, and biological networks in the cell. To summarize, we have the following goals in this book:

- A comprehensive study and a detailed study of fundamental principles of sequential graph algorithms and approaches for NP-hard problems, approximation algorithms and heuristics.
- A comparative analysis of sequential, parallel and distributed graph algorithms including algorithms for big data.
- Study of conversion principles between the three methods.

There are three parts in the book; we provide a brief background on graphs, sequential, parallel, and distributed graph algorithms in the first part. The second part forms the core of the book with a detailed analysis of sequential, parallel, and distributed algorithms for fundamental graph problems. In the last part, our focus is on algebraic and dynamic graph algorithms and graph algorithms for very large networks, which are commonly implemented using heuristics rather than exact solutions.

We review theory as much as needed for the design of sequential, parallel, and distributed graph algorithms and our emphasis for many problems is on implementation details in full. Our study of sequential graph algorithms throughout the book is comprehensive, however, we provide a comparative analysis of sequential algorithms only with the fundamental parallel and distributed graph algorithms. We kept the layout of each chapter as homogenous as possible by first describing the problem informally and then providing the basic theoretical background. We then describe fundamental algorithms by first describing the main idea of an algorithm; then giving its pseudocode; showing an example implementation and finally the analysis of its correctness and complexities. This algorithm template is repeated for all algorithms except the ones that have complex structures and phases in which case we describe the general idea and the operation of the algorithm.

The intended audience for this book is the senior/graduate students of computer science, electrical and electronic engineering, bioinformatics, and any researcher or a person with background in discrete mathematics, basic graph theory and algorithms. There is a Web page for the book to keep errata and other material at: http://ube.ege.edu.tr/~erciyes/GGA/.

I would like to thank senior/graduate students at Ege University, University of California Davis, California State University San Marcos, and senior/graduate students at Izmir University who have taken the distributed algorithms and complex networks courses, sometimes under slightly different names, for their valuable feedback when parts of the material covered in the book was presented during lectures. I would also like to thank Springer editors Wayne Wheeler and Simon Rees for their help and their faith in another book project I have proposed.

Izmir, Turkey   K. Erciyes
                                                                                          An Emeritus Professor
                                                                       International Computer Institute
                                                                            Ege (Aegean) University

Competing Interests The author has no competing interests to declare that are relevant to the content of this manuscript.

Contents

Introduction

1

1.1 Graphs

Graphs are discrete structures that are frequently used to model many real-world problems such as communication networks, social networks, and biological networks. A graph consists of *vertices* and *edges* connecting these vertices. A graph is shown as $G = (V, E)$ where V is the set of vertices and E is the set of edges it has. Figure 1.1 shows an example graph with vertices and edges between them with $V = \{a, b, c, d\}$ and $E = \{(a, b), (a, e), (a, d), (b, c), (b, d), (b, e), (c, d), (d, e)\}$, (a, b) denoting the edge between vertices a and b for example.

Graphs have numerous applications including computer science, scientific computing, chemistry, and sociology since they are simple yet effective to model real-life phenomenon. A vertex of a graph represents some entity such as a person in a social network or a protein in a biological network. An edge in such a network corresponds to a social interaction such as friendship in a social network or a biochemical interaction between two proteins in the cell.

Study of graphs has both theoretical and practical implications. In this chapter, we describe the main goal of the book which is to provide a unified view of graph algorithms in terms of sequential, parallel, and distributed graph algorithms with emphasis on sequential graph algorithms. We describe a simple graph problem from these three views and then review challenges in graph algorithms by finally outlining the contents of the book.

1.2 Graph Algorithms

An *algorithm* consists of a sequence of instructions processed by a computer to solve a problem. A *graph algorithm* works on graphs to find a solution to a problem represented by the graphs. We can classify the graph problems as sequential,

K. Erciyes, *Guide to Graph Algorithms*, Texts in Computer Science,
https://doi.org/10.1007/978-3-032-05294-0_1

Fig. 1.1 An example graph
consisting of vertices
$\{a, b, \ldots, h\}$

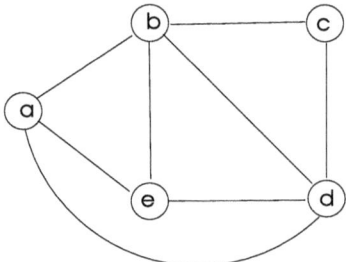

parallel, and distributed, based on the mode of running these algorithms on the computing environment.

1.2.1   Sequential Graph Algorithms

A *sequential algorithm* has a single flow of control and is executed sequentially. It accepts an input, works on the input, and provides an output. For example, reading two integers, adding them and printing the sum is a sequential algorithm that consists of three steps.

Let us assume the graph of Fig. 1.2 represents a small network with vertices labeled $\{v_1, v_2, \ldots, v_8\}$ and having integers 1, …, 13 associated with each edge. The integer value for each edge may denote the cost of sending a message over that edge. Commonly, the edge values are called the *weights* of edges. Let us assume our aim is to design a sequential algorithm to find the edge with the maximum value. This may have some practical usage, as we may need to find the highest cost edge in the network to avoid that link while some communication or transportation is done.

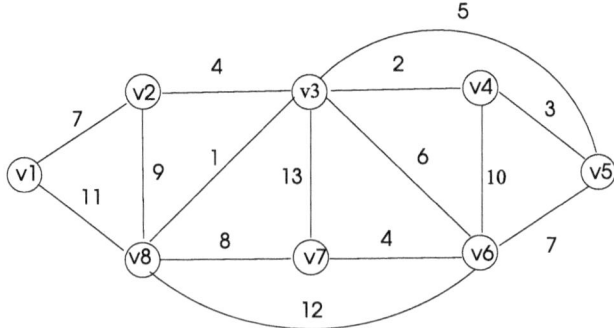

Fig. 1.2 An example graph

Let us form a distance matrix D for this graph, which has entries $d(i, j)$ showing the weights of edges between vertices v_i and v_j as below:

$$D = \begin{bmatrix} 0 & 7 & 0 & 0 & 0 & 0 & 0 & 11 \\ 7 & 0 & 4 & 0 & 0 & 0 & 0 & 9 \\ 0 & 4 & 0 & 2 & 5 & 6 & 13 & 1 \\ 0 & 0 & 2 & 0 & 3 & 10 & 0 & 0 \\ 0 & 0 & 5 & 3 & 0 & 7 & 0 & 0 \\ 0 & 0 & 6 & 10 & 7 & 0 & 4 & 12 \\ 0 & 0 & 13 & 0 & 0 & 4 & 0 & 8 \\ 11 & 9 & 1 & 0 & 0 & 12 & 8 & 0 \end{bmatrix}.$$

If two vertices v_i and v_j are not connected, we insert a zero for that entry in D to exclude that edge from computation. We can have a simple sequential algorithm that finds the largest value in each row of this matrix first and then the maximum value of all these largest values in the second step as shown in Algorithm 1.1.

Algorithm 1.1 *Sequential_graph*

1: **int** $D[n, n] \leftarrow$ edge weights
2: **int** $max[8], maximum$
3: **for** $i=1$ to 8 **do**
4:     $max[i] \leftarrow D[i, 1]$
5:     **for** $j=2$ to 8 **do**
6:         **if** $D[i, j] > max[i]$ **then**
7:             $max[i] \leftarrow D[i, j]$
8:         **end if**
9:     **end for**
10: **end for**
11: $maximum \leftarrow max[1]$
12: **for** $i=2$ to 8 **do**
13:     **if** $max[i] > maximum$ **then**
14:         $maximum \leftarrow max[i]$
15:     **end if**
16: **end for**
17: **output** $maximum$

This algorithm requires 7 comparisons for each row and then 7 for array, for a total of 63 comparisons. It needs n^2 comparisons in general for a graph with n vertices.

1.2.2   Parallel Graph Algorithms

Parallel graph algorithms aim for performance as all parallel algorithms. This way of speeding up programs is needed especially for very large graphs representing complex networks such as biological or social networks which consist of huge number of nodes and edges. We have a number of processors working in parallel on the same problem and the results are commonly gathered at a single processor for output. Parallel algorithms may synchronize and communicate through shared memory or they run as distributed memory algorithms communicating by the transfer of messages only. The latter mode of communication is a more common practice in parallel computing due to its versatility to realize in general network architectures.

We can attempt to parallelize an existing sequential algorithm or design a new parallel algorithm from scratch. A common approach in parallel computing is the partitioning of data to a number of processors so that each computing element works on a particular partition. Another fundamental approach is the partitioning of computation across the processors as we will investigate in Chap. 4. We will see some graph problems are difficult to partition into data or computation.

Let us reconsider the sequential algorithm in the previous section and attempt to parallelize it using data partitioning. Since graph data is represented by the distance matrix, the first thing to consider would be the partitioning of this matrix. Indeed, row-wise or column-wise partitioning of a matrix representing a graph is commonly used in parallel graph algorithms. Let us have a controlling processor we will call the *supervisor* or the *root* and two *worker* processors to do the actual work. This mode of operation, sometimes called *supervisor/worker model*, is also a common practice in the design of parallel algorithms. Processors are commonly called *processes* to mean the actual processor may also be doing some other work. We now have three processes p_0, p_1, and p_2, and p_0 is the supervisor. The process p_0 has the distance matrix initially, and it partitions and sends the first half of the rows from 1 to 4 to p_1 and 5 to 8 to p_2 as shown below:

$$D = \begin{bmatrix} 0 & 7 & 0 & 0 & 0 & 0 & 0 & 11 & p_1 \\ 7 & 0 & 4 & 0 & 0 & 0 & 0 & 9 & \\ 0 & 4 & 0 & 2 & 5 & 6 & 13 & 1 & \\ 0 & 0 & 2 & 0 & 3 & 10 & 0 & 0 & \\ 0 & 0 & 5 & 3 & 0 & 7 & 0 & 0 & p_2 \\ 0 & 0 & 6 & 10 & 7 & 0 & 4 & 12 & \\ 0 & 0 & 13 & 0 & 0 & 4 & 0 & 8 & \\ 11 & 9 & 1 & 0 & 0 & 12 & 8 & 0 & \end{bmatrix}.$$

Each worker now finds the heaviest edge incident to the vertices in the rows it is assigned using the sequential algorithm described and sends this result to the supervisor p_0 which finds the maximum of these two values and outputs it. A more general form of this algorithm with k worker processes is shown in Algorithm 1.2. Since data is partitioned to two processes now, we would expect to have a significant decrease in the runtime of the sequential algorithm. However, we have communication costs between the supervisor and the workers now which may not be trivial for large data transfers.

Algorithm 1.2 *Parallel_graph*

1: **int** $D[n, n]$ ← edge weights of graph G
2: **int** $Max[n]$, $E[n/k, n]maximum$
3: **if** $i = root$ **then**
4:     row-wise partition distance matrix D of graph into $D_1, ..., D_k$
5:     **for** i=1 to k **do**
6:         send D_i to p_i
7:     **end for**
8:     **for** i=1 to k **do**
9:         **receive** $largest_i$ from p_i into $max[i]$
10:     **end for**
11:     **find** the maximum value of max using the sequential algorithm
12:     **output** $maximum$
13: **else**
14:     **receive** my rows into E
15:     **find** the maximum value in E using sequential algorithm
16:     **send** $root$ my maximum value
17: **end if**

Designing parallel graph algorithms may not be trivial as in this example, in general, we need more sophisticated methods. The operation we need to do may depend largely on what was done before which means significant communications and synchronization may be needed between the workers. The inter-process communication across the network connecting the computational nodes may be costly and we may end up designing a parallel graph algorithm that is not efficient.

1.2.3 Distributed Graph Algorithms

Distributed graph algorithms are a class of graph algorithms in which we have a computational node represented by a vertex of the graph. The problems to be solved with such algorithms are related to the network they represent; for example, it may be required to find the shortest distance between any two nodes in the network so that whenever a data packet comes to a node in the network, it forwards the packet to one of its neighbors that is on the least cost path to the destination. In such algorithms, each node typically runs the same algorithm but has different neighbors to communicate and transfer its local result. In essence, our aim is to solve an overall problem related to the graph representing the network by the cooperation of the nodes in the network. Note that the nodes in the network can only communicate with their neighbors and this is the reason these algorithms are sometimes referred to as *local* algorithms.

In the distributed version of our sample maximum weight edge finding algorithm, we have computational nodes of a computer network as the vertices of the graph, and our aim is that each node in the network modeled by the graph should receive the largest weight edge of the graph in the end. We will attempt to solve this problem using *rounds* for the synchronization of the nodes. Each node starts the round, performs some function in the round, and does not start the next round until all other nodes have also finished execution of the round. This model is widely used for distributed algorithms as we will describe in Chap. 5 and there is no other central control other than the synchronization of the rounds. Each node starts by broadcasting the largest weight it is incident to all of its neighbors and receiving the largest weight values from neighbors. In the following rounds, a node broadcasts the largest weight it has seen so far and after a certain number of steps, the largest value will be propagated to all nodes of the graph as shown in Algorithm 1.3. The number of steps is the *diameter* of the graph which is the maximum number of edges between any two vertices.

Algorithm 1.3 *Distributed_graph*

1: **boolean** $finished, round\_over \leftarrow false$
2: **message type** $start, result, stop$
3: **while** $count \leq diam(G)$ **do**
4:     **receive** $max(j)$ from all neighbors
5:     **find** the maximum of all received values
6:     **send** the maximum value to all neighbors
7:     $count \leftarrow count + 1$
8: **end while**

We now can see fundamental differences between parallel and distributed graph algorithms using this example as follows.

- Parallel graph algorithms are needed mainly for the speedup they provide. There are a number of processing elements that work in parallel which cooperate to finish an overall task. The main relation between the number of processes and the size of the graph is that we would prefer to use more processes for large graphs. We assume each processing element can communicate with each other in general although there are some special parallel computing architectures such as processors forming a cubic architecture of communication as in the *hypercube*.
- In distributed graph algorithms, computational nodes are the vertices of the graph under consideration and communicate with their neighbors only to solve a problem related to the network represented by the graph. Note that the process number is the number of vertices of the graph for these algorithms.

One important goal of this book is to provide a unified view of graph algorithms from these three different angles. There are cases we may want to solve a network problem on parallel processing environment, for example, all shortest paths between any two nodes in the network may need to be stored in a central server to be transferred to individual nodes or for statistical purposes. In this case, we run a parallel algorithm for the network using a number of processing elements. In a network setting, we need each node to work to know the shortest paths from it to other nodes.

A general approach is to derive parallel and distributed graph algorithms from a sequential one but there are ways of converting a parallel graph algorithm to distributed one or vice versa for some problems. For the example problem we have, we can have each row of the distance matrix D assigned to a single process. This way, each process can be represented by a network node provided that it communicates with its neighbors only. Conversions as such are useful in many cases since we do not design a new algorithm from scratch.

1.2.4 Algorithms for Large Graphs

Recent technical advancements in the last few decades have resulted in the availability of data of very large networks. These networks are commonly called *complex networks* and consist of tens of thousands of nodes and hundreds of thousands of links between the nodes. One such type of networks is the biological networks within the cell of living organisms. A protein–protein interaction (PPI) network is a biological network formed with interacting proteins outside the nucleus in the cell.

A social network consisting of individuals interacting over the Internet may again be a very large network. These complex networks can be modeled by graphs with vertices representing the nodes and edges the interaction between the nodes like any other network. However, these networks are different than a small network modeled by a graph in few respects. First of all, they have very small diameters meaning the shortest distance between any two vertices is small when compared to their sizes. For example, various PPI networks consisting of thousands of nodes are found to have a diameter of only several units. Similarly, social networks and technological networks such as the Internet also have small diameters. This state is known as *small-world* property. Second, empirical studies suggest these networks have very few nodes with very high number of connections; and most of the other nodes have few connections to neighbors. This so-called *scale-free* property is exhibited again in most of the complex networks. Lastly, the size of these networks being large requires efficient algorithms for their analysis. In summary, we need efficient and possibly parallel algorithms that exploit various properties such as small-world and scale-free features of these networks.

1.3 Challenges in Graph Algorithms

There are numerous challenges in graphs to be solved by graph algorithms.

- *Complexity of graph algorithms*: A polynomial time algorithm has a complexity that can be expressed by a polynomial function. There are very few polynomial time algorithms for the majority of problems related to graphs. The algorithms at hand typically have exponential time complexities which means even for moderate size graphs, the execution times are significant. For example, assume an algorithm A to solve a graph problem P has time complexity 2^n, n being the number of vertices in the graph. We can see that A may have poor performance even for graphs with $n > 20$ vertices. We then have the following choices:

- *Approximation Algorithms*: Search for an approximation algorithm that finds a suboptimal solution rather than an optimal one. In this case, we need to prove that the approximation algorithm always provides a solution within an *approximation ratio* to the optimal solution. Various proof techniques can be employed and there is no need to experiment the approximation algorithm other than statistical purposes. Finding and proving approximation algorithms are difficult for many graph problems.
- *Randomized Algorithms*: These algorithms decide on the course of execution based on some random choice, for example, selection of an edge at random. The output is presented typically as *expected* or *with high probability* meaning there is a chance, if even slightly, that the output may not be correct. However, the randomized algorithms provide polynomial solutions to many difficult graph problems.
- *Heuristics*: In many cases, our only choice is the use of common sense approaches called *heuristics* in search of a solution. Choice of a heuristic is commonly pursued by intuition and we need to experiment the algorithm with the heuristic for a wide range of inputs to show it works *experimentally*.

There are other methods such as *backtracking* and *branch-and-bound* which work only for a subset of the search space and therefore have less time complexities. However, these approaches can be applied to only a subset of problems and are not general. Let us exemplify these concepts by case analysis. A *clique* in a graph is a subgraph such that each vertex in this subgraph has connections to all other vertices in the subgraph as shown in Fig. 1.3. Finding cliques in a graph has many implications as these exhibit dense regions of activity. Finding the largest clique of a graph G with n vertices, which is the clique with the maximum number of vertices in the graph, cannot be performed in polynomial time. A *brute force algorithm*, which is typically the first algorithm that comes to mind, will enumerate all 2^n subgraphs of G and check the clique condition from the largest to the smallest. Instead of searching for an approximation algorithm, we could do the following by intuition: start with the vertex that has the highest number of connections called its degree; check whether all of its neighbors have the same number of connections and if all have, then we have a clique. If this fails, continue with the next highest degree vertex. This heuristic will work fine but in general, we need to show experimentally that a heuristic works for most of the input variations, for 90% for example but an algorithm that works fine for 60% of the time with diverse inputs would not be a favorable heuristic.

- *Performance*: Even with polynomial time graph algorithms, the size of the graph may restrict its use for large graphs. Recent interest in large graphs representing large real-life networks demands high-performance algorithms which are commonly realized by parallel computing. Biological networks and social networks

Fig. 1.3 The maximum clique of a sample graph is shown by dark vertices. All vertices in this clique are connected to each other

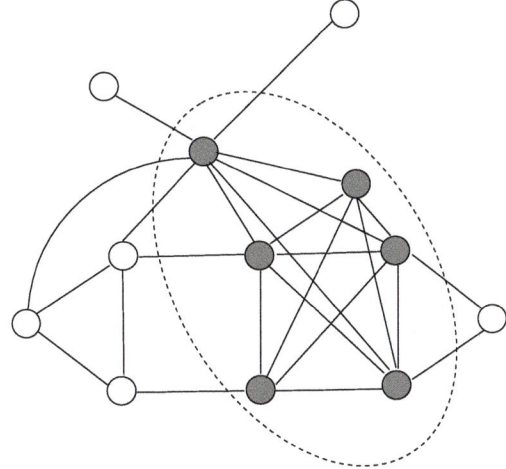

are examples of such networks. Therefore, there is a need for efficient parallel algorithms to be implemented in these large graphs. However, some graph problems are difficult to parallelize due to the structure of the procedures used.

- *Distribution*: Several large real networks are distributed in a sense that each node of the network is an autonomous computing element. The Internet, the Web, mobile ad hoc networks, and wireless sensor networks are examples of such networks which can be termed as computer networks in general. These networks can again be modeled conveniently by graphs. However, the nodes of the network now actively participate in the execution of the graph algorithm. This type of algorithms is termed *distributed algorithms*.

The main goal of this book is the study of graph algorithms from three angles: sequential, parallel, and distributed algorithms. We think this approach will provide a better understanding of the problem at hand and its solution by also showing its possible application areas. We will be as comprehensive as possible in the study of sequential graph algorithms but will only present representative graph algorithms for parallel and distributed cases. We will see some graph problems have complicated parallel algorithmic solutions reported in research studies and we will provide a contemporary research survey of the topics in these cases.

1.4   Outline of the Book

We have divided the book into three parts as follows.

- *Fundamentals*: This part has four chapters; the first chapter contains a dense review of basic graph theory concepts. Some of these concepts are detailed in individual chapters. We then describe sequential, parallel, and distributed graph algorithms in sequence in three chapters. In each chapter, we first provide the main concepts about the algorithm method and then provide a number of examples on graphs using the method mentioned. For example, in the sequential algorithm methods, we give a greedy graph algorithm while describing greedy algorithms. This part basically forms the background for parts II and III.

- *Basic Graph Algorithms*: This part contains the core material of the book. We look at the main topics in graph theory at each chapter which are trees and graph traversals; weighted graphs; connectivity; matching; subgraphs; and coloring. Here, we leave out some theoretical topics of graph theory which do not have significant algorithms. The topics we investigate in the book allow algorithmic methods conveniently and we start each chapter with brief theoretical background for algorithmic analysis. In other words, our treatment of related graph theoretical concepts is not comprehensive as our main goal is the study of graph algorithms rather than graph theory on its own. In each chapter, we first describe sequential algorithms and this part is one place in the book that we try to be as comprehensive as possible by describing most of the well-established algorithms of the topic. We then provide only sample parallel and distributed algorithms on the topic investigated. These are typically one or two well-known algorithms rather than a comprehensive list. In some cases, the parallel or distributed algorithms at hand are complicated. For such problems, we give a survey of algorithms with short descriptions.

- *Advanced Topics*: We present recent and more advanced topics in graph algorithms than Part II in this section of the book starting with algebraic and dynamic graph algorithms. Algebraic graph algorithms commonly make use of the matrices associated with a graph and operations on them while solving a graph problem. Dynamic graphs represent real networks where edges are inserted and removed from a graph in time. Algorithms for such graphs, called *dynamic graph algorithms*, aim to provide solutions in shorter time than running the static algorithm from scratch.

  Large graphs representing real-life networks such as biological and social networks tend to have interesting and unexpected properties as we have outlined. Study of such graphs has become a major research direction in network science recently. We therefore considered it to be appropriate to have two chapters of the book dedicated for this purpose. Algorithms for these large graphs have somehow different goals, and community detection which is finding dense regions in these graphs has become one of the main topics of research. We first provide a chapter on general description and analysis of these large graphs along with algorithms to compute some important parameters. We then review basic complex network types with algorithms used to solve fundamental problems in these networks. The final chapter is about describing general guidelines on how to search a graph algorithm for the problem at hand.

We conclude this chapter by emphasizing the main goals of the book once more. First, it would be proper to state what this book is not. This book is not intended as a graph theory book, or a parallel computing book or a distributed algorithms book on graphs. We assume basic familiarity with these areas although we provide a brief and dense review of these topics as related to graph problems in Part I. We describe basic graph theory including the notation and basic theorems related to the topic at the beginning of each chapter. Our emphasis is again on graph theory that is related to the graph algorithm we intend to review. We try to be as comprehensive as possible in the analysis of sequential graph algorithms but we review only exemplary parallel and distributed graph algorithms. Our main focus is guiding the reader to graph algorithms by investigating and studying the same problem from three different views: a thorough sequential, typical parallel, and distributed algorithmic approaches. Such an approach is effective and beneficial not only because it helps to understand the problem at hand better but also it is possible to convert from one approach to another saving significant amount of time compared to designing a completely new algorithm.

Part I
Fundamentals

Introduction to Graphs

2

2.1 Introduction

Objects and connections between them occur in a variety of applications such as roadways, computer networks, and electrical circuits. Graphs are used to model such applications with vertices of a graph representing the objects or nodes and the edges showing the connections between the nodes.

We review the basic graph theoretical concepts in a rather dense form in this chapter. This review includes notations used, basic definitions, vertex degrees, subgraphs, graph isomorphism, graph operations, directed graphs, distance, graph representations and matrices related to graphs. We leave discussion of more advanced properties of graphs such as matching, connectivity, special subgraphs, and coloring to Part II when we review sequential, parallel, and distributed algorithms for these problems. We also delay review of methods and parameters for the analysis of large graphs to Part III. These large graphs are used to model complex networks such as the Internet or biological networks, which consist of a huge number of vertices and edges. We will see there is a need for new parameters and analysis methods for the investigation of these large graphs.

2.2 Notations and Definitions

A graph is a set of points and a set of lines in a plane or a 3-D space. A graph can be formally defined as follows.

Definition 2.1 (*graph*) A graph $G = (V, E, g)$ or $G = (V(G), E(G), g)$ is a discrete structure consisting of a vertex set V and an edge set E and a relation g that associates each edge with two vertices of the set V.

© The Author(s), under exclusive license to Springer Nature Switzerland AG 2026   15
K. Erciyes, *Guide to Graph Algorithms*, Texts in Computer Science,
https://doi.org/10.1007/978-3-032-05294-0_2

The vertex set consists of vertices also called *nodes*; and an edge in the edge set is incident between two vertices called its *endpoints*. The vertex set of a graph G is shown as $V(G)$ and the edge set as $E(G)$. We will use V for $V(G)$ and E for $E(G)$ when the graph under consideration is known. A *trivial graph* has one vertex and no edges. A *null graph* has an empty vertex set and an empty edge set. A graph is called *finite* if both $V(G)$ and $E(G)$ are finite. We will consider only simple and finite graphs in this book, unless stated otherwise. The number of vertices of a graph G is called its *order* and we will use the literal n for this parameter. The number of edges of G is called its *size* and we will show this parameter by the literal m. An edge of a graph G between its vertices u and v is commonly shown as (u, v), uv or sometimes $\{u, v\}$; we will adopt the first one. The vertices at the ends of an edge are called its *endpoints* or *end vertices* or simply *ends*. For an edge (u, v) between vertices u and v, we say u and v are incident to the edge (u, v).

Definition 2.2 (*self-loop, multiple edge*) A *self-loop* is an edge with the same endpoints. *Multiple edges* have the same pair of endpoints.

An edge that is not a self-loop is called a *proper edge*. A *simple graph* does not have any self-loops or multiple edges. A graph containing multiple edges is called a *multigraph*. An *underlying graph* of a multigraph is obtained by substituting a single edge for each multiple edge. An example multigraph is depicted in Fig. 2.1.

Definition 2.3 (*complement of a graph*) The complement of a graph $G(V, E)$ is the graph $\overline{G}(V, E')$ with the same vertex set as G and any edge $(u, v) \in E'$ if and only if $(u, v) \notin E$.

Informally, we have the same vertex set in the complement of a graph G but only have edges that do not exist in G. Complements of two graphs are shown in Fig. 2.2.

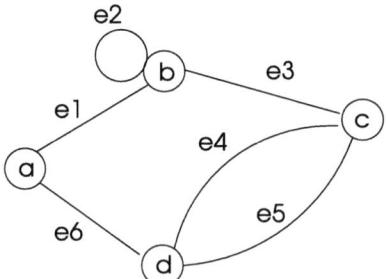

Fig. 2.1 A graph with $V(G) = \{a, b, c, d\}$ and $E(G) = \{e_1, e_2, e_3, e_4, e_5, e_6\}$. Edge e_2 is a self-loop and edges e_4 and e_5 are multiple edges

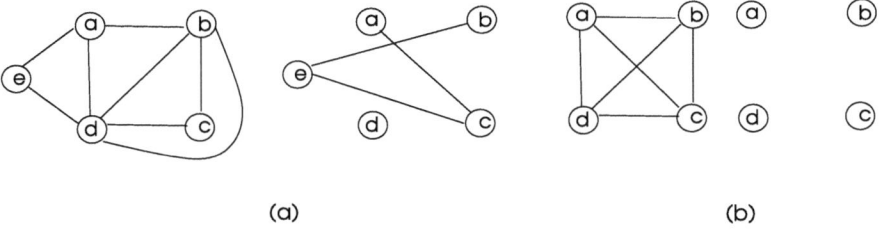

(a)   (b)

Fig. 2.2 **a** Complement of a sample graph, **b** complement of a completely connected graph in which every vertex is connected to all other vertices

2.2.1 Vertex Degrees

The degree of a vertex in a graph is a useful attribute of a vertex as defined below.

Definition 2.4 (*degree of a vertex*) The sum of the number of proper edges and twice the number of self-loops incident on a vertex v of a graph G is called its *degree* and is shown by $deg(v)$.

A vertex that has a degree of 0 is called an *isolated* vertex and a vertex of degree 1 is called a *pendant* vertex. The minimum degree of a graph G is denoted by $\delta(G)$ and the maximum degree by $\Delta(G)$. The following relation between the degree of a vertex v in G and these parameter holds:

$$0 \leq \delta(G) \leq deg(v) \leq \Delta(G) \leq n - 1. \tag{2.1}$$

Since the maximum number of edges in a simple undirected graph is $n(n-1)/2$, for any such graph,

$$0 \leq m \leq \frac{n(n-1)}{2} = \binom{n}{2}.$$

We can, therefore, conclude there are at most $2^{\binom{n}{2}}$ possible simple undirected graphs having n vertices. The *first theorem of graph theory*, which is commonly refered to as the *handshaking lemma* is as follows.

Theorem 2.1 (Euler) *The sum of the degrees of a simple undirected graph $G = (V, E)$ is twice the number of its edges shown below.*

$$\sum_{v \in V} deg(v) = 2m \tag{2.2}$$

Proof is trivial as each edge is counted twice to find the sum. A vertex in a graph with n vertices can have a maximum degree of $n - 1$. Hence, the sum of the degrees in a *complete graph* where every vertex is connected to all others is $n(n - 1)$. The total number of edges is $n(n - 1)/2$ in such a graph. In a meeting of n people, if everyone shook hands with each other, the total number of handshakes would be $n(n - 1)/2$ and hence the name of lemma. The average degree of a graph is

$$\frac{\sum_{v \in V} deg(v)}{n} = 2m/n. \tag{2.3}$$

A vertex is called *odd* or *even* depending on whether its degree is odd or even.

Corollary 2.1 *The number of odd-degree vertices of a graph is an even number.*

Proof The vertices of a graph $G = (V, E)$ may be divided into the even-degree (v_e) and odd-degree (v_o) vertices. The sum of degrees can then be stated as

$$\sum_{v \in V} deg(v) = \sum_{v_e \in V} deg(v_e) + \sum_{v_o \in V} deg(v_o)$$

Since the sum is even by Theorem 2.1, the sum of the odd-degree vertices should also be even which means there must be an even number of odd-degree vertices. $\square$

Theorem 2.2 *Every graph with at least two vertices has at least two vertices that have the same degree.*

Proof We will prove this theorem using contradiction. Suppose there is no such graph; for a graph with n vertices, this implies the vertex degrees are unique, from 0 to $n - 1$. We cannot have a vertex u with degree of $n - 1$ and a vertex v with 0 degree in the same graph G as former implies u is connected to all other vertices in G and therefore a contradiction. $\square$

This theorem can be put in practice in a gathering of people where some know each other and rest are not acquainted. If persons are represented by the vertices of a graph where an edge between two individuals who know each other is represented by an edge we can say there are at least two persons that have the same number of acquaintances in the meeting.

2.2.1.1 Degree Sequences

The *degree sequence* of a graph is obtained when the degrees of its vertices are listed in some order.

Definition 2.5 (*degree sequence*) The degree sequence of a graph *G* is the list of the degrees of its vertices in nondecreasing or nonincreasing, more commonly in nonincreasing order. The degree sequence of a digraph is the list consisting of its in-degree, out-degree pairs.

The degree sequence of the graph in Fig. 2.1 is {4, 3, 3, 2} for vertices *b*, *d*, *c*, *a* in sequence since a loop counts two. Given a degree sequence $D = (d_1, d_2, \ldots, d_n)$, which consists of a finite set of nonnegative integers, *D* is called *graphical* if it represents a degree sequence of some graph *G*. We may need to check whether a given degree sequence is graphical. The condition that $deg(v) < n - 1$, $\forall v \in V$ is the first condition and also $\sum_{v \in V} deg(v)$ should be even. However, these are necessary but not sufficient and an efficient method is proposed in the theorem first proved by Havel [7] and later by Hakimi using a more complicated method [5].

Theorem 2.3 (Havel–Hakimi) *Let D be a nonincreasing sequence $d_1, d_2, \ldots, d_n$ with $n \geq 2$. Let D' be the sequence derived from D by deleting d_1 and subtracting 1 from each of the first d_1 elements of the remaining sequence. Then D is graphical if and only if D' is graphical.*

This means if we come across a degree sequence which is graphical during this process, the initial degree sequence is graphical. Let us see the implementation of this theorem to a degree sequence by analyzing the graph of Fig. 2.3a.

The degree sequence for this graph is {4, 3, 3, 3, 2, 1}. We can now iterate as follows starting with the initial sequence. Deleting 4 and subtracting 1 from the first 4 of the remaining elements gives

$$\{2, 2, 2, 1, 1\}$$

continuing similarly, we obtain

$$\{1, 1, 0, 0\}.$$

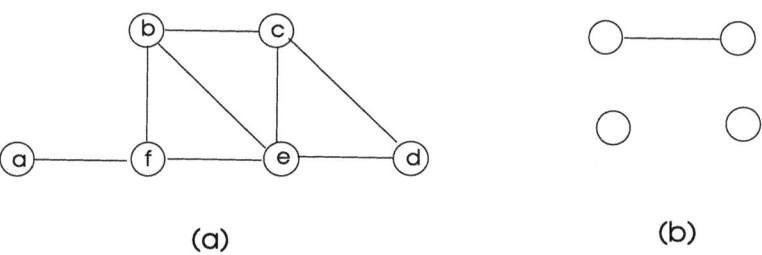

(a)   (b)

Fig. 2.3 a A sample graph to implement Havel–Hakimi theorem. **b** A graph representing graphical sequence {1, 1, 1, 1}. **c** A graph representing graphical sequence {0, 1, 1}

The last sequence is graphical since it can be realized as shown in Fig. 2.3b; thus, the initial degree sequence is graphical. This theorem can be conveniently implemented using a recursive algorithm due to its recursive structure.

2.2.2 Subgraphs

In many cases, we would be interested in part of a graph rather than the graph as a whole. A subgraph G' of a graph G has a subset of vertices of G and a subset of its edges. We may need to search for a subgraph of a graph that meets some condition, for example, our aim may be to find dense subgraphs which may indicate an increased relatedness or activity in that part of the network represented by the graph.

Definition 2.6 (*subgraph, induced subgraph*) $G' = (V', E')$ is a subgraph of $G = (V, E)$ if $V' \subseteq V$ and $E' \subseteq E$. A subgraph $G' = (V', E')$ of a graph $G = (V, E)$ is called an *induced subgraph* of G if E' contains all edges in G that have both ends in V'.

When $G' \neq G$, G' is called a *proper subgraph* of G; when G' is a subgraph of G, G is called a *supergraph* of G'. A *spanning subgraph* G' of G is its subgraph with $V(G) = V(G')$. Similarly, a *spanning supergraph* G of G' has the same vertex set as G'. A spanning subgraph and an induced subgraph of a graph are shown in Fig. 2.4.

Given a vertex v of a graph G, the subgraph of G shown by $G - v$ is formed by deleting the vertex v and all of its incident edges from G. The subgraph $G - e$ is obtained by deleting the edge e from G. The induced subgraph of G by the vertex set V' is shown by $G[V']$. The subgraph $G[V \setminus V']$ is denoted by $G - V'$.

Vertices in a *regular graph* all have the same degree. For a graph G, we can obtain a regular graph H which contains G as an induced subgraph. We simply duplicate G next to itself and join each corresponding pair of vertices by an edge if this vertex does not have a degree of $\Delta(G)$ as shown in Fig. 2.5. If the new graph G' is not $\Delta(G)$-regular, we continue this process by duplicating G' until the regular graph is obtained. This result is due to Konig who stated that for every graph of maximum degree r, there exists an r-regular graph that contains G as an induced subgraph.

2.2.3 Isomorphism

Definition 2.7 (*graph isomorphism*) An isomorphism from a graph G_1 to another graph G_2 is a bijection $f : V(G_1) \rightarrow V(G_2)$ in which any edge $(u, v) \in E(G_1)$ if and only if $f(u)f(v) \in E(G_2)$.

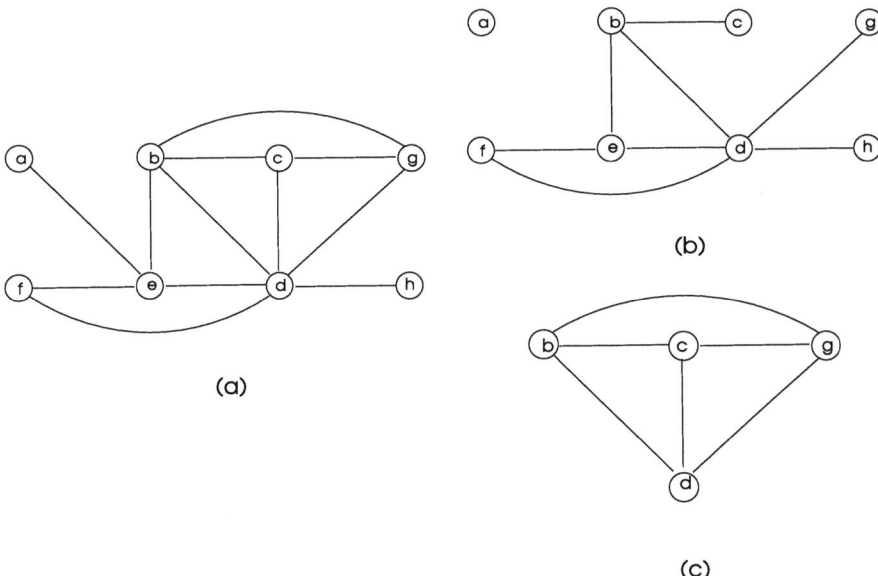

Fig. 2.4 **a** A sample graph *G*. **b** A spanning subgraph of *G*. **c** An induced subgraph of *G*

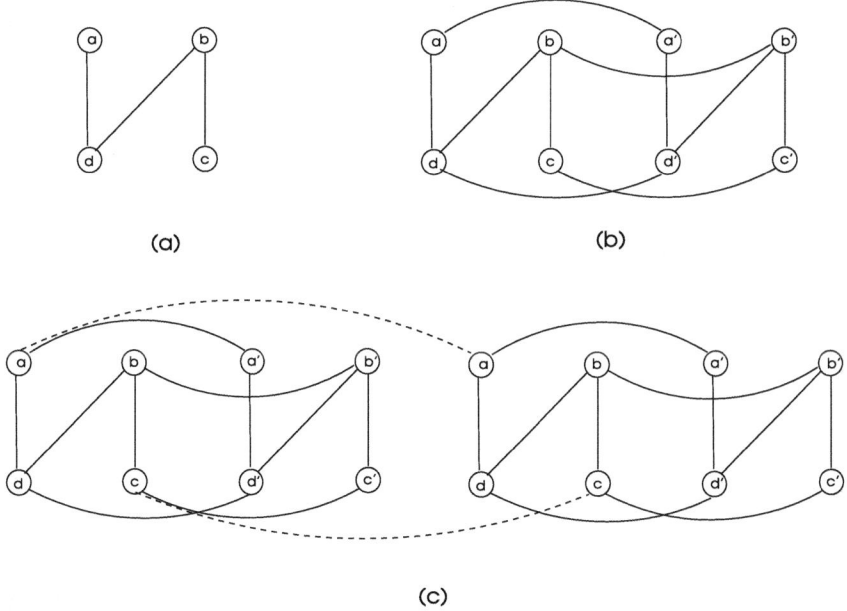

Fig. 2.5 Obtaining a regular graph. **a** The graph. **b** The first iteration. **c** The 3-regular graph obtained in the second iteration shown by dashed lines

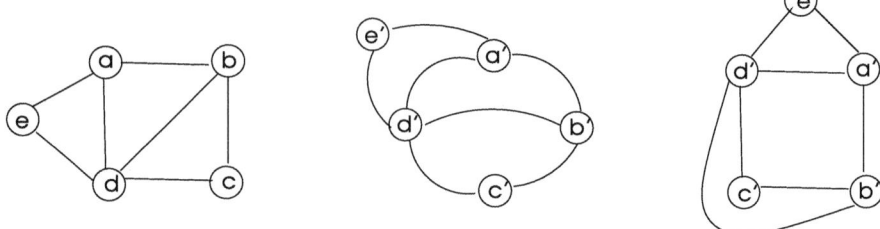

Fig. 2.6 Three isomorphic graphs. Vertex x is mapped to vertex x'

When this condition holds, G_1 is said to be *isomorphic* to G_2 or, G_1 and G_2 are isomorphic. Three isomorphic graphs are depicted in Fig. 2.6. Testing whether two graphs are isomorphic is a difficult problem and cannot be performed in polynomial time. An isomorphism of a graph to itself is called an *automorphism*. A *graph invariant* is a property of a graph that is equal in its isomorphic graphs. Given two isomorphic graphs G_1 and G_2, their orders and sizes are the same and their corresponding vertices have the same degrees. Thus, we can say that the number of vertices, the number of edges and the degree sequences are isomorphism invariants, that is, they do not change in isomorphic graphs.

2.3   Graph Operations

We may need to generate new graphs from a set of input graphs by using certain operations. These operations are uniting and finding intersection of two graphs and finding their product as described below.

2.3.1   Union and Intersection

Definition 2.8 (*union and intersection of two graphs*) The *union* of two graphs $G_1 = (V_1, E_1)$ and $G_2 = (V_2, E_2)$ is a graph $G_3 = (V_3, E_3)$ in which $V_3 = V_1 \cup V_2$ and $E_3 = E_1 \cup E_2$. This operation is shown as $G_3 = G_1 \cup G_2$. The *intersection* of two graphs $G_1 = (V_1, E_1)$ and $G_2 = (V_2, E_2)$ is a graph $G_3 = (V_3, E_3)$ in which $V_3 = V_1 \cap V_2$ and $E_3 = E_1 \cap E_2$. This is shown as $G_3 = G_1 \cap G_2$.

Figure 2.7 depicts these concepts.

Definition 2.9 (*join of two graphs*) The *join* of two graphs $G_1 = (V_1, E_1)$ and $G_2 = (V_2, E_2)$ is a graph $G_3 = (V_3, E_3)$ in which $V_3 = V_1 \cup V_2$ and $E_3 = E_1 \cup E_2 \cup \{(u, v) : u \in V_1 \text{ and } v \in V_2\}$. This operation is shown as $G_3 = G_1 \vee G_2$.

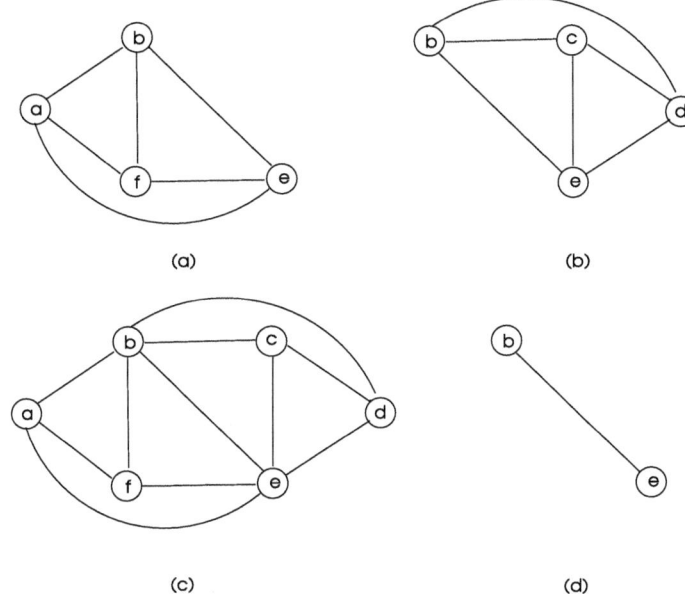

(a)   (b)

(c)   (d)

Fig. 2.7 Union and intersection of two graphs. The graph in **c** is the union of the graphs in **a** and **b** and the graph in **d** is their intersection

The join operation of two graphs creates new edges between each vertex pairs, one from each of the two graphs. Figure 2.8 displays the join of two graphs. All of the union, intersection, and join operations are commutative, that is, $G_1 \cup G_2 = G_2 \cup G_1$, $G_1 \cap G_2 = G_2 \cap G_1$, and $G_1 \vee G_2 = G_2 \vee G_1$.

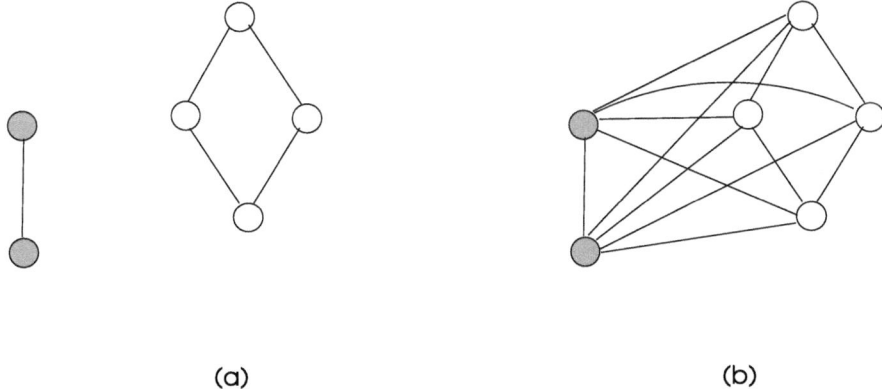

(a)   (b)

Fig. 2.8 Join of two graphs. **a** Two graphs. **b** Their join

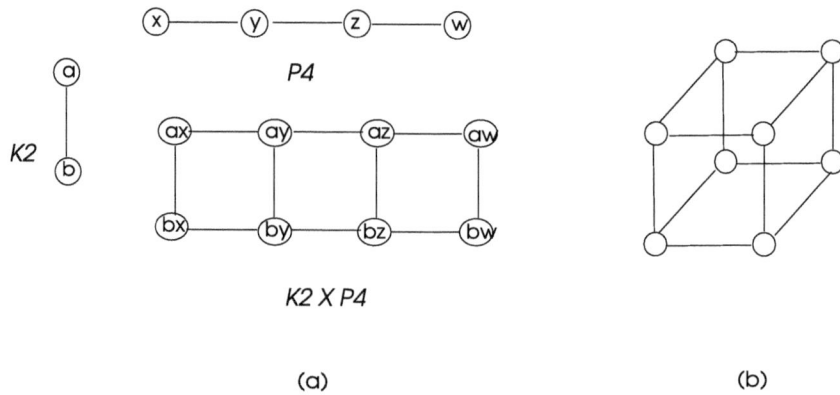

Fig. 2.9 **a** Graph product of K_2 and P_4. **b** Hypercube of dimension 3

2.3.2 Cartesian Product

Definition 2.10 (*Cartesian product*) The Cartesian product or simply the product of two graphs $G_1 = (V_1, E_1)$ and $G_2 = (V_2, E_2)$, shown by $G_1 \square G_2$ or $G_1 \times G_2$ is a graph $G_3 = (V_3, E_3)$ in which $V_3 = V_1 \times V_2$ and an edge $((u_i, v_j), (u_p, j_q))$ is in $G_1 \times G_2$ if one of the following conditions holds:

1. $i = p$ and $(v_j, v_q) \in E_2$
2. $j = q$ and $(u_i, u_p) \in E_1$.

   Informally, the vertices we have in the product are the Cartesian product of vertices of the graphs and hence each represents two vertices, one from each graph. Figure 2.9a displays the product of complete graph K_2 and the path graph with 4 vertices, P_4. Graph product is useful in various cases, for example, the hypercube of dimension n, Q_n, is a special graph that is the graph product of K_2 by itself n times. It can be described recursively as $Q_n = K_2 \times Q_{n-1}$. A hypercube of dimension 3 is depicted in Fig. 2.9b.

2.4   Types of Graphs

We review main types of graphs that have various applications in this section.

2.4.1 Complete Graphs

Definition 2.11 (*complete graph*) In a complete simple graph $G(V, E)$, each vertex $v \in V$ is connected to all other vertices in V.

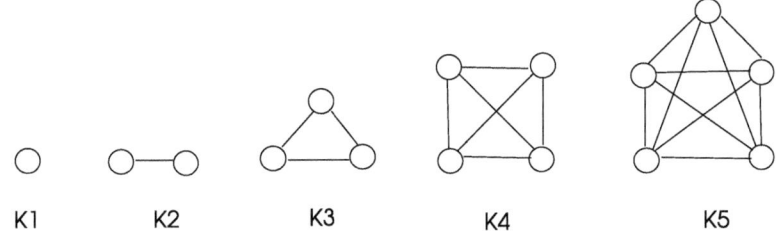

Fig. 2.10 Complete graphs of sizes 1–5

Searching for complete subgraphs of a graph G provides dense regions in G which may mean some important functionality in that region. A complete graph is denoted by K_n where n is the number of vertices. Figure 2.10 depicts $K_1, \ldots, K_5$. The complete graph with three vertices, K_3, is called a *triangle*.

The size of a simple undirected complete graph K_n is $n(n - 1)/2$ since the degree of every vertex in K_n is $n - 1$, there are n such vertices, and we need to divide by two as each edge is counted twice for both vertices in its endpoints.

2.4.2 Directed Graphs

A *directed edge* or an *arc* has an orientation from its head endpoint to its tail endpoint shown by an arrow. Directed graphs consist of directed edges.

Definition 2.12 (*directed graph*) A *directed graph* or a *digraph* consists of a set of vertices and a set of ordered pairs of vertices called *directed edges*. A partially directed graph has both directed and undirected edges.

If an edge $e = (u, v)$ is a directed edge in a directed graph G, we say e begins at u and ends at v, or u is the origin of e and v is its destination, or e is directed from u to v. The underlying graph of a directed or partially directed graph is obtained by removing the directions in all edges and replacing each multiple edge with a single edge. A directed graph is shown in Fig. 2.11. Unless stated otherwise, what we state for graphs will be valid for directed and undirected graphs. In a *complete simple digraph*; there is a pair of arcs, one in each direction between any two vertices.

Definition 2.13 (*in-degree, out-degree*) The in-degree of a vertex v in a digraph is the number of edges directed to v and the out-degree of v is the number of edges originating from it. The degree of v is the sum of its in-degree and its out-degree.

The sum of the in-degrees of the vertices in a digraph is equal to the sum of the out-degrees which are both equal to the sum of the number of edges. A directed graph that has no cycles is called a *directed acyclic graph* (DAG).

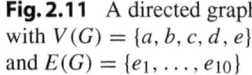

Fig. 2.11 A directed graph
with $V(G) = \{a, b, c, d, e\}$
and $E(G) = \{e_1, \ldots, e_{10}\}$

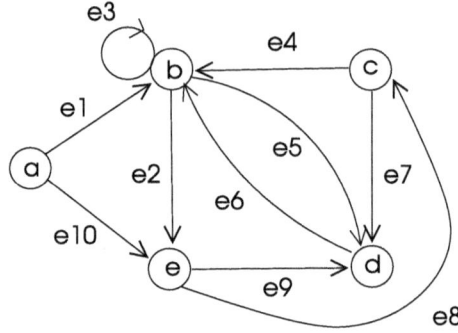

2.4.3   Weighted Graphs

We have considered unweighted graphs up to this point. Weighted graphs have edges and vertices labeled with real numbers representing weights.

Definition 2.14 (*edge-weighted, vertex-weighted graphs*) An edge-weighted graph $G(V, E, w)$, $w : E \rightarrow \mathbb{R}$ has weights consisting of real numbers associated with its edges. Similarly, a vertex-weighted graph $G(V, E, w)$, $w : V \rightarrow \mathbb{R}$ has weights of real numbers associated with its vertices.

Weighted graphs find many real applications, for example, weight of an edge (u, v) may represent the cost of moving from u to v as in a roadway or cost of sending a message between two routers u and v in a computer network. The weight of a vertex v may be associated with capacity stored at v which may be used to represent a property such as the storage volume of a router in a computer network.

2.4.4   Bipartite Graphs

Definition 2.15 (*bipartite graph*) A graph $G = (V, E)$ is called a bipartite graph if the vertex set V can be partitioned into two disjoint subsets V_1 and V_2 such that any edge of G connects a vertex in V_1 to a vertex in V_2. That is, $\forall (u, v) \in E$, $u \in V_1 \wedge v \in V_2$, or $u \in V_2 \wedge v \in V_1$.

In a *complete bipartite graph*, each vertex of V_1 is connected to each vertex of V_2 and such a graph is denoted by $K_{m,n}$, where m is the number of vertices in V_1 and n is the number of vertices in V_2. The complete bipartite graph $K_{m,n}$ has mn edges and $m + n$ vertices. A weighted complete bipartite graph is depicted in Fig. 2.12.

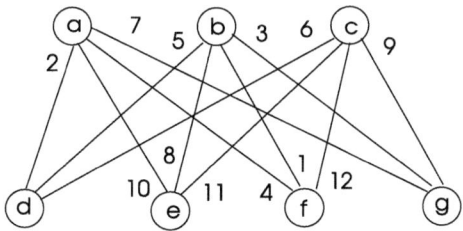

Fig. 2.12 A weighted complete bipartite graph $K_{3,4}$ with $V_1 = \{a, b, c\}$ and $V_2 = \{d, e, f, g\}$

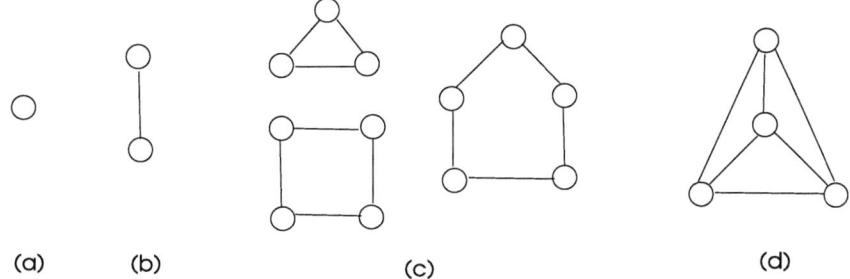

(a)   (b)   (c)   (d)

Fig. 2.13 **a** A 0-regular graph. **b** A 1-regular graph. **c** 2-regular graphs. **d** A 3-regular graph

2.4.5 Regular Graphs

In a *regular graph*, each vertex has the same degree. Each vertex of a k-regular graph has a degree of k. Every k-complete graph is a $k - 1$ regular graph but the latter does not imply the former. For example, a d-hypercube is a d-regular graph but it is not a d-complete graph. Examples of regular graphs are shown in Fig. 2.13. Any single n-cycle graph is a 2-regular graph. Any regular graph with odd-degree vertices must have an even number of such vertices to have an even number sum of vertices.

2.4.6 Line Graphs

In order to construct a line graph $L(G)$ of a simple graph G, each vertex of L representing an edge of G is formed. Then, two vertices u and v of L are connected if the edges represented by them are adjacent in G as shown in Fig. 2.14 where v_i represents e_i.

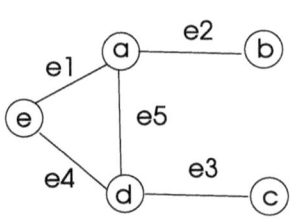
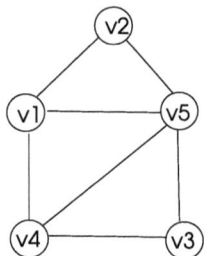

Fig. 2.14 a A simple graph G. **b** Line graph of G

2.5   Walks, Paths, and Cycles

We need few definitions to specify traversing the edges and vertices of a graph.

Definition 2.16 (*walk*) A *walk* W between two vertices v_0 and v_n of a graph G is an alternating sequence of $n + 1$ vertices and n edges shown as $W = (v_0, e_1, v_1, e_2, \ldots, v_{n-1}, e_n, v_n)$, where e_i is incident to vertices v_{i-1} and v_i. The vertex v_0 is called the *initial* vertex and v_n is called the *terminating* vertex of the walk W.

In a directed graph, a directed walk can be defined similarly. The length of a walk W is the number of edges (arcs in digraphs) included in it. A walk can have repeated edges and vertices. A walk is *closed* if it starts and ends at the same vertex and *open* otherwise. A walk is shown in Fig. 2.15.

A graph is *connected* if there is a walk between any pair of its vertices. Connectivity is an important concept that finds many applications in computer networks and we will review algorithms for connectivity in Chap. 8.

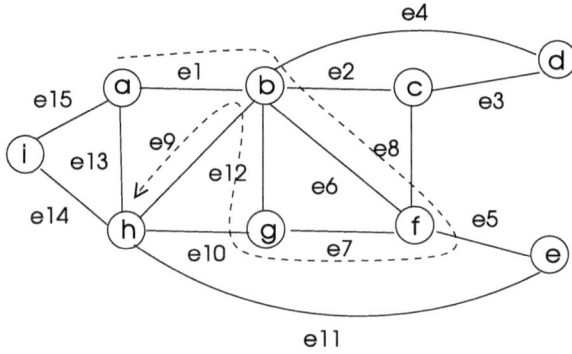

Fig. 2.15 A walk $(a, e_1, b, e_6, f, e_7, g, e_{12}, b, e_9, h)$ in a sample graph shown by a dashed curve

Table 2.1 Properties of simple graph traversals

	Repeated vertices	Repeated edges
Walks	Yes	Yes
Trails	Yes	No
Paths	No except initial and terminal vertex	No
Circuits	Yes, starts and ends at the same vertex	No
Cycles	No except initial and term	No

Definition 2.17 (*trail*) A *trail* is a walk that does not have any repeated edges.

Definition 2.18 (*path*) A *path* is a trail that does not have any repeated vertices with the exception of initial and terminal vertices.

Paths are shown by the vertices only. For example, (i, a, b, h, g) is a path in Fig. 2.15.

Definition 2.19 (*cycle*) A closed path which starts and ends at the same vertex is called a *cycle*.

The length of a cycle can be an odd integer in which case it is called an *odd cycle*. Otherwise, it is called an *even cycle*.

Definition 2.20 (*circuit*) A closed trail which starts and ends at the same vertex is called a *circuit*.

All of these concepts are summarized in Table 2.1.

A trail in Fig. 2.15 is $(h, e_9, b, e_6, f, e_5, e)$. When e_{11} and h are added to this trail, it becomes a cycle. An *Eulerian tour* is a closed Eulerian trail and an *Eulerian graph* is a graph that has an Eulerian tour. The number of edges contained in a cycle is denoted its *length l* and the cycle is shown as C_l. For example, C_3 is a triangle.

Definition 2.21 (*Hamiltonian cycle*) A cycle that includes all of the vertices in a graph is called a *Hamiltonian cycle* and such a graph is called *Hamiltonian*. A Hamiltonian path of a graph G passes through every vertex of G.

2.5.1   Connectivity and Distance

We would be interested to find if we can reach a vertex v from a vertex u. In a *connected graph G*, there is a path between every pair of vertices. Otherwise, G is *disconnected*. The connected subgraphs of a disconnected graph are called *components*. A connected graph itself is the only component it has. If the underlying graph of a digraph G is *connected*, G is connected. If there is a directed walk between each pair of vertices, G is *strongly connected*.

In a connected graph, it is of interest to find how easy it is to reach one vertex from another. The distance parameter defined below provides this information.

Definition 2.22 (*distance*) The distance $d(u, v)$ between two vertices u and v in a (directed) graph G is the length of the shortest path between them.

In an unweighted (directed) graph, $d(u, v)$ is the number of edges of the shortest path between them. In a weighted graph, this distance is the minimum sum of the weights of the edges of a path out of all possible paths between these vertices. The *shortest path* between two vertices u and v in a graph G is another term used instead of distance between the vertices u and v to have the same meaning. The shortest paths between vertices h and e are h, b, f, e and h, g, f, e both with a distance of 3 in Fig. 2.15. Similarly, shortest paths between vertices i and b are i, a, b and i, h, b both with a distance of 2. The *directed distance* from a vertex u to v in a digraph is the length of the shortest walk from u to v. In an undirected simple (weighted) graph $G(V, E, w)$, the following can be stated:

1. $d(u, v) = d(v, u)$
2. $d(u, v) \leq d(u, w) + d(w, v), \forall w \in V$

Definition 2.23 (*eccentricity*) The eccentricity of a vertex v in a connected graph G is its maximum distance to any vertex in G.

The maximum eccentricity is called the *diameter* and the minimum value of this parameter is called the *radius* of the graph. The vertex v of a graph G with minimum eccentricity in a connected graph G is called the *central vertex* of G. Finding central vertex of a graph has practical implications, for example, we may want to place a resource center at a central location in a geographical area where cities are represented by the vertices of a graph and the roads by its edges. There may be more than one central vertex.

2.6   Graph Representations

We need to represent graphs in suitable forms to be able to perform computations on them. Two widely used ways of representation are the *adjacency matrix* and the *adjacency list* methods.

2.6.1   Adjacency Matrix

An adjacency matrix of a simple graph or a digraph is a matrix $A[n, n]$ where an element $a_{ij} = 1$ if there is an edge joining vertex i to j and $a_{ij} = 0$ otherwise. For multigraphs, the entry a_{ij} equals the number of edges between the vertices i and j. For a digraph, a_{ij} shows the number of arcs from the vertex i to vertex j. The adjacency matrix is symmetric for an undirected graph and is asymmetric for a digraph in general. A digraph and its adjacency matrix are displayed in Fig. 2.16. An adjacency matrix requires $O(n^2)$ space.

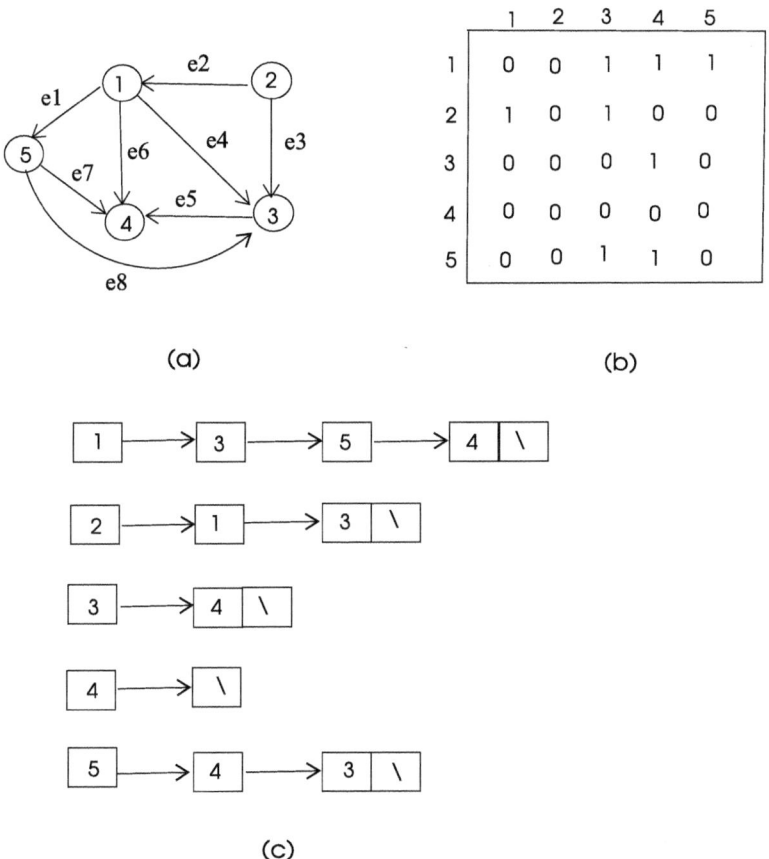

(a)   (b)

(c)

Fig. 2.16 a A digraph. **b** Its adjacency matrix. **c** Its adjacency list. The end of the list entries are marked by a backslash

Table 2.2 Predecessor lists and successor lists of vertices	v	P_v	S_v
	1	$\{2\}$	$\{3, 4, 5\}$
	2	$\{\emptyset\}$	$\{1, 3\}$
	3	$\{1, 2, 5\}$	$\{4\}$
	4	$\{1, 3, 5\}$	$\{\emptyset\}$
	5	$\{1\}$	$\{3, 4\}$

2.6.2 Adjacency List

An *adjacency list* of a simple graph (or a digraph) is an array of lists with each list representing a vertex and its (out) neighbors in a linked list. The end of the list is marked by a NULL pointer. The adjacency list of a graph is depicted in Fig. 2.16c. The adjacency matrix requires $O(n + m)$ space. For sparse graphs, adjacency list is preferred due to the space dependence on the number of vertices and edges. For dense graphs, adjacency matrix is commonly used as searching the existence of an edge in this matrix can be done in constant time. With the adjacency list of a graph, the time required for the same operation is $O(n)$.

In a digraph $G = (V, E)$, the *predecessor list* $P_v \subseteq V$ of a vertex v is defined as follows.

$$P_v = \{u \in V : (u, v) \in E\}$$

and the *successor list* of v, $S_v \subseteq V$ is,

$$S_v = \{u \in V : (v, u) \in E\}.$$

The predecessor and successor lists of the vertices of the graph of Fig. 2.16a are listed in Table 2.2.

2.6.3 Incidence Matrix

An *incident matrix* $B[n, m]$ of a simple graph has elements $b_{ij} = 1$ if edge j is incident to vertex i and $b_{ij} = 0$. otherwise. The incidence matrix for a digraph is defined differently as below.

$$b_{ve} = \begin{cases} -1 & \text{if arc } e \text{ ends at vertex } v \\ 1 & \text{if arc } e \text{ starts at vertex } v \\ 0 & \text{otherwise} \end{cases}$$

The incidence matrix of the graph of Fig. 2.16a is as below:

$$B = \begin{pmatrix} 1 & -1 & 0 & 1 & 0 & 1 & 0 & 0 \\ 0 & 1 & 1 & 0 & 0 & 0 & 0 & 0 \\ 0 & 0 & -1 & -1 & 1 & 0 & 0 & -1 \\ 0 & 0 & 0 & 0 & -1 & -1 & -1 & 0 \\ -1 & 0 & 0 & 0 & 0 & 0 & 1 & 1 \end{pmatrix}.$$

In the *edge list* representation of a graph, all of its edges are included in the list.

2.7   Trees

A graph is called a *tree* if it is connected and does not contain any cycles. The following statements equally define a tree T:

1. T is connected and has $n - 1$ edges.
2. Any two vertices of T are connected by exactly one path.
3. T is connected and each edge is a *cut-edge* removal of which disconnects T.

In a *rooted-tree* T, there is a special vertex r called the *root* and every other vertex of T has a directed path to r; the tree is unrooted otherwise. A rooted tree is depicted in Fig. 2.17. A *spanning-tree* of a graph G is a tree that covers all vertices of G. A *minimum spanning tree* of a weighted graph G is a spanning tree of G with the minimum total weight among all spanning trees of G. We will be investigating tree structures and algorithms in more detail in Chap. 6.

Fig. 2.17 A general tree structure

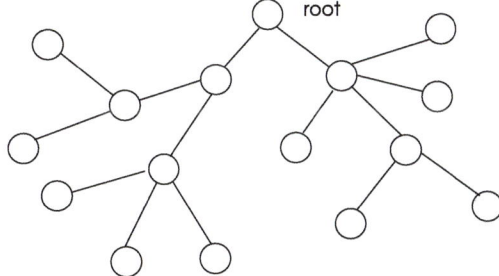

2.8   Graphs and Matrices

The spectral analysis of graphs involves operations on the matrices related to graphs.

2.8.1   Eigenvalues

Multiplying an $n \times n$ matrix A by an $n \times 1$ vector x results in another $n \times 1$ vector y which can be written as $Ax = y$ in equation form. Instead of getting a new vector y, it would be interesting to know if the Ax product equals a vector which is a constant multiplied by the vector x as shown below.

$$Ax = \lambda x. \tag{2.4}$$

If we can find values of x and λ for this equation to hold, λ is called an *eigenvalue* of A and x as an *eigenvector* of A. There will be a number of eigenvalues and a set of eigenvectors corresponding to these eigenvalues in general. Rewriting Eq. 2.4,

$$Ax - \lambda x = 0 \tag{2.5}$$

$$(A - \lambda I)x = 0$$

$$det(A - \lambda I)x = 0. \tag{2.6}$$

For $A[n, n]$, $det(A - \lambda I) = 0$ is called the *characteristic polynomial* which has a degree of n and therefore has n roots. Solving this polynomial provides us with eigenvalues and substituting these in Eq. 2.4 provides the eigenvectors of matrix A.

2.8.2   Laplacian Matrix

Definition 2.24 (*degree matrix*) The degree matrix D of a graph G is the diagonal matrix with elements $d_1, \ldots, d_n$ where d_i is the degree of vertex i.

Definition 2.25 (*Laplacian matrix*) The Laplacian matrix L of a simple graph is obtained by subtracting its adjacency matrix from its degree matrix.

The entries of the Laplacian matrix L are then as follows.

$$l_{ij} = \begin{cases} d_i & \text{if } i = j \\ -1 & \text{if } i \text{ and } j \text{ are neighbors} \\ 0 & \text{otherwise.} \end{cases}$$

The Laplacian matrix L is sometimes referred to as the combinatorial Laplacian. The set of all eigenvalues of the Laplacian matrix of a graph G is called the *Laplacian spectrum* of G. The *normalized Laplacian* matrix $\mathcal{L}$ is closely related to the combinatorial Laplacian. The relation between these two matrices is as follows:

$$\mathcal{L} = D^{-1/2}LD^{-1/2} = D^{-1/2}(D-A)D^{-1/2} = I - D^{-1/2}AD^{-1/2}. \qquad (2.7)$$

The entries of the normalized Laplacian matrix $\mathcal{L}$ can then be specified as below.

$$\mathcal{L}_{ij} = \begin{cases} 1 & \text{if } i = j \\ \dfrac{-1}{\sqrt{d_i d_j}} & \text{if } i \text{ and } j \text{ are neighbors} \\ 0 & \text{otherwise.} \end{cases}$$

The Laplacian matrix and adjacency matrix of a graph G are commonly used to analyze the spectral properties of G and design algebraic graph algorithm to solve various graph problems.

2.9 Chapter Notes

We have reviewed the basic concepts in graph theory leaving the study of some of the related background including trees, connectivity, matching, network flows and coloring to Part II when we discuss algorithms for these problems. The main graph theory background is presented in a number of books including books by Harary [6], Bondy and Murty [1], and West [8]. Graph theory with applications is studied in a book edited by Gross et al. [4]. Algorithmic graph theory focusses more on the algorithmic aspects of graph theory and books available in this topic include the books by Gibbons [2] and Golumbic [3].

Exercises

1. Show that the relation between the size and the order of a simple graph is given by $m \leq (n/2)$ and decide when the equality holds.
2. Find the order of a 4-regular graph that has a size of 28.
3. Show that the minimum and maximum degrees of a graph G are related by $\delta(G) \leq 2m/n \leq \Delta(G)$ inequality.
4. Show that for a regular bipartite graph $G = (V1, V2, E)$, $|V1| = |V2|$.
5. Let $G = (V_1, V_2, E)$ be a bipartite graph with vertex partitions V_1 and V_2. Show that

$$\sum_{u \in V_1} deg(u) = \sum_{v \in V_2} deg(v)$$

6. A graph G has a degree sequence $D = (d_1, d_2, \ldots, d_n)$. Find the degree sequence of the complement $\overline{G}$ of this graph.

7. Show that the join of the complements of a complete graph K_p and another complete graph K_q is a complete bipartite graph $K_{p,q}$.
8. Draw the line graph of K_3.
9. Let $G = (V, E)$ be a graph where $\forall v \in V$, $deg(v) \geq 2$. Show that G contains a cycle.
10. Show that if a simple graph G is disconnected, its complement $\overline{G}$ is connected.

References

1. Bondy AB, Murty USR (2008) Graph theory. In: Graduate texts in mathematics, 1st corrected edition 2008, 3rd printing 2008 edition. Springer, Berlin. ISBN-10: 1846289696, ISBN-13: 978-1846289699
2. Gibbons A (1985) Algorithmic graph theory, 1st edn. Cambridge University Press, Cambridge. ISBN-10: 0521288819, ISBN-13: 978-0521288811
3. Golumbic MC, Rheinboldt W (2004) Algorithmic graph theory and perfect graphs. In: Annals of discrete mathematics, vol 57, 2nd edn. North Holland, New York. ISBN-10: 0444515305, ISBN-13: 978-0444515308
4. Gross JL, Yellen J, Zhang P (eds) (2013) Handbook of graph theory, 2nd edn. CRC Press, Boca Raton
5. Hakimi SL (1962) On the realizability of a set of integers as degrees of the vertices of a graph. SIAM J Appl Math 10(1962):496–506
6. Harary F (1969) Graph theory. In: Addison Wesley series in mathematics. Addison–Wesley, Reading
7. Havel V (1955) A remark on the existence of finite graphs. Casopis Pest Mat 890(1955):477–480
8. West D (2000) Introduction to graph theory, 2nd edn. PHI Learning, Prentice Hall, Englewood Cliffs

Graph Algorithms

3

3.1 Introduction

An algorithm is a finite set of instructions to solve a given problem. It receives a set of *inputs* and computes a set of *outputs* using the inputs. Design and analysis of algorithms has been a key subject in any Computer Science Curriculum.

There has been a growing interest in the study of algorithms for graph theoretical problems in the last few decades mainly because of numerous increasing applications of graphs in real life. Also, recent technological advancements provided data of very large networks such as biological networks, social networks, the Web, and the Internet, which are commonly called *complex networks* which can be represented by graphs. The problems encountered in these networks can be quite different than the ones studied in classical graph theory as these networks have large sizes and do not have random structures. Hence, there is a need for new methods and algorithms in these networks.

Study of graph algorithms is reported in various headings including algorithmic graph theory, graphs, and applications. Our main goal in this chapter is to provide a brief and dense review of the main principles of sequential algorithm design and analysis with focus on graph algorithms. We first describe the basic concepts of algorithms and then review the mathematics behind the analysis of algorithms. We then provide a short survey of NP-completeness with focus on NP-hard graph problems. When we are dealing with such difficult graph problems, we may use approximation algorithms which provide suboptimal solutions with proven approximation ratios. In many cases, however, our only choice is to use heuristic algorithms that work for most input combinations as we describe. Finally, we briefly review the major algorithm design methods showing their implementations for graph problems.

© The Author(s), under exclusive license to Springer Nature Switzerland AG 2026   37
K. Erciyes, *Guide to Graph Algorithms*, Texts in Computer Science,
https://doi.org/10.1007/978-3-032-05294-0_3

3.2   Basics

An algorithm is a set of instructions working on some input to produce some useful output. The fundamental properties of any algorithm as well as any graph algorithm are as follows:

- It accepts a set of *inputs*, processes these inputs, and produces some useful *outputs*.
- It should provide *correct* output. Correctness is a fundamental requirement of any algorithm.
- It should execute a *finite* number of steps to produce the output. In other words, we do not want the algorithm to run forever without producing any output. It is possible to have algorithms running infinitely such as server programs, but these produce outputs while running.
- An algorithm should be *effective*. It should perform the required task using a minimum number of steps. It does not make sense to have an algorithm that runs 2 days to estimate the weather for tomorrow since we know what it would be by then. Given two algorithms that perform the same task, we would prefer the one with less number of steps.

When presenting an algorithm in this book and in general, we first need to provide a simple description of the main idea of the algorithm. We then would need to detail its operation using *pseudocode* syntax which shows its running using basic structures as described next. Pseudocode is the description of an algorithm in a more formal and structured way than verbally describing it but less formal than a programming language. We then typically show an example operation of the algorithm in a sample graph. The second fundamental thing to do is to prove that the algorithm works correctly which is self-evident in many cases, trivial in some cases and needs rigorous proof techniques for various others as we describe in this chapter. Finally, we should present its worst-case analysis which shows its performance as the number of steps required in the worst case.

3.2.1   Structures

Three fundamental structures used in algorithms are the *assignment, decision,* and *loops.* An assignment provides assigning a value to a variable as shown below:

$$b \leftarrow 3$$

$$a \leftarrow b^2.$$

Here we assign an integer value of 3 to a variable b and then assign the square of b to a. The final value of a is 9. Decisions are the key structures in algorithms as in daily life. We need to branch to some part of the algorithm based on some condition. The following example uses *if...then...else* structure to determine the larger one of two input numbers:

```
input a,b
if a > b then print a
        else if a=b then print ''they are equal''
        else print b
end if
```

Yet another fundamental structure in algorithms is the *loop* structure to perform an action. There are three main ways to perform loops in an algorithm as follows.

- *for* loops: The for loops are commonly used when we know how many times the loop should execute before we start with the loop. There is a *loop variable* and test condition. The loop variable is modified at the end of the loop according to the starting line and tested against the condition specified in the testing line. If this condition yields a true value, the loop is entered. In the following example, i is the loop variable and it is incremented by 1 at the end of each loop iteration and checked against the boundary value of 10. This simple loop prints the squares of integers between 1 and 10.

```
for i=1 to 10 step 1
    print i * i
end for
```

- *while* loops: In case we do not know how many times the loop will be executed, *while* structure may be used. We have a test condition at the beginning of the loop and if this succeeds, the loop is entered. The following example illustrates the use of *while* loop where we input Q for quitting by the user and otherwise add the two numbers given by the user. We do not know when the user may want to stop, so the use of *while* is appropriate here. Also note that we need two input statements for control, one outside the loop to be executed once and another one inside the loop to test iteratively since check is at the beginning of the loop.

```
input chr
while chr <> 'Q'
    input a,b
    print a+b
    input chr
end while
```

- *repeat ... until* loops: This structure is used in similar situations to *while* loops when we do not know how many times the loop will be executed. The main difference is that we do the test at the end of the loop which means this loop is executed at least once whereas the *while* loop may not be executed even once. We will write the previous example with the *repeat...until* (or *loop...until*) structure with a clearly shorter code.

```
repeat
    input a,b
    print a+b
until chr <> 'Q'
```

3.2.2   Procedures and Functions

A procedure or a function is a self-contained block of code to perform a specific task. These modules are provided in programming languages to break large programs or tasks into smaller ones to ease debugging and maintenance. Also, a procedure or a function can be used more than once, resulting in simplified code with less space requirements. A procedure or a function is implemented using the call/return method. They are called from the main program by the *call* routine. A procedure or a function may input parameters to work on, and they end by a return statement that takes the running program back to the point after they are called from. A function always returns a value whereas a procedure may not. Algorithm 3.1 displays a procedure called *Count* that is called from the main program with the parameter k. It displays all integers between 1 and k. Running this algorithm will display 1, 1 2, 1 2 3, and 1 2 3 4 at the output calling the procedure four times.

Algorithm 3.1 *Proc_Example*

1: **Input** : **int** $n = 4$
2: **int** i ▷ algorithm variable i
3: **procedure** COUNT(k)   ▷ procedure input variable k
4:     **int** $j = 1$ ▷ procedure local variable j
5:     **while** $j \leq k$ **do**
6:         **output** j
7:         $j \leftarrow j + 1$
8:     **end while**
9: **end procedure**
10: **for** i=1 to n **do**   ▷ main body of the algorithm
11:     Count(i)   ▷ procedure call
12: **end for**

3.3   Asymptotic Analysis

We need to assess the running time of algorithms for various input sizes to evaluate their performances. We can assess the behavior of an algorithm experimentally but theoretical determination of the required number of steps as a function of input size is needed in practice. Let us illustrate these concepts by writing an algorithm that searches for a key integer value in an integer array and returns its first occurrence as the index of the array as shown in Algorithm 3.2. We want to find out the execution time of this algorithm as the number of steps executed and if we can find an algorithm that has less number of steps for the same process, we would prefer that algorithm. In such analysis, we are not interested in the constant number of steps but rather, we need to find the number of steps required as the size of the input grows. For example, initializing the variable i takes constant time but it is performed only once therefore can be neglected, and this is more meaningful when $n \gg 1$. The number of times the loop is executed is important as it affects the performance of the algorithm significantly. However, the number of steps, say 2 or 3, inside the loop is insignificant again since $2n$ or $3n$ is invariable when n is very large.

Algorithm 3.2 *Search_Key*

1: **Input** : array $A[n]$ of n integers, integer *key*
2: **Output** : the first location of *key* or NOT_FOUND if it is not in $A[n]$
3: **int** i
4: **for** i=1 to n **do**
5:     **if** $key = A[i]$ **then**
6:         **return** i
7:     **end if**
8: **end for**
9: **return** NOT_FOUND

When we run this algorithm, it is possible that the key value is found in the first array entry in which case the algorithm completes in one step. This will be the lowest running time of the algorithm. In the worst case, we need to check each entry of the array A, running the loop n times. We are mostly interested in the worst execution time of an algorithm as this is what can be expected as the worst case.

We need to analyze the running time and space requirement of an algorithm as functions of the input size. Our interest is to determine the asymptotic behavior of these functions when input size gets very large. The number of steps required to run an algorithm is termed its *time complexity*. This parameter can be specified in three ways: the best-case, average-case, and worst-case complexities described as follows, assuming f and g are functions from $\mathbb{N}$ to $\mathbb{R}^+$ and n is the input size.

The Worst-Case Analysis
The worst running time of an algorithm is $f(n) = O(g(n))$, if there exists a constant $c > 0$ such that $\forall n_0 \geq n$, $f(n) \leq cg(n)$. This is also called the *big-Oh notation* and states that the running time is bounded by a function $g(n)$ multiplied by a constant when the input size is greater than or equal to a threshold input value. There are many $O(g(n))$ functions for $f(n)$ but we search for the smallest possible value to select a tight upper bound on $f(n)$.

Example 3.1 Let $f(n) = 5n + 3$ for an algorithm, which means its running time is this linear function of its input size n. We can have a function $g(n) = n^2$ and $n_0 = 6$, and hence claim $cg(n) \geq f(n)$, $\forall n \geq n_0$. Therefore, $5n + 3 \in O(n^2)$ which means

$f(n)$ has a worst-time complexity of $O(n^2)$. Note that any complexity greater than n^2, for example, $O(n^3)$, is also a valid worst-time complexity for $f(n)$. In fact $O(n)$ is a closer complexity for the worst case for this algorithm as this function approaches n in the limit when n is very large. We would normally guess this and proceed as follows:

$$5n + 3 \leq cn$$

$$(c - 5)n \geq 3$$

$$n \geq 3/(c - 5)$$

We can select $c = 6$ and $n_0 = 4$. for this inequality to hold and hence complete the proof that $5n + 3 \in O(n)$. As another example, consider the time complexity of $4 \log n + 7$. We claim this is $O(\log n)$ and need to find c and n_0 values such that $4 \log n + 7 \leq c \log n$ for $n \geq n_0$ which holds for $c = 12$ and $n_0 = 2$.

The Best-Case Analysis
The best or the lowest running time of an algorithm which is also the minimum number of steps to execute it is $f(n) = \Omega(g(n))$, if there exists a constant $c > 0$ such that $\forall n_0 \geq n$, $f(n) \geq cg(n)$. Informally, this is the best running time of the algorithm among all inputs of size n. In general, this parameter does not yield much information about the general performance of the algorithm as the algorithm may be slow on various other input combinations. Hence, it may not be reliable to compare algorithms based on their best running times but this parameter still gives us an idea of what to expect best.

Example 3.2 Let $f(n) = 3 \log n + 2$ for an algorithm, and let us consider the function $g(n) = \log n$ to be a lower bound on the running time of the algorithm. In this case, we need to verify $3 \log n + 2 \geq c \log n$ for some constant c for all $n \geq n_0$ values for a threshold n_0 value.

$$3 \log n + 2 \geq c \log n$$

$$(3 - c) \log n \geq -2$$

$$\log n \geq -2/(3 - c)$$

and for $n_0 = 2$ and $c = 4$, this equation holds and hence claims $3 \log n + 2 \in \Omega(\log n)$. The key point here was guessing that this function grows at least as $\log n$.

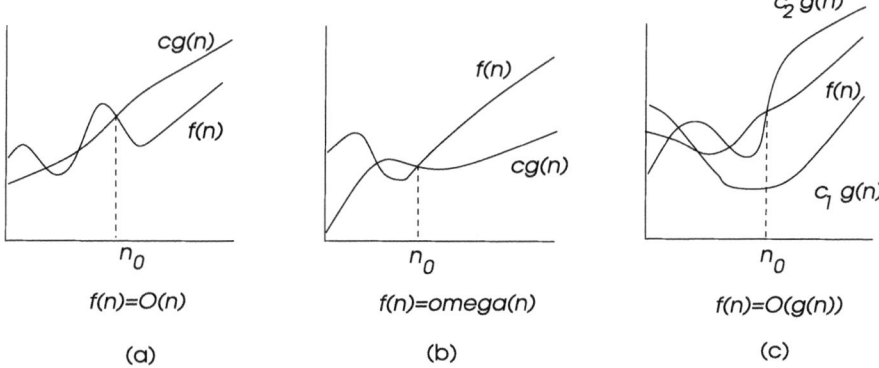

Fig. 3.1 The growth rate of a function

Theta Notation

$\Theta(n)$ is the set of functions that grow at the same rate as $f(n)$ and is considered as a tight bound for $f(n)$. These functions are both in $= O(g(n))$ and $\Omega(g(n))$. Formally, $g(n) \in \Theta(n)$ if there exists constants n_0, c_1 and c_2 such that $\forall n \geq n_0$, $c_1 g(n) \leq |f(n)| \leq c_2 g(n)$. The relation between the growth rate of a function $f(n)$, $O(n)$, $\Theta(n)$, and $\Omega(n)$ is depicted in Fig. 3.1.

The Average-Case Analysis

Our aim in determining the average case is to find the expected running time of the algorithm using a randomly selected input, assuming a probability distribution over inputs of size n. This method in general is more difficult to assess than the worst or best cases as it requires probabilistic analysis, but it can provide more meaningful results. Another point of concern is the memory space needed by an algorithm. This is specified as the maximum number of bits required by the algorithm and called the *space complexity* of the algorithm.

General Rules

- The order of growth of commonly found worst cases in increasing order is as follows:

$$O(1) \subset O(\log n) \subset O(n) \subset O(n \log n) \subset O(n^2) \subset O(n^3),$$

$$\ldots, O(n^k)n \subset 2^n \subset O(n!) \subset O(n^n)$$

- When the running time of an algorithm is determined as a polynomial, ignore low-order terms and the constant factor of the largest term as these will have a very little effect on the running time when input size is very large. For example, if the running time of an algorithm is $5n^4 + 3n^2 + 2n + 5$, its worst-case time complexity is $O(n^4)$.
- $O(cf(n)) = O(f(n))$ for any constant c.

- The sum of the worst-case time complexities of two algorithms is the worst-case complexities of the sum of the two algorithms as follows.

$$O(f(n)) + O(g(n)) = O(f(n) + g(n))$$

- Similarly, $O(f(n))O(g(n)) = O(f(n)g(n))$.
- $\log_{10} n \in \Theta(\log_2 n)$ since $\log_{10} n = \log_2 n / \log_2 10$ which is $\log_2 n / 3.32$. In general, $\log_a n \in \Theta(\log_b n)$.

Although asymptotic analysis shows the running time of the algorithm as the size of the input is increased to very large values, we may have to work with only small input sizes, which means low-order terms and constants may not be ignored. Also, given two algorithms, choice of the one with better average complexity rather than the one with better worst-case complexity would be more sensible as this would cover most of the cases.

3.4   Recursive Algorithms and Recurrences

We have two main methods of algorithm design and implementation: *recursion* and *iteration*. A recursive algorithm is the one that calls itself and these algorithms are commonly employed to break a large problem into smaller parts, solve the smaller parts, and then combine the results as we will see in this chapter. Iterative solutions keep repeating the same procedure until the desired condition is met. Recursive algorithms commonly provide shorter solutions but they may be more difficult to design and analyze. Let us consider a recursive function *Power* to find the nth power of an integer x. The iterative solution would involve n times multiplication of x by itself in a *for* or another type of loop. In the recursive algorithm, we have the function calling itself with decremented values of n each time until the *base case* is encountered when $n = 0$. The nested calls to this function start returning to the caller after this point, each time multiplying x with the returned value. The first returned value from the last call is 1 followed by x, x^2, x^3 until x^n as shown in Algorithm 3.3.

Algorithm 3.3 *Power*

1: **function** POWER(x,n)
2:    **if** $n = 0$ **then**   ▷ base case
3:        **return** 1
4:    **else**
5:        **return** $x \times power(x, n - 1)$ ▷ recursive call
6:    **end if**
7: **end function**

In order to analyze the running time of recursive algorithms, we need to define *recurrence relations* which are relations defined recursively in terms of themselves.

Let us attempt to form the recurrence relation for the recursive power algorithm defined above. Let $T(x)$ denote the time spent when x is input to this algorithm. Considering two constants c_1 and c_2 show constant time at each step of the algorithm for the base case and recursion case, respectively, we can form the following recurrence equations:

$$T(0) = c_1$$

$$T(1) = c_2 + T(1)$$

$$T(n) = c_2 + T(n - 1)$$

$$T(n) = 2c_2 + T(n - 2)$$

$$T(n) = 3c_2 + T(n - 3)$$

$$\dots$$

$$T(n) = kc_2 + T(n - k)$$

when $k = n$,

$$T(n) = nc_2 + T(n - n) = c_1 + nc_2 \in \Theta(n).$$

We have thus shown this algorithm takes n steps. Intuitively, we substituted the recurrence for lower values of n until we saw a pattern which is called the *iteration method* for solving recurrences. However, solving recurrences may involve more complicated procedures than this simple example. A commonly used approach to solve recurrence relations is by guessing a solution and proving the solution by induction. For example, let us assume the recurrence function below:

$$T(n) = 2T(n - 1) + 1$$

with

$$T(0) = 0$$

and guess the solution is $T(n) = 2^n - 1$ simply by looking at the values of this function for the first few values of n which are 1, 3, 7, 15, 31, and 63 for inputs 1, 2, 3, 4, and 5. Considering the base case, we find it holds.

$$T(1) = 2^1 - 1 = 1$$

Substitution for the general case yields

$$T(n) = 2T(n-1) = 2(2^{n-1} - 1) + 1 = 2^n - 1.$$

Therefore, we conclude our guess was correct.

The Master Method
The Master method is used to solve recurrence relations that can be expressed as follows:

$$T(n) = aT(n/b) + f(n)$$

where $a \geq 1$ and $b > 1$ are constants with a function $f(n)$ of positive n. There are three cases to consider:

1. $f(n) = O(n^{\log a - \varepsilon})$ for some $\varepsilon > 0$: $T(n) = \Theta(n^{\log a})$.
2. $f(n) = \Theta(n^{\log a})$: $T(n) = \Theta(n^{\log a} \log n)$.
3. $f(n) = \Omega(n^{\log a + \varepsilon})$ for some $\varepsilon > 0$, and $af(n/b) \leq cf(n)$ for some $c < 1$ and $\forall n > n_0$: $T(n) = \Theta(n^{\log a})$.

The proof can be found in [1].

3.5   Proving Correctness of Algorithms

Correctness is a fundamental requirement for any algorithm. It may be easy in many cases to determine that an algorithm works correctly for any input but in many other cases, we need to prove formally that the algorithm works. We can use various proving methods such as direct method, contraposition, contradiction; but mathematical induction method is used more frequently than others.

A logical statement or a *proposition* "*if p then q*" can be written as $p \rightarrow q$ where p is called the *premise* and q is the *conclusion*. We need to show that the conclusion holds based on the premise while proving a proposition. The direct proof involves arriving at the conclusion directly from the premise.

Example 3.3 If a and b are two even integers, their product ab is even.

Proof We can write $a = 2m$ and $b = 2n$ for some integers m and n since they are even, and therefore are divisible by 2. The product $ab = 2m \cdot 2n = 4mn = 2(2mn)$ is an even number since it is divisible by 2.   □

The proofs may be as simple as in this example. In many cases, however, a proof involves more sophisticated reasoning to arrive at the conclusion. Let us look at another example that involves a direct but not so easy to derive proof.

Example 3.4 Given two integers a and b, if $a + b$ is even then $a - b$ is even.

Proof Since $a + b$ is even, we can write $a + b = 2m$ for some integer m. Then, substitution for b yields:

$$a + b = 2m$$

$$b = 2m - a$$

$$a - b = a - 2m + a$$

$$= 2a - 2m$$

$$= 2(a - m)$$

which shows that the difference is an even number and completes the proof. □

3.5.1   Contraposition

A *contrapositive* of a logical statement $p \rightarrow q$ is $\neg q \rightarrow \neg p$. The contrapositive of such a statement is equivalent to the statement, and hence, we can prove the contrapositive to verify the original statement. This method is also called *indirect proof* and can be applied to a variety of problems.

Example 3.5 For any integer $a > 2$, if a is a prime number, then a is an odd number.

Proof Let us assume the opposite of the conclusion, a is even. We can then write $a = 2n$ for some integer n. However, this implies a is divisible by 2, and hence, it cannot be a prime number which contradicts the premise. □

3.5.2   Contradiction

In this proof method, we assume the premise p is true and the conclusion q is not true $(p \wedge \neg q)$ and try to find a contradiction. This contradiction can be against what we assume as hypothesis or simply be something against we know to be true such as $1 = 0$. In this case, if we find $(p \wedge \neg q)$ is false, it means either p is false or $\neg q$ is false. Since we assume p is true as it is the premise, $\neg q$ must be false which means q is true if there is a contradiction and that completes the proof.

Example 3.6 Let us prove the statement: $3n + 2$ is even when n is even.

Proof Here $p =$ "$3n + 2$ is even" and $q =$ "n is even". We will assume $\neg q$ which is "n is odd" and thus $n = 2k + 1$ for some integer k. Substituting in p yields $3(2k + 1) + 2 = 6k + 5 = 2(3k + 2) + 1$. Substituting $t = (3k + 2)$ which an integer, results in $3n + 2 = 2t + 1$ which is odd by the definition of an odd integer. Hence, we arrive at $\neg p$.   □

3.5.3   Induction

In induction, we are given a sequence of propositions in the form $P(1)$, ..., $P(n)$ and we perform two steps:

1. *Basis step*: Establish $P(1)$ is true.
2. *Induction step*: If $P(k)$ is true for any given k, then establish $P(k + 1)$ is also true.

If these two steps provide true results, we can conclude $P(n)$ is true for any n. It is one of the most commonly used methods to prove sequential, parallel, and distributed algorithms.

Example 3.7 Let us illustrate this method by proving that the sum S of the first n odd numbers $1 + 3 + 5 \ldots$ is n^2.

Proof

1. *Basis step*: $P(1) = 1 = 1^2$, so the basis step yields a true answer.
2. *Induction step*: Assuming $P(k) = k^2$, we need to show that $P(k + 1) = (k + 1)^2$. Since the kth element of $P(n)$ is expressed as $2k - 1$ as it is an odd number, the following can be stated:

$$P(k + 1) = P(k) + 2(k + 1) - 1 = k^2 + 2k + 1 = (k + 1)^2$$

Therefore, this proposition is true for all positive integers.   □

3.5.4   Strong Induction

In induction, we attempted to prove the proposition $P(k + 1)$ assuming the statement $P(k)$ is true. In strong induction, we have the following steps:

1. *Basis step*: $P(1)$ is proven.
2. *Strong Induction step*: Assuming $P(1)$, ..., $P(k)$ are all true, we need to establish that $P(k + 1)$ is true.

This proof method is useful when $P(k + 1)$ does not depend on $P(k)$ but on some smaller values of k. In fact, the two induction methods are equivalent.

Example 3.8 Every integer greater than 1 is either a prime number or can be written as the product of prime numbers.

Proof We need to consider the base case and the strong induction case.

- *Base case*: $P(2) = 2$ is prime so the base case is true.
- *Strong Induction step*: Assuming each integer n with $2 \leq n \leq k$ is either a prime or a product of prime numbers, we need to prove $(k + 1)$ is either prime or a product of prime numbers. We have two cases as follows:
  - $(k + 1)$ is a prime number: Then $P(k + 1)$ is true.
  - $(k+1)$ is a composite number and not a prime: Then $\exists\, a$ and b with $2 \leq a, b \leq k$ such that $k + 1 = a \cdot b$. By the strong induction step, a and b are either prime numbers or product of prime numbers. Therefore, $k + 1 = a \cdot b$ is a product of prime numbers. $\qquad\square$

3.5.5 Loop Invariants

An algorithm with a loop starts by initializing some variables based on some inputs, executes a loop, and produces some output based on the values of its variables. A *loop invariant* is an assertion about the value of a variable after each iteration of a particular loop, and the final value of this variable is used to determine the correctness of the algorithm.

A *precondition* is a set of statements that are true before the algorithm executes which is commonly represented as a set of inputs, and a *postcondition* is a set of statements that remain true after the algorithm executes which are the outputs. We use loop invariants to help us understand why an algorithm is correct. We must show three things about a loop invariant:

1. *Initialization*: The loop invariant should be true before the first iteration of the loop.
2. *Maintenance*: If the loop invariant is true for the nth iteration, it should be true for $(n + 1)$th iteration.
3. *Termination*: The invariant is true when the loop terminates.

The first two properties of a loop variant assert that the variant is true before each loop iteration, similar to the induction method. Initialization is like the base case of induction, and maintenance is similar to the inductive step. There are no definite rules to choose a loop variant and we proceed by commonsense in most cases. Let us consider the following *while* loop. Our aim is to prove this loop works correctly and terminates.

```
a > 0;
b = 0;
while ( a != b)
    b = b + 1;
```

We will choose $a \geq b$ as the loop variant L. We need to show the three conditions: L is true before the loop starts; if L is true before an iteration, it remains true after the iteration and lastly, it should establish the postcondition. We can now check these conditions as follows and can determine that this loop works correctly and terminates:

1. *Initialization*: $a \geq 0$ and $b = 0$ before the loop starts is *true*.
2. *Maintenance*: $(a \geq 0) \wedge (a \neq b) \to b = b + 1$.
3. *Termination*: $(a \geq 0) \wedge \neg(a \neq b) \to a = b$.

3.6   Reductions

We may want to prove that some computational problems are difficult to solve. In order to verify this, we need to show some problem X is at least as hard as a known problem Y. An elegant way of proving this assertion is to reduce problem Y to X. Let us assume we have a problem P_1 that has an algorithmic solution A_1 and a similar problem P_2 that does not have any solution. Similarity may imply we can borrow some of the solutions found for P_1 if we can find a *reduction* of problem P_2 to P_1.

Definition 3.1 (*reduction*) A reduction of a problem P_2 to P_1 transforms the inputs of P_2 to the inputs of P_1, obtains outputs from P_1 by running algorithm A_1 and depicts these outputs as the solutions from P_1. A problem P_2 is reducible to problem P_1 if there is a function f that takes an arbitrary input to P_2, transforms it to an input to P_1, solves there and obtains the solution to x. This is shown as $P_2 \leq P_1$.

Specifying an upper bound on the transformation of the input is sensible as this process itself may be time-consuming, for example, it may be exponential. If transferring of any input to problem P_2 to problem P_1 can be performed in polynomial time, P_2 is said to be *polynomial-time reducible* to P_1 and shown as $P_2 \leq_P P_1$. Our interest is mainly on polynomial-time reductions when solving similar problems.

3.6.1   Difficult Graph Problems

Let us consider the vertex cover problem. A *vertex cover* (VCOV) of a graph
$G = (V, E)$ is a subset V' of its vertices such that any edge $e \in E$ is incident
to at least one vertex $v \in V'$. Informally, we try to *cover* all edges of G by the
vertices in this subset. The decision form of this problem (VCOV) asks: Given a
graph $G = (V, E)$ and an integer k, does G have a vertex cover of size at most
k? The optimization VCOV problem is to find the vertex cover with the minimum
number of vertices among all vertex covers of a graph.

An independent set of a graph G is a subset V' of its vertices such that no vertex
in V' is adjacent to any other vertex in V'. The decision form of this problem (IND)
seeks to answer the question: Given graph G and an integer k, does G contain an
independent set of at least k vertices? The optimization IND problem is to find the
independent set with the maximum number of vertices among all independent sets
of a graph.

A related graph problem is finding the dominating set (DOM) of a graph
$G = (V, E)$ which is a subset V' of its vertices such that any $v \in V \setminus V'$ is
adjacent to at least one vertex in V'. The decision form of DOM seeks to answer
the question: Given graph G and an integer k, does G contain a dominating set of
at most k vertices? These subgraphs are displayed in a sample graph in Fig. 3.2.
The optimization DOM problem is to find the dominating set with the minimum
number of vertices among all dominating sets of a graph. Minimal or maximal versions of all of these problems are to find vertex subsets that cannot be reduced or
enlarged by removal/addition of any other vertices. We will review these problems
in more detail in Chap. 10.

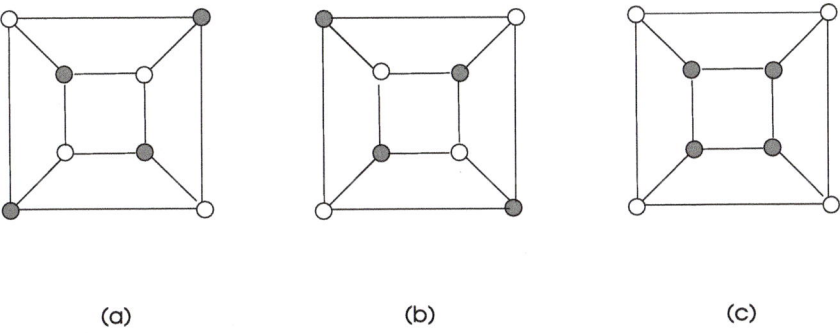

    (a)   (b)   (c)

Fig. 3.2 Some difficult graph problems. **a** A minimum vertex cover, **b** a maximum independent
set, and **c** a minimum dominating set. Note that **a** is also a maximum independent set and a minimum dominating set, **b** is also a minimum vertex cover and a minimum dominating set but **c** is
only a minimum dominating set for this graph

3.6.2   Independent Set to Vertex Cover Reduction

We will show that an independent set of a graph can be reduced to a vertex cover by first considering the theorem below.

Theorem 3.1 *Given a graph $G = (V, E)$, $S \subset V$ is an independent set of G if and only if $V - S$ is a vertex cover.*

Proof Let us consider an edge $(u, v) \in E$. If $S \subset V$ is an independent set in G, either $u \in S$ or $v \in S$ but not both. This means at least one of the endpoints of (u, v) is in $V - S$, hence $V - S$ is a vertex cover. We need to prove the theorem in the other direction; if $V - S$ is a vertex cover, let us consider two vertices $u \in S$ and $v \in S$. Then, $(u, v) \notin E$ since if such an edge existed, it would not be covered in $V - S$, and hence, $V - S$ would not be a vertex cover.

Figure 3.3 shows the equivalence of these two problems in a sample graph. Given the five vertices in (a) with $k = 5$, we can see these form an independent set. We now transform this input to an input of the vertex cover problem in polynomial time, which are the *white* vertices in (a). We check whether these form a vertex cover in (b) again in polynomial time and conclude they do. Our test algorithm simply marks incident edges to *black* vertices in $O(k)$ time and checks whether any edges are left unmarked in the end. All edges are covered by these four vertices in this case. Hence, we can deduce the five *black* vertices in (a) are indeed a solution to the independent set decision problem for this graph. We have shown an example of *IND* $\leq_P$ *VCOV*.

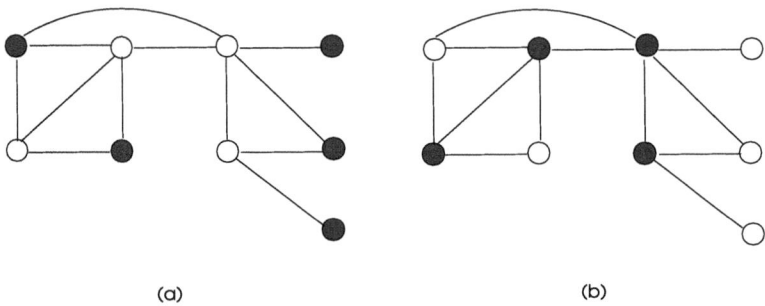

(a)   (b)

Fig. 3.3 A sample graph with an independent set in **a** and a vertex cover in **b** shown by *black* circles in both cases. The independent set in **b** is formed by the *white* vertices in **a**

3.7   NP-Completeness

The problems we face can be classified based on the time it takes to solve them. A *polynomial function* $O(n^k)$, with n as the variable and k as the constant, is bounded by n^k as we saw. The *exponential functions* refer to functions such as $O(2^n)$ or $O(n^n)$ which grow very rapidly with the increased input size n. A polynomial-time algorithm has a running time bounded by a polynomial function of its input, and an exponential algorithm is the one which does not have a time performance bounded by a polynomial function of n. A *tractable problem* is solvable by a polynomial-time algorithm, and an *intractable problem* cannot be solved by a polynomial-time algorithm. Searching for a key value in a list can be performed by a polynomial-time algorithm as we need to check each entry of a list of size n in n time and listing all permutations of n numbers is an example of an exponential algorithm. In fact, the third class of problems have no known polynomial-time algorithms but they are not proven to be intractable either. When we are presented with an intractable problem, we can do one of the following:

- Attempt to solve a simpler or a restricted version of a problem which can be accomplished in polynomial time.
- Implement a polynomial-time probabilistic algorithm which provides the correct solution only with very high probability. This means it may fail on rare occasions.
- Use a polynomial-time approximation algorithm with a proven approximation ratio. In this case, we have a suboptimal solution to the problem which may be acceptable in many applications.
- When all fails, we can use some heuristics which are commonsense rules to design polynomial-time algorithms. We need to show experimentally that the heuristic algorithm works fine with a wide spectrum of input combinations. Many problems in graphs are intractable and this method is frequently employed especially in large graphs, as we shall see.

3.7.1   Complexity Classes

The tractable problems have polynomial-time algorithms to solve them. The intractable problems, on the other hand, can be further divided into two subclasses; the ones proven to have no polynomial-time algorithms and others that have exponential time solution algorithms. At this point, we will need to classify the problems based on the expected output from them as *optimization problems* or *decision problems*. In an optimization problem, we attempt to maximize or minimize a particular objective function and a decision problem returns a *yes* or a *no* as an answer to a given input. Let us consider the IND problem as both optimization and a decision problem. The optimization problem asks to find the independent set of a graph $G = (V, E)$ with the largest order. The decision problem we saw seeks an answer to the question: Given an integer $k \leq |V|$, does G have an independent set with at least k vertices? Dealing with decision problems is advantageous not only because these problems are in general easier than their optimization versions but they also may provide a transfer to the optimization problem. In the IND decision problem, we can try all possible k values and find the largest one that provides an independent set to find a solution to the IND optimization problem.

Complexity Class P

The first complexity class we will consider is P which contains problems that can be solved in polynomial time.

Definition 3.2 (*class P*) The complexity class P refers to decision problems that can be solved in polynomial time.

For a problem A to be in P, there has to be an algorithm with worst execution time $O(n^k)$ for an input size n and a constant k that solves A. For example, finding the largest value of an integer array can be performed in $O(n)$ time for an array of size n, and hence is a problem in P.

Complexity Class NP

In cases where we do not know a polynomial-time algorithm to solve a given decision problem, a search can be made for an algorithm that solves an input instance of a problem in polynomial time. The specific input is called the *certificate* and the polynomial-time algorithm that checks whether the certificate is acceptable, meaning it is a solution to the decision problem, is called the *certifier*. We can now define a new complexity class as follows.

Definition 3.3 (*class NP*) The complexity class Nondeterministic Polynomial (NP) is the set of decision problems that can be verified by a polynomial algorithm.

The NP class includes P (P $\subset$ NP) class since all of the problems in P have certifiers in polynomial time but whether P $=$ NP has not been determined and remains a grand challenge in Computer Science. Many problems are difficult to solve, but an input instance can be verified in polynomial time whether it yields a solution or not. For example, given a graph $G = (V, E)$ and a certificate $S \in V$, we can check in polynomial time whether S is an independent set in G. The certifier program is shown in Algorithm 3.4 where we simply check whether any two vertices in S are adjacent. If any such two vertices are found, the answer is NO and the input is rejected. The algorithm runs two nested loops in $O(n^2)$ time, and hence, we have a polynomial-time certifier which shows IND $\in$ NP.

Algorithm 3.4 *IS_Certifier*

1: **Input** : $G = (V, E), S \in V$
2: **Output** : *yes* or *no*
3: **for all** $u \in S$ **do**
4:     **for all** $v \in S$ **do**
5:         **if** $(u, v) \in E$ **then**
6:             **return** *No*
7:         **end if**
8:     **end for**
9: **end for**
10: **return** *Yes*

NP-Hard Problems

Definition 3.4 (*class NP-Hard*) A decision problem P_i is NP-hard if every problem in NP is polynomial time reducible to P_i. In other words, if we can solve P_i in polynomial time, we can solve all NP problems in polynomial time.

An NP-hard problem P_i does not need to be in NP. An NP-hard problem P_i means P_i is as hard as any problem in NP.

NP-Complete Problems
A problem P_i is NP-complete if it is NP-hard and it is also a member of the NP class.

Definition 3.5 (*class NP-Complete*) A decision problem P_i is NP-Complete if it is in NP and NP-hard.

Figure 3.4 displays the relation between the complexity classes. Class P is contained in class NP as every problem in P has a certifier, and NP-complete problems are in the intersection of NP problems and NP-hard problems.

In order to show that a problem P_i is NP-Complete, we need to do the following:

1. Prove P_i is a decision problem.
2. Show $P_i \in$ NP.
3. Show P_i is as hard as any problem that is NP-hard.

3.7.2 The First NP-Hard Problem: Satisfiability

The *satisfiability problem* (SAT) states that given a Boolean formula, is there a way to assign truth values to the variables in the formula such that the formula evaluates to true value? Let us consider a set of logical variables $x_1, x_2, \ldots, x_n$

Fig. 3.4 Relation between
complexity classes

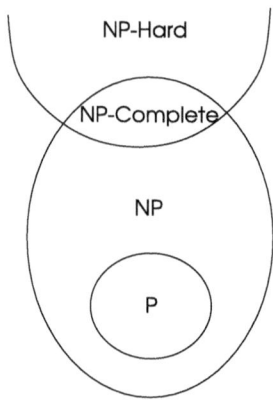

each of which can be *true* or *false*. A clause is formed by disjunction of logical variables such as $(x_1 \lor \overline{x_2} \lor x_3)$. A conjunctive normal form (CNF) formula is a conjunction of the clauses as $C_1 \land C_2, \ldots \land C_k$ such as below:

$$(x_1 \lor \overline{x_2} \lor x_3) \land (\overline{x_1} \lor x_2 \lor \overline{x_3}) \land (\overline{x_1} \lor \overline{x_2} \lor x_3) \tag{3.1}$$

The CNF formula is satisfied if every clause in it yields a *true* value. The satisfiability problem searches for an assignment to variables $x_1, x_2, \ldots, x_n$ such that CNF formula is satisfied. 3-SAT problem requires each clause to be of length 3 over variables $x_1, x_2, \ldots, x_n$. The SAT problem has no known polynomial-time algorithm but we cannot conclude it is intractable either. We can try all combinations of the input in 2^n time to find the solution. However, we can have a distinct input and check whether this input is accepted by the SAT circuit, and hence conclude SAT is in NP. This problem was shown to be NP-hard and therefore to be NP-complete by Cook in 1970 [3]. We can, therefore, use 3-SAT problem as basis to prove other problems to be NP-complete or not. The relationships between various problems is depicted in Fig. 3.5.

Let us show how to reduce the 3-SAT problem to IND problem. In the former, we know that we have to set at least one term in each clause to be true and we cannot set both x_i and $\overline{x_i}$ to be true at the same time. We first draw triangles for each clause of 3-SAT with each vertex representing the term inside the clause. A true literal from each clause suffices to obtain a true value for the 3-SAT formula. We add lines between a term and its inverse as we do not want to include both in the solution. The graph drawn this way for Eq. 3.1 is shown in Fig. 3.6.

We now claim the following.

Theorem 3.2 *The 3-SAT formula F with k clauses is satisfiable if and only if the graph formed this way has an independent set of size k.*

Proof Let the graph formed this way be $G = (V, E)$. If formula F is satisfiable, we need to have at least one *true* literal from each clause. We form the vertex set

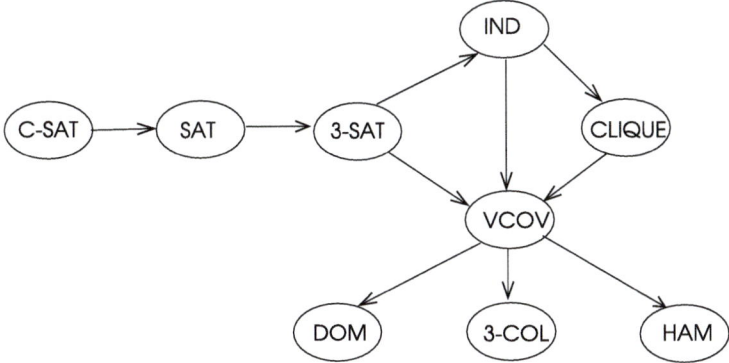

Fig. 3.5 The reduction relationship between various NP-hard problems

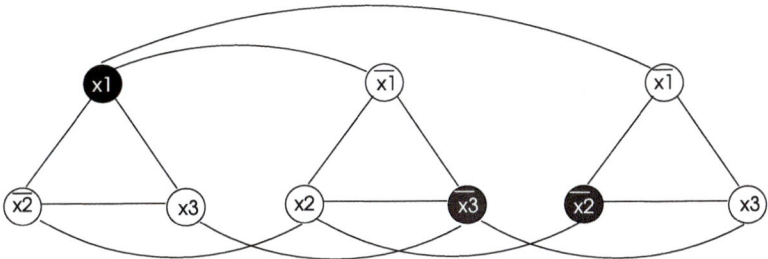

Fig. 3.6 The graph for the 3-SAT equation of Eq. 3.1. The black vertices x_1, $\overline{x_3}$ and $\overline{x_2}$ represent the independent set of this graph which is also the solution to the 3-SAT of Eq. 3.1 with $x_1 = 1$, $x_2 = 0$ and $x_3 = 0$ values

$V' \subset V$ by selecting a vertex from each triangle, and also by not selecting a variable x and its complement $\bar{x}$ at the same time since a variable and its complement cannot be true at the same time. V' is an independent set since there are no edges between the vertices selected. To prove the claim in the reverse direction, let us consider G has an independent set of size k. The set V' cannot have two vertices from the same cluster and it will not have a variable and its complement at the same time since it is an independent set. Moreover, when we set *true* values to all variables in V', we will have satisfied the SAT formula F. Transformation of the 3-SAT problem to IND problem can be performed in polynomial time; hence, we deduce these two problems are equivalent, and finding a solution to one means solution to the other one is also discovered. □

3.8   Coping with NP-Completeness

Many of the graph optimization problems are NP-hard with no known solutions in polynomial time. However, methods to result in solutions *most of the time* within a specified bound of probability (*randomized algorithms*), or results that are close to the exact solution within a specified margin to the exact result (*approximation algorithms*); or methods that eliminate some of the unwanted intermediate results to achieve improvement in the performance (*backtracking* and *branch and bound*) are the topics we will review in this section.

3.8.1   Randomized Algorithms

Randomized algorithms are frequently used for some of the difficult graph problems as they are simple and provide efficient solutions. Randomly generated numbers or random choices are typically used in these algorithms to decide on the courses of computation. The output from a randomized algorithm and its running time varies for different inputs and even for the same input. Two classes of randomized algorithms are *Las Vegas* and *Monte Carlo* algorithms. The former always returns a correct answer but the runtimes of such algorithms depend on the random choices made. The algorithm runs a constant amount of time but the answer may or may not be correct in Monte Carlo algorithms.

The average cost of the algorithm over all random choices gives us its expected bounds and a randomized algorithm is commonly specified in terms of its *expected running time* for all inputs. On the other hand, when we say an algorithm runs in $O(x)$ time with *high probability*, it means the runtime of this algorithm will not be above the value of x with high probability. Randomized algorithms are commonly used in two cases: when an initial random configuration is to be chosen and to decide on a local solution when there are several options. The randomized choice may be repeated with different seeds and then the best solution is returned [2].

Karger's Minimum Cut Algorithm
We will describe how randomization helps to find the minimum cut (*mincut* henceforth) of a graph. Given a graph $G = (V, E)$, finding a mincut of G is to partition the vertices of the graph into two disjoint sets V_1 and V_2 such that the number of edges between V_1 and V_2 is minimum. There is a solution to this problem using the maximum flow as we will see in Chap. 8, here we will describe a randomized algorithm due to Karger [4].

This simple algorithm selects an edge at random, makes a supervertex from the endpoints of the selected edge using *contraction* and continues until there are exactly two supervertices left as shown in Algorithm 3.5. The vertices in each final supervertex are the vertices of the partitions.

Algorithm 3.5 *Karger_mincut*

1: **Input** : $G(V, E)$
2: **Output** mincut V_1 and V_2 of G
3:
4: $G' = (V', E') \leftarrow G(V, E)$
5: **repeat**
6:     **select** an edge $(u, v) \in E'$ at random
7:     **contract** vertices u and v into a super vertex uv
8: **until** there are only two super vertices u_1 and u_2 left
9: $V_1 \leftarrow$ all of the vertices in u_1
10: $V_2 \leftarrow$ all of the vertices in u_2

Contracting two vertices u and v is done as follows:

- Delete all edges between u and v.
- Replace u and v with a supervertex uv.
- Connect all edges incident to u and v to supervertex uv.

Let us see how this algorithm works in the simple graph of Fig. 3.7. The edges picked at random are shown inside dashed regions and the final cut consists of three edges between $V_1 = \{b, h, a\}$ and $V_2 = \{c, f, d, e, g\}$ as shown in (h). This is not the mincut however, the minimum cut consists of edges (b, c) and (g, f) as depicted in (i).

Karger's algorithm will find the correct minimum cut if it never selects an edge that belongs to the minimum cut. In our example, we selected the edge (g, f) that belongs to the mincut in Fig. 3.7e deliberately to result in cut that is not minimum. On the other hand, the mincut edges have the lowest probability to be selected by this algorithm since they have fewer edges than all edges that do not belong to the mincut. Before its analysis, let us state few observations about the mincut of a graph.

Remark 1 The size of a mincut of a graph G is at most the minimum degree $\delta(G)$ of G.

This is valid since the mincut is not larger than any cut of G. Therefore, $\delta(G)$ sets an upper bound on the size of the mincut. Since we cannot determine $\delta(G)$ easily, let us check whether an upper bound in terms of the size and order of G exists.

Theorem 3.3 *The size of the mincut is at most 2m/n.*

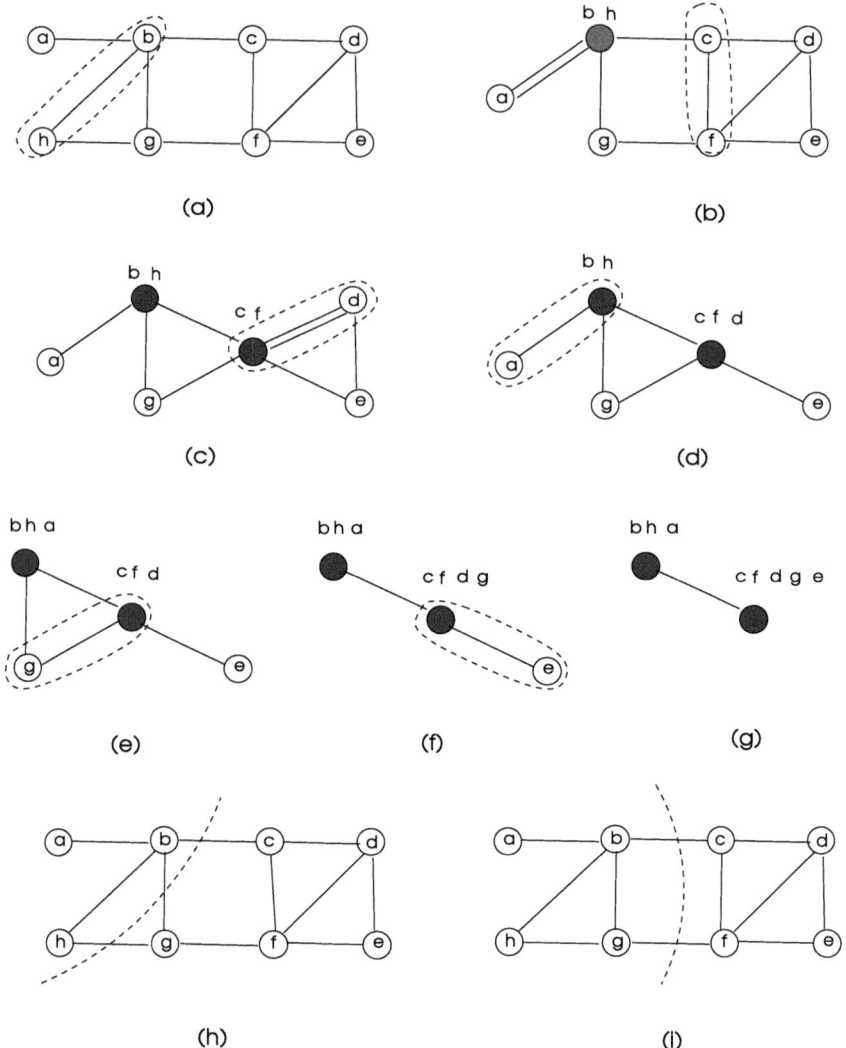

Fig. 3.7 Running of Kargel's algorithm in a simple graph

Proof Assume the size of the mincut is k. Then, every vertex of G must have at least a degree of k. Therefore, by Euler theorem (handshaking lemma),

$$m = \frac{\sum_{v \in V} deg(v)}{2} \geq \frac{\sum_{v \in V} k}{2} = \frac{nk}{2} \qquad (3.2)$$

which means $k \leq 2m/n$. $\qquad\square$

Corollary 3.1 *Given a graph G with a mincut C, the probability of selecting an edge $(u, v) \in C$, $P(\varepsilon_1)$ is at most 2/n.*

Proof There are m edges and at most $2m/n$ are in the mincut by Theorem 3.3, so $P(\varepsilon_1) = m/(2m/n) = 2/n$. $\qquad\square$

Remark 2 Given a graph G with a mincut C, the algorithm must not select any edge $(u, v) \in C$.

Therefore, probability of not selecting the first edge of the mincut is

$$P(\overline{\varepsilon_1}) = 1 - P(\varepsilon_1) \geq 1 - \frac{2}{n} \geq \frac{n-2}{n}. \tag{3.3}$$

Let us choose a minimum cut C with size k and find the probability that an edge $(u, v) \in C$ is not contracted by the algorithm which will give us the probability that the algorithm finds the correct result. We will first evaluate the probability of selecting an edge $(u, v) \in C$ in the first round, $P(\varepsilon_1)$ which is k/m for a mincut with size k. Therefore,

$$P(\varepsilon_1) = \frac{k}{m} \leq \frac{k}{nk/2} = \frac{2}{n} \tag{3.4}$$

Let $P(\varepsilon_C)$ be the probability that the final cut obtained s minimum. This probability is the product of the probabilities of the probability the first selected edge is not in mincut, probability the second selected edge is not in mincut, etc., until two last supervertices are formed. A contraction of an edge results in one less vertex in the new graph. Therefore,

$$P(\varepsilon_C) \geq \left(1 - \frac{2}{n}\right)\left(1 - \frac{2}{n-1}\right)\left(1 - \frac{2}{n-2}\right)\left(1 - \frac{2}{n-3}\right) \cdots \left(1 - \frac{2}{3}\right) \tag{3.5}$$

since nominators and denominators cancel in every two terms except the two first denominators,

$$= \left(\frac{n-2}{n}\right)\left(\frac{n-3}{n-1}\right)\left(\frac{n-4}{n-2}\right) \cdots \left(\frac{1}{3}\right)$$

$$= \frac{2}{n(n-1)}.$$

Hence, the probability that the algorithm returns the mincut C is at least $\frac{2}{n(n-1)}$. We can therefore conclude this algorithm succeeds with probability $p \geq 2/n^2$ and running it $O(n^2 \log n)$ time provides a minimum cut with high probability. Using the adjacency matrix of the graph, we can run each iteration in $O(n^2)$ time, and the total time is $O(n^4 \log n)$.

3.8.2 Approximation Algorithms

A common approach in search of a solution to NP-hard graph problems is to relax the requirements and inspect *approximation algorithms* which provide suboptimal solutions.

Definition 3.6 (*approximation ratio*) Let A be an approximation algorithm for problem P, I be an instant of the problem P, and OPT(I) the value of the optimum cost of solution for I. The approximation ratio of algorithm A is defined as

$$\alpha_A = \max_I \frac{A(I)}{\text{OPT}(I)}. \tag{3.6}$$

Example 3.9 We had already investigated the vertex cover problem in Sect. 3.6. Finding the minimum vertex cover which has the least number of vertices among all vertex covers of a graph is NP-hard. Since our aim is to cover all edges of the graph by a subset of vertices, we can design an algorithm that picks each edge in a random order and since we cannot determine which vertex will be used to cover the edge, we include both ends of the edge in the cover as shown in Algorithm 3.6. For each selected edge (u, v), we need to delete edges incident to u or v from graph since these edges are covered.

Algorithm 3.6 *Approx_Vertex_Cover*

1: Input $G(V, E)$
2: $E' \leftarrow E, V' \leftarrow \emptyset$
3: **while** $E' \neq \emptyset$ **do**
4:    **select** randomly an edge $(u, v) \in E'$
5:    $V' \leftarrow V' \cup \{u, v\}$
6:    $E' \leftarrow E' \setminus \{$ all edges incident to either u or v $\}$
7: **end while**

The iterations of this algorithm in a sample graph are shown in Fig. 3.8 which provides a vertex cover with an approximation of 2.

Theorem 3.4 *Algorithm 3.6 provides a vertex cover in $O(m)$ time and the size of MVC is $2\,|MinVC|$.*

Proof Since the algorithm continues until there are no more edges left, every edge is covered, therefore the output from *Seq1_MVC* is an MVC, taking $O(m)$ time. The set of edges picked by this algorithm is a matching, as edges chosen are disjoint and it is maximal as addition of another edge is not possible. Since two vertices are covered for each matched edge, the approximation ratio for this algorithm is 2. $\quad\square$

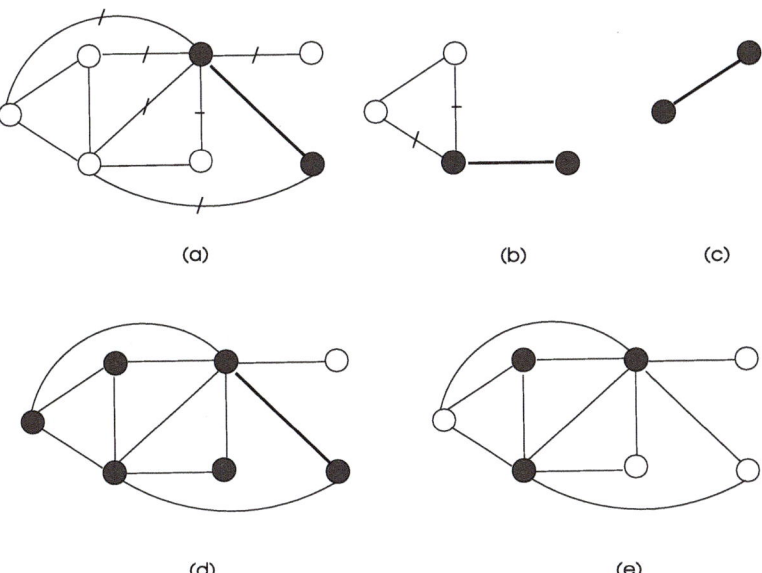

Fig. 3.8 A possible iteration of Algorithm 3.6 in a sample graph, showing the selected edge in bold and the vertices included at the endpoints of this edge as black at each step, from **a–c**. The final vertex cover has six vertices as shown in **d** and the minimum vertex cover for this graph has three vertices as shown in **e** resulting in the worst approximation ratio of 2

3.8.3 Backtracking

In many cases of algorithm design, we are faced with a search space that grows exponentially with the size of the input. The *brute-force* or *exhaustive search* approach searches all available options. *Backtracking* is a convenient way of searching the available options while looking for a solution. In this method, we look at a partial solution and if we can pursue it further, we do. Otherwise, we *backtrack* to the previous state and proceed from there since proceeding from current state violates the requirements. This way, we save some of the operations needed from the current state onwards. The choices that can be made are placed in a *state-search tree* where nodes except the leaves correspond to partial solutions and edges are used to expand the partial solutions. A subtree which does not lead to a solution is not searched and we backtrack to the parent of such a node. Backtracking can be conveniently implemented by recursion since we need to get back to the previous choice if we find the current choice does not lead to a solution.

Let us see how this method works using an example. In the *subset sum problem*, we are given a set of S of n distinct integers and are asked to find possibly more than one subsets of S sum of which equals a given integer M. For example, if $S = \{1, 3, 5, 7, 8\}$ and $M = 11$, then $S_1 = \{1, 3, 7\}$ and $S_2 = \{3, 8\}$ are the

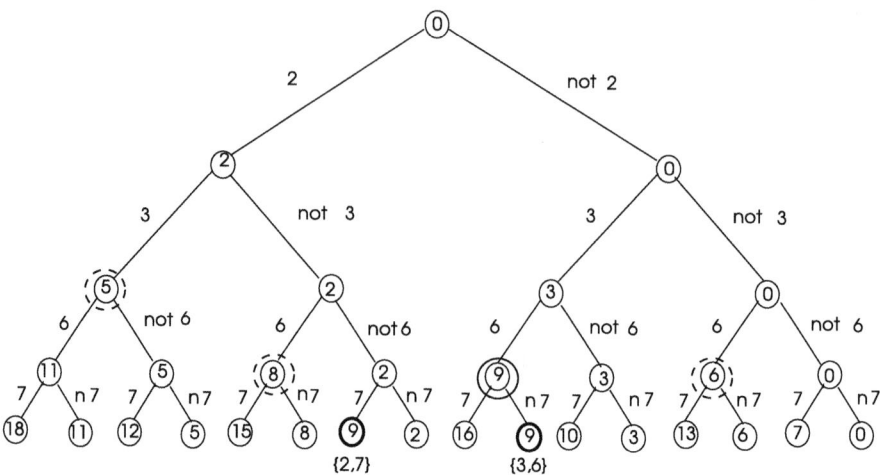

Fig. 3.9 State-space tree for the subset sum example

solutions. We have 2^n possible subsets and a binary tree representing the state-space tree will have 2^n leaves with one or more leaves providing the solutions if they exist.

Given $S = \{2, 3, 6, 7\}$ and $M = 9$, a state-space tree can be formed as shown in Fig. 3.9. The nodes of the tree show the sum accumulated up to that point from the root down the tree and we start with 0 sum. We consider each element of the set S in increasing order and at length i from the root, the element considered is the ith element of S. At each node, we have the left branch showing the decision if we include the element and the right branch proceeds to the subtree when we do not include that element in the search. The nodes shown in dashed circles are the points in the search where we do not need to proceed any further since the requirement can not be met if we do. For example, when we include the first two elements of the set, we have 5 as the accumulated sum and adding the third element, 6 will give 11 which is more than M. So we do not take the subtree rooted at left branch of node 5. Similarly, selecting 3 and 6 gives us the solution and we report it but still backtrack since all solutions are required. If only one solution was required, we would have stopped there.

3.8.4   Branch and Bound

Branch and bound method is similar to backtracking approach used for decision problems, which is modified for optimization problems. The aim in solving an optimization problem is to maximize or minimize an objective function and the result found is the *optimal solution* to the problem. Branch and bound algorithms employ the state-space trees as in backtracking with the addition of the record of the best solution *best* found up to that point in the execution. Moreover, as the execution progresses, we need the limit on the best value $next_i$ that can be obtained for each node i of the tree if we continue processing from that node. This way, we can compare these values and if $next_i$ is no better than *best*, there is no need to process the subtree rooted at node i.

We will illustrate the general idea of this method by the traveling salesperson problem (TSP) in which a salesperson starts her journey from a city, visits each city, and returns to the original city using a minimal total distance. This, in fact, is the Hamiltonian cycle problem with edge weights. Let $G = (V, E)$ be the undirected graph that represents the cities and roads between them. For the sample graph of Fig. 3.10, we can see there are six possible weighted Hamiltonian cycles and only two provides the optimal route. The routes in both are the same but the order of visits are reversed. Note that starting from any vertex would provide the same routes. In fact, there are $(n - 1)!$ possible routes in a fully connected graph with n vertices.

A brute-force approach would start from a vertex a, for example, and search all possible Hamiltonian cycles using the state-space tree and then record the best route found. We need to define a lower bound value (*lb*) for the branch and bound algorithm. The *lb* value is calculated as the total sum of the two minimum weight edges from each vertex divided by two to get the average value. For the graph depicted in Fig. 3.10, this value is

$$((2 + 4) + (3 + 1) + (3 + 2) + (4 + 1))/2 = 10.$$

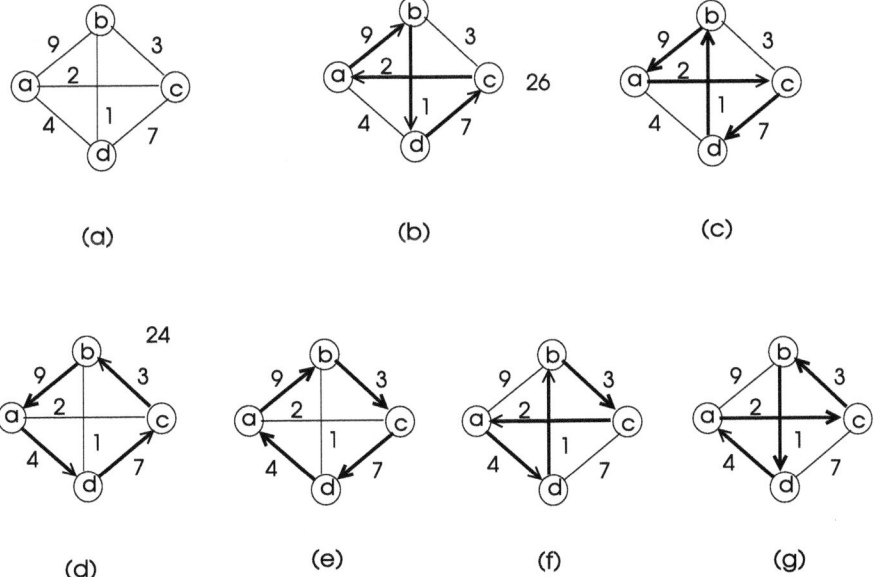

(a)　　　　　　　　(b)　　　　　　　　(c)

(d)　　　　　　(e)　　　　　　(f)　　　　　　(g)

Fig. 3.10 Hamiltonian cycles of a simple weighted graph

We can now start to build the state-space tree and every time we consider adding an edge to the existing path, we will modify the values in the lower bound function as affected by the selection of that edge and calculate a new lower bound. Then, we will select the edge with the minimum lower bound value among all possible edges. The state-space tree of the branch and bound algorithm for TSP in the graph of Fig. 3.10 is depicted in Fig. 3.11. The optimal Hamiltonian cycles are a, c, b, d, a and a, d, b, c, a. These paths correspond to paths (f) and (g) of Fig. 3.10.

Key to the operation of any branch and bound algorithm is the specification of the lower bound. When search space is large, we need this parameter to be easily computed at each step yet to be selective enough to prune the unwanted nodes of the state-space tree early. For the TSP example, another lower bound is the sum of the minimum entries in the adjacency matrix A of the graph G. This is a solid lower bound since we are considering the lightest weight edge from each vertex and we know we can not do better than this. For our example of Fig. 3.10, the lower bound calculated this way is 9. Computing the lower bound in each step would then involve deleting the row a and column b from A when the edge (a, b) is considered and then including lightest edges from each remaining vertices that are not connected to a or b to compute the lower bound for including the edge (a, b) in the path.

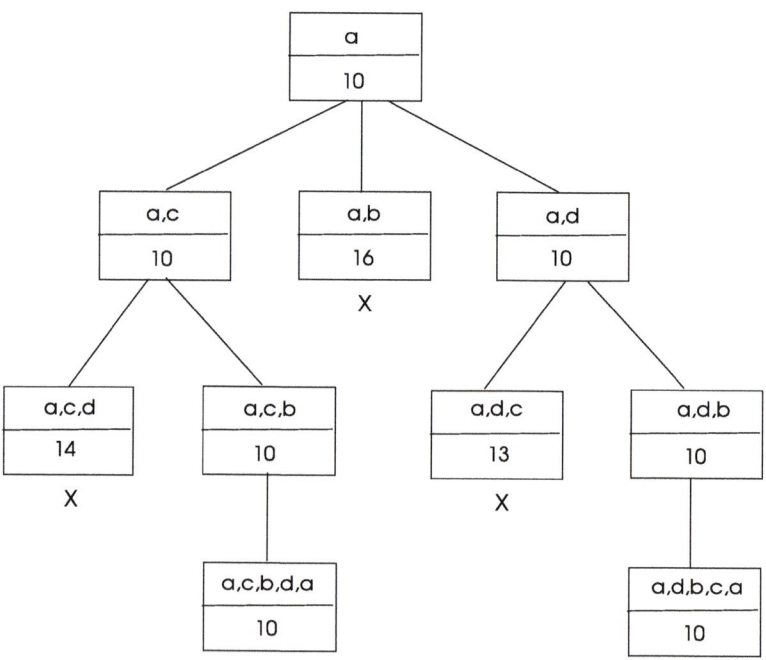

Fig. 3.11 The state-space tree for the graph of Fig. 3.10. Each tree node has the path and the lower bound value using this path

3.9 Major Design Methods

Given a problem to be solved by an algorithm, we can approach the design using various methods. One basic approach is the *brute-force* method in which we basically evaluate all possibilities. It is simple and easy to apply but usually does not have favorable performance. Other commonly employed techniques are the greedy method, divide and conquer method, and dynamic programming.

3.9.1 Greedy Algorithms

The *greedy method* searches for solutions that are locally optimal based on the current state. It chooses the best alternative available using the currently available information. A real-life example is the change provided by a cashier in a supermarket. Commonly, the cashier will select the largest coin to result in least number of coins in the next step which is optimal in some coin combinations such as in the U.S. In many cases, following the locally best solution at each step will not yield an overall optimal solution. However, in certain problems, the greedy method provides optimal solutions. Finding shortest paths between the nodes of a weighted graph and constructing a minimum spanning tree of a graph are examples of greedy method that can be used efficiently to find the solutions. Greedy algorithms can also be used to find approximate solutions to some problems. We will describe Kruskal's algorithm to find the minimum spanning tree (MST) as an example greedy graph algorithm.

3.9.1.1 Kruskal's Algorithm

Kruskal's MST algorithm orders the edges of the weighted graph $G(V, E, w)$ in nondecreasing weights. Then, starting from the lightest weight edge, edges are included in the partial MST T' as long as they do not form cycles with the edges already contained in T'. This process continues until all edges in the queue are processed. This algorithm consists of the following steps:

1. **Input**: A weighted undirected graph $G = (V, E, w)$
2. **Output**: MST T of G
3. Sort edges of G in nondecreasing order and place them in Q
4. $T \leftarrow \emptyset$
5. **while** Q is not empty
6.   Pick the first element (u, v) from Q that does not form a cycle with any edge of T
7. $T \leftarrow T \cup \{(u, v)\}$.

Figure 3.12 shows the iterations of Kruskal's algorithm in a graph.

The complexity of this algorithm is $O(n^2)$ as the *while* loop is executed for all vertices and the search for the minimum weight edge for each vertex will take $O(n)$ time. This complexity can be reduced to $O(m \log n)$ time by using binary

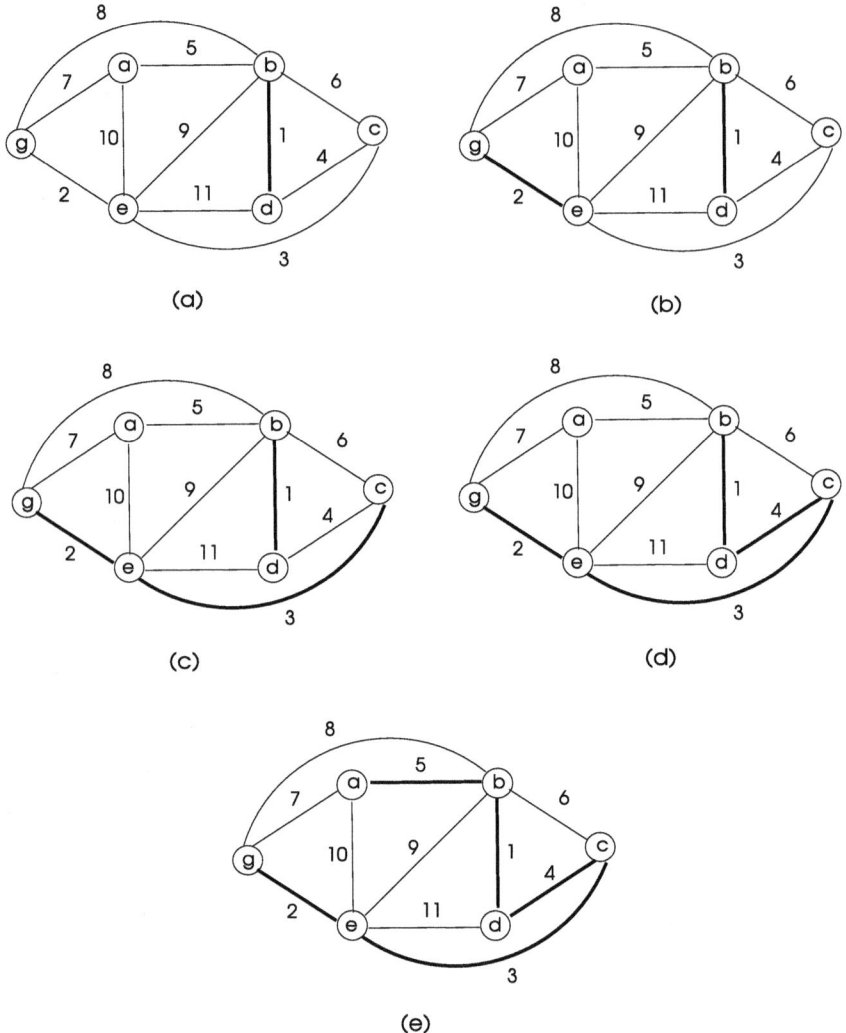

Fig. 3.12 Running of Kruskal's MST algorithm for a sample graph

heaps and adjacency lists. We will review this algorithm in more detail when we investigate MST algorithms for graphs in Chap. 7.

Analysis
The weights of edges of the graph G can be sorted in $O(m \log m)$ time. We can use the data structure *union-find* described alternatively in Chap. 7 which results in $\log n$ time for searching and testing for m edges can then be done in $O(m \log n)$ time.

3.9.2   Divide and Conquer

The *divide and conquer strategy* involves dividing a problem instance into several smaller instances of possibly similar sizes. The smaller instances are then solved and the solutions to the smaller instances are then typically combined to provide the overall solution. Solving the smaller instances and combining these solutions is performed recursively.

3.9.2.1   Taking Power with Divide and Conquer

We have seen the recursive algorithm to find the nth power of an integer in Sect. 3.4. Let us attempt to find the nth power of an integer using the divide and conquer method this time. Dividing step involves halving the problem until the base case is encountered. Each call to the function results in calling the same function with half of the power for even n and half of one less than n for odd integers as shown in Algorithm 3.7.

Algorithm 3.7 *Power_DivideConq*

1: **function** POWER(**int** a, **int** n)
2:    **if** $n = 1$ **then**   ▷ base case
3:        return a
4:    **else**
5:        **if** n MOD $2 = 0$ **then**
6:            **return** $Power(a, n/2) \times Power(a, n/2)$
7:            **else return** $Power(a, (n-1)/2) \times Power(a, (n-1)/2) \times a$
8:        **end if**
9:    **end if**
10: **end function**

For example, in order to compute 2^9, we make the following calls shown in Fig. 3.13.

This algorithm has the following recurrence relation:

$$T(n) = T(n/2) + \Theta(1)$$

which has a solution as $O(\log n)$.

3.9.2.2   Fibonacci Series with Divide and Conquer

Let us test another example for this method; we want to find Fibonacci series which has 0 and 1 as the first two elements, and each term thereafter is the sum of the two previous terms with 0, 1, 1, 2, 3, 5, ..., $F(n-1) + F(n-2)$ as the first n terms. This simple series has numerous diverse applications. We can form this algorithm recursively in a straightforward manner as shown in Algorithm 3.8 using the divide and conquer method.

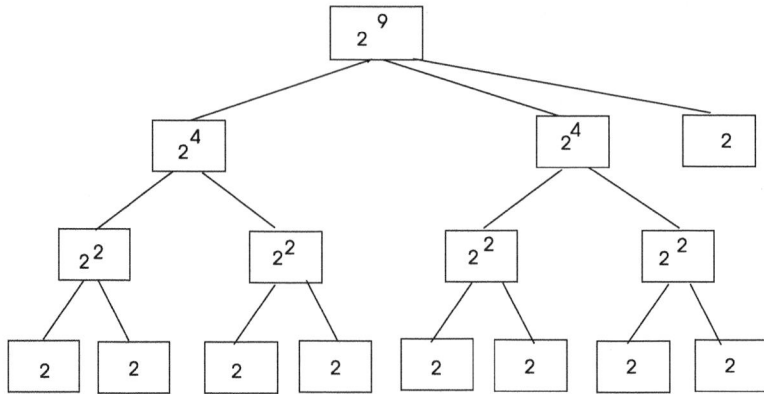

Fig. 3.13 Recursive calls to compute 2^9 by Algorithm 3.7

Algorithm 3.8 *Fibo_DivideConq*

1: **function** FIBO(**int** n)
2:     **if** $n \leq 1$ **then**   ▷ base case
3:         return 1
4:     **else**
5:         **return** $Fibo(n-1) + Fibo(n-2)$ ▷ recursive call
6:     **end if**
7: **end function**

Let us try to analyze the recursive calls to this function. For example, $F(9)$ is $F(8) + F(7)$; $F(8)$ is $F(7) + F(6)$ and moving in this direction, we will reach the base values of $F(1)$ and $F(0)$. Note the value of $F(7)$ is calculated twice to find $F(9)$. This recurrence relation has exponential time complexity; however, we can see some of the calls are repeated, for example, $F(3)$ has to be calculated twice and deduce this is not the best way to solve this problem. Dynamic programming solution provides a solution with better complexity as we will see next.

3.9.3   Dynamic Programming

Dynamic programming is an algorithmic method to solve problems that have overlapping subproblems. The word *programming* means a tabular method rather than actual computer programming. The choices made are based on the current state, and hence, the word dynamic is used. A dynamic programming algorithm requires a recurrence relation; the tabular computation and traceback to provide the optimal solution.

In this method, the problem is divided into smaller instances first, the small problems are solved, and the results are stored to be used in the next stage. It is similar to divide and conquer method in a way as it recursively divides the instance of the problem into smaller instances. However, it computes the solution to smaller instances, records them, and does not recalculate these solutions as in divide and conquer.

3.9.3.1 Fibonacci Series with Dynamic Programming

Let us attempt to form the Fibonacci series using dynamic programming this time. We still divide the problem into smaller instances but we save the intermediate results to be used later. The dynamic programming solution to Fibonacci series is shown in Algorithm 3.9 where the intermediate results are stored in a table and are used subsequently resulting in $O(n)$ time complexity.

Algorithm 3.9 *Dynamic_Fibo*

1: **Input** : int n
2: **Output** : array $F[n]$
3: **int** i
4: $F[0] \leftarrow 0$; $F[1] \leftarrow 1$
5: **for** $i=2$ to n **do**
6:     $F[i] \leftarrow F[i-1] + F[i-2]$
7: **end for**

3.9.3.2 Bellman–Ford Algorithm

As an example of dynamic programming application in graphs, we will consider the shortest path problem. We are given a weighted graph that has edges labeled with some real numbers, showing costs associated with edges. Given a weighted $G = (V, E, w)$ with $w : E \rightarrow \mathbb{R}$ and a source vertex s, our aim is to find the shortest path from vertex s to all other vertices, the path having the minimum total weight between the source and a destination vertex. Bellman–Ford provided a dynamic programming algorithm for this purpose which can work with negative weight edges. It provides all shortest paths and detects negative cycles if they exist. The main idea of this algorithm is to use the previously calculated shortest paths and if the currently calculated path has a smaller weight, update the shortest path with the current one. It is a dynamic programming algorithm as we use the prior results without recalculating them. At each iteration, we extend the reachable vertices from the source by one more hop as depicted in Algorithm 3.10.

Algorithm 3.10 $BellFord\_SSSP$

1: **Input** : $G = (V, E)$, s ▷ a directed or undirected edge-weighted graph, a source vertex s
2: **Output** : $d_u, \forall u \in V$ ▷ distances to s
3: $d_s \leftarrow 0$;
4: **for all** $i \neq s$ **do**   ▷ initialize distances and predecessors
5:     $d_i \leftarrow \infty$
6: **end for**
7: **for** $i = 1$ to $n - 1$ **do**
8:     **for all** $(u, v) \in E$ **do**   ▷ update distances
9:        $d_u \leftarrow \min\{d_u, d_u + w(u, v)\}$
10:     **end for**
11: **end for**
12: **for all** $(u, v) \in E$ **do**   ▷ report negative cycle
13:     **if** $d_u + w(u, v) > d_v$ **then**
14:        **report** *"Graph contains a negative cycle"*
15:     **end if**
16: **end for**

Example

An undirected and edge-weighted sample graph is depicted in Fig. 3.14 where we implement this algorithm with source vertex g. The maximum number of changes of the shortest path for a vertex is $n - 1$ requiring $n - 1$ iterations of the outer loop at line 8. Each loop requires at most m edge checking resulting in $O(nm)$ time complexity for this algorithm. We will see a more detailed version of this algorithm that also provides a tree structure in Chap. 7.

3.10   Chapter Notes

We provided a dense review of basic algorithmic methods with focus on graph algorithms in this chapter. We first reviewed basic algorithm structures and described asymptotic analysis of time complexity of general algorithms. The worst-case, best-case, and average-case analysis of an algorithm gives us insight into the time it requires when input size is large. We noted worst-case time complexity called *big-Oh notation* is commonly required for algorithms to predict their running time. Our approach in presenting algorithms is to first describe the main idea with key points of the algorithm verbally. We then provide the pseudocode of the algorithm to detail its operation in a programming language like syntax which is accompanied by an example operation in a small sample graph in many cases. We then show that the algorithm works correctly using various proof methods. We also provide the worst-case timing analysis of the graph algorithm under consideration. We reviewed the necessary background for all these steps including main proof strategies such as loop variants and induction which are commonly used.

The second topic we investigated was NP-completeness. We classified problems as the ones in P, in NP, and the problems that are NP-hard and NP-complete. A problem that has a solution in polynomial time is in P; a decision problem that

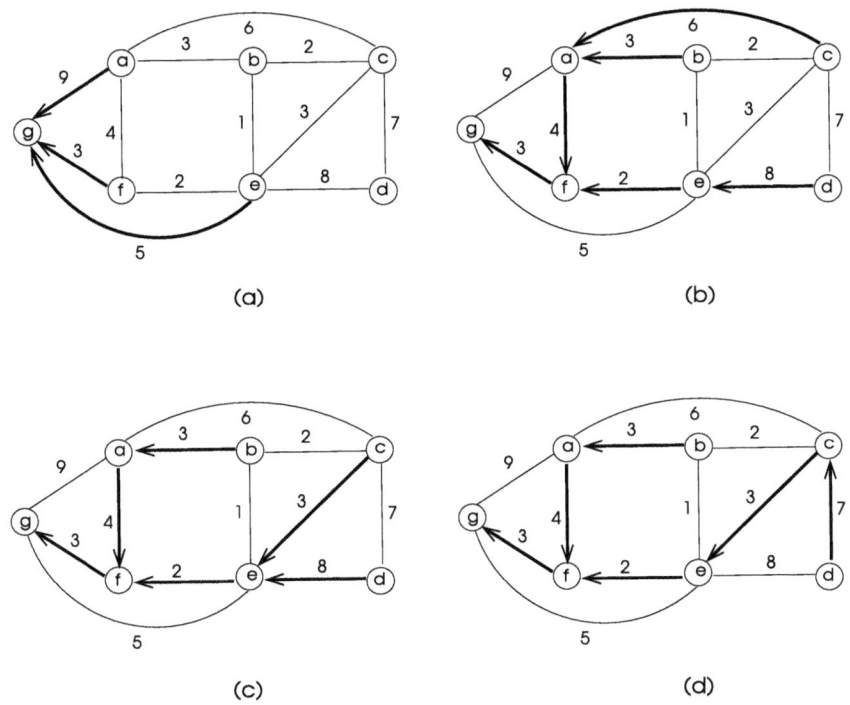

Fig. 3.14 An example running of Bellman–Ford algorithm in a sample graph for the source vertex g. The first reachable vertices are a, f, and e which are included in the shortest path tree T. This tree is updated at each iteration when less cost paths compared to the previous ones are found

has a certifier which can provide a yes or no answer to a certificate is in NP. An NP-complete problem is in NP and is as hard as any other problem that is NP-hard. The problems C-SAT and SAT were the first problems that were shown to be NP-hard and reductions are used to prove that a problem is as hard as another problem. Once we know that a problem is NP-hard, we may do one of the following. We can search an approximation algorithm that provides a suboptimal solution in polynomial time. An approximation algorithm needs to be proven to have an approximation ratio to the optimal solution of the problem. Proofs of such algorithms may be complex as we do not know the optimal algorithm itself to show how the approximation algorithm converges to it. We may use heuristics which are commonsense rules that are shown to work with a wide range of inputs in practice. However, there is no guarantee they will work for every input. Nevertheless, for many graph problems in applications such as in bioinformatics, using heuristics continues to be the only viable method.

In the final part, we described fundamental algorithmic methods which are the greedy method, divide and conquer strategy, and dynamic programming. Greedy methods attempt to find an optimal global solution by always selecting the local optimum solutions. These local solution choices are based on what is known so far and may not lead to an optimal solution in general. However, we saw greedy algorithms provide optimal solutions in a few graph problems including shortest paths and minimum spanning trees. In the divide and conquer method, the problem at hand is divided into a number of smaller problems which are solved and solutions are merged to find the final solution. These algorithms often employ recursion due to the nature of their operation. Dynamic programming also divides the problem into smaller parts but makes use of the partial solutions found to obtain the general solution. The background we have reviewed is related mainly to sequential graph algorithms and we will see that further background and considerations are needed for parallel and distributed graph algorithms in the next chapters. There are a number of algorithm books which provide the basic background about algorithms including the one by Skiena [6], Cormen et al. [1], and by Tardos and Kleinberg [5].

Exercises

1. Work out by proofs the worst-case running times for the following:
   a. $f(n) = 4n^3 + 5n^2 - 4$.
   b. $f(n) = 2^n + n^7 + 23$.
   c. $f(n) = 2n \log n + n + 8$.
2. Prove $3n^4$ is not $O(n^3)$. Note that you need to show there are no valid constant c and a threshold n_0 values for this worst case to hold.
3. Write the pseudocode of a recursive algorithm to find the sum of first n positive integers. Form and solve the recurrence relation for this algorithm to find its worst-time complexity.
4. Prove $n! \leq n^n$ by the induction method.
5. A clique is the subgraph of a graph in which every vertex is adjacent to all other vertices in the clique. Given a graph $G = (V, E)$ and its subgraph V', V' is a maximal clique of G if and only if $G - V'$ is a maximal independent set. Prove that the decision problem of finding whether a graph has a maximal clique of size k is NP-complete by reduction from IND.
6. Find all minimum vertex covers, maximum independent sets, and minimum dominating sets of the sample graph shown in Fig. 3.15.
7. Work out the vertex cover for the sample graph in Fig. 3.16 by first finding its maximal independent set and then reduction to vertex cover.

Fig. 3.15 Sample graph for Exercise 8

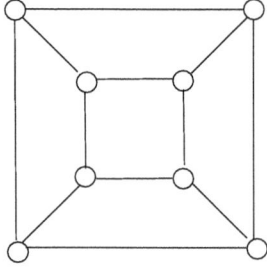

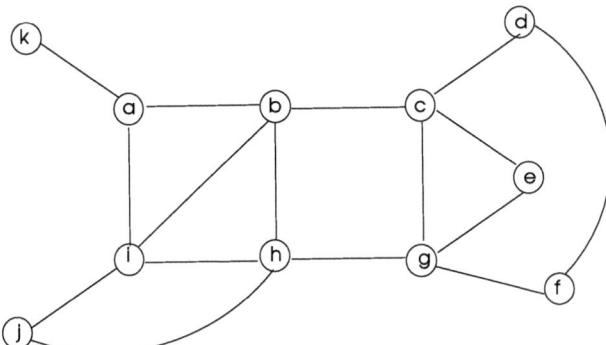

Fig. 3.16 Sample graph for Exercise 9

Fig. 3.17 Sample graph for
Exercise 11 and 12

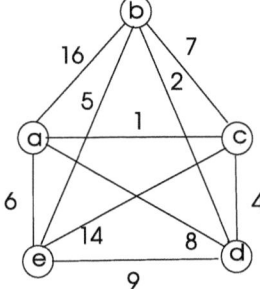

8. Find the solution to the following 3-SAT formula using reduction to independent set:

$$(x_1 \vee \overline{x_2} \vee x_3) \wedge (\overline{x_1} \vee x_2 \vee \overline{x_3} \wedge (\overline{x_1} \vee \overline{x_2} \vee x_3)$$

9. Work out the optimal path for the TSP problem in the graph depicted in Fig. 3.17 using the lower bound calculated as the sum of the mean weights of two lightest weight edges from each vertex. Draw the state-space tree by showing all the pruned nodes.

10. Find the optimal TSP tour for the graph of Fig. 3.17 this time setting the lower bound as the sum of the weights of the lightest weight edges at each vertex.

11. Find the minimal vertex cover for the sample graph in Fig. 3.18 using the 2-approximation algorithm. Show each iteration of this algorithm and work out the approximation achieved by your iterations.

12. Find the MST of the sample graph of Fig. 3.19 using Kruskal's algorithm by showing each iteration.

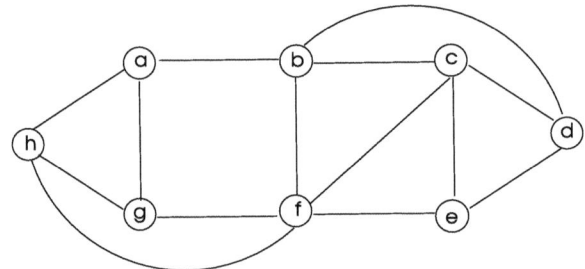

Fig. 3.18 Sample graph for Exercise 13

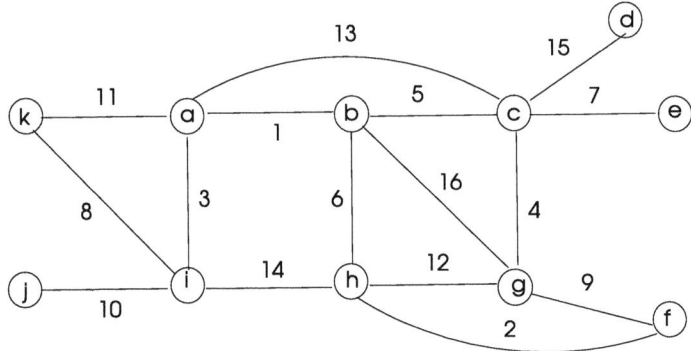

Fig. 3.19 Sample graph for Exercise 14

References

1. Cormen TH, Stein C, Rivest RL, Leiserson CE (2009) Introduction to algorithms, 3rd edn. MIT Press, Cambridge. ISBN-13: 978-0262033848
2. Dasgupta S, Papadimitriou CH, Vazirani UV (2006) Algorithms. McGraw-Hill, New York
3. Garey MR, Johnson DS (1979) Computers and intractability: a guide to the theory of NP completeness. W.H Freeman, New York
4. Karger D (1993) Global min-cuts in RNC and other ramifications of a simple mincut algorithm. In: Proceedings of the 4th annual ACM-SIAM symposium on discrete algorithms
5. Kleinberg J, Tardos E (2005) Algorithm design, 1st edn. Pearson, London. ISBN-13: 978-032129535
6. Skiena S (2008) The algorithm design manual. Springer, Berlin. ISBN-10: 1849967202

Parallel Graph Algorithms

4

4.1   Introduction

We have reviewed basic concepts in algorithms and main methods to design sequential algorithms with emphasis on graph algorithms in the previous chapter. Our aim in this chapter is to investigate methods for parallel algorithm design with emphasis on graph algorithms again. Parallel processing is commonly used to solve computationally large and data-intensive tasks on a number of computational nodes. The main goal in using this method is to obtain results much faster than would be acquired by sequential processing and hence improve the system performance. Parallel processing requires design of efficient parallel algorithms and this is not a trivial task as we will see.

There are various tasks involved in parallel running of algorithms; we first need to identify the subtasks that can be executed in parallel. For some problems, this step can be performed conveniently; however, many problems are inherently sequential and we will see a number of graph algorithms fall in this category. Assuming a task T can be divided into n parallel subtasks $t_1, \ldots, t_n$, the next problem is to assign these subtasks to processors of the parallel system. This process is called *mapping* and is denoted as a function from the task set T to the processor set P as $M : T \rightarrow P$. Subtasks may have dependencies so that a subtask t_j may not start before a preceding subtask t_i finishes. This is indeed the case when t_i sends some data to t_j in which case starting t_j without this data would be meaningless. If we know all of the task dependencies and also characteristics such as the execution times, we can distribute tasks evenly to the processors before running them and this process is termed as *static scheduling*. In many cases, we do not have this information beforehand and *dynamic load balancing* is used to provide each processor with fair share of the workload at runtime based on the load variation of the processes.

We can have shared memory parallel processing in which computational nodes communicate via a shared memory and in this case, the global memory should be

© The Author(s), under exclusive license to Springer Nature Switzerland AG 2026
K. Erciyes, *Guide to Graph Algorithms*, Texts in Computer Science,
https://doi.org/10.1007/978-3-032-05294-0_4

protected against concurrent accesses. In distributed memory parallel processing, communication and synchronization among parallel tasks are handled by sending and receiving messages over a communication network without any shared memory. Parallel programming implicates writing the actual parallel code that will run on the parallel processors. For shared memory parallel processing, lightweight processes called *threads* are widely used and for parallel processing applications in distributed memory architectures; the Message Passing Interface (MPI) is a commonly implemented interface standard. We will see that we can have different models of parallel processing at hardware, algorithm, and programming modes. These models are related to each other to some extent as will be described. For example, message passing model at algorithmic level requires distributed memory at hardware level which can be implemented by a message passing programming model that runs the same code with different data or different codes possibly with different data.

We start this chapter by describing fundamental concepts of parallel processing followed by the specification of models of parallel computing. We then investigate parallel algorithm design methods focusing on parallel graph algorithms which require specific techniques. Static and dynamic load balancing methods to evenly distribute parallel tasks to processors are outlined and we conclude by illustrating the parallel programming environments.

4.2   Concepts and Terminology

Parallel computing involves simultaneous employment of multiple computational resources to solve a computational problem. Commonly applied steps of parallel computing are partitioning the overall task into a number of parallel subtasks: designing inter-task communication and mapping of subtasks to processors. These steps are dependent on selecting certain criteria in a step that affect the other processes. We will investigate these tracks in detail later in this chapter. Some of the common terminology used in parallel computing can be described as follows.

- *Parallelism versus concurrency*: *Parallelism* indicates at least two tasks are running physically at the same time on different processors. *Concurrency* is achieved when two or more tasks run at the same time frame but they may not run at the same physical time. Tasks that communicate using shared memory in a single-processor system are concurrent but are not parallel. Concurrency is more general than parallelism and encompasses parallelism.
- *Fine-grain versus coarse-grain parallelism*: When the computation is partitioned into number of tasks, the size of tasks as well as the size of data they work on affects their running time. In *fine-grain* parallelism, tasks communicate and synchronize frequently and coarse-grain parallelism involves tasks with larger computation times that communicate and synchronize much less frequently.

- *Embarrassingly parallel*: The parallel computation consists of independent tasks that have no inter-dependencies in this mode. In other words, there are no precedence relationships or data communications among them. The speedup achieved by these algorithms may be close to the number of processors used in the parallel processing system.
- *Multi-core computing*: A multi-core processor contains more than one of the processing elements called *cores*. Most contemporary processors are multi-core and *multi-core computing* is running on programs in parallel on multi-core processors. The parallel algorithm should make use of the multi-core architecture effectively and the operating system should provide effective scheduling of tasks to cores in these systems.
- *Symmetric multiprocessing*: A symmetric multiprocessor (SMP) contains a number of identical processors that communicate via a shared memory. Note that SMPs are organized on a coarser scale than multi-core processors which contain cores in a single integrated circuit package.
- *Multiprocessors*: A multiprocessor consists of a number of processors which communicate through a shared memory. We typically have a set of microprocessors connected by a high-speed parallel bus to a global memory. Memory arbitration at hardware level is needed in these systems.
- *Multicomputer*: Each processor has private memory and typically communicates with other microcomputers by sending and receiving messages. There is no global memory in general.
- *Cluster computing*: A cluster is a set of connected computers that communicate and synchronize using messages over a network to finish a common task. A cluster is envisioned as a single computer by the user and it acts as a single computer by the use of suitable software. Note that a cluster is a more abstract view of a multiprocessor system, possibly with added software capabilities such as dynamic load balancing.
- *Grid computing*: A grid is a large number of geographically distributed computers that work and cooperate to achieve a common goal. Grid computing provides a platform of parallel computing mostly for embarrassingly parallel applications due to unpredictable delays in communication.
- *Cloud computing*: Cloud computing enables sharing of networked computing resources for various applications using the Internet. It provides delivery of services such as online storage, computing power, and specialized user applications to the user.
- *Parallel processing with GPUs*: A graphical processing unit (GPU) is a co-processor with enhanced graphic processing capabilities. CUDA is a parallel computing platform that uses GPUs for parallel processing formed by NVIDIA [9].

4.3   Parallel Architectures

A processing unit has a processor, a memory, and an input/output unit in its most basic form. We need a number of processors that should execute in parallel to perform subtasks of a larger task. These subtasks need two basic operations: communication to transfer data produced and synchronization. We can have these operations in shared memory or distributed memory configurations as described next. A general trend in parallel computing is to employ general-purpose off-the-shelf processors connected by a network due to the simplicity and the scalability of such configurations.

4.3.1   Shared Memory Architectures

In a shared memory architecture as shown in Fig. 4.1a, each processor has some local memory, and interprocess communication and synchronization are performed using a shared memory that provides access to all processors. Data is read from and written to the shared memory locations; however, we need to provide some form of control on access to this memory to prevent race conditions. We can have a number of shared memory modules as shown in Fig. 4.1b with the network interface to these modules providing concurrent accesses to different modules by different processors.

The main advantage of shared memory parallel processors is fast data access to memory. However, the shared memory should be protected against concurrent accesses by the parallel tasks and this process should be controlled by the programmer in many cases. Another major disadvantage of shared memory approach is the limited number of processors that can be connected due to the bottleneck over the bus while accessing the shared memory. In conclusion, we can state shared memory systems are not easily scalable.

4.3.2   Distributed Memory Architectures

There is no shared memory in distributed memory architectures, and the communication and synchronization are performed exclusively by sending and receiving of messages. For this reason, these systems are also called *message passing* systems. Each processor has its own local memory, and address space is not shared among processors. One important advantage of distributed memory systems is we can use off-the-shelf computers and connect them using a network to have a parallel computing system. Access to local memories is faster than global memory access; however, the algorithm designer should take the responsibility of how and when to synchronize the processor and transfer of data. The main advantage of distributed memory systems is their scalability and use of the off-the-shelf computers. Also, there are no overheads in memory management as in shared memory. As the

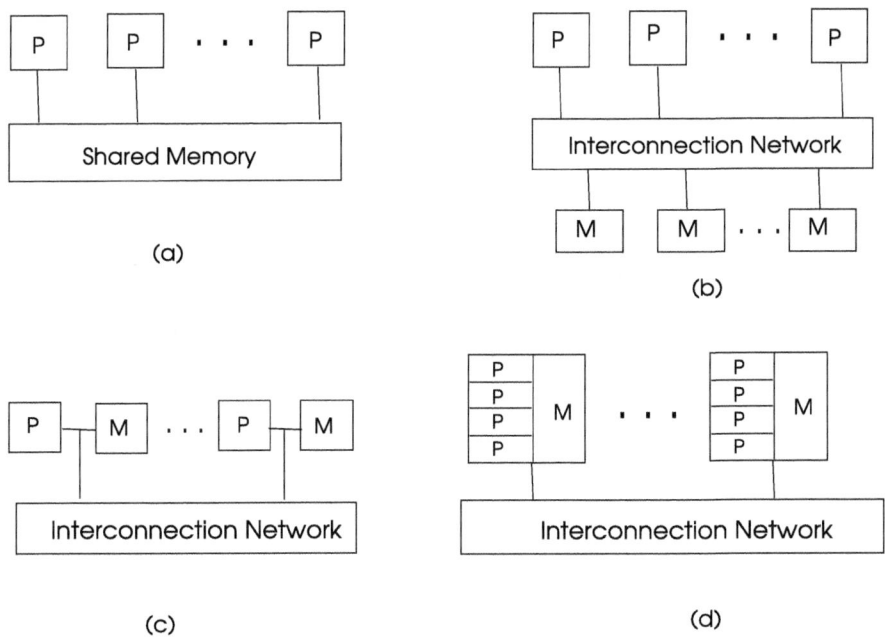

Fig. 4.1 Parallel computing architectures, **a** shared memory architecture with a single global memory, **b** shared memory architecture with a number of memory modules, **c** distributed memory architecture, **d** distributed and shared memory architecture

main disadvantage, it is the task of the algorithm designer to manage data communication using messages; and communication over the network is commonly serial which is slower than the parallel communication of the shared memory system. Also, some algorithms are based on sharing of global data converting and mapping of which to distributed memory may not be a trivial task.

In many cases, contemporary parallel computing applications use both shared memory and distributed memory architectures. The nodes of the parallel computing system are symmetric multiple processors (SMPs) that are connected via a network to other SMPs in many cases. Each SMP node works in shared memory mode to run its tasks but communicates with other nodes in distributed memory mode as shown in Fig. 4.1d.

4.3.3   Flynn's Taxonomy

Flynn's taxonomy of parallel computers classifies parallel computers based on how they process instructions and data as follows [2]:

- *Single-Instruction Single-Data* (SISD) computers: These computers execute one instruction per unit time on one data item. First, computers with one processor operated in this mode. Most modern processors have a number of cores in their processing unit.

- *Single-Instruction Multiple-Data* (SIMD) computers: We have synchronous parallel processors executing the same instruction on different data in this architecture. Processor arrays and vector pipelines are the two main examples in this model. These parallel processors are suitable for scientific applications that deal with large data.
- *Multiple-Instruction Single-Data* (MISD) computers: These processors execute different instructions on the same data. This model is not practical except the case of pipelining stages within the processor.
- *Multiple-Instruction Multiple-Data* (MIMD) computers: We have processors running different instructions on different data in this case. This model exhibits the most common and versatile model of parallel computing. Parallel computer clusters with network communications, grids, and some supercomputers fall in this category.

Special architectures provide communication links that can transfer data between multiple source–destination pair of nodes in parallel. In a *hypercube*, processors are connected as the vertices of a cube as shown in Fig. 4.2a for a hypercube of size 4. The largest distance between any two processors is $\log n$ in a hypercube with n processors, and a hypercube of size d has $n = 2^d$ processors. Each node has an integer label which has a difference of one bit from any of its neighbors providing a convenient way of detecting neighbors while designing parallel algorithms on the hypercube.

A *linear array* consists of connected processors connected in a line each having a left and right neighbor except the starting and terminating processor as depicted in Fig. 4.2b. A *ring network* has processors connected in a cycle as in Fig. 4.2c. The *mesh architecture* has a 2D array of processors connected as a matrix and the balanced tree architecture is a tree with nodes as processors each of which has two children except the leaves as shown in Fig. 4.2c, d. Few parallel computers including Cray T3D, SGI, and IBM Blue Gene have mesh structures.

4.4   Models

We need a model of parallel computing which specifies what can be done in parallel and how these operations can be performed. Two basic models based on architectural constraints are described next.

4.4.1   PRAM Model

The parallel random access memory (PRAM) extends the basic RAM model to parallel computing. The processors are identical with some local memory, and there is a global shared memory used for communication and synchronization.

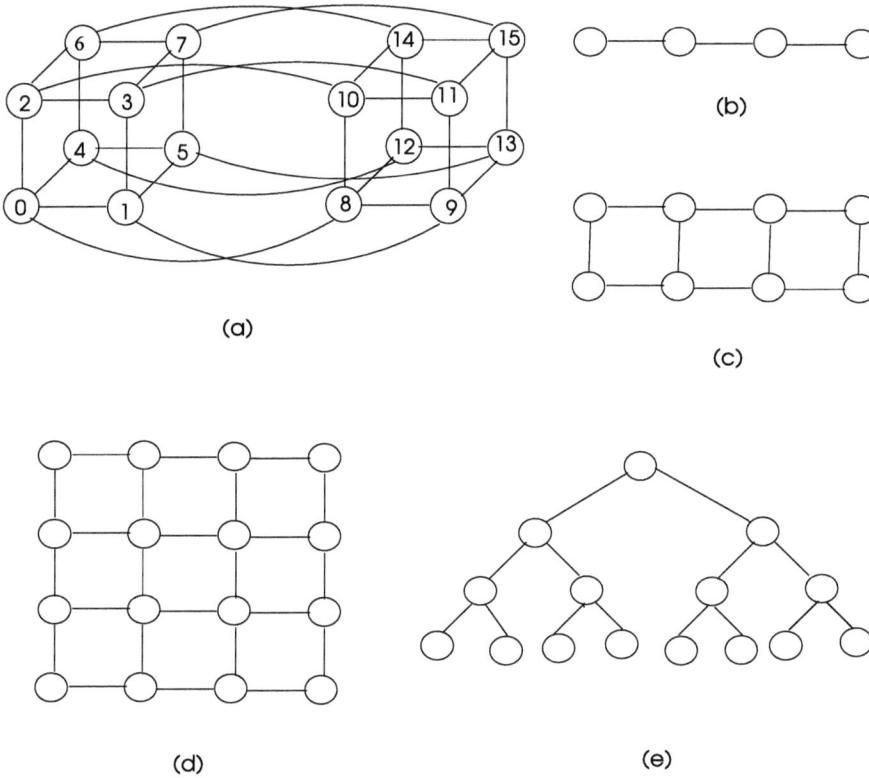

(a)

(b)

(c)

(d)

(e)

Fig. 4.2 Special parallel processing architectures, **a** hypercube of dimension 4, **b** linear array, **c** mesh of two dimensions, **d** balanced binary tree

Therefore, it assumes the shared memory architecture described in Sect. 4.3.1. Processors work synchronously using a global clock and at each time unit, a processor can perform a read from a global or local memory location; execute a single RAM operation, and write to one global or local memory location. PRAM models are classified according to the read or write access rights to the global memory as follows.

- The exclusive-read-exclusive-write (EREW) model requires each processor to read from or to write to global memory locations exclusively, one processor at a time.
- The concurrent-read-exclusive-write (CREW) model allows concurrent reading of the same global memory location by more than one processor; however, writing to a global memory location has to be done exclusively.

- The concurrent-read-concurrent-write (CRCW) model is the most versatile of all as it allows both concurrent reads and concurrent writes. In case of concurrent writes, a mechanism is needed to decide what is to be written as last to the memory location. There may be arbitrary writes or writing with priority.

The PRAM model is idealistic as it assumes unbounded number of processors and unbounded size of shared memory. However, it simplifies parallel algorithm design greatly by abstracting communication and synchronization details. We have n processors that work synchronously and we can have at most n operations in any unit time. We can therefore use this model to compare various parallel algorithms to solve a given problem. We will see examples of PRAM graph algorithms in Sect. 4.8.

4.4.2   Message Passing Model

Message passing model is based on the distributed memory architecture. There is no shared memory, and the parallel tasks that run on different processors communicate and synchronize using two basic primitives: *send* and *receive*. Calls to these routines may be *blocking* or *non-blocking*. A blocking send will stop the caller until an acknowledgement from the receiver is received. A blocking receive on the other hand will prevent the caller from continuing until a specific/general message is received. We can assume that the network is reliable and transfers the messages to its destination with high reliability and the receiving task needs the data it receives before continuing. It is therefore general practice to have a non-blocking send and a blocking receive pair as shown in Fig. 4.3a. We may need the blocking send and receive pair of routines as shown in Fig. 4.3b for applications requiring high reliability such as in parallel hard real-time systems.

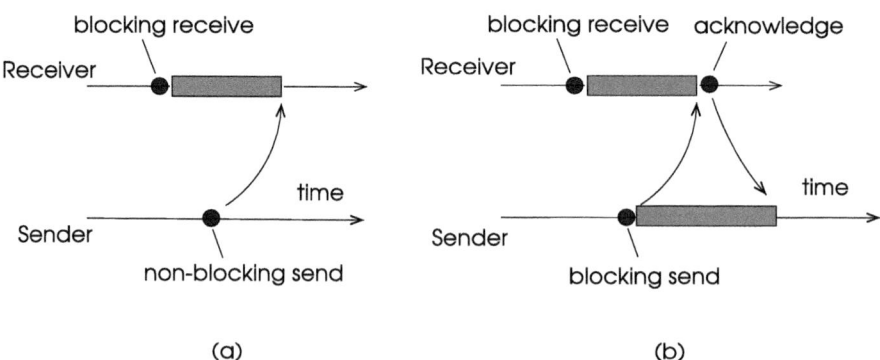

Fig. 4.3 Blocking and non-blocking communication modes, **a** non-blocking *send* and a blocking *receive*, **b** blocking *send* and a blocking *receive*. Blocked times are shown by gray rectangles. Network delays cause the duration between sending and receiving of a message

4.5 Analysis of Parallel Algorithms

We need to assess the efficiency of a parallel algorithm to decide its goodness. The running time of an algorithm, the number of processors it uses, and its cost are used to determine the efficiency of a parallel algorithm. The running time of a parallel algorithm T_p can be specified as

$$T_p = t_{fin} - t_{st}$$

where t_{st} is the start time of the algorithm on the first (earliest) processor and t_{fin} is the finishing time of the algorithm in the last (latest) processor. The depth D_p of a parallel algorithm is the largest number of dependent steps performed by the algorithm. The dependency in this context means a step cannot be performed before a previous one finishes since it needs the output from this previous step. If T_p is the worst-case running time of a particular algorithm A for a problem Q using p identical processors and T_s is the worst-case running time of the fastest known sequential algorithm to solve Q, the *speedup* S_p is defined as below:

$$S_p = \frac{T_s}{T_p}. \tag{4.1}$$

We need the speedup to be as large as possible for efficiency. The parallel processing time T_p increases with increased interprocess communication costs resulting in a lower speedup. Placing parallel tasks in fewer processors to reduce network traffic decreases parallelism and these two contradicting approaches should be considered carefully while allocating parallel tasks to processors. Efficiency of a parallel algorithm is defined as

$$E_p = \frac{S_p}{p} \tag{4.2}$$

where p is the number of processors. A parallel algorithm is said to be *scalable* if its efficiency remains almost constant when both the number of processors and the size of the problem are increased. Efficiency of a parallel algorithm is between 0 and 1. A program that is scalable with speedup approaching p has efficiency approaching 1. Let us analyze adding n numbers using k processors assuming n/k elements are distributed to each processor. Each p_i, $0 \le i < k$, finds its local sum in $\Theta(n/k)$ time. Then, the partial sums are added in $\log(k)$ time by k processors resulting in a total parallel time $T_p = \Theta(n/k + \log k)$. The sequential algorithm has a time complexity of $T_s = \Theta(n)$. Therefore, efficiency of this algorithm is

$$E_p = \frac{T_s}{kT_p} = \frac{n}{n + k \log k}. \tag{4.3}$$

We need E_p to remain as a constant c and solving for n yields

$$n = \frac{c}{1-c} k \log k. \tag{4.4}$$

Listing the efficiency values against the size of the problem n and the number of processors k, we can see for (n, k) values of $(64, 4)$, $(192, 8)$, and $(512, 16)$, the efficiency is 80% [5] for a maximum of $(512, 32)$ value; all other efficiency values are lower. Total cost or simply the *cost* of a parallel algorithm is the collective time taken by all processors to solve the problem. The cost on p parallel computers is the time spent multiplied by the number of processor as pT_p, and hence

$$T_p \geq \frac{T_s}{p}$$

which is called the *work law*. We are interested in the number of steps taken rather than physical time duration of the parallel algorithm. A parallel algorithm is *cost-optimal* if the total work done W is

$$W = pT_p = \Theta(T_s).$$

In other words, when its cost is similar to the cost of best-known sequential algorithm for the same problem, the parallelism achieved $\mathcal{P}$ can be specified in terms of these parameters as below:

$$\mathcal{P} = \frac{W}{S}$$

Let us illustrate these concepts by another parallel algorithm to add 8 numbers; this time each processor does one addition only. A possible implementation of this algorithm using four processors is shown in Fig. 4.4. We can see the number of dependent steps which is the depth of this algorithm is 3 and the total work done is 7 as 4, 2, and 1 additions are done in steps 1, 2, and 3. These problems are termed as *circuit problems* which include finding minimum/maximum values in an array and their depth is $\log n$ as can be seen. Total work done in these algorithms is $n - 1$ (Fig. 4.4).

4.6   Basic Communication Modes

Processes residing on different computing nodes of the parallel system need to communicate to finish an overall task. We can distinguish the two basic modes of communication between the processes: either all processes involved in the parallel task communicate or a group of processes communicate with each other. Viewed from another angle, there is also the architecture of the hardware that needs to be considered when specifying these communication modes. The following are the basic communication between all n processes of the system or the group [5].

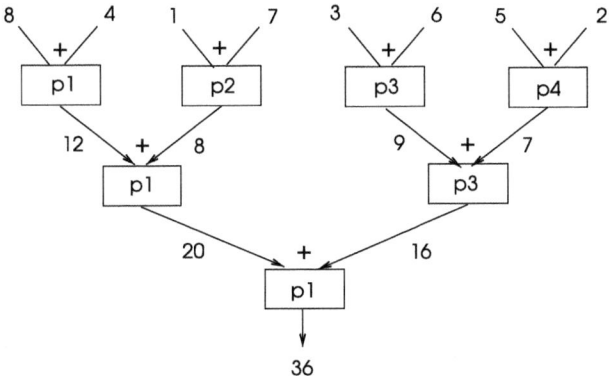

Fig. 4.4 Parallel summation algorithm with four processors $p_1, \ldots, p_4$ to add 8 numbers

- *One-to-All Broadcast*: A process p_i broadcasts a message m to all $n-1$ other processes. The message m is copied to the private memories of all processes in the distributed memory model and to the global memory in the shared memory model. In a tree-structured network, this mode can be realized by the root sending a message m to its children which transfer m to their children until leaves of the tree receive it. This transfer is achieved in $\Theta(\log n)$ time by the transfer of $O(n)$ messages.
- *All-to-One Reduction*: The data of each process of the parallel system is combined using an associative operator and the result is stored in a single memory location of a specific process.
- *All-to-All Broadcast and All-to-All Reduction*: We have each process p_i, $0 \le i < n$ of n processes sending message m_i to all other processes simultaneously. Each process stores messages m_i, $0 \le i < n$ at the end of this communication. In the *all-to-all reduction* mode is the reverse operation of all-to-all broadcast in which each process p_i stores n distinct messages sent by other processes in the end.
- *Scatter and Gather Operation*: A personalized unique message for each process is sent from a process in the *scatter* operation. The source process p_i has messages m_j, $0 \le j < n, j \neq i$ for each of the processes and each process $p_j, j \neq i$, receives its message m_j in the end. The reverse procedure is performed in the *gather* operation with each process sending its unique message to be gathered at one specific process.
- *All-to-All Personalized Communications*: In this mode, each process p_i has a distinct message m_i and sends this message to all other processes. All processes in the system receive messages m_i, $0 \le i < n$ at the end of this communication.

Let us try to implement all-to-all-broadcast communication in a hypercube of dimension d. Each node i has a distinct message m_i in the beginning and we require each node to have all messages m_i, $0 \le i < n$, with $n = 2^d$ at the end of the algorithm. We can have all neighbors exchange their messages along x-axis in the first step, then exchange the obtained result along y-axis, and finally along z-axis in the last step for a 3D hypercube as shown in Fig. 4.2. Note that the size of messages transmitted is doubled at each step and the total number of steps is d (Fig. 4.5).

The problem here is to write an algorithm that provides the data transfer specified above. We can make use of the labeling of the nodes in a hypercube which differ in one bit from any neighbor. The least significant bit (LSB) difference of a node from a neighbor is in x direction, the second LSB difference is in y direction, and so on. We can therefore bitwise exclusive-OR (XOR) of an identity of a node to find the identity of its neighbor. For example, node 5 (0101B) when XORed with 0001B results in 4 which is the neighbor of node 5 in x direction. Algorithm 4.1 makes use of this property and selects neighbors (*neigh*) for transfer which have

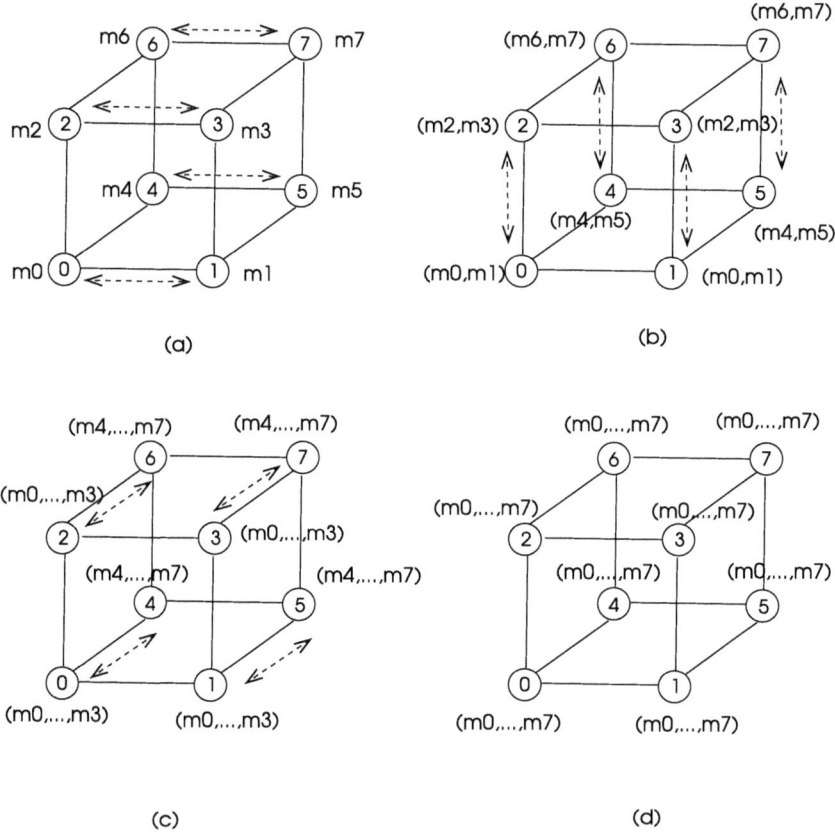

Fig. 4.5 All-to-all communication in a 3D hypercube. The first data exchange between neighbors is in x direction, then y and finally in z directions in **a**, **b** and **c** consecutively

1-bit difference at each iteration. Total number of steps is the dimension d of the hypercube.

Algorithm 4.1 *All-to-All Broadcast on Hypercube*

1: **procedure** ATA_BCAST_HC($my\_id, my\_msg, d, all\_msgs$)
2:    **for** $i \leftarrow 0$ to $d - 1$ **do**
3:       $neigh \leftarrow my\_id \oplus 2^i$
4:       **send** all_msgs to $neigh$
5:       **receive** msg from $neigh$
6:       $all\_msgs \leftarrow all\_msgs \cup msg$
7:    **end for**
8: **end procedure**

4.7 Parallel Algorithm Design Methods

We can employ one of the following strategies when designing a parallel algorithm; modify an existing sequential algorithm by detecting subtasks that can be performed in parallel which is by far one of the most commonly used approaches. Alternatively, we can design a parallel algorithm from scratch or we can start the same sequential algorithm on a number of processors but with different, possibly random initial conditions and the first one that finishes becomes the winner. Foster proposed a four-step design approach for parallel processing stated below [3]. We will look into these steps in more detail in the following sections.

1. *Partitioning*: Data, the overall task or both, can be partitioned into a number of processors. Partitioning of data is called *data* or *domain decomposition* and partitioning of code is termed *functional decomposition*.
2. *Communication*: The amount of data and the sending and receiving parallel subtasks are determined in this step.
3. *Agglomeration*: The subtasks determined in the first two steps are arranged into larger groups with the aim of reducing communication among them.
4. *Mapping*: The formed groups are allocated to the processors of the parallel system. When the task graph that depicts subtasks and their communication is constructed, the last two steps of this methodology are reduced to graph partitioning problem as we will wee.

4.7.1 Data Parallelism

Data parallelism or *data decomposition* method comprises simultaneous execution of the same function on the elements of a dataset. Dividing data to k processors $p_1, \ldots, p_k$, each of which working on its partition can be applied to both PRAM and message passing models and this method is widely implemented in many parallel computing problems. Let us illustrate this approach by the matrix multiplication. We need to form the product C of two $n \times n$ matrices A and B and we partition these matrices as $n/2$, $n/2$ sub-matrices for four processes as follows:

$$\begin{pmatrix} C_1 & C_2 \\ C_3 & C_4 \end{pmatrix} = \begin{pmatrix} A_1 & A_2 \\ A_3 & A_4 \end{pmatrix} \times \begin{pmatrix} B_1 & B_2 \\ B_3 & B_4 \end{pmatrix}.$$

The tasks to be performed by each process $p_1, \ldots, p_4$ can now be stated as below:

$$C_1 = (A_1 \times B_1) + (A_2 \times B_3) \rightarrow p_1$$

$$C_2 = (A_1 \times B_2) + (A_2 \times B_4) \rightarrow p_2$$

$$C_3 = (A_3 \times B_1) + (A_4 \times B_3) \rightarrow p_3$$

$$C_4 = (A_3 \times B_2) + (A_4 \times B_4) \rightarrow p_4.$$

We can simply distribute the associated partitions of matrices to each process in the message passing model, for example A_1, A_2, B_1, and B_3 to p_1, or have the processes work on their related partitions in shared memory in the PRAM model. In the *supervisor/worker model* of parallel processing, we have one process, say p_1, that has all the inputs which are matrices A and B in this case. This supervisor node is responsible for the distribution of the initial data to worker processes and then collecting the results. It may also be involved in computing the results if there is a load unbalance such that the supervisor remains idle when other processes are involved in computation. Intuitively, when there are a large number of processes with needed dense communication, the role of the supervisor can be confined to manage basic dataflow and provide the output. Alternatively, in the fully distributed model, all processes are equal with input data provided to all. Each node in the network works in its partition but exchanges messages to have the total result stored in them. This mode may be used in the first step of a parallel task if total results are needed by each process as input data to the next step of individual computations. We have used *block partitioning* of the input matrices in this example where the matrix is partitioned into blocks of equal size as shown in Fig. 4.6 for an 8×8 matrix where we have 16 processes, p_1 to p_16 each having a 2×2 partition of the matrix.

Row-Wise Array Partitioning

Let us consider a matrix $A[n, n]$ with n rows and n columns. In row partitioning, we simply partition the matrix A to k parts such that p_i gets $((i-1)n/k) + 1$ to in/k rows. Such a partitioning is depicted in Fig. 4.7a for an 8×8 matrix with four processes p_1, p_2, p_3, and p_4.

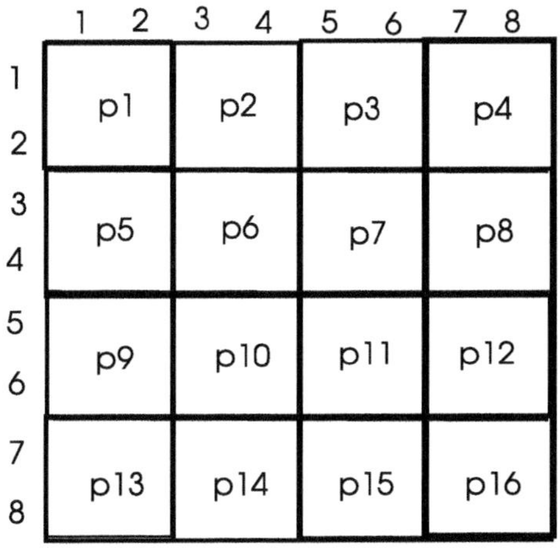

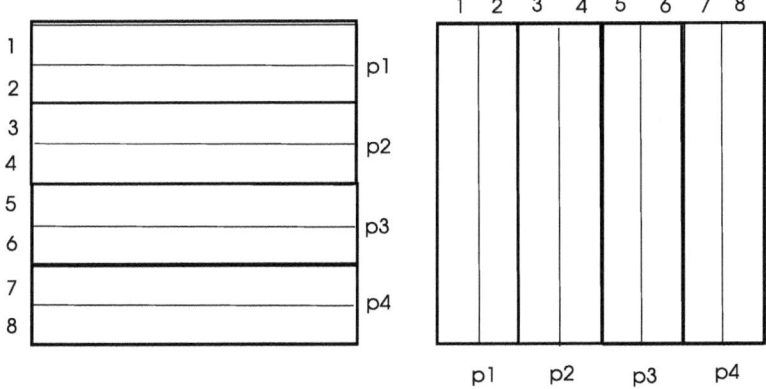

Fig. 4.6 Block partitioning of a matrix

Fig. 4.7 Row-wise and column-wise partitioning of a matrix

Column-Wise Array Partitioning

In column-wise partitioning of a matrix $A[n, n]$, each process now has n/k consecutive columns of A. Column-wise partitioning of an 8×8 matrix with four processes is shown in Fig. 4.7b. Row-wise and column-wise partitioning of a matrix are called 1D *partitioning* and the block partitioning is commonly called 2D *partitioning* of a matrix.

4.7.2   Functional Parallelism

Functional parallelism or *functional decomposition* approach involves applying different operations to possibly different data simultaneously. Also called *task parallelism*, this model involves simultaneous execution of various functions on multiple processors on the same or different datasets. The divide and conquer algorithmic method can be efficiently implemented using this model. We have reviewed the sequential divide and conquer method in which we recursively divided the problem into smaller instances, solved the smaller cases, and merged the results recursively to obtain the final result. Divide and conquer algorithms are good candidates for parallelization since the recursions can be done in parallel. Let us consider adding n numbers using divide and conquer as shown in Algorithm 4.2.

Algorithm 4.2 *Recursive Sum*

```
1: function SUM(A[1..n])
2:    if n = 1 then                                        ▷ base case
3:        return A[1]
4:    else
5:        x ← Sum(A[1..n/2])
6:        y ← Sum(A[n/2 + 1..n])
7:        return x + y                                     ▷ recursive call
8:    end if
9: end function
```

Analysis of this algorithm shows it obeys the recurrence relation $T(n) = 2T(n/2) + 1$ which has solution $T(n) = O(n)$. In order to have a parallel version of this algorithm, we note the recursive calls are independent as they operate on different data partitions; hence, we can perform these calls in parallel simply by performing the operations within the *else* statement between lines 4 and 8 of Algorithm 4.2 in parallel. The recurrence relation for this algorithm in this case is $T(n) = T(n/2) + 1$ which has solution $T(n) = O(\log n)$.

4.8   Parallel Algorithm Methods for Graphs

The divide and conquer method requires graph structure to be partitioned into smaller graphs and this is not an easy task due to the irregular structures of graphs. Partitioning of data for parallel graph algorithms means balanced partitioning of graph among the processors which is an NP-hard problem. We need radically different methods for parallel graph algorithms, and graph contraction, pointer jumping, and randomization are the three fundamental approaches for this purpose.

4.8.1   Randomization and Symmetry Breaking

A *randomized algorithm* makes certain decisions based on the result of coin flips during the execution of the algorithm as we reviewed in Chap. 3. These algorithms assume any input combination is possible. Two main classes of randomized algorithms are *Las Vegas* and *Monte Carlo* algorithms as we have outlined.

Randomized algorithms can be used effectively for parallel solution of various graph problems. Discovering connected components of a graph, finding maximal independent sets, and constructing minimum spanning trees of a graph can all be performed by parallel randomized algorithms as we will see in Part II.

Symmetry breaking in a parallel graph algorithm involves selection of a subset from a large set of independent operations using some property of the graph. For example, finding all candidate vertices for the maximal independent set (MIS) of a graph can be done in parallel. However, we cannot have both of adjacent vertices included in the MIS as this violates the definition of MIS. A symmetry breaking procedure may select the vertex with a lower identifier or a lower degree. In general, symmetry breaking may be employed to correct the output when independent parallel operations on the vertices or edges of a graph produce a large and possibly incorrect result.

4.8.2   Graph Partitioning

Given an unweighted undirected graph $G = (V, E)$, graph partitioning task is dividing the vertex set V into disjoint vertex sets $V_1, \ldots, V_k$ such that the number of vertices in each partition is approximately equal and the number of edges between the subgraphs induced by the vertices in the partition is minimal. This process is depicted in Fig. 4.8 where a graph is partitioned to three balanced subgraphs. Vertex and edges may have weights associated with them representing some physical parameter related to the network represented by a graph. In such a case, our aim in partitioning is to have approximately equal sum of weights of vertices in each partition with a total minimum sum of edge weights between the partitions.

In PRAM and distributed memory model, each process p_i works on its partition. Assuming work done by a processor is a function of the number of vertices and edges in its partition, the load is evenly distributed. However, inter-partition edges and border vertices should be handled with care when obtaining the overall solution to the problem. The duplicated border vertices in partitions they do not belong are called *ghost vertices* and using these nodes helps to overcome the difficulties encountered in the partition boundaries. A simple and effective algorithm was proposed rather early in 1970 by Kernighan and Lin to partition a graph recursively [13]. It is basically used to improve an existing partition by swapping vertices between the partitions to reduce the cost of inter-partition edges. In the *multi-level graph partitioning* method, the graph $G = (V, E)$ is coarsened to a small graph $G' = (V', E')$ using suitable heuristics, a k-way partition of G' is computed, and

Fig. 4.8 Partitioning of a
sample graph into three
disjoint vertex sets;
$V_1 = \{a, b, c, d\}$,
$V_2 = \{f, g, h, e\}$ and
$V_3 = \{i, j, k\}$. A number of
vertices in these partitions are
4, 4, and 3, respectively, and
there is a total of three edges
between the partitions

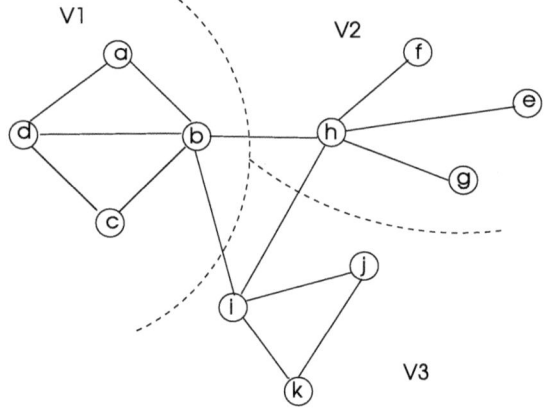

the partition obtained is projected back to the original graph G [11]. A parallel
formation of this method that uses maximal matching during coarsening phase is
presented in [12].

4.8.3 Graph Contraction

Graph contraction method involves obtaining smaller graphs by shrinking the original graph at each step. This scheme is useful in designing efficient parallel graph
algorithms in two respects. It can be conveniently performed in $O(\log n)$ steps as
the size of the graph is reduced by a constant factor at each step. Therefore, if we
can find some way of contracting a graph in parallel while simultaneously solving
a graph problem during contraction, then we have a suitable parallel graph algorithm. Searching for a solution to some graph problems such as minimum spanning
trees during contraction is possible as we will see. Moreover, careful selection of
contraction method maintains basic graph properties, and hence we can solve the
problem on a smaller graph with much ease in parallel and then combine these
solutions to find the solution for the original graph.

Let us assume we have an input graph G and obtain $G_1, \ldots, G_k$ small graphs
after contraction. We can solve the problem in parallel in these small graphs and
then merge the solutions. A graph contraction algorithm template shown in Algorithm 4.3 follows a typical recursive algorithm structure with the base case and
an inductive case. When we reach the base case, we start computing the required
function on the small graph and then recurse on this small graph. Note that the
vertex partition should be disjoint.

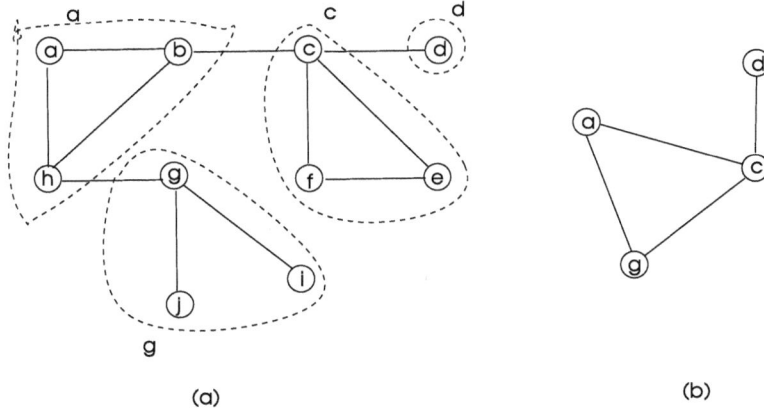

Fig. 4.9 Contraction of a sample graph. The vertex set in **a** is partitioned into four subsets and each subset is represented by one of the vertices in the subset to get the contracted graph in **b**

Algorithm 4.3 *Graph Contraction*

1: **procedure** CONTRACT_GRAPH($G = (V, E)$)
2:    $G \leftarrow G_i$
3:    **if** *graph G_i is small enough* **then**   ▷ base case
4:       **compute** the required task on G_i
5:       **return**
6:    **else**
7:       **compute** vertex partitioning $V_i = V_1, ..., V_k$ of G_i
8:       **contract** each $V_j \in V_i$ into a supervertex V_k
9:       **remove** edges inside each supervertex V_k
10:       **merge** multiple edges between supervertices
11:       $G_j = (V_j, E_j) \leftarrow$ contracted graph
12:       **return** *Contract_Graph*($G_j = (V_j, E_j)$)   ▷ recursive call
13:    **end if**
14: **end procedure**

Let us consider the example in Fig. 4.10 where we partition the vertex set into four subsets which have a representative vertex as shown. Assuming the newly formed graph is small enough, we can now solve the original problem in the small graph which is the base case and recurse to obtain the full solution (Fig. 4.9).

4.8.3.1 Edge Contraction

The *edge partitioning* of a graph involves selecting distinct edges or isolated vertices which will form the partitions. We then contract vertices pairwise that are incident to these selected edges in this method. Since two vertices and the edge between them are contracted to have a single vertex, the selected edges must not share endpoints which in fact is the graph matching problem described as follows.

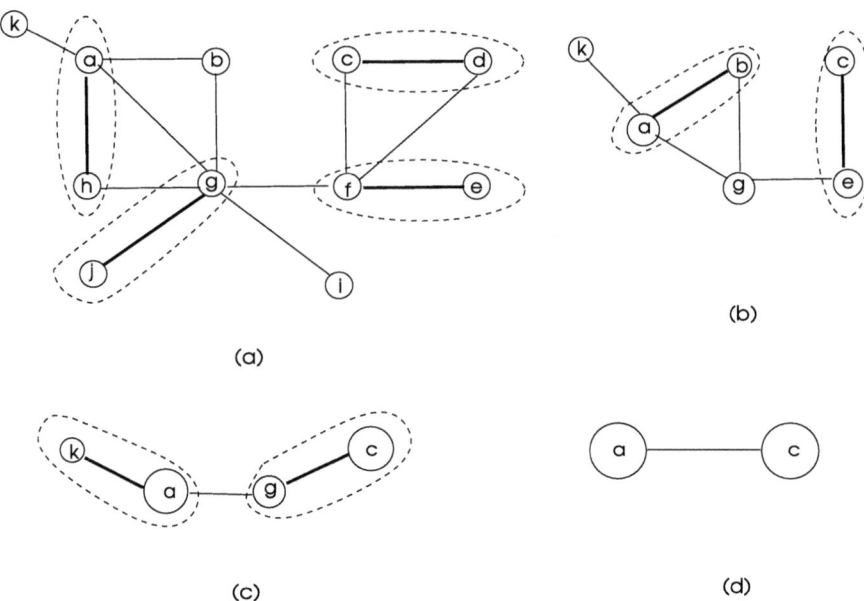

(a)

(b)

(c)

(d)

Fig. 4.10 Edge contraction of a sample graph in **a**. A maximal matching is found and graph is contracted to obtain the graph in **b**. The size of partitions is enlarged as more vertices are included and the label of a partition is the smallest label it has lexicographically. Final graph has two supervertices

Definition 4.1 A matching in a graph $G = (V, E)$ is a set $E' \subset E$ of its edges such that any $v \in V$ is incident to at most one edge in E'.

In other words, edges in the matching are disjoint with no shared vertices. A *maximal matching* in a graph G cannot be enlarged by the addition of new edges and the *maximum matching* in G is the matching with the maximum size among all matchings in G. We can therefore view edge partitioning and contraction problem as recursively finding maximal matching E' in a graph G, edge contraction in E to obtain G' and continuing with finding maximal matching in G', and so on. For graph contraction, a sufficiently large matching rather than a maximal matching can be used. We will search the graph matching problem in more detail in Chap. 9.

In order to find an edge contraction of a graph, we can use a simple greedy algorithm that picks an edge (u, v) arbitrarily in G, deletes u, v and all edges incident to u and v from G, and repeats until no edges are left. Performing parallel matching is not so simple, however, since more than one vertex may select the same vertex for matching. One way to overcome this problem is to use randomization for symmetry breaking. We can select each edge to be included in matching M with 0.5 probability. If an edge (u, v) is selected to be in M and all other edges incident to u or v are not included in M, then (u, v) is decided to be in M [1]. This scheme ensures finding correct matching in parallel.

4.8.3.2 Star Contraction

A star graph is an undirected graph with a center vertex and vertices that are connected directly by vertices to the center as shown in Fig. 4.11. In *star contraction*, we select a vertex in each component as the *center* of the star and all other vertices connected directly to this center called *satellites* are contracted to form a supervertex.

In a sequential setting, we can select a vertex arbitrarily as a center and contract all its neighbors, remove the star from the graph, and continue until (there are no vertices left). Figure 4.12 depicts the sequential star contraction.

A parallel star contraction algorithm can make use of randomization and symmetry breaking as in edge contraction. This time, each vertex selects to be center or a satellite with probability 0.5. A vertex u that selects to be a center becomes a center; however, a vertex v that selects to be a satellite searches for a neighbor that has become a center. If such a neighbor exists, it becomes a satellite of that neighbor. Otherwise, the vertex v becomes a center. If there are more than one center neighbors of v, it selects one arbitrarily to be its center [1].

4.8.4 Pointer Jumping

Let us consider a linked list of n elements with each element pointing to the next element in the list. The *pointer jumping* method provides each element to point to the end of the list after $\log n$ steps. At each step of the algorithm, each element points to the element pointed by its successor as shown in Algorithm 6.1. This algorithm can run in parallel as each pointer update can be performed in parallel.

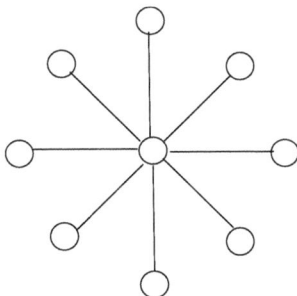

Fig. 4.11 Star network with a center and eight satellites

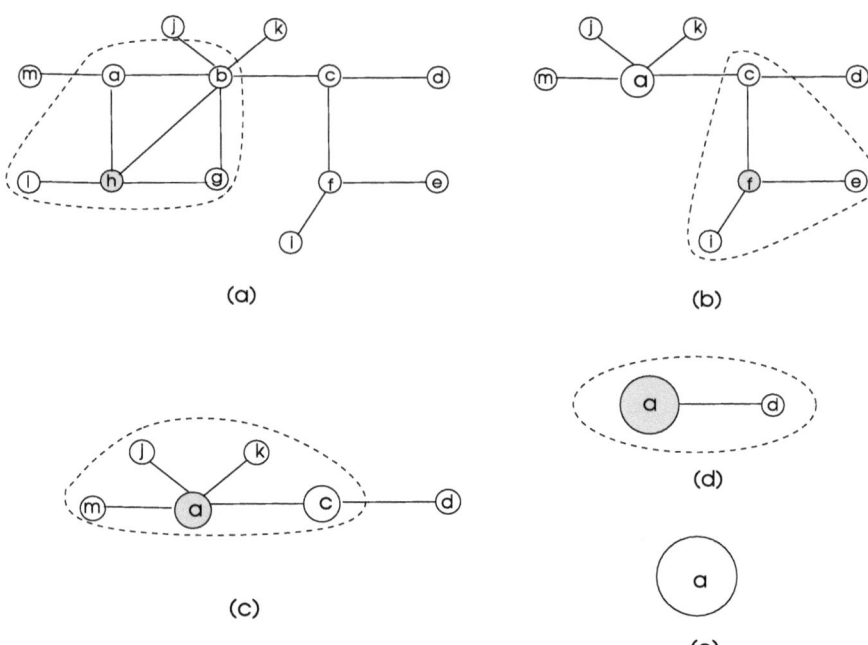

Fig. 4.12 Iterations of star contraction of a sample graph. The contracted vertices are shown as enlarged circles and centers as shaded

---

Algorithm 4.4 *Pointer Jumping*

1: **Input** : a linked list L with n elements
2: **Output** : modified L
3: **for** $i = 1$ to $\lceil \log n \rceil$ **do**
4:     **for** each list element a **in parallel do**
5:         $a.next \leftarrow (a.next).next$
6:     **end for**
7: **end for**

The operation of this algorithm for a linked list of eight elements is shown in Fig. 4.13. We can use this template for parallel graph algorithm design that uses linked lists as graphs can be represented by adjacency or edge lists. Pointer jumping method is suitable for PRAM model with shared memory.

List Ranking
Given a linked list L, finding the distance from each node of L to the terminal node is called *list ranking*. Algorithm 4.4 can be modified to compute these distances as shown in Algorithm 4.5.

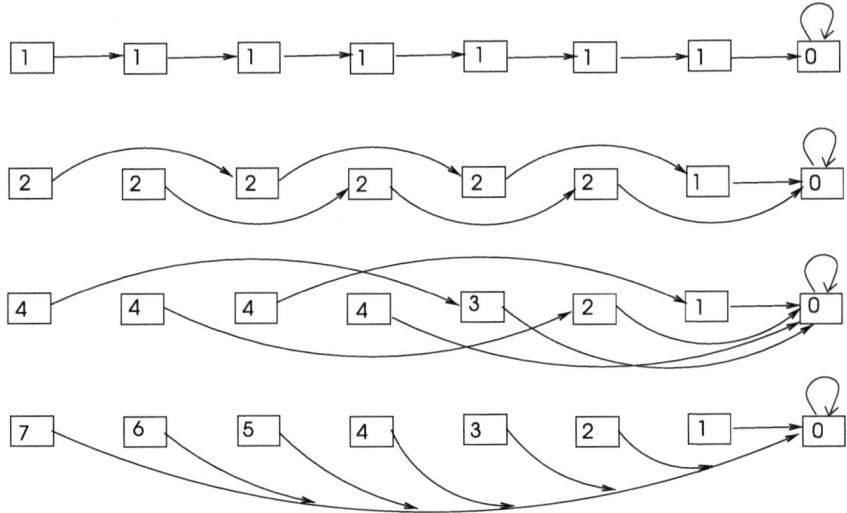

Fig. 4.13 Pointer jumping method in a linked list of eight elements. After three steps, all of the elements point to the end of the list. List ranking algorithm is also depicted in this figure at the end of which all nodes have distances to the head stored

---

Algorithm 4.5 *List Ranking*

1: **Input** : a linked list L with n elements as $(dist, next)$
2: **Output** : distances of list elements to the list head
3: **for** each list element a **in parallel do**   ▷ initialization
4:     **if** $a.next = \emptyset$ **then**
5:         $a.dist \leftarrow 0$
6:     **else** $a.dist \leftarrow 1$
7:     **end if**
8: **end for**
9: **for** i=1 to $\lceil \log n \rceil$ **do**
10:        **for** each list element a **in parallel do**
11:            **if** $a.next \neq \emptyset$ **then**
12:                $a.dist \leftarrow a.dist + (a.next).dist$
13:                $a.next \leftarrow (a.next).next$
14:            **end if**
15:        **end for**
16: **end for**

Line 12 of this algorithm for a node a involves reading the distance of the next node of a and then adding this distance to own and writing the sum as the new distance to a. These two consecutive operations can be done in constant time in only CRCW PRAM model. In order to provide EREW version of this algorithm, we need to replace line 12 by a read line and a write line below, both of which should be executed in EREW mode. The time complexity of this algorithm is $O(\log n)$.

Fig. 4.14 Ear decomposition
of a graph

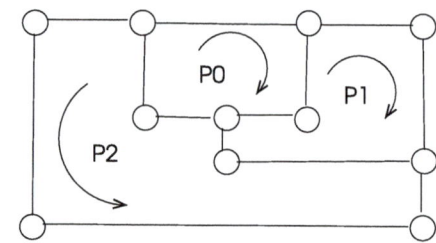

```
temp = (a.next).d
a.d = temp
```

4.8.5   Ear Decomposition

An *ear decomposition* of a graph is defined as follows.

Definition 4.2 (*Ear Decomposition*) An ear decomposition of a graph $G = (V, E)$ is the union of paths $P_0, P_1, \ldots, P_n$ with P_0 being a simple cycle and P_i ($1 \leq I \leq n$) is a path with both endpoints in $P_0 \cup P_1 \cup \ldots \cup P_{I-1}$.

A ear decomposition of a graph is depicted in Fig. 4.14. A graph has an ear decomposition if and only if it has no bridges. An ear decomposition of a graph can be used to determine if two edges are part of a common cycle which can be used to resolve some graph properties such as connectivity and planarity. A ear decomposition of a graph can be found in logarithmic time in parallel and hence these problems can be solved in parallel conveniently using this method.

4.9   Processor Allocation

Allocation of subtasks of a task to processors such that each processor is utilized efficiently during computation is one of the important tasks in parallel computing. We may know the subtask characteristics such as its computation time, communication times and durations, and the subtasks it communicates beforehand in which case we may apply static scheduling strategies. If these parameters are not known, dynamic load balancing methods are commonly used to keep all processors busy at all times.

4.9.1   Static Allocation

Static allocation of tasks to processors is performed before the execution of the algorithm. We need to know the execution time of tasks, their interaction, and the size of data transferred between tasks to be able to perform this allocation. We will assume the computation time of a subtask t_i; its predecessors which are subtasks that have to finish before t_i and its successors which can start when t_i finishes are known in advance with the duration of each communication. In this case, we can draw a *task dependency graph* that illustrates these parameters as shown in Fig. 4.15 for six subtasks of a task. Assuming we have a task set T consisting of n subtasks $t_1, \ldots, t_n$ and a processor set $P = p_1, \ldots, p_k$ of k elements, this problem is reduced to finding the optimal function $F : T \rightarrow P$. This process is commonly called *mapping* or *static scheduling*. Unfortunately, the solution set for F grows exponentially with increased n and heuristic methods for static task allocation are frequently employed. The chart in Fig. 4.15b which shows the allocation of tasks to processors against time is called a *Gantt chart* designed by Henry Gantt in 1910 [4].

4.9.2   Dynamic Load Balancing

In dynamic load balancing, we allocate the tasks to the processor during the execution of the parallel algorithm. This method is needed when task characteristics are not known *en priori*. We may have a centralized load balancing scheme in which a central process commonly called the *supervisor* manages load distribution. Whenever a processor becomes idle, supervisor is informed which provides a new subtask/data to the idle worker. Fully distributed approaches have no supervisor processes and monitoring of load at each processor is performed by equal workers which exchange their states during execution of the parallel program. The overloaded process may start transfer of work to a lightly loaded process or work transfer may be initiated by the receiving process. The first approach is called *sender-initiated* and the latter *receiver-initiated* dynamic load balancing.

4.10   Parallel Programming

Parallel programming involves writing the actual code that will run on the parallel machine. We will review parallel programming with examples in shared memory and message passing distributed memory models. This task can be modeled as *single-program multiple-data* (SPMD) paradigm in which all processes execute the same code on different data or *multiple-program multiple-data* (MPMD) model with each process running different codes on different data in both of these models.

4.10.1   Shared Memory Programming with Threads

Operating systems are built around the concept of a *process* which is the basic unit of code to be scheduled that has data such as registers, stack, and private memory.

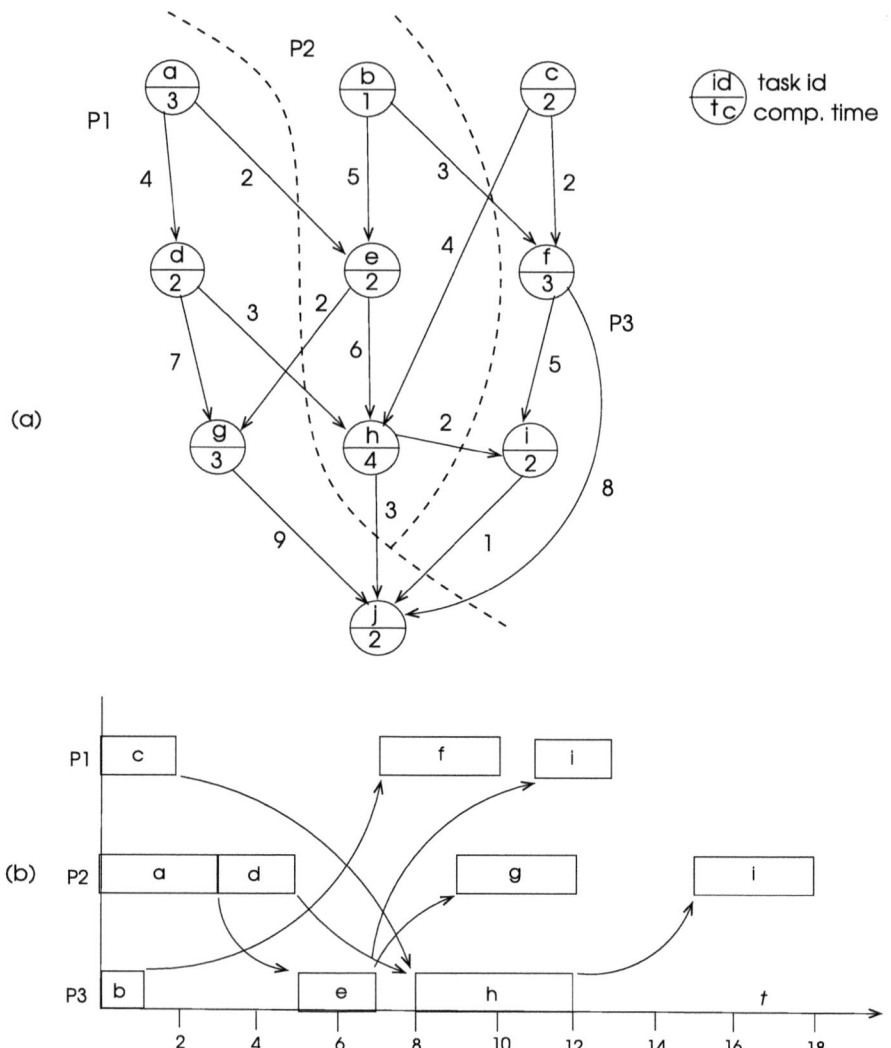

Fig. 4.15 Allocation of static tasks using the task dependency graph. We have a graph of nine tasks in **a** and a possible allocation to three processors is shown in **b**. The partitioning here attempts to put tasks that communicate heavily to the same processor by also trying to keep the workload in each processor similar

Organizing the main functions of an operating system which are resource management and convenient user interface around this perception has many advantages. In the very basic sense, many processes can be scheduled independently in multitasking operating systems preventing the unnecessary waits due to slow input/output devices such as disks. A process can be in one of the three basic states at any time:

running when it is executing, *blocked* when it cannot execute due to the unavailability of a resource, or *ready* when the only resource it needs is the processor. Using processes provides the capability to switch the processor among different processes. The current environment of the running process such as its registers, file pointers, and local data is stored and the saved environment of the new process to run is restored in *context switching*. Another problem encountered when using this model of computing is the protection of the shared memory among processes when data in this area needs to be read or written. The code of a process or the operating system that performs access to shared memory with other processes is called the *critical section*. Although in theory we can use processes for parallel processing in shared memory environment, two main difficulties are the costly overhead of context switching and protection of shared memory segments against concurrent read/write operations by the processes.

4.10.1.1 Threads
Modern operating systems support *threads* which are lightweight processes within a process. A thread has program counter, registers, stack, and a small local memory making the context switching at least an order of less costly than switching processes. There is the global area of the process which needs to be protected since threads need to access this area often. There are two main types of threads: *kernel threads* and *user threads*. A kernel thread is known by the kernel of an operating system and hence can be scheduled independently, contrary to the user threads which are only identified in user space. The user threads are managed by the runtime thread management system, resulting in an order of decrease in their context switch when compared to kernel threads. However, a user thread blocked on an input/output operation blocks the whole process since these threads are not identified by the kernel.

4.10.1.2 POSIX Threads
Portable operating system interface (POSIX) is a set of standards which grew out of necessity as joint efforts by IEEE and ISO to provide compatible software for different operating systems [14]. POSIX threads standard is an application programming interface (API) specifying a number of routines for thread management. The fundamental thread management functions in this library can be classified as below.

- *Thread function*: The thread itself is declared as a procedure with input parameters and also possible return parameters to be invoked by the main thread.
- *Thread creation*: This system call creates a thread and starts running it.

```
int pthread_create(&thread_id,&attributes,start_function,
arguments);
```

where *thread_id* is the variable; the created thread identifier will be stored after this system call, certain properties of a thread can be initialized by the *attributes* variable, *start_function* is the address of the thread code, and the *arguments* are the variables passed to the created thread.

- *Waiting for thread termination*: The main thread waits for the threads it has created using this function call.

```
int pthread_join(pthread_t thread, void **status);
```

where *thread* is the identifier of the thread to wait and *status* is used for the return status or passing back a variable to the main thread.

- *Thread synchronization*: Threads need to synchronize for critical sections and also for notifying events to each other using data structures such as *mutual exclusion variables, condition variables*, and *semaphores*.

Mutual Exclusion

Protection of global variables using POSIX threads is provided by mutual exclusion variables. In the sample C code using POSIX threads below, we have two threads $T1$ and $T2$, a global shared variable *data* between them, and a mutual exclusion variable m which is initialized by the main thread which also activates threads and finally waits for them to finish. Each thread locks m before entering its critical section preventing interruption in this section. Upon exit, it unlocks m to enable any other thread enter its critical section protected by m. The operating system ensures that *lock* and *unlock* operations on the mutual exclusion variables are executed atomic and hence cannot be interrupted.

```
#include <pthread.h>

int data;
pthread_t thread1, thread2;
pthread_mutex_t m;
```

```
T1 () {                                    T2 () {
   . . .                                       . . .
   pthread_mutex_lock(&m);                     pthread_mutex_lock(&m);
   data=data+1;                                data=data*4;
   pthread_mutex_unlock(&m);                   pthread_mutex_unlock(&m);
   . . . }                                     . . .  }
```

```
main() {
   pthread_mutex_init(&m);
   pthread_create(&thread1,NULL,T1,*void);
   pthread_create(&thread2,NULL,T2,*void);
   . . .
   pthread_join(thread1,NULL);
   pthread_join(thread2,NULL);
}
```

Synchronization

Threads, as processes, need to synchronize on conditions. Let us assume two
threads one of which produces some data (*producer*) and needs to inform another
thread that it has finished this task so that the second thread can retrieve and
process this data (*consumer*). The consumer thread cannot proceed before the
first producer declares the availability of data using some signaling method.
Semaphores are data structures consisting of an integer and commonly a process
queue associated with them. Processes and threads can perform two main atomic
actions on semaphores: *wait* in which the caller may wait or continue depending
on the condition it intends to wait, and *signal* call provides signaling the completion of the waited event by also possibly freeing any waiting process for that
event.

The C code provided below shows how two threads synchronize using two
semaphores *sema1* and *sema2*. We have a thread *producer* which inputs some
data and writes this data to shared memory location *data*. It then signals the other
thread *consumer* which reads this data and processes it. Synchronization is needed
so that unread data is not overwritten by the *producer* and also data is not read and
processed more than once by the *consumer*. The semaphore *sema1* is initialized
to *true* value in the main thread thereby allowing thread *producer* that executes a
wait on it to continue without waiting in the first instance, since it has nothing to
wait initially. The second semaphore *sema2* is initialized to *false* value since we
do not want *consumer* to proceed before *producer* signals the availability of data

for *producer*. Semaphores can also be used for mutual exclusion but employment of mutual exclusion variables for mutual exclusion should be preferred to enhance readability of the program and also for performance.

```
#include <pthread.h> pthread_t t1, t2;
sem_t sema1,sema2;
int data;

void producer()              void  consumer()
{ while(true) {                { while(true){
      wait(sema1);                 wait(sema2);
      input some_data;             my_data = data;
      data = some_data;            signal(sema1);
      signal(sema2);               process my_data;
  }                              }
}                              }

main() {
    pthread_create(&t1,NULL,producer,*void);
    pthread_create(&t2,NULL,consumer,*void);
    sem_init(&sema1,1,0);
    sem_init(&sema2,1,0);
    ...
}
```

A multithreaded program with the POSIX thread library (*lpthread*) can be compiled and linked as follows in UNIX environment:

```
cc -o sum sum.c -lpthread
```

A Parallel Algorithm with POSIX Threads

Let us use POSIX threads to calculate the PI number using shared memory parallel processing. This number can be defined by various methods and one such technique is taking the integral of $4/(1 + x^2)$ between 0 and 1 to find an approximate solution. In our design, we will divide the area under this curve into n number of strips; calculate the area of each strip in parallel using POSIX threads and add these areas to find the whole area under the curve between 0 and 1, using the fact that integral of a function between two limits is the size of the area with it within these limits and the x-axis. Each thread has a unique thread identifier given by the operating system at runtime; however, we pass an increasing sequential integer starting from 1 to each one to define the area of the curve the thread will work. The global memory variable *total_area* is to be updated by each thread and therefore needs to be protected against concurrent accesses by a mutual exclusion lock named $m1$.

```
#include <stdio.h>
#include <pthread.h>

#define n 100
#define n_threads 1024

pthread_t threads[n_threads];
pthread_mutex_t m1;

int total_area;// global variable to store PI

/*********************************************************
       thread code to be invoked n_threads times
*********************************************************/

void *worker(void *id)
{ int me=*((int *)id);
  int i, n_slice;
  double width, my_sum;

  width = 1/(double) n_threads;
  x = my_id * width;
  y = 4.0 / (double)(1+(x * x));
  my_area = x * y;             // find my_area
  pthread_mutex_lock(&m1); // update total_area
  total_area = total_area + my_area;
  pthread_mutex_unlock(&m1);
}

/*********************************************************
                    main thread
*********************************************************/

main()
{ pthread_t threads[n_threads];
  int i;
  pthread_mutex_init(&m1);
  for(i=1; i<=n_threads; i++)
    pthread_create(&threads[i],NULL,worker,i);
  for(i=1; i<=n_threads; i++)
    pthread_join(threads[i],NULL);
    printf("Approximate PI is:
```

Figure 4.16 displays the PI curve between 0 and 1 x-axis values where we assume five parallel threads for simplicity. Precision can be improved by either

Fig. 4.16 PI curve between 0 and 1 x-axis values. The area under the curve can be better approximated if each thread calculates the average of the areas of two rectangles formed by border values as shown for T_3 which should find areas for $x = 0.4$ and $x = 0.6$ values and average them

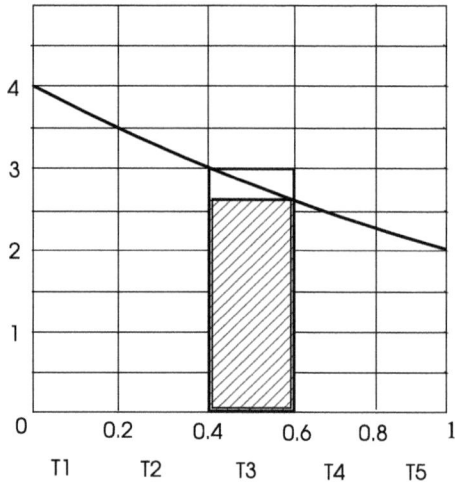

increasing the number of threads and/or calculating the area as the mean value of the border area values as shown in the figure.

Threads can be used conveniently for parallel processing in modern multi-core processors. They communicate using shared memory and hence do not need to send messages thereby preventing data communication delays at the expense of overheads caused for the protection of shared memory. OpenMP is a widely used parallel processing platform that uses multithreading [6, 15].

4.10.2 Distributed Memory Programming with MPI

Message passing interface (MPI) standard specifies a message passing library of routines to provide a portable, flexible, and efficient method to write message passing programs [8, 10]. Although it is primarily targeted for distributed memory programs, the later developments and versions of MPI provide distributed memory, shared memory, or hybrid implementations. Parallel Virtual Machine (PVM) is another tool widely used for parallel processing [7]. MPI consists of routines to send and receive data among processes and various other modes of communication such as broadcasting a message to all processes in the system or multicasting in which a message is sent to a group of messages. MPI programs start by initializing the environment, performing parallel computations by sending and receiving messages among processes, and then terminating. In order to define the set of processes that will run, objects called *communicators* are employed. Each process in MPI belongs to a communicator and inside a communicator, each process has a unique identifier called its *rank*. The following C routines are used to initialize the parallel computing environment and then terminate.

- **MPI_Init**(int *argc*, char **argv*): Inputs a pointer to the number of arguments and a pointer to the argument vector. These parameters are passed from the command line to specify the number of processes to be invoked.
- int **MPI_Comm_size**(MPI_Comm *comm*, int *size*): The number of parallel processes in the communicator *comm* is returned in *size* in the communicator *comm*.
- int **MPI_Comm_rank**(MPI_Comm *comm*, int *rank*): Returns the rank of a process in the group *comm*. The ranks are ordered 0, …, *size* −1.
- int **MPI_Finalize**(*void*): This routine is called by each process before exiting to clean up the library and terminate.

The main procedures for data transfer are the send and receive routines with many variations. The blocking send and the blocking receive are specified below:

```
int MPI_Send(const void *buf, int count, MPI_Datatype datatype,
             int dest, int tag, MPI_Comm comm)

int MPI_Recv(void *buf, int count, MPI_Datatype datatype,
             int source, int tag, MPI_Comm comm, MPI_Status *status)
```

where *buf* is the address of send/receive buffer, *count* is the number of elements in the buffer, *datatype* is the data type of each send/receive buffer element, *dest/source* is the integer rank of destination/source, *tag* is the type of message, *comm* is the communicator, and *status* is the status object to be examined. Note that two processes may use message tag to perform different actions by different tags. We can now write a simple MPI application of two processes sending and receiving messages in C programming language as below:

```
#include <mpi.h>

int main(int argc, char** argv) {
  int my_rank, size, data;

  // Initialize the MPI environment
  MPI_Init(argc, argv);
  MPI_Comm_rank(MPI_COMM_WORLD, &my_rank);
  MPI_Comm_size(MPI_COMM_WORLD, &size);

  if (my_rank == 0) { // rank 0 is the sender
    data = 23;
    MPI_Send(&data, 1, MPI_INT, 1, 0, MPI_COMM_WORLD);
    printf("This is process %d",my_rank);
  } else if (my_rank == 1) { // rank 1 is the receiver
    MPI_Recv(&data, 1, MPI_INT, 0, 0, MPI_COMM_WORLD,
                                    MPI_STATUS_IGNORE);
    printf("This is process %d",my_rank);
  }
  MPI_Finalize();
}
```

The same code is run on all processors in this SPMD model which is very common in MPI applications due to the difficulty in writing different codes for different processes. The instructions to be run by each process are separated by the use of process identifiers. In this example, rank 0 is the sender and rank 1 is the receiver. We may compile and run this code named *mess.c* for eight processes in UNIX environment as follows:

```
mpicc -o mess mess.c
mpirun -np 8 mess
```

The first line is the compiling and linking command using the *mpicc* compiler/ linker and the second line starts running the executable program with eight processes. Note that we pass this argument of eight processes to the main program in which MPI_Init uses to initialize the environment and starts running these identical processes in a hardware environment we do not know. In fact, we could have installed MPI on a single computer and eight processes could run on the same computer. The following example displays the use of MPI to calculate PI using the same method of finding the area under the curve $4/(1 + x^2)$ between $x = 0$ and $x = 1$ as we did with POSIX threads.

```
#include <mpi.h>
#include <math.h>

int main(int argc,char **argv)

{
    int i, n, my_id, n_slices, n_procs, n_sl;
    double my_area, total_area, width, x, my_sum;

    MPI_Init(&argc,&argv);
    MPI_Comm_size(MPI_COMM_WORLD,&n_procs);
    MPI_Comm_rank(MPI_COMM_WORLD,&my_id);

    if (myid == 0) {
        printf("Enter the number of slices for each process:");
        scanf("%d",&n_slices);
        MPI_Bcast(&n_slices, 1, MPI_INT, 0, MPI_COMM_WORLD);
        for(i= 1; i < n_procs; i++) {
          MPI_Recv( &part_area, 1, MPI_DOUBLE, MPI_ANY_SOURCE,
                        1, MPI_COMM_WORLD, &status);
             total_area += part_area; }
        printf("Approximate PI is: %f", total_area);
    }
    else {
      MPI_Recv(&n_sl, 1, MPI_INT, 0, 0, MPI_COMM_WORLD,
                    MPI_STATUS_IGNORE);
        my_sum = 0.0;
        width = 1.0 / (double) (n_sl + 1);
        for (i = my_id + 1; i <= n_sl; i++) {
        x = width * (double)i ;
        my_area += 4.0 / (1.0 + x*x);
        }
      MPI_Send(&my_area, 1, MPI_DOUBLE, 1, 0, MPI_COMM_WORLD);
    }

    MPI_Finalize();
    return 0;
}
```

Note that this is a supervisor/worker model of message passing parallel processing that uses SPMD paradigm. The supervisor process, sometimes called the *root*,
has rank 0 and starts by initialization as all other processes followed by asking the
user with the number of slices to be processed by each process under the curve.
It then broadcasts this value to all processes. Each process then computes the area
under its portion of the curve for the number of slices and sends this partial area
to the supervisor which adds them and outputs. We could have the supervisor also
involved in the computation of PI (see Exercise 6).

4.11   Conclusions

We have reviewed the parallel computing fundamental concepts in this chapter with emphasis on design methods for parallel graph algorithms. Parallel algorithms may use the shared memory or the message passing model in a general sense. The PRAM model is an idealistic method to design parallel algorithms in the shared memory platform; moreover, it provides an abstract model hiding details of implementation and hence can be used for high-level design and comparison of shared memory parallel algorithms. Access mode to shared memory is important in this model, and reads and writes can be performed in concurrent or exclusive modes. Message passing model is suitable for distributed memory processors which communicate and synchronize by sending and receiving messages only. Basic communication modes in a parallel computing system maybe classified based on the source and destination of the data transfer. Grouping the communications under operations such as *one-to-all* or *all-to-all* modes eases the burden of writing a parallel algorithm since we can simply specify the needed mode rather than writing the actual algorithm for communications. Design methods for parallel algorithms may be broadly classified as data or functional decomposition. Data is decomposed into a number of sets to be processed by parallel processors using the same algorithm in the first and different tasks are allocated to different processors in the second method.

We saw graphs require special methods of parallel computing and graph contraction is a commonly used approach to enable parallel graph operations. The graph under consideration is made smaller at each step, using methods such as edge or star contraction. Graph contraction can be performed in parallel and also solving the problem in a smaller graph can be done more conveniently. Some graph problems can be solved efficiently using sequential algorithms; however, the same problems do not have simple parallel algorithmic solutions due to the dependencies involved. Randomization and symmetry breaking methods provide simple and elegant parallel graph algorithms for various graph problems such as maximal independent sets and minimum spanning trees. Data partitioning for graphs frequently involves dividing the adjacency matrix row-wise, column-wise, or block-wise to a number of processors. Parallel tasks relations can be depicted by a task dependency graph and allocation of these tasks to processors is a variation of graph partitioning problem. This approach requires task computation times and their interaction to be known in advance which may not be realistic in many real-life applications. Dynamic load balancing involves keeping the loads on processes even at runtime. Finally, we reviewed two commonly used platforms for shared memory and distributed memory programming: POSIX threads provide a convenient API to implement shared memory parallel algorithms and MPI is widely used for distributed memory programming.

At a more abstract level, we can view the modeling of the whole process of parallel computing at four related levels: hardware, operating system, programming, and algorithmic levels as shown in Fig. 4.17. In all these layers of design, the main distinction is whether shared or distributed memory is used. We find

Fig. 4.17 Hierarchy of
shared and distributed
memory parallel processing
models

PRAM	Message Passing	ALGORITHM MODEL
Threads OpenMP	MPI, PVM	PROGRAMMING
locks, semaphores	messages (send, receive)	OPERATING SYSTEM
Shared Memory	Distributed Memory	HARDWARE

operating system and middleware should provide different services at these levels. The main problem with shared memory approach is the protection of memory during concurrent accesses and this is provided by the operating system constructs such as semaphores and locks. At the programming level, threads which are lightweight processes are widely used for shared memory programming. The POSIX thread library provides all necessary routines for thread synchronization and mutual exclusion. The MPI standard is widely used for distributed memory parallel computing with a wide range of needed message passing procedures. At algorithmic level, the PRAM model which assumes shared memory is not practical as it assumes infinitely large shared memory and infinite number of processors; however, it is used to compare various parallel algorithms for the same problem. We will mostly consider distributed memory platforms in our analysis of parallel graph algorithms in this book to provide implementable solutions, except in few places where we describe PRAM algorithms.

Exercises

1. Write the pseudocode of an EREW PRAM algorithm that uses p processors to find the sum of p values in $O(\log p)$ time.
2. The prefix sum of an array $A[1 \ldots n]$ is an array $B[1 \ldots n]$ such that $B[j] = \sum_{i=1}^{j} A[i]$. Design a parallel EREW PRAM algorithm to find the prefix sum of an input array.
3. Write the pseudocode of *one-to-all* broadcast communication routine in a hypercube of dimension d. This routine inputs the dimension d and the identifier i of the node and the message. Assume node 0 is the source of the message and it is first transmitted in x direction (1), then in y direction, and finally in z direction (3) as shown in Fig. 4.18. Work out the number of steps and work done.
4. We need to partition the task graph of Fig. 4.19 to two parallel processors. We want to place processes that have significant communication to the same processor to keep interprocess communications as small as possible and also

Fig. 4.18 Hypercube
one-to-all communication

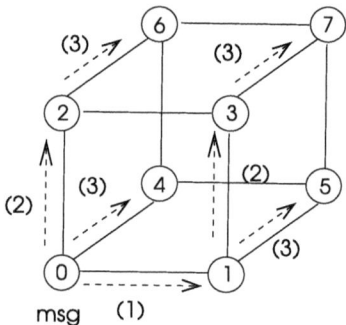

want to keep the total load (execution time of tasks) on each processor similar. The method to be employed is to find a matching of the graph such that the heaviest disjoint edges are selected at each iteration. Then, the vertices at each end of selected edges are contracted to obtain supervertices. Use graph contraction until only two supervertices remain each of which can be assigned to a single processor. Work out the total parallel execution time by drawing a Gantt chart of the schedule of processes. Workout also the total interprocess communication cost ignoring the communication costs of tasks assigned to the same processor.

5. Write a C program using POSIX threads API that finds the sum of an array of size n stored in global memory. Use data parallelism to decompose data equally to k threads. The total sum is kept in global memory and should be protected against concurrent accesses.

6. Modify the C code for PI computation using MPI of Sect. 4.10.2 so that the supervisor also computes the area under its portion of curve.

7. Eight processors with labels in increasing order are connected in a unidirectional ring structure as shown in Fig. 4.20. Each process holds an integer value and we need to sum these values and output the sum by process p_0. Write an

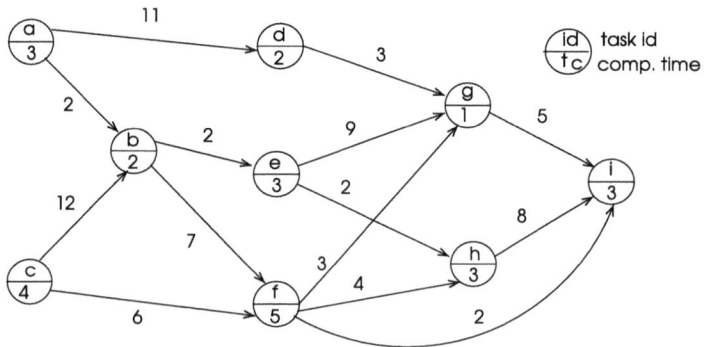

Fig. 4.19 Task dependency graph for Exercise 4

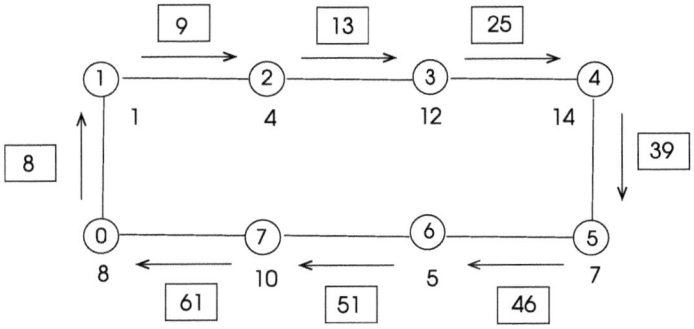

Fig. 4.20 Unidirectional ring of eight processes used to calculate sum of integers stored at each node. The integers stored are shown next to nodes and the messages contain the partial sums transferred

MPI program in C language which starts by p_0 sending its value to its next neighbor which adds the value it receives with the value it has and sends the sum to its next neighbor. The partial sums are propagated in this manner until p_0 receives the total sum and outputs it as shown in the figure. Modify this program such that p_0 has an integer array of size n and distributes the portions of this array to processes using data decomposition which perform as above to find the total sum of array.

8. A multithreaded file server is to be designed which receives a message by the front end thread, and this thread invokes one of the *open*, *read*, and *write* threads depending on the action required in the incoming message. The *read* thread reads the number of bytes from the file which is sent to the sender, the *write* thread writes the specified number of contained bytes in the message to the specified file. Write this program in C with POSIX threads.

References

1. Belloch G (2015) Algorithm design: parallel and sequential. Draft book
2. Flynn MJ (1972) Some computer organizations and their effectiveness. IEEE Trans. Comput. C21(9):948960
3. Foster I (1995) Designing and building parallel programs: concepts and tools for parallel software engineering. Addison-Wesley, Boston
4. Gantt HL (1910) Work, wages and profit, The engineering magazine, New York
5. Grama A, Karypis G, Kumar V, Gupta A (2003) Introduction to parallel computing, 2nd edn. Addison-Wesley, Boston
6. http://openmp.org/wp/
7. http://www.csm.ornl.gov/pvm/
8. http://www.mcs.anl.gov/research/projects/mpi/
9. http://www.nvidia.com/object/cuda_home_new.html
10. http://www.open-mpi.org/

11. Karypis G, Kumar V (1995) Multilevel k-way partitioning scheme for irregular graphs. Technical Report TR 95-064, Department of Computer Science, University of Minnesota
12. Karypis G, Kumar V (1997) A coarse-grain parallel formulation of multilevel k-way graph partitioning algorithm. In: 8th SIAM conference on parallel processing for scientific computing
13. Kernighan BW, Lin S (1970) An efficient heuristic procedure for partitioning graphs. Bell Syst Tech J 49:291307
14. POSIX.1 FAQ. The open group, 5 Oct 2011
15. Quinn M (2003) Parallel programming in C with MPI and OpenMP, 1st edn. McGraw-Hill Science/Engineering/Math, New York

Distributed Graph Algorithms

5

5.1   Introduction

A distributed computing system or a distributed system as more commonly termed, consists of a number of computational nodes connected by a communication network. Computing nodes are autonomous and the network can be a wired medium, a wireless communication channel or both. The nodes of a distributed system cooperate and communicate to achieve a common goal. It is evident that synchronization among computations at the nodes of such system is needed to provide this coordination.

A distributed system appears to users as a single computing system. In that respect, a cloud is a distributed system since there are numerous computing elements and databases in a cloud, yet it appears as a single system to a user. Distributed systems are needed because they provide convenient access to remote resources for users and applications. In many cases, the application itself is inherently distributed. For example, an airline reservation system is used by many users and provides all necessary communication and synchronization. Distributed systems provide fault tolerance in which case failure of a node or a link does not harm the operation of the system as these are replaced by other nodes or links. Distributed systems are commonly dynamic in which nodes and links may be inserted to or deleted from the network due to failures or movement of the nodes as in the case of a mobile network. A rescue operation consisting of moving nodes is an example of a mobile network.

A distributed system can be conveniently modeled by a graph in which the vertices of the graph represent the computational nodes and an edge between two nodes represents a communication facility between them. The algorithms running at the nodes of a graph representing the distributed system are commonly termed *distributed graph* or *network algorithms*. Note that distributed memory-employing algorithms in a parallel processing environment are also called distributed algorithms in the literature but in the context of this book, we will use distributed

© The Author(s), under exclusive license to Springer Nature Switzerland AG 2026   117
K. Erciyes, *Guide to Graph Algorithms*, Texts in Computer Science,
https://doi.org/10.1007/978-3-032-05294-0_5

(graph) algorithms to mean algorithms running in a network represented by a graph. We will see designing a distributed version of a sequential graph algorithm is not a trivial task. We start this chapter by describing common distributed system platforms. We then investigate distributed graph algorithms, classify them, and show the operation of some basic distributed graph algorithms.

5.2   Types of Distributed Systems

Distributed system applications vary from clusters of computers to networks of embedded systems. We can classify distributed systems as distributed computing systems, distributed information systems, and distributed pervasive systems [5]. Distributed computing systems typically consist of a cluster of homogenous computers connected by a local area network. The *Grid* is also a distributed computing system, which consists of numerous heterogeneous computing systems with many different users that cooperate to achieve a common goal [3]. *Cloud computing* is more general than grid computing and provides users with various resources such as storage, data management, web site hosting, and computation [4]. Fault tolerance due to failing nodes and links, and load balancing are the main issues to be handled in both Grid and a cloud.

Distributed information systems commonly involve large database applications such as a transaction processing system. An online banking system with millions of users is an example of such system. Distributed pervasive systems typically consist of small and sometimes mobile computers that communicate using wireless medium. We will take a closer look at these systems since unlike Grid or a cloud, these can be modeled conveniently by a graph. The Internet is the largest network in the world connecting personal, infrastructured, wireless, or any other type of network.

Wireless networks communicate using wireless communication and networking medium. They can be broadly classified as *infrastructured* and *ad hoc*. An infrastructured wireless network has a fixed wired backbone consisting of routers and access points to provide communication among hosts of the network such as a cellular network. In contrast, ad hoc wireless networks do not have this structure and each node in such a network acts as a router for the transfer of messages. Ad hoc networks are widely used due to easiness and speed in their deployment. Two types of wireless networks have gained importance recently; *mobile ad hoc networks* and *wireless sensor networks*.

5.2.1   Mobile Ad Hoc Networks

A mobile ad hoc network (MANET) is an infrastructure-less wireless network consisting of nodes that move dynamically. Vehicular ad hoc networks (VANETs) that provide communication between moving vehicles, military MANETs used by military, and MANETs used in rescue operations are examples of such systems.

Each node in a MANET acts as a router for *multihop communications* between hosts in which a message is transferred between a number of host pairs before it reaches its destination.

One of the main challenges in a MANET is *routing*, which is the process of transferring a message between a sender and a receiver in the most efficient way. Nodes in a MANET are mobile which means routes have to be computed dynamically requiring efficient routing algorithms. Staying connected in a MANET is also another problem that needs to be solved. A robot network is another example of a MANET and keeping the network connected at all times is needed for the coordinated operations of robots in such a network.

5.2.2   Wireless Sensor Networks

A wireless sensor network (WSN) consists of a network of sensors with radio transceivers and controllers. These networks of physically tiny nodes in most cases have many applications including environmental control, e-health, and intelligent buildings. A sensor node has a very limited power and sensors are typically controlled by a central node called the *sink* with more computational capabilities. Data recorded by sensor nodes is collected at the sink for further processing. Routing of data messages to the sink efficiently using network protocols as well as keeping the network connected are the main issues to be addressed in WSNs. Sensor networks are mostly stationary and require low-power operation, implementation of which is more critical than managing power in MANETs.

A MANET or a WSN can be conveniently modeled by a graph and the problems such as routing, connectivity can then be transferred to graph domain to be solved with methods developed for graphs. For example, efficient routing problem can be solved with the aid of the method of finding the shortest distance between two nodes of a weighted graph. However, these problems should now be solved in a distributed manner without any global knowledge, which makes the problem harder than an ordinary graph problem. A node in a graph representing a WSN can only communicate with its neighbors, but we need to have a global decision using the collected data from all of the sensors. Figure 5.1 displays a wireless network with nodes that can transmit and receive radio signals within a radius of r meters. We can then connect the nodes that are within transmission ranges of each other by an edge and obtain the graph shown.

5.3   Models

Messages are crucial for the correct operation of a distributed algorithm. We can define the widely accepted *message-passing model* of a distributed system formally as follows [1, 6]:

Fig. 5.1 Graph
representation of a wireless
network. Transmission range
of a node is shown by a
dashed circle centered at that
node

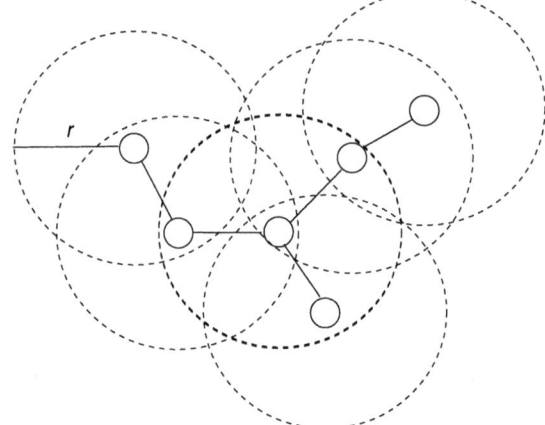

- A process p_i at a node i communicates with other processes by exchanging messages only.
- Each process p_i has a state $s_i \in S$, where S is the set of all possible states that a process p_i can be.
- A configuration of a system consists of a vector of states as $C = [s_1, \ldots, s_n]$
- The configuration of a system may change by either a *message delivery event* or a *computation event*.
- A distributed system continuously goes through executions as $C_0, \phi_1, C_1, \phi_2, \ldots$ where ϕ_i is either a computation or a message delivery event.

A *finite-state machine* (*FSM*) or *finite-state automaton* is a mathematical model to represent a complex system. An FSM consists of states, inputs, and outputs. It may change its state based on its current state and the input it receives. FSMs are widely used to design algorithms, network protocols, and sequence analysis in bioinformatics. Formally, a deterministic FSM is a quintuple (I, S, S_0, δ, F) where

- I is a set of input signals.
- S is a finite nonempty set of states.
- $s_0 \in S$ is the initial start state.
- δ is the state transition function such that $\delta : S \times I \to S$.
- $O \in S$ is the set of output states.

The next state of an FSM is determined by its current state and the input it receives. The same input may cause different actions in different states. As an everyday example, let us consider students in a school who for simplicity can have only two states: *in_class* or *out_class* meaning that they can be either in the class or out of the class. When the bell rings in *in_class* state, it means they can go out and the bell ringing in *out_class* state means they should go in the class. An *FSM diagram* or a *state transition diagram* is a visual aid to understand the behavior of an FSM. The circles in such a diagram denote its states and transitions between states are shown by directed arcs which are labeled as *a/b* where *a* is the set of inputs received and *b* is the set of outputs produced when these inputs are received. A double circle denotes the accept (the final) state of the FSM.

A *state table* provides an alternative way of representing a FSM. It has states of the FSM as rows and inputs as columns and the elements of the table can be the next FSM state and actions to be taken when the input is received. The output of a *Moore Machine* type of FSM is the next state, whereas the output in a *Mealy Machine* type of FSM contains outputs as well as the next state.

Example 5.1 We will design a simple FSM for an elevator that can only go to floors 0, 1, and 2. There are two buttons in the elevator: *up* and *down* which take the elevator up and down respectively. We can associate the current state of the elevator with the floor it currently stays; therefore, we have three states 0, 1, and 2. At each state, the *up* or *down* button can be pressed represented by two inputs *up* by 0 and *down* by 1. The FSM diagram for this example is shown in Fig. 5.2 which shows all state transitions, considering there will be two inputs at each state. We cannot go down from 0 state and also going up from second floor is not allowed shown by loops at these states.

We can now form the state table for this FSM with entries showing the next state of the FSM when the input shown in columns is received at state shown in rows as shown in Table 5.1. This way of expressing an FSM provides a very convenient way of writing its algorithm. We can form a 2D array with each element being a function pointer. We then define functions to be performed for each table entry; for example, receiving "0" (up) at "1" (first floor) state should cause a transition to state 2 (elevator should move to second floor) which is realized by changing the current state to "2." The running of the algorithm is then straightforward; every time an input is received, we activate the function shown by the FSM table entry as shown by the C programming language code below.

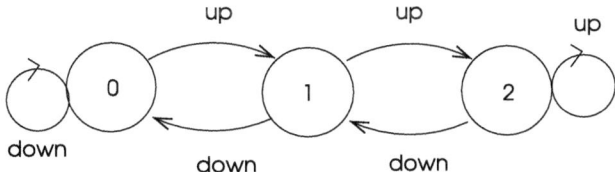

Fig. 5.2 FSM of the elevator example

Table 5.1 Elevator state table

State	0 (up)	1 (down)
0	1	0
1	2	0
2	2	1

```
#include <stdio.h>
    # define UP     0
    # define DOWN   1

    void *fsm_tab[3][2]();
    int  input;

    void act00(){curr_state=1;}
    void act01(){curr_state=0;}
    void act10(){curr_state=2;}
    void act11(){curr_state=0;}
    void act20(){curr_state=2;}
    void act21(){curr_state=1;}

main()
    { curr_state=0;              // initialize curr_state
      fsm_tab[0][0]=act00;       // initialize FSM table
      fsm_tab[0][1]=act01;
      fsm_tab[0][2]=act02;
      fsm_tab[1][0]=act10;
      fsm_tab[1][1]=act11;
      fsm_tab[1][2]=act12;

      while (true)
        { printf("Type 0 for up, and 1 for down \n");
          scanf("%d", &input);
          *fsm_tab[curr_state][input];
          printf("now at floor \%d", curr_state)
        }
    }
```

5.4   Communication and Synchronization

The algorithms that run at the nodes of a distributed system need to synchronize to accomplish a common goal. This process can be performed at various levels. Let us see how synchronization can be handled locally at three main levels of hierarchy: the hardware, the operating system, and the application. At the lowest level, hardware may provide synchronization at a certain number of clock ticks periodically. At a higher level, one of the main tasks of local operating systems at each node is the synchronization of the processes residing at that node. Moreover, this function can be extended to processes running at the nodes of the distributed system at the application level.

However, we need a mechanism to provide synchronization among the nodes which should be translated to local synchronization mechanisms described above. A very commonly used method in a distributed system is synchronization via messages. In this so-called *message-passing* model, each local operating system or middleware provides two basic primitives; *send* and *receive* for sending and receiving messages. These procedures can be executed in *blocking* or *non-blocking* fashion. A blocking *send* stops the caller until an acknowledgment from the receiver is received. A blocking *receive* means the receiver should wait until a message is received. The blocking *receive* maybe *selective* in which a message from a particular sender is waited and execution is resumed only after this happens. It is common practice to employ a non-blocking *send* with a blocking *receive* since the sent message is assumed to be delivered correctly while the actions of a receiver depend on whether the message is received and also its contents and thus a blocking *receive* is frequently used.

Sending and receiving are commonly employed indirectly using data structures called *ports* or *mailboxes*. These are depository places for messages, and placement or removal of messages can be performed asynchronously from these structures. In a distributed system, the locally executed *send* procedure typically deposits the message in the mailbox of the network process which appends the necessary network headers and transfers the message through lower network layer software to the network. The receiving network process removes network headers and deposits the message in the mailbox of the receiver which takes it from there as shown in simplified form in Fig. 5.3. There are three main software modules at each node of a distributed system: system (OS), network protocol stack (N/W), and the application (APP) as shown in this figure.

In summary, operating system and network processes provide synchrony between two processes p_i and p_j which execute distributed algorithms at nodes i and j of the network. At a higher level of abstraction at application, synchronization among distributed algorithms at different nodes may be achieved by *rounds* which are executed in lock-step fashion. In this case, a special process at a node starts each round and each process waits for all other processes to finish execution before the next round starts. Synchronization at the beginning of a round can be achieved by broadcasting a special message and end of a round can be identified by the convergence of messages which are two basic communication operations as we will see shortly. Commonly, a process p_i of a distributed system performs the following steps at each round:

1. *Send* message.
2. *Receive* message.

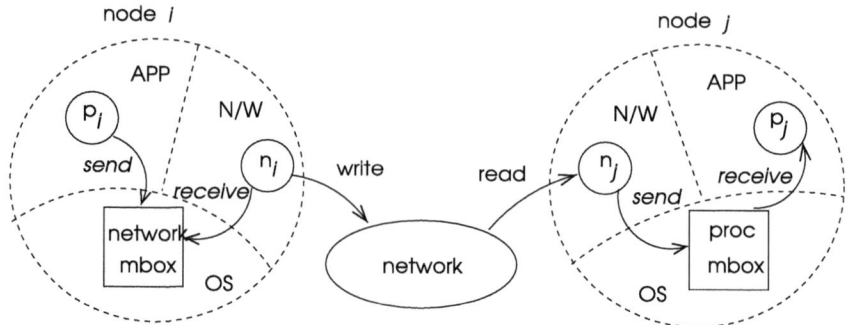

Fig. 5.3 Distributed communication via mailboxes. Process p_i at node i sends a message to process p_j at node j using mailboxes via network processes n_i and n_j

3. Do some computation.

We assume here that a process sends the results of its computation from round k-1 in kth round. This order is not strict; however, we could have *compute-send-receive* sequence which would mean each process now computes new results in round k based on what it has received in the previous round and sends the new results in the current round. Distributed algorithms that work asynchronously and do not have this synchronously executing rounds are called *asynchronous algorithms*. Detecting the termination of distributed algorithms is needed to stop the algorithm when a certain condition is met which is not a trivial task. Although starting and ending a round cause overhead in terms of needed extra messages, designing synchronous distributed algorithms is more straightforward than asynchronous algorithms in general. The asynchronous algorithms require more complex control logic and detection of termination in such algorithms is also more difficult.

Yet another distinction is whether a single initiator starts the distributed algorithm or there exists more than one initiators. A single initiator that also controls the overall running of the algorithm means a single point of supervision which is easier to manage than individually controlled processes. We can, hence, classify distributed algorithms based on synchronization at application or algorithmic level as follows.

- Synchronous Single Initiator (SSI) Algorithms: There is a single initiator which starts the algorithm, synchronizing start and end of each round, and initiating termination. These algorithms are simpler to design than others since there is a single process that controls the operation.
- Asynchronous Single Initiator (ASI) Algorithms: This type of algorithms has a single initiator but activity at each node is performed asynchronously from the other nodes. Synchronization and termination detection are more difficult for such an algorithm than a synchronous algorithm.

- Synchronous Concurrent Initiator (SCI) Algorithms: These algorithms execute synchronously but may be started by concurrent initiators.
- Asynchronous Concurrent Initiator (ACI) Algorithms: There are more than one initiators in this case and the activities are asynchronous. This mode of operation is the most general case but may require complex control.

In the case of an SSI algorithm, a previously built spanning tree to transfer control messages can be conveniently used. Based on foregoing, a possible SSI algorithm round template is sketched in Algorithm 5.1. All processes start the kth round when they receive the *start* message over the spanning tree T, which is basically a broadcast operation over T as we will see shortly. The three actions in the round are sending results of the previous round to all neighbors, receiving results of the previous round from neighbors, and prepare new results for the next round. When a process finishes executing a round, it waits for all of its children in T to finish before it can send the *stop* message to its parent. When the root of the spanning tree T receives *stop* message from all of its children, the round k is over and the root can now start the round $k + 1$. We will use this structure frequently while designing distributed graph algorithms.

Algorithm 5.1 *SSI_template*

1: **boolean** *round_over* ← *false*
2: **message type** *start, result, stop*
3: **while** ¬*round_over* **do**
4:     **receive** *msg(j)*
5:     **case** *msg(j).type* **of**
6:             *start* : **send** *result(i)* to all neighbors
7:                       **receive** *result(j)*, from each neighbor j
8:                       **compute** *result(i), finished* ← *true*
9:             *stop* : **if** *stop* received from all children **and** *finished* **then**
10:                       **send** *stop* to *parent*
11:                       *round_over* ← *true*
12: **end while**

5.5   Performance Criteria

The performance of a distributed algorithm is evaluated in terms of time, message, space, and bit complexities outlined below:

- *Time Complexity*: Time complexity is the number of steps required for the distributed algorithm to finish as in a sequential algorithm. For synchronous distributed algorithms, we would be mostly interested in the number of rounds as time complexity.

- *Message Complexity*: This parameter is commonly considered as the dominant cost of a distributed algorithm since it directly shows the utilization of the network and indicates synchronization costs among the nodes of the network. Transferring a message over a network is magnitudes of orders more costly than doing local computations.
- *Bit Complexity*: The length of a message may also affect the performance of a distributed algorithm, especially if message length increases as the message traverses the network. For a large network modeled by a graph with many vertices and edges, bit complexity may be significant which directly affects the network performance.
- *Space Complexity*: This is the required storage at a node of the distributed system for the algorithm under consideration.

5.6 Distributed Graph Algorithm Examples

We are now ready to design and implement simple distributed graph algorithms. We will describe sample basic algorithms which follow a logical sequence. The first algorithm uses a technique named *flooding* to send a message from a node of the graph representing the network to all other nodes. We then make use of this algorithm to build a spanning tree of the network which can be used for efficient broadcast and convergecast of messages in the network as described next.

5.6.1 Flooding Algorithm

Our aim is to send a message from a single node to all nodes in the graph. This operation is called *broadcast* and has many applications in real networks, for example to inform all nodes of an alarm condition that occurs at a node. In the simplest case, we can have the following rules as a first attempt to solve this problem:

1. The initiator i sends a message $msg(i)$ to all of its neighbors.
2. Any node that receives message msg sends it to its neighbors except the one it received it from.

This algorithm works fine and all nodes will receive message msg sent by p_i eventually. However, we can obtain a more efficient algorithm with less messages transferred between the nodes by a simple modification: A node sends msg to its neighbors only when it receives it for the first time. This way, duplicate transmission along an edge of the graph in the same direction is prevented. We now need a way to detect whether a message is received first time or not which can be implemented simply using a variable such as *visited* which is false initially and becomes true when msg arrives for the first time. Nevertheless, this modified algorithm is

Fig. 5.4 FSM of the *Flooding* algorithm

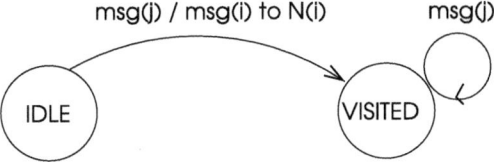

simple to implement by an FSM having two states as shown in Fig. 5.4, which will also aid us to understand the use of FSMs in distributed algorithms.

We can implement this algorithm based on the FSM as shown in Algorithm 5.2. When the message *msg* arrives for the first time, the VISITED state is entered and any further receptions of *msg* are ignored.

Algorithm 5.2 *Flooding*

1: { code for process i, message received from process j }
2: *currstate* ← IDLE   ▷ start with IDLE state
3: **if** $i = initiator$ **then**
4:     **send** $msg(i)$ to $N(i)$
5:     *currstate* ← VISITED
6: **end if**
7:
8: **while** *true* **do**   ▷ all nodes execute this part
9:     **receive**($msg(j)$)
10:     **case** *currstate* **of**
11:         IDLE: **send** $msg(i)$ to $N(i) \backslash j$
12:                 *currstate* ← VISITED
13:         VISITED: ▷ do nothing
14: **end while**

We can have some improvements to this algorithm as follows.

- Instead of waiting forever at line 8, we can have a terminating condition. A process i can wait certain times such as the diameter $diam(G)$ of the graph. This is logical since $diam(G)$ is the longest path that can be traversed by the message *msg*. Once the message *msg* is received $diam(G)$ times, process i terminates.
- We can have a counter commonly named *time-to-live* (TTL) contained in the message. Each time it is received, TTL is decremented and a message with 0 TTL value is not forwarded to neighbors.

Analysis

A careful look at this algorithm reveals that each edge of the graph will be traversed at most twice, once in each direction when both nodes at the ends of an edge start sending the message *msg* concurrently. Therefore, message complexity is $O(m)$. Assuming there is at least one message transfer at each time unit, time

Fig. 5.5 FSM of spanning tree construction using flooding

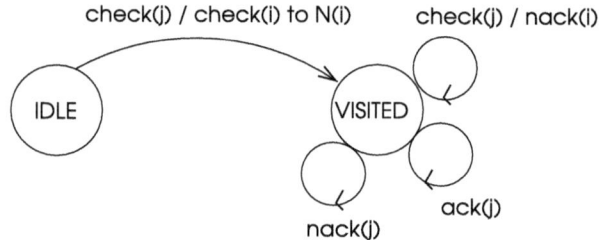

taken by this algorithm is the longest distance between any two vertices of the graph which is its diameter and thus, time complexity is $\Theta(diam(G))$.

5.6.2   Spanning Tree Construction Using Flooding

We can design a spanning tree construction of a network using the *Flooding* algorithm with few modifications. Building a spanning tree in a network environment means each node knows its parent and its children in the general sense. We will not attempt to store all of the tree structure at a special node or at each node of the graph since parent/children relationship at each node is adequate for transferring messages over the spanning tree. We have a single initiator as in the *Flooding* algorithm and this node becomes the root of the spanning tree to be formed. The first modification we have is to assign the sender j of the message $msg(j)$ as the *parent* of the receiver i if $msg(j)$ is received for the first time. Since we also require the parent to be aware of its children, node i should send an acknowledgment message $ack(i)$ to j to inform j of this situation. Otherwise, if node i already has a parent, meaning it has been visited before, it sends back a negative acknowledgment message $nack(i)$ to node j. We have, therefore, three types of messages: *check*, *ack*, and *nack*. Determining the types of messages is crucial in the design of distributed graph algorithms; moreover, determination of states is performed by messages if we are to use a FSM. Let us modify the FSM of Fig. 5.4 to reflect what we have been discussing. We can see that the states may remain the same since a node can be either in IDLE or VISITED state as before. Based on its state and the type of the message, we may need to take different actions. The modified FSM is shown in Fig. 5.5 with the VISITED state having all possible message types as input now.

This FSM can be directly translated to a distributed algorithm as shown in Algorithm 5.3 where additionally, a termination condition for a node is also specified. The activity of any node is finished when it has received *ack* or *nack* messages from all of its neighbors except the sender of the message it has received for the first time.

Algorithm 5.3 *Flooding2*

1: **int** *parent* ← Ø
2: **set of int** *childs* ← {Ø} , *others* ← {Ø}
3: **message types** *check, ack, nack*
4:
5: **if** *i* = *initiator* **then**
6:     **send** *check* to *N(i)*
7:     *currstate* ← VISITED
8: **end if**
9:
10: **while** (*childs* ∪ *others*) ≠ (*N(i)*\{*parent*}) **do**   ▷ all nodes execute this part
11:     **receive**(*msg(j)*)
12:     **case** *currstate* ∧ *msg(j).type* **of**
13:         IDLE ∧ *check*:   *parent* ← *j*
14:                       **send** *check* to *N(i)**j*
15:                       **send** *ack* to *j*
16:                       *currstate* ←VISITED
17:         VISITED ∧ *check*: **send** *nack* to *j*
18:         VISITED ∧ *ack*:   *childs* ← *childs* ∪ {*j*}   ▷ *j* is now a child
19:         VISITED ∧ *nack*:   *others* ← *childs* ∪ {*j*}   ▷ *j* is not a child
20: **end while**

The operation of this algorithm is shown in Fig. 5.6.

We could have easily implemented this algorithm without using an FSM, a node having a parent or not basically shows its state as IDLE or VISITED. With this in mind, this algorithm is shown in Algorithm 5.4 as in [2]. However, for complicated distributed algorithms, using FSMs would ease the design and implementation.

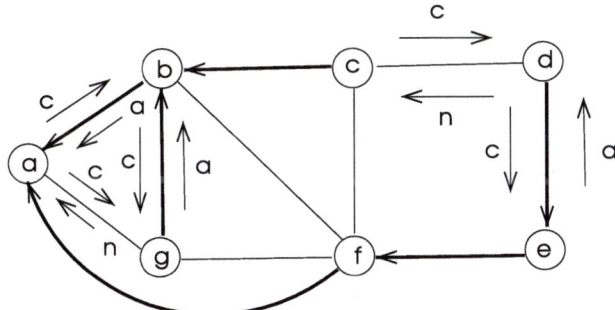

Fig. 5.6 A spanning tree constructed in a graph using flooding. The branch (*g, b*) is on tree but (*g, a*) is not since *check* message (*c*) from node *a* arrives at *g* later than *c* from *b*, which is replied by a *nack* (*n*) message. A similar situation is depicted for branch (*e, d*) where message *c* from node *d* is replied by an *ack* (*a*) message and (*d, e*) is included in the tree

---

Algorithm 5.4 *Flooding*3

1: **int** *parent* ← ⊥
2: **set of int** *childs* ← {∅} , *others* ← {∅}
3: **message types** *check, ack, nack*
4:
5: **if** *i* = *initiator* **then**   ▷ root initiates tree construction
6:     **send** *check* to *N*(*i*)
7:        *parent* ← *i*
8: **end if**
9:
10: **while** (*childs* ∪ *others*) ≠ (*N*(*i*)\{*parent*}) **do**
11:     **receive** *msg*(*j*)
12:     **case** *msg*(*j*).*type* **of**
13:             *check* :   **if** *parent* = ∅ **then**   ▷ *check* received first time
14:                         *parent* ← *j*
15:                         **send** *ack* to *j*
16:                         **send** *msg*(*i*).*check* to *N*(*i*) \ {*j*}
17:                      **else**   ▷ *check* received before
18:                         **send** *msg*(*i*).*reject* to *j*
19:             *ack*   :   *childs* ← *childs* ∪ {*j*}   ▷ *j* is a child
20:             *nack* :   *others* ← *others* ∪ {*j*}   ▷ *j* is not a child
21: **end while**

Analysis

Each edge of the graph will be traversed at least twice by *check/ack* or *check/nack* message pairs and at most four times when two nodes start to send *check* messages to each other simultaneously. Therefore, message complexity of this algorithm is $O(m)$. The depth of the tree constructed will be at most $n - 1$, assuming a linear network is built. If there is at least one message transfer per unit time, time complexity is $O(n)$.

5.6.3   Basic Communications

There are a number of basic communication operations performed in a distributed system. One such process is the *broadcast* which is initiated by a node by sending a message, and all of the nodes in the distributed system have a copy of the message at the end of the broadcast operation. Another fundamental primitive is the *convergecast* where data from each node is collected at a special node in the system. We will look into these two operations in this section. One other activity is the multicast sending of messages in which a message is delivered to only a specified subset of processes.

Broadcast over a Spanning Tree

For the broadcast operation, we will assume a graph represents the network of the distributed system and a spanning tree T is already built by an algorithm similar to what we have discussed. The broadcast is initiated by a node by sending *msg* to all of its children. Any node on the tree T that has children simply forwards *msg* to all of its children. Since *msg* is transferred only over tree edges, the number of messages will be $n - 1$ for a graph with n vertices. Time taken will be the depth of T, assuming concurrent sending of messages at each level. Depth of T can be a maximum of $n - 1$ assuming a linear network.

Convergecast over a Spanning Tree

In certain networks, data from all nodes are to be collected at a node with more capabilities and this special node can then analyze and evaluate all of the data, provide reports containing statistics which can be transferred to more advanced computation centers or users for further processing. This situation is commonly encountered in wireless sensor networks where data sensed needs to go through these steps of operation. Collecting data is very much simplified when a spanning tree constructed beforehand is used. In this case, the leaves of the tree send their data to their parents, the parents combine their own data with those of leaves, and send these to their parents. An intermediate node may in fact perform some simple operation on data such as taking average or finding extreme values. This way, data sent upwards in the tree does not have to get much larger at each level. This process of gathering called *convergecast* continues until all data is collected at the special node, commonly called the *sink* in sensor networks. Algorithm 5.5 shows the pseudocode for the convergecast process over a spanning tree. Leaves of the tree start the algorithm and any intermediate node in the tree should wait until data from all of its children are received before combining these data with its own to be sent to its parent as realized at line 8 of the algorithm. The termination condition for the root of the tree is met when it receives the convergecast messages from all of its children at line 12. For all others, termination is on line 17 when they send their data to their parents.

Algorithm 5.5 *Convergecast*

1: **int** *parent*
2: **set of int** *childs, received* ← {∅} , *data* ← {∅}
3: **message types** *convcast*
4:
5: **if** *childs* = {∅} **then**   ▷ *leaf* nodes start convergecast
6:    **send** *convcast* to *parent*
7: **else**   ▷ any intermediate node or root
8:    **while** *childs* ≠ *received* **do**   ▷ wait for convergecast messages from all children
9:       **receive** *convcast(j)*
10:       *received* ← *received* ∪ {*j*}
11:       *data* ← *data* ∪ *convcast(j)*
12:    **end while**
13: **end if**
14: **if** *i* ≠ root **then**
15:    **combine** *data* into *convcast*
16:    **send** *convcast* to *parent*
17: **end if**

Fig. 5.7 Convergecast over
the spanning tree of Fig. 5.6.
Message label values show
concurrent transfer of siblings
and the duration of messages

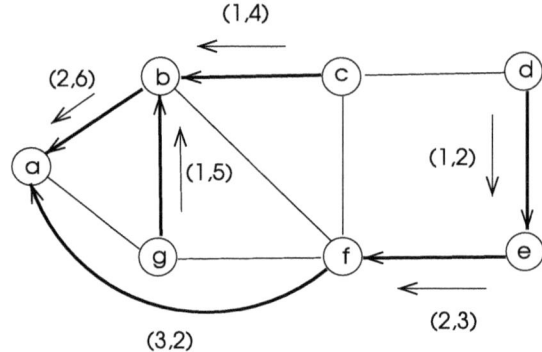

Message and time complexities for this algorithm are the same as the *Broadcast*
algorithm using similar reasoning. Figure 5.7 shows the operation of the *Convergecast* algorithm using the spanning tree built in Fig. 5.6. The messages are labeled
with pair (a, b); a showing the time frame and b is the duration of the message.
We can see the highest level of tree finishes convergecast in 7 time units, followed
by 11 and 10 units at level 2 and 6 units at level 1; total time is 11 units as the
longest interval over g, b, and a.

5.6.4   Leader Election in a Ring

Leader or coordinator election is needed in distributed systems as this special
node can initiate an algorithm and supervise the overall execution of the algorithm
as in a SSI algorithm. The leader may also take remedy actions when failures
are encountered in the execution of an algorithm. Nodes and communication links
may physically fail and although we can initially assign a node as the leader of the
network, we need to elect a new leader when failure happens. *Election algorithms*
provide ways of assigning a new leader in the network when the current leader
fails.

There are many leader election algorithms in literature for distributed systems.
As an introductory distributed algorithm example, we will consider leader election
in a ring with nodes having unique identifiers. The transfer of messages is in one
direction only. This example can be conveniently described by a simple FSM with
the following states:

- LEAD: The nodes have a leader in this stable state.
- ELECT: Election is going on when a node is in this state.

The main idea of this algorithm is that any node detecting the failure of the current leader initiates the algorithm by sending an *election* message containing its
identifier to its neighbor at its right assuming a clockwise unidirectional ring. A
node that receives this message changes its state to ELECT. If the identifier in the

Fig. 5.8 FSM of the ring leader election algorithm

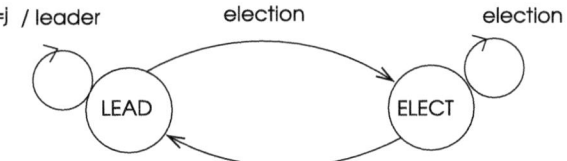

message is greater than its own identifier, it simply passes the election message to its next neighbor. Otherwise, it inserts its identifier which is greater than the identifier in the incoming message and sends it to the neighbor. We have two messages in this example:

- *election*: Sent by any node that detects leader failure. This message may be sent by more than one initiator.
- *leader*: The new leader broadcasts this message to notify all nodes that election is over.

When a process with identifier i receives a message with an identifier j in it, it checks and does one of the following:

- $i > j$: Process i replaces j with i in message and passes it to the next node.
- $i < j$: Process i simply passes message to next node.
- $i = j$: Process i becomes the leader and sends the *leader* message to its next neighbor.

In the last case, the election message originating from node i has returned to itself meaning that it has the highest identifier among all active processes. Basically, the highest identifier is transferred between all functioning nodes, and when the originator receives its own message, it determines it is the leader and sends the *leader* message to its neighbor which is then broadcast to all nodes by neighbor transfers. The FSM for this algorithm is depicted in Fig. 5.8.

Analysis
The worst case happens when the nodes are ordered from smallest to largest with respect to their identifiers in clockwise direction and start election concurrently in anticlockwise direction. The largest identifier message travels through all nodes n times, the second largest identifier is transferred $n - 1$ times, and in total, there will be $\sum_{i=1}^{n} = n(n+1)/2$ messages as shown in Fig. 5.9a. The best case occurs for a total of $2n$-1 messages when messages are transmitted in clockwise direction as in Fig. 5.9b. In this case, even if all nodes start election concurrently, their messages will be purged by the next nodes for $n - 1$ times and only the message of the highest identifier node, which is 7 in this case, will traverse the ring all the way back to the originator at n step. Total number of steps will then be $2n - 1$, excluding the declaration message sent by the leader.

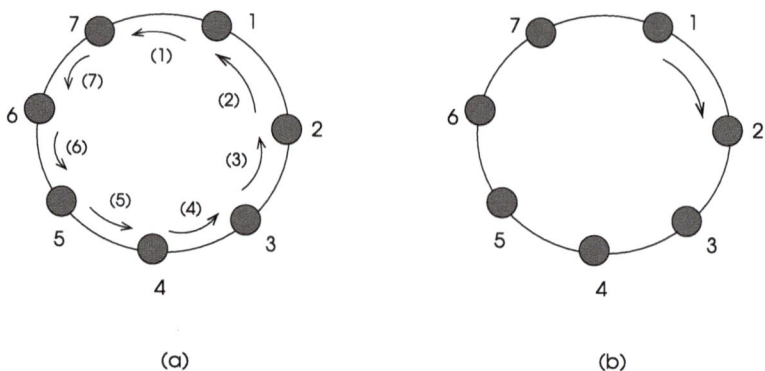

(a)   (b)

Fig. 5.9 Ring leader election algorithm: worst and best scenarios. In **a**, each message by the originator is tagged with the number of links it travels. For example, message originating at node 7 is tagged with 7 since it goes through 7 edges back to node 7. The best case is depicted in **b**

5.7   Chapter Notes

We have described distributed systems and the fundamental problems in designing algorithms for such systems in this chapter. Distributed systems are needed since they provide sharing of resources, fault tolerance and in various implementations; the application is inherently distributed such as a factory control system. The common platforms to implement distributed algorithms are the Internet, mobile ad hoc networks, wireless sensor networks, the Grid, and the Cloud. In many cases, we can model these networks suitably by graphs with vertices of a graph representing computational nodes and edges showing the communication links between them.

Distributed systems require distributed algorithms that run at the nodes of such a system. Synchronization and communication are two basic requirements in efficient design of such algorithms. Synchronization may be realized at various levels: hardware, operating system/middleware, and at application level. We see synchronization at application level using messages is commonly used due to versatility and easiness in implementation which may then be transferred to local synchronization primitives. In this so-called *message-passing* distributed systems, the main communication and synchronization are achieved by messages only. The receiver of a message decides on what to do next mainly by the type of the arriving message. A synchronous distributed algorithm typically runs in rounds and the next round is not started until all nodes finish executing the current round. The synchronization at the beginning and end of round is commonly realized by special messages sent by a distinct node.

Distributed algorithms can be modeled by FSMs which are mathematical models which include states and transitions between states as we have outlined. We can design a distributed algorithm without a FSM but for complicated algorithms, FSMs provide a neater algorithm with visual aid and less error-prone than algorithms which otherwise could involve many decision-making statements.

Fig. 5.10 Ring structure for
Exercises 4

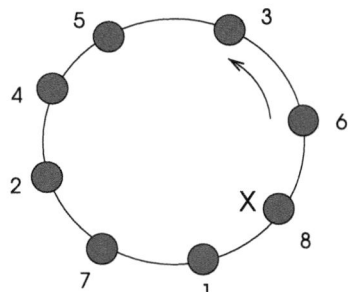

We then described some sample distributed graph algorithms which include building a spanning tree of the graph, broadcast and convergecast operations over a spanning tree, and a leader election algorithm to find the new coordinator of nodes in a ring when leader fails. We need to prove that a distributed algorithm correctly achieves what it is intended for; and time, message, bit, and space complexities of a distributed algorithm are used to evaluate its performance. In general, message complexity is considered as the dominant cost of a distributed algorithm.

Exercises

1. The elevator algorithm of Example 5.1 is to be modified so that a green light showing moving up and a yellow light for downwards movement are added. Provide the necessary modifications to the FSM diagram, FSM table, and C code to incorporate these two outputs.
2. A binary bit string S has even parity if the number of bits in S is an even number and has odd parity otherwise. Provide the FSM diagram, FSM table, and the C code of an algorithm that reads a binary string bit-by-bit and decides to be in either even or odd state after each read. Use the programming style shown in the C code of Example 5.1.
3. We need to modify the broadcast algorithm over a spanning tree so that the initiator becomes aware that each node has received the broadcast message. This can be realized simply by each node deferring to send an acknowledgment to the sender of the message until it receives acknowledgments from all of its children, similar to the convergecast operation which should be started by the leaves of the spanning tree once they receive the broadcast message. Write the pseudocode for this algorithm with comments and work out its time and message complexities.
4. Show the execution of the ring election algorithm for the nodes shown in Fig. 5.10. Assume nodes 2 and 5 find concurrently that the leader 8 is not working and decide to run an election.
5. In a fully connected graph with each node having unique identifiers, *bully algorithm* may be used to elect a new leader. A node u that finds that the leader is not functioning may start this algorithm by sending an *election* message to all nodes that have higher identifiers than itself. Any node v that receives this

message sends back and *ack* message to the node u which then leaves the election. The node v now starts an election and this process continues until there is one winner which is the active node with the highest identifier. The new leader broadcasts it is the winner by a special message to all nodes. Write the pseudocode for this algorithm and find its time and message complexities. Show its operation in a complete graph of 8 nodes where nodes 4 and 6 find simultaneously the leader 8 is down.

References

1. Attiya H, Welch J (2004) Distributed computing: fundamentals, simulations, and advanced topics, 2nd edn. Wiley, New York
2. Erciyes K (2013) Distributed graph algorithms for computer networks. Springer computer communications and networks series. Springer, Berlin. ISBN- 10:1447151720 (May 16, 2013)
3. Foster I, Kesselman C (2004) The grid: blueprint for a new computing infrastructure. Morgan Kaufmann, San Mateo
4. Mell P, Grance T (2011) The NIST definition of cloud computing. National institute of standards and technology, US department of commerce, special publication, 800145
5. Tanenbaum AS, Steen MV (2007) Distributed systems, principles and paradigms, 2nd edn. Pearson-Prentice Hall, Upper Saddle River. ISBN 0-13-239227-5
6. Tel G (2000) Introduction to distributed algorithms, 2nd edn. Cambridge University Press, Cambridge

Trees and Graph Traversals

6

6.1 Introduction

A *tree* is a connected acyclic graph and a *forest* consists of trees. Trees find many applications in computer science and real-life situations. For example, the organization of a university or any establishment is typically shown by a tree, and family trees illustrate the parental relationships between the individuals. In computer science, trees are used for efficient data storage and tree-based algorithms find a wide range of applications. A spanning tree of a graph is its tree subgraph that includes all of the vertices of the graph. A graph may have a number of spanning trees. We have described trees briefly in Chap. 2; now, we provide a more detailed analysis of trees with related algorithms in this chapter. We start by defining the tree structure and stating its properties. We then describe algorithms for constructing spanning trees and tree traversals and briefly review special tree types.

Traversing all vertices or all edges of a graph in some order is required in various applications, for example, to find all reachable vertices in a graph. The algorithms that perform traversals may also be used as the building blocks of more complicated graph algorithms. In an undirected graph traversal, all edges are considered whereas only the outgoing edges from a node are considered in a directed graph. Two main methods of graph traversal are the *depth-first search* and *breadth-first search*. In the first method, we start from any vertex of a graph and go as deep as we can by visiting neighbors of each visited vertex. The breadth-first search involves visiting all neighbors of a vertex first, then visiting all neighbors of these neighbors and proceeding in this manner until all vertices are visited. Both these methods produce spanning trees rooted at the starting vertex. We describe sequential, parallel, and distributed algorithms for both of these approaches with their possible applications in this chapter.

© The Author(s), under exclusive license to Springer Nature Switzerland AG 2026
K. Erciyes, *Guide to Graph Algorithms*, Texts in Computer Science,
https://doi.org/10.1007/978-3-032-05294-0_6

6.2   Trees

A graph is a *tree* if it is connected and does not contain any cycles. A *forest* is a graph with no cycles. Every path is a tree and a tree T is path if and only if the maximum degree of T is 2. A tree can be *rooted* or *unrooted*. A designated node called the *root* in a rooted tree is at the top of the hierarchy and every other vertex of T has a path to the root; the tree is *unrooted* otherwise. A binary tree consists of nodes that have at most two children. The following statements equally define a tree T:

1. T is connected and has $n - 1$ edges.
2. Any two vertices of T are connected by exactly one path.
3. T is connected and a cut-edge removal of each edge disconnects T.

Definition 6.1 (*Level*) The *level* of a vertex in a rooted tree is its distance to the root. The level of a vertex is also called its *depth* in the tree.

Definition 6.2 (*Parent, Child*) A vertex v that is connected to vertex u (the predecessor of u) on the path to the root is called the *parent* of u and v is called the child of u.

A vertex v can have only one parent as having more than one parent produces a cycle and the resulting structure therefore will not be a tree.

Definition 6.3 (*Leaf, Internal vertex, Siblings*) A *leaf* is a vertex of the tree that has no children. An *internal vertex* of a tree has a parent and one or more children. *Siblings* in a tree have the same parent.

The maximum level of a leaf is the *height* (or *depth*) of the tree as it is the farthest vertex to the root.

Definition 6.4 (*Spanning Tree*) A spanning tree $T = (V, E')$ of a graph $G = (V, E)$ has the same vertex set of G with an edge set that is a subset of the edges of G.

A minimum spanning tree MST(G) of a weighted, undirected graph G is a spanning tree of G with minimum total edge weight cost among all spanning trees of G. MSTs find numerous applications and we will investigate sequential, parallel, and distributed algorithms for MSTs in Chap. 8.

Definition 6.5 (*m-ary tree* ($m \geq 2$)) An *m*-ary tree is a rooted tree in which every vertex other than the leaves has at most m children. In a *binary tree*, $m = 2$.

Definition 6.6 (*Complete m-ary tree*) A complete *m*-ary tree is an *m*-ary tree in which every internal vertex of the tree has exactly *m* children and all leaves have the same depth.

Definition 6.7 (*Ordered Tree*) In an ordered tree, there is a linear ordering of the children of each node. This means we can identify the children as first, second, etc.

Figure 6.1 displays these concepts.

Definition 6.8 (*Center of a Tree*) A *center* of a tree T is either a vertex v such that $\max(d(u, v)) \forall u \in V$ is minimum or there may be two adjacent centers with this property.

We can find center(s) of a tree by recursively removing its leaves until there are one or two centers left. This procedure is illustrated in Fig. 6.2.

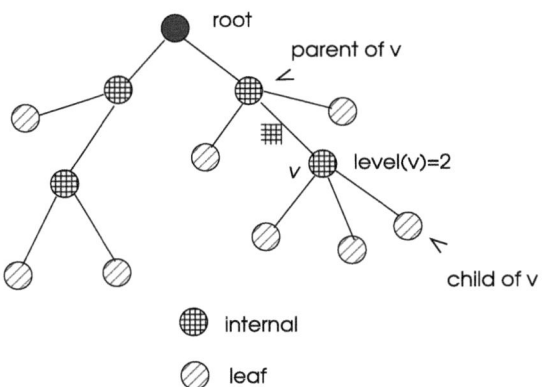

Fig. 6.1 Vertices of a sample tree which is a 3-ary tree since all vertices other than the leaves have at most three children

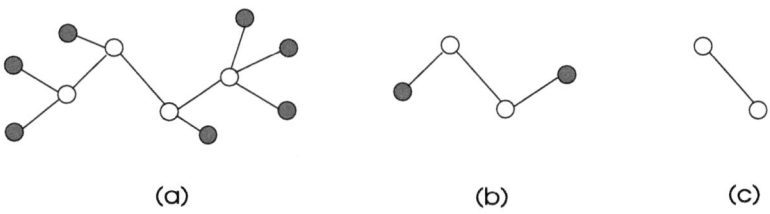

(a) (b) (c)

Fig. 6.2 Finding two centers of an unrooted tree. The leaves are recursively removed from **a–c** to obtain the two centers

6.2.1   Properties of Trees

Theorem 6.1 *An undirected graph G is a tree if and only if there is a unique simple path between any two vertices of G.*

Proof We will first assume G is a tree which means it has no cycles. Now, if there are two simple paths between any vertex pair (u, v) in G, the total path from u to v and then back to u would form a cycle; however, from the definition of a tree, we know G does not have a cycle and therefore a contradiction. In the other direction of the statement, let us assume G is a graph in which any two vertices u and v are connected by a unique path. If there were two distinct paths between a vertex pair u and v, these paths would form a cycle and since G is a tree and does not contain a cycle, we have a contradiction.   □

Theorem 6.2 *Every tree of order n has size $m = n - 1$.*

Proof We will prove this theorem by induction n. The induction hypothesis is that a tree with n nodes has $n - 1$ edges. For the base case, when $n = 1$, the trivial graph has no edges; therefore, the base case holds. Let us consider a tree T with $n + 1$ nodes. Removing a leaf node v with its incident edge from T leaves a tree $T' = (V', E')$. Since we have not created a cycle in doing so, T' is also a tree, say with p edges and q vertices. With the inductive hypothesis, $p = q - 1$. Since we removed one edge and one node from T, $p = m - 1$, $q = n - 1$. Substitution yields $p = n - 2$, when $q = n - 1$, hence $m = n - 1$.   □

Theorem 6.3 *Any connected graph G with n vertices and $n - 1$ edges is a tree.*

Proof We need to show G is acyclic. Let us assume the contrary that G has at least one cycle, and iteratively remove edges from cycles until we have a graph G' which is acyclic and therefore is a tree. We can conclude G' has $n - 1$ edges by Theorem 6.2. Since we have removed at least one edge to obtain G', G had a size at least n which is one greater than the size of G' and hence a contradiction.   □

6.2.2   Finding Root of a Tree by Pointer Jumping

In pointer jumping method, we would have each element of a linked list point to the link of the element it points to in each step. This method can be conveniently used to find the root of a rooted tree as shown in Algorithm 6.1. After $\lceil \log depth \rceil$ steps, all vertices of the tree will point to the root. Note that this method is inherently parallel.

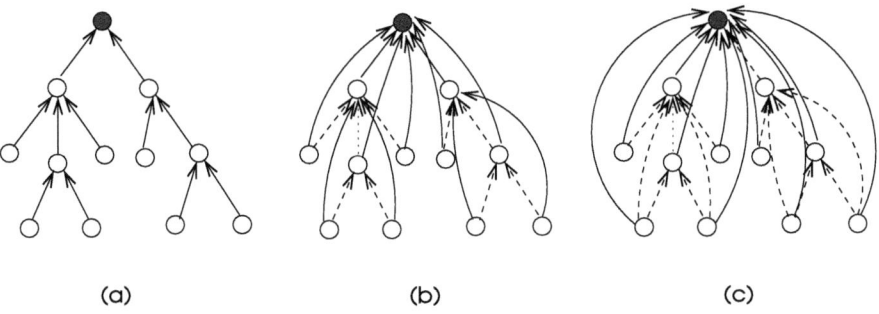

(a)   (b)   (c)

Fig. 6.3 Root of a tree is found in two iterations by all nodes

---

Algorithm 6.1 *Finding Root of a Tree*

1: **Input**: a rooted tree $T = (V, E)$ with n elements
2: **Output**: each vertex v points to the root
3: **for** i=1 to $\lceil \log depth \rceil$ **do**
4:     **for** each tree vertex $v \in V$ **in parallel do**
5:         $(v- > parent) \leftarrow ((v- > parent)- > parent)$
6:     **end for**
7: **end for**

Finding the root of a tree with depth 3 is shown in Fig. 6.3. It takes three iterations for all nodes to point to the root.

6.2.3 Counting Spanning Trees

A spanning tree of a graph G is its subgraph that is a tree and contains all vertices of G. Every connected and undirected graph has at least one spanning tree. Spanning trees find various applications such as providing a communication infrastructure in computer networks and cluster analysis. Let us denote the number of spanning trees of a graph G by $\tau(G)$. Cayley provided a formula to find the number of spanning trees of a labeled complete graph K_n which has a unique identifier for each of its vertices as follows [4]:

$$\tau(K_n) = n^{n-2}. \tag{6.1}$$

The 3 spanning trees of K_3 and 16 spanning trees of labeled K_4 are shown in Figs. 6.4 and 6.5.

Fig. 6.4 All possible labeled three spanning trees of K_3

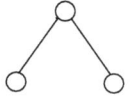 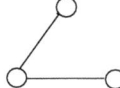 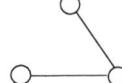

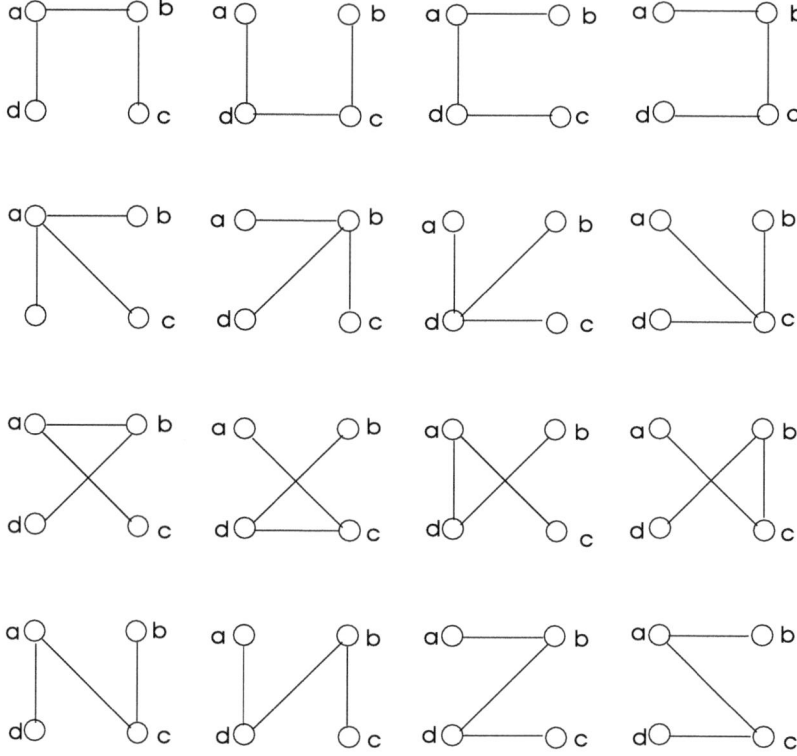

Fig. 6.5 All possible labeled 16 spanning trees of K_4

6.2.4 Constructing Spanning Trees

We describe three simple algorithms to construct a spanning tree of an undirected, simple, and connected graph in this section.

6.2.4.1 The First Algorithm

As a simple approach, we will start with the all edges of a graph belonging to T which may not be a tree in fact, and then iteratively remove edges from T which will not result in a disconnected graph G. We continue until T has one of the basic tree properties described above; that is, it has $n - 1$ edges, it is acyclic, etc. The pseudocode of this algorithm is given in Algorithm 6.2 where we check the tree property that each edge of a tree is a cut-edge, that is, removal of any edge leaves a tree disconnected.

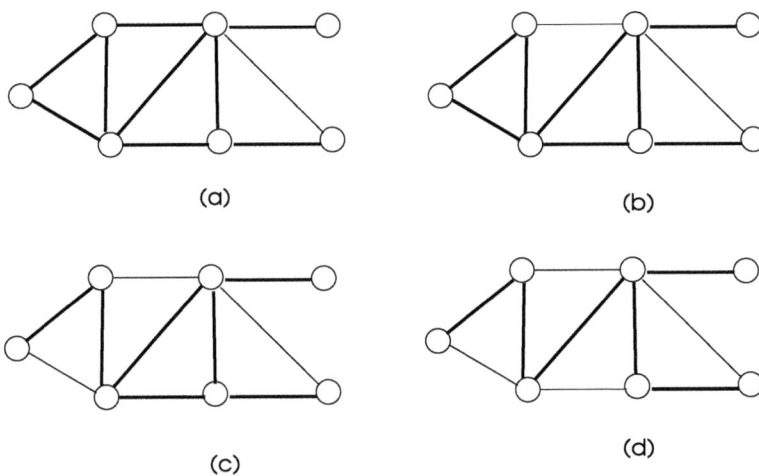

Fig. 6.6 Steps of Algorithm 6.2 in a sample graph. All of the graph edges shown in bold are included in the spanning tree initially. Then, an edge removal which does not disconnect the tree is removed at each iteration starting from **a** until this is not possible

---

Algorithm 6.2 *ST_Alg1*

1: **Input**: $G = (V, E)$
2: **Output**: Spanning Tree T of G
3: $T \leftarrow E$
4: **repeat**
5:    **pick** any edge $e \in T$ removal of which does not disconnect T
6:    $T \leftarrow T - \{e\}$
7: **until** T any edge removal leaves T disconnected

The steps of operation of this algorithm are depicted in Fig. 6.6. We remove an edge in each iteration, and hence the *repeat − until* loop is executed $O(m)$ times. We check that the resulting structure after each edge removal has $n-1$ edges and is connected. Checking the former is simple by keeping a counter and decrementing it after each edge deletion. However, checking the connectedness of the graph is not trivial as we will see in the second part of this chapter. Note that we require both properties by Theorem 6.3. We will see we can construct spanning trees with special properties in linear time in Chap. 7.

6.2.4.2 The Second Algorithm

An alternative algorithm to build a spanning tree T of a graph $G = (V, E)$ can be constructed as follows. This time, we start with an empty tree T and pick an arbitrary vertex u, and include u in T. Thereafter, we arbitrarily select any *outgoing edge* (u, v) which is an edge with one endpoint say u in T and the other endpoint v outside T, and include (u, v) in T. We proceed in this manner until all vertices are processed as shown in Algorithm 6.3.

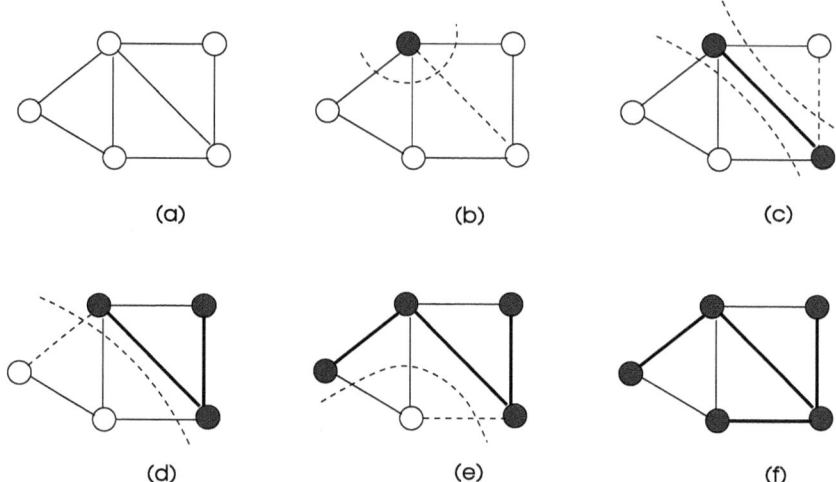

(a)   (b)   (c)

(d)   (e)   (f)

Fig. 6.7 Construction of a spanning tree of a small sample graph using outgoing edge concept. Selected outgoing edge at each iteration is shown by a dashed line

---

Algorithm 6.3 *ST_Alg2*

1: **Input:** $G = (V, E)$
2: **Output:** Spanning Tree T of G
3: $T \leftarrow$ an arbitrary vertex u
4: $V' \leftarrow \emptyset$
5: **while** $V' \neq V$ **do**
6:     **select** any outgoing edge (u, v) from T vertices with $u \in T \wedge v \notin T$
7:     $T \leftarrow T \cup \{(u, v)\}$
8:     $V' \leftarrow V' \cup \{v\}$
9: **end while**

Correctness is evident since any outgoing edge will not produce a cycle with edges already included in the tree. Hence, the resulting structure will be cycle free and therefore a tree. A possible operation of this algorithm in a small sample graph is shown in Fig. 6.7. The time complexity is $O(n)$ which is the number of times the *while* loop is executed. Note that we do not need any extra processing such as checking tree property or connectivity as in the previous algorithm. Selecting minimum weight outgoing edges from the set of edges already included in the tree will be useful in forming minimum spanning trees as we will see in the next chapter.

6.2.4.3 The Third Algorithm

Another approach to build a spanning tree is to first start with each vertex being a tree and merge any two trees until this is not possible. The main idea and the related correctness argument here is that merging of two trees will always produce another tree simply because each tree before the merge is acyclic and an edge

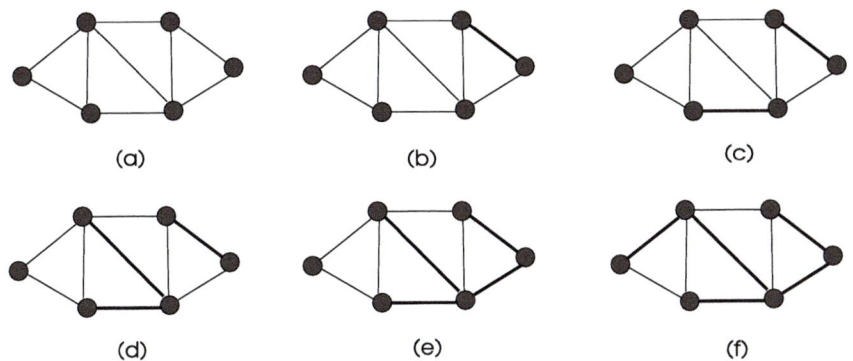

Fig. 6.8 Running of the third spanning tree construction algorithm in a sample graph. Spanning tree edges are shown in bold and each vertex is a tree initially

joining them will not produce a cycle with the smaller trees. We have again $O(n)$ steps of this algorithm to form the spanning tree as in the linear graph case with iteratively merging each vertex with the neighbor vertex. The operation of this algorithm in a sample graph is depicted in Fig. 6.8.

This method lends itself to parallel processing since independent subtree formations are possible. In fact, it is also suitable for distributed processing in a network environment. Each vertex is a network node and requests merge operation from a neighbor node. We need to be careful as not to form cycles when concurrent requests are made by two nodes from the same subtree to two nodes that coexist in another neighbor subtree. This problem can be handled by selecting a leader for each subtree which controls the requests to merge by a suitable protocol.

6.2.5 Tree Traversals

Tree traversal is the process of recursively visiting each node of the tree exactly once. Traversing trees in some determined sequence is useful in many graph applications. We can classify tree traversals by the order in which the vertices are visited as preorder and postorder for general trees.

Preorder Traversal

In *preorder traversal* of a rooted tree, a vertex is visited before its descendants as shown in Algorithm 6.4. The time complexity of this traversal is $O(n)$ for a tree with n vertices.

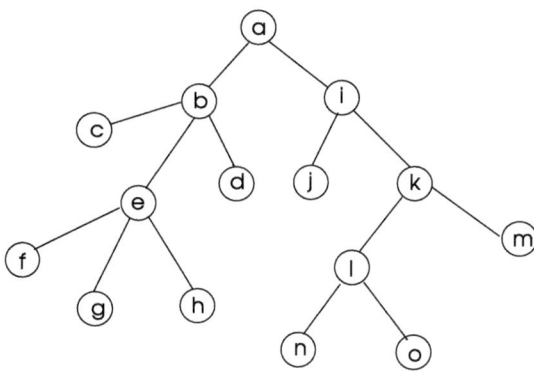

Fig. 6.9 Sample tree for tree traversals

Algorithm 6.4 *Preorder*

1: **procedure** PREORDER(*root*)
2:     **Input**: tree *T* and its root *root*
3:     **Output**: preorder traversal of *T*
4:     **if** *root* ≠ Ø **then**
5:         *process(root)*
6:         *Preorder(root → left)*
7:         *Preorder(root → right)*
8:     **end if**
9: **end procedure**

The preorder traversal of the sample tree of Fig. 6.9 results in vertex processing sequence of *a, b, c, d, e, f, g, h, I, j, k, l, n, o, m*.

Postorder Traversal

Postorder traversal of a tree involves visiting a tree vertex after visiting its descendants as depicted in Algorithm 6.5. The postorder traversal of the sample tree of Fig. 6.5 provides vertex processing sequence of *c, f, g, h, e, d, b, j, n, o, l, m, k, I, a*. Since each vertex is visited exactly once, time complexity of this method is $O(n)$.

Algorithm 6.5 *Postorder Traversal*

1: **procedure** POSTORDER(*root*)
2:     **Input**: tree *T* and its root *root*
3:     **Output**: postorder traversal of *T*
4:     **if** *root* ≠ Ø **then**
5:         *Postorder(root → left)*
6:         *Postorder(root → right)*
7:         *process(root)*
8:     **end if**
9: **end procedure**

6.2.6   Binary Trees

Definition 6.9 (*Binary Tree*) A *binary tree* is a rooted tree in which every vertex has at most two children and each child of a vertex v is *left child* or *right child* of v.

A complete binary tree of depth d has $2^{d+1} - 1$ vertices which is the maximum order of any binary tree. Binary trees can be traversed as preorder or postorder as in general trees. An additional traversal method for binary trees is *inorder* tree traversal.

Inorder Traversal

In this mode of binary tree traversal, the vertices at each left subtree of a vertex are processed first, the vertex is processed second, and the right subtree vertices of the vertex are processed finally. The pseudocode for this operation is shown in Algorithm 6.6. The time complexity of this algorithm is also $O(n)$.

Algorithm 6.6 *Inorder Traversal*

```
1: procedure INORDER(root)
2:     Input: tree T and its root root
3:     Output: inorder traversal of T
4:     if root ≠ Ø then
5:         inorder(root → left)
6:         process(root)
7:         inorder(root → right)
8:     end if
9: end procedure
```

Python code that provides the traversals of the tree depicted in Fig. 6.10 is shown in Listing 6.1 with the output in Listing 6.2. The tree is constructed as a list with each element as [node, left, right].

Listing 6.1 Tree traversals

```
Tree = [[0,1,2],[1,3,4],[2,5,6],[3,None,None],[4,None,None],[5,
    None,None],[6,None,None]]

def preorder(root):
    if root == None:
        return
    print(Tree[root][0],end=" ")
    preorder(Tree[root][1])
    preorder(Tree[root][2])

def inorder(root):
    if root == None:
        return
    inorder(Tree[root][1])
    print(Tree[root][0],end=" ")
    inorder(Tree[root][2])

def postorder(root):
    if root == None:
        return
    postorder(Tree[root][1])
    postorder(Tree[root][2])
    print(Tree[root][0],end=" ")
print("Preorder Traversal:", end=" ")
preorder(0)
print("\n")
print("Postorder Traversal:", end=" ")
postorder(0)
print("\n")
print("Inorder Traversal:", end=" ")
inorder(0)
```

Listing 6.2 Output of KG Analysis

```
Preorder Traversal: 0 1 3 4 2 5 6
Postorder Traversal: 3 4 1 5 6 2 0
Inorder Traversal: 3 1 4 0 5 2 6
```

Fig. 6.10 Binary tree

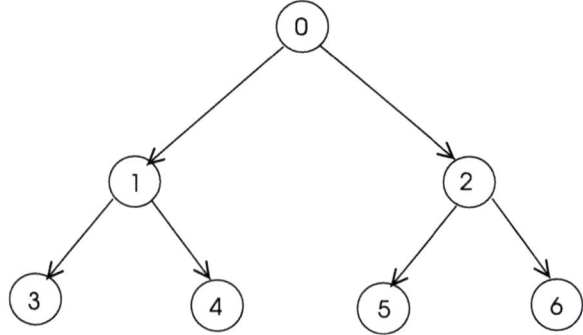

Fig. 6.11 Evaluation of an
expression using a binary tree $((a*b)+c)-(d*((f*g)+e))$

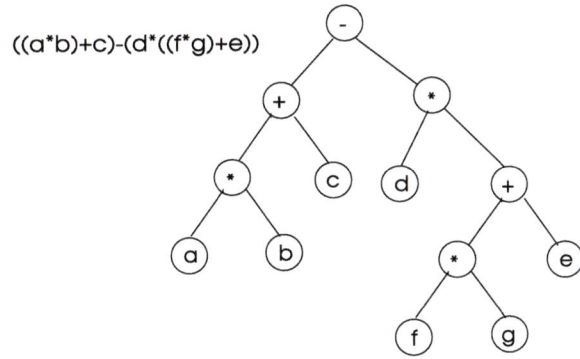

Fig. 6.12 Example of a
binary search tree

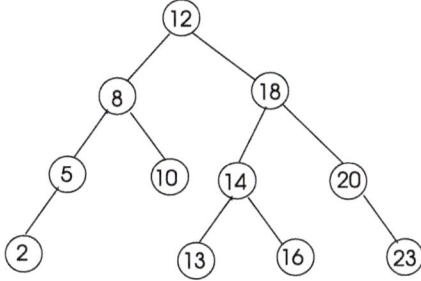

We can store arithmetic or logical expressions in a binary tree where leaves of the tree are operands and internal nodes are unary or binary operators as shown in Fig. 6.11. Such a binary tree is commonly referred to as an *expression tree*. This method is used by compilers to parse and evaluate various expressions. We can see that inorder traversal of an expression tree yields the expression.

A *binary search tree* (BST) is a binary tree in which the value in every node is greater than all values in the left subtree of the node and less than all the values in the right subtree of the node as shown in Fig. 6.12. BSTs are used in various applications such as sorting or data search.

6.2.7   Priority Queues and Heaps

A *priority queue* is a data structure to store a set of elements each having a value called a *key*. This data structure is useful in implementing various graph algorithms as we will see when reviewing algorithms for weighted graphs in Chap. 7. A min-priority queue provides the following operations.

- *Insert(S,x)*: Inserts the element x into the set S.
- *Minimum(S)*: The element of S with the largest key is returned.

Fig. 6.13 Example of a
heap. The dashed path from
the root to a leaf shows the
monotonically increasing key
values

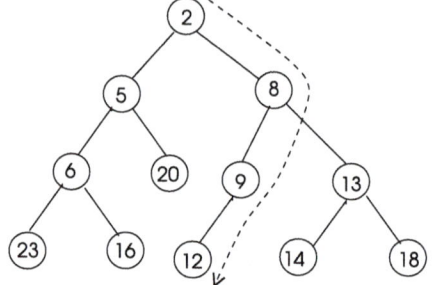

- *ExtractMin(S)*: Similar to *Minimum(x)* but the element *x* is deleted from *S*.
- *DecreaseKey(S,x,k)*: The value of the key of *x* is decreased to *k*.

When we are dealing with a max-priority queue, the operations in such a queue
are *Maximum(S)*, *ExtractMax(S)*, and *IncreaseKey(S,x,k)* which find the maximum
element of the queue, extract this value, and increase the value of the key of an
element in turn.

Heap as a Priority Queue

A *min-binary heap* is a complete binary tree except possibly the leaves in which
the keys of the children of any vertex *u* are greater than or equal to the key of
u. Therefore, along each path from the root, keys monotonically increase and the
root has the minimum key value as shown in Fig. 6.13. We can have *max-binary
heap* in which the key values decrease from the root downward.

All of the priority queue operations defined above can be implemented with
heaps as *HeapInsert*, *HeapExtract*, *HeapMin*, and *HeapDecrease* procedures in
$O(\log n)$ time and building the binary heap takes $O(n)$ time. A detailed description
of the heap structure is provided in [5].

6.3 Depth-First Search

Depth-first search (DFS) is a basic method to traverse all nodes of a graph. It
basically traverses a graph by going as deep as possible from a given vertex and
hence the name. We explore a path as far as we can go by marking vertices we
visit along the path and when we cannot go any further since either we encounter
a vertex with a degree of 1 or all neighbors of a vertex are visited, we return
to where we come from. This method can be best described by a person in a
maze carrying a chalk and a string. Each room (vertex) has a number of doors
(neighbors) and the person after entering the room selects one of the unmarked
doors and marks it with the chalk and goes through that door to another room. If
that has no unmarked doors (all neighbors visited) or no doors other than the one

she came from (a vertex with one edge), she returns to where she came from. The string is used to keep track of where she came from. The DFS algorithm has two versions as recursive and iterative described below.

6.3.1   A Recursive Algorithm

We assume the graph is connected and if it is not, DFS algorithm is performed on each component of the graph. The latter version of the algorithms is called *DFS_ Forest*. We select any vertex u of the graph G and this vertex is the root of the DFS tree to be formed. We then select an edge (u, v) that is incident to u. This edge is a *tree edge* and is included in the DFS tree T, and the vertex u is the parent of vertex v in T. We proceed in this manner by always selecting unexplored edges that are incident to vertices that are not visited. If all the edges incident on a vertex u are explored meaning all of its neighbors are visited, we return to the parent of u and continue the search from there. When we find an unexplored edge (u, v) incident on a vertex v, the edge (u, v) is traversed, it is included in T, and u becomes the parent of v as in the root case. If v has been visited before, and (u, v) is unexplored, (u, v) is a non-tree edge and is not included in T. The pseudocode of an algorithm that performs the described procedure is given in Algorithm 6.7. We also record the visit times for each vertex; the first time of visit when a vertex is discovered is in $d[v]$ and the final time when we return from the recursive call to v is stored in $f[v]$.

Algorithm 6.7 *DFS_Recursive_Forest*

1: **Input**: $G(V, E)$, directed or undirected graph
2: **Output**: $Pred[n]$; $d[n]$, $f[n]$ ▷ place of a vertex in DFS tree and its visit times
3: **int** *time* ← 0; **boolean** $Marked[1..n]$
4: **for all** $u \in V$ **do**   ▷ initialize
5:     $Marked[u]$ ← $false$
6:     $Pred[u]$ ←⊥
7: **end for**
8: **for all** $u \in V$ **do**
9:     **if** $Marked[u] = false$ **then**
10:        $DFS(u)$ ▷ call for each connected component
11:     **end if**
12: **end for**
13:
14: **procedure** $DFS(u)$
15:     $Marked[u]$ ← $true$
16:     *time* ← *time* + 1; $d[u]$ ← *time*   ▷ first visit
17:     **for all** $(u, v) \in E$ **do**   ▷ visit neighbors
18:        **if** $Marked[v] = false$ **then**
19:           $Pred[v]$ ← u
20:           $DFS(v)$
21:        **end if**
22:     **end for**
23:     *time* ← *time* + 1
24:     $f[u]$ ← *time*   ▷ return visit
25: **end procedure**

The output of the algorithm is a DFS tree stored in the array *Pred* which shows the predecessors of each vertex. There are few things to note about this algorithm as follows.

- We can select the neighbors of the visited vertex arbitrarily or using some ordering in line 17. If vertices are labeled with unique integers, the ordering can be linear, from smallest to the largest vertex identifiers. When vertices are labeled with letters, we can have a lexicographically first choice. As a consequence, the DFS tree obtained as the result of this algorithm is not unique as the order of the selection of unexplored edges affects the structure of this tree.
- If the adjacency matrix representation of the graph is used, we need to check the entire row that belongs to vertex u in lines 17 and 18 for n times for a graph with n vertices. Using adjacency list means the checking in these lines will be a maximum of $\Delta(G)$ times; hence, we can deduce using adjacency list that is a better choice for this algorithm than using adjacency matrix.
- The *DFS* procedure terminates when it returns to the root vertex it is called from. The algorithm terminates when the *DFS* procedure is run on all components of the graph.
- The first visit time $d[v]$ of a vertex v is called the *depth-first number* of vertex u and corresponds to the number given to it during the preorder traversal of the

tree formed. We will see first and last visit times of vertices can be used for various DFS applications.

- If we know the graph G is connected prior to executing the algorithm, we can simply execute the procedure between the lines 14 and 25 for a vertex u to find a DFS tree rooted at u.

Analysis
The edges included in the DFS tree form a directed spanning tree of G. This is true since we never form a cycle by never selecting an edge between two marked vertices (lines 18 and 19) and all of the vertices are marked and are included in the DFS tree in the end. We need to invoke the *DFS* procedure n times, one for each vertex. Each activation of this procedure involves checking each entry in the adjacency matrix for a total of n times. The time taken using the adjacency matrix is therefore $\Theta(n^2)$. Using the adjacency list means checking each edge in G twice for each vertex in its ends plus the time taken for initialization resulting in a total time of $\Theta(n+m)$.

Example
The running of *DFS_Forest* algorithm on a sample graph with two components is shown in Fig. 6.14. We can see two DFS trees which are directed spanning trees of two components are formed.

For any two vertices u and v on the DFS tree formed with u as the ancestor of v,

$$d[u] < d[v] < f[v] < f[u]$$

since vertex u is discovered before vertex v and we return to vertex u after we complete processing vertex v. When these vertices are on different trees or on the same tree but do not have parental relationship, then either $f[u] < d[v]$ or $f[v] < d[u]$.

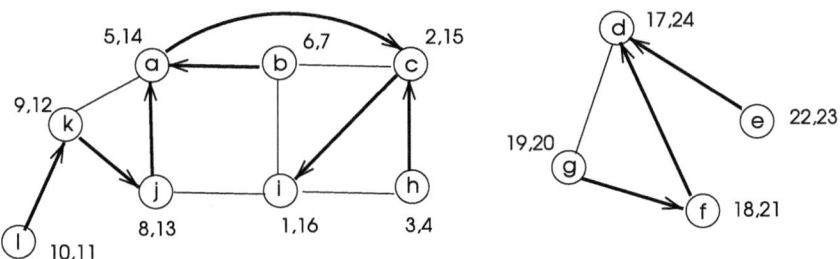

Fig. 6.14 Running of *DFS_Recursive_Forest* algorithm in a sample disconnected graph with first and last visit times shown next to the vertices. The source vertices are I and d in the components and the arrows point to parents in the tree

Edge Properties

The DFS run on a directed graph works similarly except that we have to consider only the outgoing edges from a vertex when searching its neighbors. The DFS algorithm working in this manner partitions the edges of a directed graph into the following types.

- *Tree Edges*: These are the edges on the tree formed. An edge (u, v) belongs to DFS tree if DFS(u) calls DFS(v).
- *Back Edges*: When vertex u of an edge (u, v) is a descendant other than its children of vertex v in tree, (u, v) is a *back edge*.
- *Front Edges*: When vertex u of an edge (u, v) is an ancestor of vertex v other than its parent in tree, (u, v) is a *front edge*.
- *Cross Edges*: Any edge that is not a tree, back, or front edge is called a *cross edge*.

There is also the following relationship between the discovery time d and finish time f of vertices in a graph G, commonly called the *parenthesis theorem*. Other orderings of discovery and finish times are not possible.

- Vertex u is a descendant of v in the DFS forest if and only if $[d(u), f(u)]$ is a subinterval of $[d(v), f(v)]$.
- Vertex v is a descendant of u in the DFS forest if and only if $[d(v), f(v)]$ is a subinterval of $[d(u), f(u)]$.
- Vertex u is not related to v in the DFS forest if and only if $[d(u), f(u)]$ and $[d(v), f(v)]$ are disjoint intervals.

Figure 6.15 displays these edges in a sample graph. Discovering a back edge on a DFS tree helps us to discover various properties of a graph. Let us modify the DFS procedure in Algorithm 6.7 for a directed graph to classify these edges using the discovery and finish times of their endpoints described above. We will use three colors for each vertex in this implementation: a vertex is *white* when it is unexplored, it is *gray* when it is explored but not finished, and it is *black* when it is finished. The array *color* is initialized to *white* for all vertices and holds the color of each vertex. The pseudocode for the modified DFS is shown in Algorithm 6.8 [5].

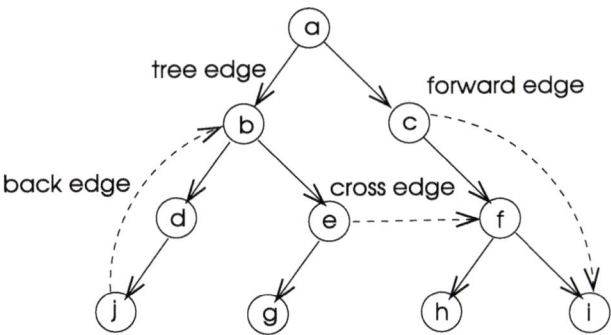

Fig. 6.15 DFS edges of a digraph; (j, b): back edge, (c, i): forward edge, (e, f): cross edge and all other edges are DFS tree edges

---

Algorithm 6.8 *DFS2_Edges*

1: **procedure** $DFS(u)$
2:     $color[u] \leftarrow grey$
3:     $time \leftarrow time + 1; d[u] \leftarrow time$ ▷ first visit
4:     **for all** $(u, v) \in E$ **do**   ▷ visit neighbors
5:         **if** $color[v] = black$ **then**
6:             **if** $d(u) < d(v)$ **then**
7:                 $type(u, v) \leftarrow forward\_edge$
8:             **else** $type(u, v) \leftarrow cross\_edge$
9:             **end if**
10:             **if** $color[v] = grey$ **then**
11:                 $type(u, v) \leftarrow back\_edge$
12:             **end if**
13:             **if** $color[v] = white$ **then**
14:                 $type(u, v) \leftarrow tree\_edge$
15:                 $DFS(v)$
16:             **end if**
17:         **end if**
18:         $color[u] \leftarrow black$
19:         $Pred[v] \leftarrow u,$
20:         $time \leftarrow time + 1$
21:         $f[u] \leftarrow time$ ▷ return visit
22:     **end for**
23: **end procedure**

We search all of the neighbors of a vertex u that the procedure takes as input; if we encounter an edge with an endpoint v that is completed before, then (u, v) is a *forward* edge if $d[u] < d[v]$ else it is a *cross* edge. The vertex v may be *gray* meaning it is explored but not finished in which case (u, v) is a back edge, otherwise it is a tree edge. In an undirected graph, there are no forward or cross edges; hence, we only need to check the color of the neighbor vertex v of the vertex u; if it is *gray*, we have a back edge and otherwise v is *white* and (u, v) is a tree edge. The running time for this edge classification algorithm is $\Theta(n+m)$ since we have only added constant time operations to the DFS procedure in Algorithm 6.7.

6.3.2   An Iterative DFS Algorithm

When we know the recursion depth of a graph is very large, for example more than few thousand, we can replace recursive calls with a stack and obtain a non-recursive DFS algorithm. The iterative DFS algorithm shown in Algorithm 6.9 starts from the source vertex and at each iteration, the neighbors of the vertex u under consideration are pushed into a stack S. This way, visiting all neighbors of u is guaranteed. Once this step is finished, a vertex w is popped from S, marked as visited and the parent of w is marked as u. This step is repeated now for vertex w. Note that this algorithm performs recursive visiting of vertices using the stack S which keeps track of vertices seen but not processed. Different than the recursive DFS algorithm, we have the last pushed vertex on to the stack being processed first. We have the same time complexity of $\Theta(n+m)$.

Algorithm 6.9 *DFS_Iterative*

1: **Input**: $G(V, E)$ and a vertex v, directed or undirected graph
2: **Output**: $Pred[n]$; $d[n]$, $f[n]$ ▷ place of a vertex in DFS tree and its visit times
3: **int** *time* ← 0; **boolean** *Marked[n]*; **stack** S ← ∅
4: **for all** $u \in V$ **do**   ▷ initialize
5:     $Marked[u]$ ← $false$
6:     $Pred[u]$ ←⊥
7: **end for**
8: $push(S,v)$ ▷ push source vertex in stack
9: **while** $S \neq \emptyset$ **do**
10:     u ← $pop(S)$
11:     **if** $visited[u] \neq true$ **then**
12:        $visited[u]$ ← $true$
13:        $Pred[u]$ ← v
14:        **for all** $(u, w) \in E$ **do**   ▷ push neighbors onto stack
15:           $push(S, w)$
16:        **end for**
17:     **end if**
18: **end while**

6.3.3   Parallel DFS

Due to the nature of its execution, DFS algorithm is difficult to parallelize. A simple way to provide parallel processing is to divide the search space among processors. However, this static allocation commonly results in poor load balance since the size of the subtrees may vary significantly. Search space in general is formed dynamically and is difficult to estimate beforehand. Dynamic load balancing for parallel DFS processing may then be used.

A simple dynamic load balancing for parallel DFS may work as follows. A process p_i works on a given search space and when it finishes its work, it requests work from other processes. This can be handled by a central process or in a truly distributed manner with no central control. In terms of implementation, the whole search space may be given to a single process and all other processes may be given empty search spaces initially as described in [8]. The search space is then divided among processes when they request work.

6.3.4 Distributed Algorithms

In a network setting, our aim is to have the nodes of the network cooperate to find the DFS of the whole network. We may use the DFS tree formed for various applications such as finding connected nodes in such a distributed environment. The DFS tree information may be gathered at the root which may then transfer the connected vertices in the network to a management utility which can take remedy actions if there are unconnected nodes. There are various algorithms for this purpose and we will review a basic one that imitates the sequential algorithm we have seen, using a special message called the *token*. This special message provides a single point of execution which is the holder of the token. Any node that possesses the token can run the algorithm while the others stay idle. The token serves a second purpose; it holds the identifiers of the nodes that are visited to prevent visiting them again.

This algorithm called the *Token_DFS* [6] which operates using the same principle as in the DFS procedure is shown in pseudocode in Algorithm 6.10 [6]. We have now a *root* node which starts the algorithm which is where we would start the sequential algorithm. A node receiving the token for the first time records the sender as its parent. It then checks whether it has an unvisited neighbor by comparing the list contained in the token with its neighbors. If such a neighbor exists, it sends the token to that node. Otherwise, token is returned to the parent which is basically imitating the return from the recursive procedure in the sequential algorithm. Correct termination of the distributed algorithm is a fundamental problem in a distributed setting. Any node other than the root terminates when it returns the token to its parent. The root has a different termination, it stops when token is returned to it, and it has no other unvisited neighbors.

Algorithm 6.10 $Token\_DFS$

1: **int** $parent \leftarrow \perp$
2: **set of int** $childs \leftarrow \{\emptyset\}$, $others \leftarrow \{\emptyset\}$, $list \leftarrow \{\emptyset\}$
3: **boolean** $used[] \leftarrow false$
4: **message types** $token$
5:
6: **if** $i = root$ **then**   ▷ root node starts the algorithm
7:     $parent \leftarrow i$, **choose** $j \in N(i)$
8:     **send** $token(\{i\})$ to j
9: **end if**
10:
11: **while** $true$ **do**
12:     **receive** $token(j, list)$
13:     **if** $parent = \perp$ **then**   ▷ token received first time
14:         $parent \leftarrow j$
15:     **end if**
16:     **if** $\exists\, j \in \{N(i) \setminus \{list\}\}$ **then**   ▷ choose an unvisited node if it exists
17:         **choose** $j \in \{N(i) \setminus \{list\}\}$
18:         **send** $token(list \cup \{i\})$ to j
19:     **else if** $i = root$ **then**   ▷ if i is the root and all visited, terminate
20:         **exit**
21:     **else**   ▷ if all visited and i is not root, return token to parent
22:         **send** $token(list \cup \{i\})$ to $parent$
23:         **exit**   ▷ all nodes except root terminate
24:     **end if**
25: **end while**

Analysis

Theorem 6.4 *The Token_DFS correctly constructs a DFS tree in* $2n - 2$ *time using* $2n - 2$ *messages.*

Proof Since the operation is basically the same of the sequential DFS algorithm, the DFS tree will be constructed correctly, that is, visiting each node in DFS manner and forming a tree without any cycles. The DFS tree constructed will have $n - 1$ edges since any tree with n vertices has $n - 1$ edges, and only the edges of this tree will have been traversed twice, once in each direction resulting in a total of $2\,n - 2$ token transfers among the nodes resulting in $2n - 2$ messages. The non-tree edges will not be traversed since we always search unvisited nodes. There is a single activity at any time dictated by the possession of the token, and each message transfer takes a single time unit resulting in $2n - 2$ time.

We followed a very simple route to design a distributed algorithm following the same logic of the sequential one. A good side of this algorithm is that the token is forwarded only along tree edges. However, a known problem with this algorithm is that the size of the token is dependent on the order of the graph. Assuming we need $\log n$ bits to hold the identifier of a node, the required size of the token would

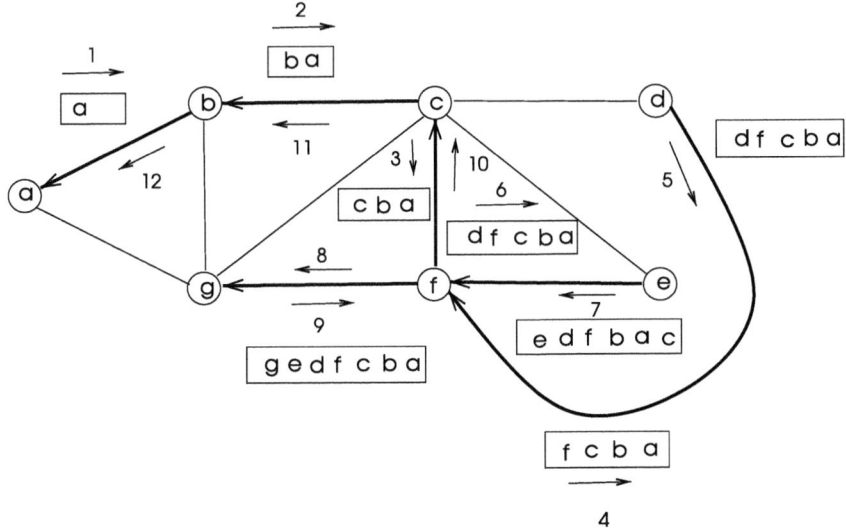

Fig. 6.16 Running of *Dist_BFS* algorithm in a sample graph. The contents of the token are shown only when there is a change. The directed tree is rooted at vertex *a* and the arrows show the sequence of execution

be $n \log n$ which means we need to transfer a message of $O(n \log n)$ size through $n - 1$ edges of the DFS tree. Moreover, the operation is sequential as there is a single activity by the holder of the token at any time.

Example
An operation of this algorithm in a sample graph is depicted in Fig. 6.16. We can see that the token is transferred 12 times which is $2 \times n - 2$, n being 7 in this case. This is also the time taken by the algorithm. The number of bits required in the token is $O(n \log n) = 21$.

As an attempt to provide an algorithm that uses a token as before but without containing the identifiers of the visited nodes, Awerbuch proposed a distributed algorithm for DFS tree construction [1]. In this algorithm, a node informs its neighbors by a special message named *vis* that is visited for the first time. The neighbors respond by *ack* messages. The nodes can keep a track of visited and unvisited neighbors at the expense of extra messages. This algorithm requires $4m$ messages in $4n - 2$ time.

The classical distributed DFS algorithm also implements a token and works using three rules as follows [6]. This algorithm needs $2m$ messages and $2m$ time to construct a DFS tree.

1. A node never forwards the token through the same edge.
2. Any node other than the root forwards the token to its parent when first rule is not applicable.

3. A node receiving the token sends it back through the same edge if this is allowed by rules 1 and 2.

6.3.5   Applications of DFS

We can use the DFS algorithm to test connectivity of undirected or directed graphs, to detect cycles in undirected or directed graphs and for topological order.

6.3.5.1   Testing Connectivity

The DFS algorithm may be used to find the connected components of a graph. If a single call to procedure DFS visits all vertices of G, then it is connected. As a consequence, we can also determine the number of connected components of G; this can be achieved simply by adding a counter to find the number of times the DFS procedure is called in line 10 of Algorithm 6.7. If count is 1, there is only one component in the graph, that is, graph is connected.

DFS was also used by Hopcroft and Karp for bipartite matching [9]. Another use of DFS method was proposed by Hopcroft and Tarjan to test the planarity of a graph in linear time [10] and Even and Tarjan provided vertex and edge connectivity algorithms based on DFS [7]. An essential DFS application is the *topological ordering* or *topological sorting* of a directed acyclic graph (DAG) as we will describe.

6.3.5.2   Detecting Cycles

A cycle in a graph is a path that starts and ends at the same vertex. We can use the DFS algorithm to detect cycles in undirected or directed graphs by the observation that there will not be any back edges such as an edge (u, v) with v being the ancestor of u in the tree in acyclic graphs. We can make use of the following theorem to detect a back edge.

Theorem 6.5 *Given an undirected or directed graph $G = (V, E)$, $(u, v) \in E$ is a back edge if and only if $d(v) < d(u) < f(u) < f(v)$.*

Proof If (u, v) is a back edge, it connects vertex u to its ancestor v; hence, vertex u is a descendant of vertex v. By the parenthesis theorem, the interval $[d(u), f(u)]$ is a subinterval of $[d(v), f(v)]$ so the forward direction of the theorem holds. For the reverse direction, let us consider the parenthesis theorem again. When $d(v) < d(u) < f(u) < f(v)$, vertex u is a descendant of v. This means edge (u, v) connects a vertex to its ancestor and hence it is a back edge. $\square$

We can therefore detect cycles in an undirected or directed graph by simply running the DFS algorithm in $O(n + m)$ time to assign discovery and finish times for each vertex and then checking the type of edges formed in $O(m)$ time for this property, resulting in a total of $O(n + m)$ time. In order to detect back edges online as we are performing a DFS, we can check the finish time $f(u)$ of a vertex

u with the discovery time of its neighbors. If there exists a neighbor v such that $d(v) < f(u)$, then v is an ancestor of u and (u, v) is a back edge. Note that this test is adequate to determine $d(v) < d(u) < f(u) < f(v)$ since we know $d(v) < f(v)$ and $d(u) < f(u)$ for any vertex pair u, v of the graph. In the example graph of Fig. 6.14, edge (h, i) is a back edge since vertex i is an ancestor of vertex h; $d(h), f(h) = 3, 4$; $d(i), f(i) = 3, 4$ and hence $d(h) < f(i)$.

As an alternative approach, we can use the coloring scheme of Algorithm 6.8 where *white* is unexplored, *gray* is discovered but not finished, and *black* means we have discovered and returned from that vertex. Whenever we consider a neighbor vertex v of a vertex u as in line 18 of Algorithm 6.8, we check its color. If it is *gray*, vertex v has to be an ancestor of vertex u on the DFS tree and therefore (u, v) is a back edge which means graph G is not acyclic.

6.3.5.3 Topological Order

Topological ordering is an important process on digraphs where we search for a linear ordering of vertices such that for any edge (u, v) in the graph, u precedes v in the ordering. Let us assume there are n tasks some of which are dependent on other tasks to start, similar to what we have in a computational task dependency graph described in Sect. 4. In order to perform these tasks, we need to arrange them in the order displaying their dependencies. Topological sorting in a digraph with a cycle is not possible and hence, we will assume the input digraph is cycle free. Basically, if a vertex v can be reached from a vertex u, then vertex u should have a lower ordering (Fig. 6.17).

Definition 6.10 (*Topological Order*) A *topological order* $\prec$ of a directed acyclic graph $G = (V, E)$ is a total order of the vertex set V such that for all edges $(u, v) \in E$, $u \prec v$.

A Simple Algorithm

We can have a simple algorithm for topological order as follows. We first find a vertex v with in-degree 0. There is always such a vertex in a DAG since it is loop-free. If there are more than one such vertices, an arbitrary selection is made. This vertex is placed in the ordered output list; and v with all of its outgoing edges is deleted from graph. This process is repeated until there are no vertices left. Correctness is ensured since if deleting outgoing edges from a vertex v placed in the list leaves a vertex u with no incoming edges, than we know $v \prec u$. Algorithm 6.11 shows this routine in pseudocode where we have L as the ordered output list. The operation of this algorithm is shown in Fig. 6.17.

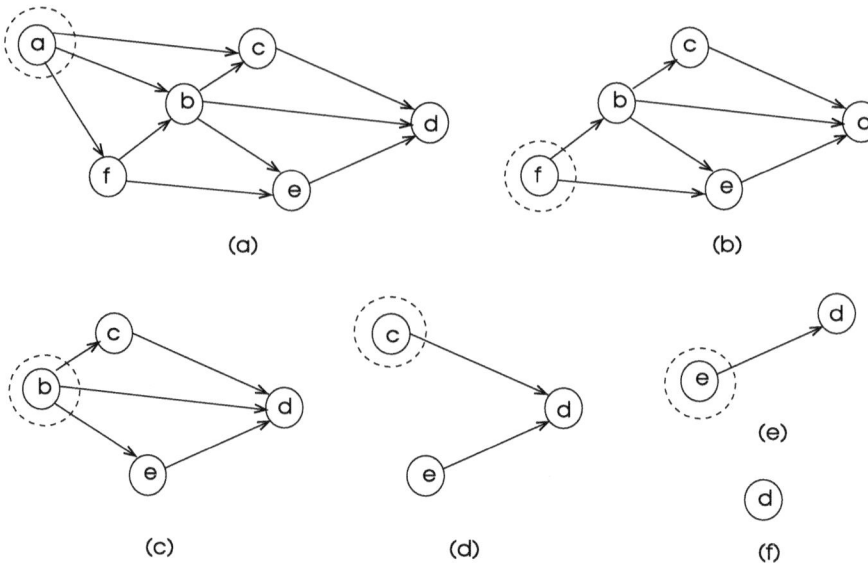

Fig. 6.17 Iterations of the simple algorithm for topological ordering in a sample graph. The selected vertex with no incoming edges at each iteration is shown inside a dashed circle. The ordering formed is $\{a, f, b, c, e, d\}$ in the order of deleted vertices

---

Algorithm 6.11 *BFS*

1: **Input** : $G(V, E)$ ▷ A directed, connected graph G
2: **Output** : L ▷ List of ordered elements
3: $G' = (V', E') \leftarrow G = (V, E)$
4: **while** $V' \neq \emptyset$ **do**   ▷ do until V' is empty
5:     **find** $v \in V'$ such that $indeg(v) = 0$
6:     $L \leftarrow L \cup \{v\}$
7:     $G' \leftarrow G' \setminus \{v\} \cup$ {all outgoing edges of v}
8: **end while**

Analysis

In terms of implementation, we can have an array A of adjacency lists of the graph and also an array D showing the in-degrees of each vertex as shown below for the example graph of Fig. 6.18. We need to check each entry of D for a 0 value and if there is more than one such vertex, we select one arbitrarily. We then remove this vertex from graph by inserting a -1 in the in-degree array D and deleting this vertex and removing it from the array A of adjacency lists of its outgoing neighbors. This process is repeated until there is one vertex with an in-degree of 0.

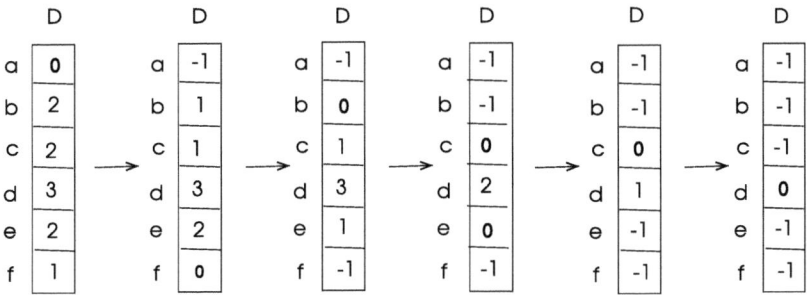

Fig. 6.18 Values of array D during iterations of the simple topological ordering algorithm for the graph of Fig. 6.18

Initializing the in-degree array D takes $O(m)$ time, searching each entry is $O(n)$ time resulting in $O(n^2)$ time for the whole array D. We then reduce in-degree of all its neighbors in $O(m)$ time resulting in a total runtime of $O(n^2 + m)$ for this algorithm.

DFS-Based Algorithm

We can make use of the edge properties discovered during the recursive DFS algorithm to perform topological order. When we run the recursive DFS algorithm on a graph G to obtain a DAG G', every edge (u, v) in G' has $f(v) < d(u)$ since there are no back edges in G'. This provides us the necessary information to form the topological order of G: simply list the vertices of G' from the largest finish time to the smallest. We can obtain this list by adding the identity of the vertex to the front of a list L when we finish with it and when DFS is completed, the list L contains the topological ordering of vertices. Total time taken therefore is $\Theta(n+m)$ as in the recursive DFS. Figure 6.19 displays a DFS tree obtained in the same graph of Fig. 6.18 and sorting the finish times of vertices provides the same topological ordering as the previous algorithm in this graph.

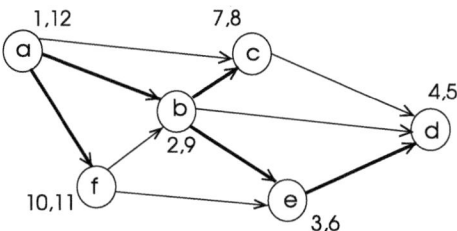

Fig. 6.19 Topological ordering by DFS on the sample graph of Fig. 6.19. The DFS tree is shown in bold and the first and last times of a visit to each vertex is shown next to it. The ordering using the last visit times of vertices found using this method is $\{a, f, b, c, e, d\}$

Analysis

Theorem 6.6 *The DFS-based topological sort algorithm correctly provides a topological sort of an acyclic digraph $G = (V, E)$.*

Proof We need to show that for any directed edge $(u, v) \in E, f(v) < f(u)$ for the correctness of this algorithm. For any edge (u, v) considered during the DFS, vertex v cannot be an ancestor of vertex u since G is given acyclic, hence $f(v) < f(u)$. In terms of our color coding of vertices, vertex v cannot be *gray*; it can either be *white* (unexplored) or *black* (explored). If vertex v is *white*, it will be processed and become *black* before u becomes *black* and hence $f(v) < f(u)$. If it is *black*, this means it has already been processed to have its finishing time $f(v)$ determined and vertex u will have a greater finish time than v in this case as well. $\square$

6.4   Breadth-First Search

The main idea of the breadth-first search (BFS) method is to visit all neighbors of a vertex first before visiting other vertices, and hence the name. Starting from the source vertex s, we first visit all neighbors of s in an arbitrary order. These neighbor vertices, $N(s)$, all have a distance of unity from the vertex s after the visit. We then visit neighbors of vertices in $N(s)$ which are labeled with distance of 2 to vertex s. This process continues until all vertices are visited.

6.4.1   The Sequential Algorithm

We can implement this algorithm by inserting the adjacent vertices of the currently visited vertex in a queue, and then removing them from the queue one by one and repeat the process. We need to keep track of the visited vertices as in the DFS algorithms to prevent a vertex to be visited again. Algorithm 6.12 shows one way of implementing the described procedure.

Algorithm 6.12 *BFS*

1: **Input** : $G(V, E), s$	▷ undirected, connected graph G and a source vertex s
2: **Output** : $D[n]$ and $Pred[n]$	▷ distances and predecessors of vertices in BFS tree
3: **for all** $v \in V \setminus \{s\}$ **do**	▷ initialize all vertices except source s
4: $D[v] \leftarrow \infty$	
5: $Pred[v] \leftarrow \perp$	
6: **end for**	
7: $D[s] \leftarrow 0$	▷ initialize source s
8: $Pred[s] \leftarrow s$	
9: $Q \leftarrow s$	
10: **while** $Q \neq \emptyset$ **do**	▷ do until Q is empty
11: $v \leftarrow deque(Q)$	▷ deque the first element u
12: **for all** $(u, v) \in E$ **do**	▷ process all neighbors of u
13: **if** $D[u] = \infty$ **then**	
14: $D[u] \leftarrow D[v] + 1$	
15: $Pred[u] \leftarrow v$	
16: $enque(Q, u)$	
17: **end if**	
18: **end for**	
19: **end while**	

Analysis

The distance and predecessor initialization for vertices takes $O(n)$ time. Each vertex is enqueued at most once and each edge is explored at most twice, once for each vertex incident to it. Total time spent to construct the BFS tree is therefore $O(n + m)$.

Example

The running of BFS algorithm on a sample graph is shown in Fig. 6.20. The BFS tree constructed shows the shortest paths from the root vertex a.

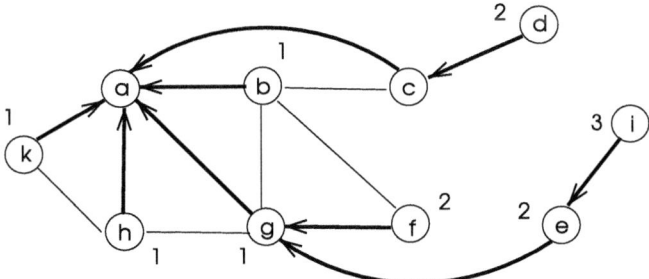

Fig. 6.20 Running of *BFS* algorithm in a sample graph from vertex a. The directed tree is rooted at a and the levels of vertices are shown next to them. Note that we mark tree edges with arrows to point to their parents along the path to the source vertex

Properties
The properties of the BFS graph traversal are as follows.

- A directed spanning tree of the graph G is obtained as a result of this algorithm. We update distances of vertices with infinity distances only in line 13 which are not visited and hence no cycles are formed. Note that we do not need a variable to check whether a vertex is visited or not as in the DFS algorithm since the distance value of infinity shows that vertex is not visited. All vertices will be processed at the end of the algorithm and therefore the output is a spanning tree of G.
- The order we select and hence enqueue the neighbors of v in line 12 of the algorithm affects the structure of the BFS tree obtained and therefore this tree is not unique.
- The BFS algorithm partitions the edges of an undirected graph into *tree edges* and *back edges*.

6.4.2 Parallel BFS

When we view the BFS procedure, we can see all the vertices at the same level can be processed in parallel. For example, we can have a parallel loop to explore all of the neighbor vertices of vertices at level i to find the vertices at level $i + 1$. However, we can have a situation in which two vertices u and v at level I will both attempt to set level of an unexplored neighbor vertex w to $i + 1$ and set themselves as the parent of w. However, this seemingly race condition does not cause any problem as it does not matter which vertex sets the level of w to $i + 1$ and it is also immaterial whether u or v is the parent of w. Moreover, we need to synchronize all of the processes at level i to ensure they all finish before starting the parallel processing at level $i + 1$ to find vertices at level $i + 2$.

This level-synchronous approach is implemented in various parallel BFS algorithms as described in [3]. A PRAM-based approach may work similar to what we have described providing parallel loop processing by a number of processing units and atomic level updates with barrier synchronization between level processing. The total execution time based on this model would then be $O(diam(G))$ since the number of levels would not exceed the diameter of the graph G. Various parallel algorithms, whether shared or distributed memory, adapt this strategy with possibly added load balancing during parallel processing of the exploration loop. A fine-grain parallel BFS algorithm running on shared memory Cray MTA-2 system is reported in [2]. In a distributed memory parallel processing system, partitioning of the graph to processing nodes is commonly pursued. A distributed memory parallel algorithm using 2D graph partitioning is implemented in BlueGene/L in [12].

6.4.3   Distributed Algorithms

We review two distributed algorithms to form a BFS tree in a network: A synchronous algorithm that works in rounds and an asynchronous algorithm. Termination should be handled carefully in both cases.

6.4.3.1   A Synchronous BFS Algorithm

Our starting point will again be the sequential BFS algorithm and we will attempt to have its distributed version. This time, however, we will assume the method of single initiator synchronous distributed (SISD) algorithm. There is a root (supervisor) node r which is where the BFS tree will be rooted and the algorithm runs in synchronous rounds [6]. The root enlarges the current BFS tree T' with a new layer at each round. The messages used in this algorithm are as follows.

- *round*: Sent by root r at the beginning of each round over the partial tree T'. Each node on T' broadcasts *round* to its children.
- *probe*: Sent by leaves of T' to all neighbor nodes except the parent.
- *ack*: A non-tree node responds to *probe* message by an *ack* message. It marks the sender of *probe* as its parent and the receiver of *ack* marks the sender as one of its children. If there are more than one *probe* messages received concurrently, the receiving node arbitrarily picks one of them.
- *nack*: If the receiver of a *probe* message already has a parent assigned, its sends the sender a *nack* message.
- *upcast*: Sent by leaves of T' to their parents to inform the round is over.

Once a new layer is formed, the leaves from the previous round collect all *ack* and *nack* messages and start a convergecast operation described in Sect. 5.6.3 over the edges of T'. When the root receives *upcast* messages from all of its children, it can start the next round. Algorithm 6.13 shows round k of the distributed synchronous BFS algorithm run by any node except the root. The set *childs* is the set of children for the root and intermediate nodes; *others* are the set of neighbors of a leaf node that are not its children and the set *collected* is used to keep track of which children have sent an *upcast* message to an intermediate node.

Algorithm 6.13 *Synchron_BFS*

1: **int** *parent* ← Ø
2: **set of int** *childs* ← {Ø}, *others* ← {Ø}, *collected* ← {Ø}
3: **message types** *round, probe, ack, nack*
4: **boolean** *visited* ← *false*; *round_over* ← *false*
5: **while** ¬*round_over* **do**
6:     **receive** *msg(j)*
7:     **case** *msg(j).type* **of**
8:         <u>*round(k)*</u> :   **if** *leaf_node* **then**   ▷ if leaf check neighbors
9:                                    **send** *probe(k)* to $N(i) \setminus \{j\}$
10:                                **else**
11:                                    **send** *probe(k)* to *childs*   ▷ else send probe to children
12:         <u>*probe(k)*</u> :   **if** ¬*visited* **then**   ▷ update distance
13:                                *parent* ← *j*; *visited* ← *true*
14:                                    **send** *ack(k)* to *j*   ▷ inform parent I am child
15:                                **else**
16:                                    **send** *nack(l)* to *j*   ▷ else reject sender
17:         <u>*ack(k)*</u> :   *childs* ← *childs* ∪ *{j}*   ▷ include sender in children
18:         <u>*nack(k)*</u> :   *others* ← *others* ∪ *{j}*   ▷ include sender in unrelated
19:         <u>*upcast(k)*</u> :   *collected* ← *collected* ∪ *{j}*   ▷ collect upcast signals
20:     **if** (*leaf_node* ∧ (*childs* ∪ *others*) = *N(i)*) ∨ (¬*leaf_node* ∧ (*collected* = *childs*))
        then
21:         **send** *upcast* to *parent*
22:         *round_over* ← *true*; *collected* ← {Ø}
23:     **end if**
24: **end while**

We can see this algorithm works rather asynchronously in one synchronous round. We do not pay attention to the order of messages received although we know each node on T' will first receive the *round* message and it will act differently depending on whether it is a leaf or an intermediate node. A leaf node searches for neighbors to be included in the next layer and an intermediate node simply acts as a gateway by sending the *round* message to its children. Upon completion of the round, an intermediate node collects *upcast* messages from its children and sends an *upcast* message to its parent. We check whether a leaf node has received *ack* or *nack* messages from all of its neighbors; or a non-leaf node has received upcast messages from all of its children to decide if the round is over. When this is decided, an *upcast* message is sent to the parent.

As with all distributed algorithms, termination detection is needed. In other words, the root should know how many rounds it has to initiate at most. We can see that the number of rounds needed is the farthest distance between the root and any node in the network which is the diameter of the network graph and hence the root needs to know the diameter prior to running the algorithm. Unfortunately, this creates another problem; however, close inspection of the algorithm may provide an easier and an elegant way out as follows. A leaf node *l* in the graph terminates when either it has a degree of unity or all of its neighbors have rejected it, meaning they are children of other intermediate tree nodes. This means there is no reason for *l* to continue as it has no children to transfer *round* or *upcast* messages nor any neighbors to probe. We can elaborate on this condition and when this happens, a special *terminate* message can be convergecast toward the root. When all of the children of an intermediate node v upcast a *terminate* message, the node v terminates and sends a *terminate* message to its parent and the root terminates when all of its children upcast a *terminate* message (see Ex. 10).

Example
An example operation of this algorithm is depicted in Fig. 6.21 where the root node g starts the algorithm by sending *probe* messages to its neighbors a and f in the first round which respond by *ack* messages and the edges (g, a) and (g, f) become BFS tree edges. Note that nodes a and f send *probe* messages to each other in the next round (round 2) which are rejected. The number of rounds started by the node g is 4 which is the diameter of the graph.

Analysis

Theorem 6.7 *Algorithm Synchron_BFS correctly constructs a BFS tree in* $O(diam^2(G))$ *time using* $O(n \cdot diam(G) + m)$ *messages.*

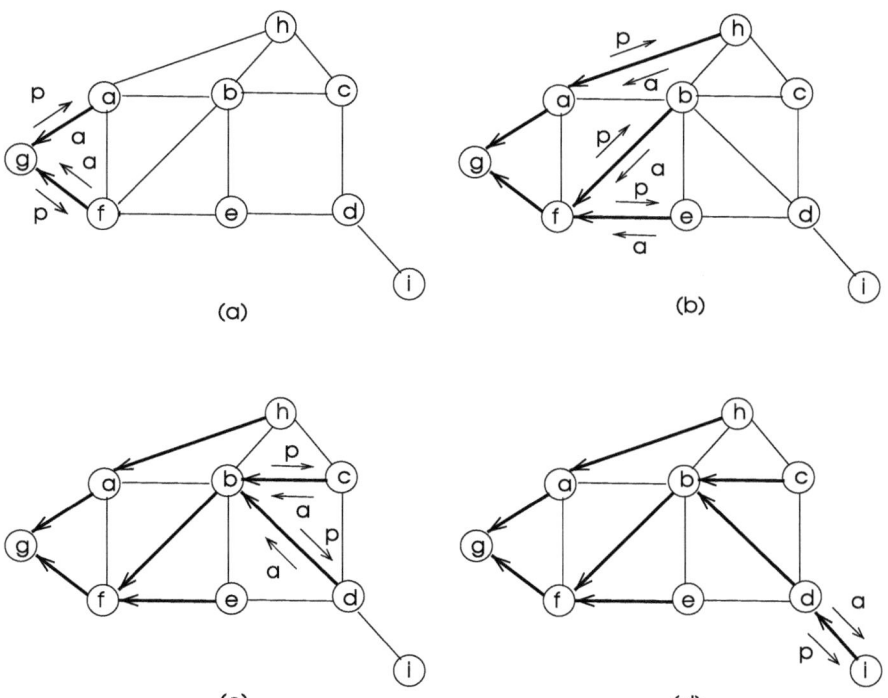

Fig. 6.21 Running of Algorithm 6.13 on a sample graph for four rounds. Only *probe* messages (*p*) that are acknowledged with *ack* (*a*) messages are shown

Proof At each step of the algorithm, only leaves at layer k of the partial tree formed will be sending *probe* messages to form layer $k + 1$ leaves which would be enlarging the partial BFS tree one layer. Hence, BFS tree property is obeyed to form the final BFS tree.

At each kth step of the algorithm, time spent will be proportional to the current level k as messages are broadcast and convergecast through k layers. We can have at most $diam(G)$ levels and hence totaling gives

$$\sum_{k=1}^{diam(G)} k = O(diam^2(G)).$$

An analysis of message complexity is as follows. There will be a maximum of n synchronization messages in each round for a total of $O(n \cdot diam(G))$ since there will be at most $diam(G)$ rounds. Also, each edge will be traversed at most twice by *probe/ack* or *probe/nack* message pairs resulting in $O(m)$ messages for queries. Total number of messages needed will be the sum of these as $O(n \cdot diam(G) + m)$. □

6.4.3.2 An Asynchronous BFS Algorithm

We will attempt to provide a distributed version of the dynamic Bellman–Ford algorithm described in Sect. 3.9.3.2. This algorithm provided shortest distances in a weighted graph; however, we will use it for an unweighted graph with the same reasoning as in [6, 11]. The distributed version is a single-source asynchronous algorithm and hence we do not know the order of the messages to be received. Therefore, we need to modify distances of the nodes to the source node at each incoming message.

We have three distinct messages $level(k)$, $ack(k)$, and $nack(k)$. The root node starts the algorithm by sending $level(0)$ message to its neighbors. All nodes initialize their distance to the root as infinity first and any node that receives this message compares its distance value to the root with the value contained in the message. If the route to the root via the neighbor sending the message is shorter, the sending neighbor is assigned as the parent and the distance is modified as shown in Algorithm 6.14. The modified distance is broadcast to neighbors so they can also update their distances. This is needed since the newly found route may affect the shortest distances of the neighbors to the root. We have the sets *childs* and *others* as in the synchronous algorithm since we need a node to be aware of the neighbors that are children or that are unrelated.

Algorithm 6.14 *Asynchron_BFS*

1: **int** *parent* ← ∅, *my_layer* ← ∞, *count*=1
2: **set of int** *childs* ← {∅} , *neighbors* ← {∅}, *others* ← {∅}
3: **message types** *level, ack, nack*
4: **if** *i* = *root* **then**
5:     **send** *level*(0) to *N(i)*   ▷ Only root executes this part
6: **end if**
7: **while** *count* ≤ *diam(G)* **do**
8:     **receive** *msg(j)*
9:     **case** *msg(j).type* **of**
10:         *layer(l)* :   **if** *my_level* > *l* + 1 **then**   ▷ update my distance
11:                              *parent* ← *j*
12:                              *my_level* ← *l* + 1
13:                              **send** *ack(l)* to *j*   ▷ inform parent i am child
14:                              **send** *my_level* to *N(i)* \ {*j*}   ▷ inform neighbors of new level
15:                      **else**
16:                              **send** *nack(l)* to *j*   ▷ else reject sender
17:         *ack(l)* :   *childs* ← *childs* ∪ {*j*}   ▷ include sender in children
18:         *nack(l)* :   *others* ← *others* ∪ {*j*}   ▷ include sender in unrelated
19:     *count* ← *count*+1
20: **end while**

It can be seen this process eventually builds a BFS tree starting from the root. The termination condition would be the traversing of the longest shortest path between any two nodes which would be the diameter of the graph G. Therefore, each node should wait a maximum of $diam(G)$ of the network messages. Unlike the synchronous algorithm, this time we do not have an easy solution for termination since we do not know the diameter *en priori*. We can include a *time-to-live* field in each message which is initialized to an upper limit of the diameter value and is decremented at each reception at a node. When this field becomes zero, the message is no longer transmitted to neighbors.

Theorem 6.8 *Algorithm Asynchron_BFS correctly constructs a BFS tree in $O(diam(G))$ time using $O(nm)$ messages.*

Proof After $diam(G)$ steps, all nodes will have received $layer(diam(G) - 1)$ message and will set its distance to this value and hence the BFS tree will be constructed. Time needed is the diameter of the network to reach the farthest node from the root node, hence time complexity is $O(diam(G))$. The longest path in the network will have a length of $n - 1$ and a node having this value for the first time may need to change it $n - 2$ times and will send at most $n \cdot deg(v)$ messages resulting in the below total number of messages [11].

$$\sum_{v=1}^{m} n \cdot deg(v) = O(nm)$$

□

6.4.4   Applications of BFS

We can find whether a graph G is connected or not using this algorithm as in the DFS algorithm. If all of the vertices of G are processed at the end, then it is connected. Finding the shortest path from the root vertex to all others in an undirected is also provided by this method. The BFS algorithm in an unweighted simple graph provides the distance between a vertex v and the source vertex; this distance is simply the level of v in the BFS tree formed.

6.4.4.1   Testing Bipartiteness

A graph $G_(V, E)$ is bipartite if its vertex set V can be partitioned into two subsets V_1 and V_2 such that every $(u, v) \in E$ has one endpoint in V_1 and the other endpoint in V_2. In other words, there are no edges between any two vertices in V_1 and no edges between any two vertices in V_2. We can use the BFS algorithm to check whether a graph G is bipartite or not as follows. An edge joining two layers of the BFS tree means G has a cycle of odd length, and hence cannot be bipartite. We give the same color to each node of G in the same layer discovered by the BFS algorithm. Clearly, we can color all of the nodes using two colors, say *white* and *gray*. Hence, if G is bipartite, there will not be any edge of G that has two vertices of the same color at its endpoints. Specifically, we have the following steps of this algorithm.

1. **Input**: $G = (V, E)$
2. **Output**: Evaluate G as bipartite or non-bipartite
3. Select $s \in V$ and set $color(s) \leftarrow white$
4. Run modified BFS algorithm starting from vertex $s \in V$ as follows.
   a. Color a vertex v at level I by the color gray if vertices at level $(i - 1)$ are white. Conversely color v white if level $(i - 1)$ vertices are gray.
   b. Let $color(v) = c$ for a vertex v colored as above.
   c. If $\exists u \in N(v)$ such that $color(v) = color(u)$
   d.   Output "G is not bipartite". Stop.
5. If all vertices are colored properly, then output "G is bipartite"

Figure 6.22 shows a sample graph partitioned into two sets by the BFS algorithm and we can see it is not bipartite as there is an edge joining two vertices of the same color.

This algorithm works correctly with the following reasoning to show only one of the cases is valid.

- If there is not an edge between two vertices in the same layer, vertices in adjacent layers will be colored with opposite colors. Therefore, G is bipartite in this case.
- Let us assume there is an edge (u, v) between vertices u and v of the same layer L_j. Let w be the least common ancestor of u and v in the BFS tree T at layer

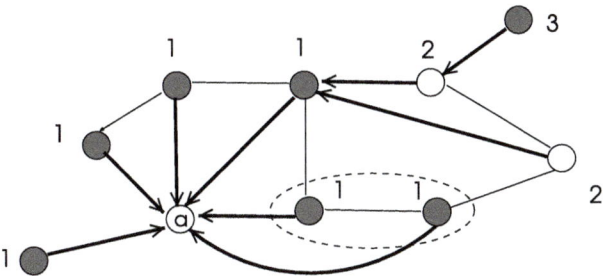

Fig. 6.22 Graph that is partitioned into white and gray vertices by the BFS algorithm from a source vertex a. Levels of vertices are shown next to them. The edge enclosed in the dashed ellipse is between two vertices that are gray, therefore this graph is not bipartite

L_i. Then, $w - u - v - w$ is a cycle of the graph having length $2(j - i) + 1$ which is an odd cycle. Hence, G is not bipartite.

BFS is accomplished in $O(n + m)$ time and scanning the edges can be performed in $O(m)$ time resulting in a total time of $O(n + m)$ for this algorithm.

6.5 Chapter Notes

We described tree structure in graphs in the first part of this chapter. Trees have numerous implementations in computer science and in real-life applications. We reviewed algorithms to construct spanning trees in a graph and then tree traversal algorithms.

In the second part of the chapter, we reviewed two fundamental graph traversal methods: DFS and BFS. We saw DFS can be implemented as an effective recursive algorithm with $O(n+m)$ time complexity and it also has an iterative version using a stack with the same time complexity. It can be used to test connectivity and to find the number of components of a graph. An important DFS implementation is to find the topological order of a directed acyclic graph in which vertices have precedence relationships. DFS algorithm is difficult to parallelize due to its sequential and dependent nature of execution between each algorithm step. However, the graph contraction method in which we obtained a coarser graph of a previous step can be used for parallel DFS construction. Distributed DFS tree building involves nodes of a communication network cooperating to construct this tree. We can convert the sequential DFS algorithm to a distributed one using a special message called *token* between the nodes. Any node that possesses the token is allowed to execute, and hence we have in fact a sequential algorithm running in a distributed manner. There are various other distributed DFS algorithms which achieve better parallelism at the expense of increased number of messages as we have reviewed.

The BFS algorithm visits vertices of a graph using layer-by-layer search and it can be used to find distances from a source vertex to all other vertices in an undirected, unweighted graph. For a weighted graph, we need to modify this algorithm to find distances as we will see in the next chapter. BFS algorithm can be used to test bipartiteness of an undirected graph as we saw. The parallel version of BFS algorithm uses graph contraction as in the parallel DFS algorithm. There are few distributed BFS algorithms and one such algorithm works synchronously in rounds under the control of a special node called the *supervisor*. This node enlarges the BFS tree layer-by-layer at each round and in fact imitates the sequential BFS algorithm in a distributed setting. Other than solving explicit problems such as connectivity, topological order, and bipartiteness, these two basic methods of graph traversals provide building blocks of various more complex graph algorithms as we will see. The implementations of these algorithms in directed graphs are similar, and we should consider only the outgoing edges from a vertex in a digraph.

Exercises

1. Write the pseudocode of the recursive tree center finding algorithm of Sect. 6.2 and show step-by-step execution of this algorithm in the sample tree depicted in Fig. 6.23.
2. Construct a possible spanning tree of the graph depicted in Fig. 6.24 using the second spanning tree algorithm of Sect. 6.2.4.2.
3. Design and form the pseudocode of a distributed algorithm that forms a spanning tree based on the third algorithm of Sect. 6.2.4.3 for spanning tree

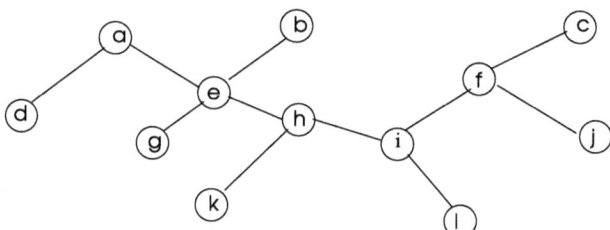

Fig. 6.23 Sample graph for Exercise 1

Fig. 6.24 Sample graph for
Exercise 2

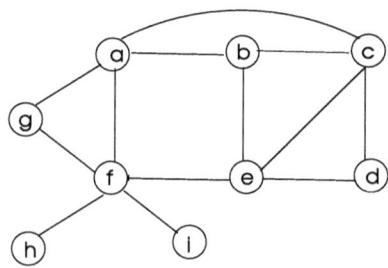

Fig. 6.25 Sample graph for
Exercise 3

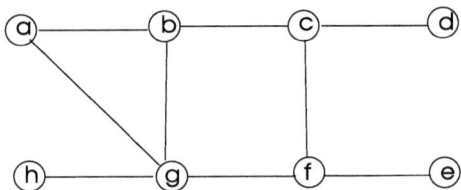

construction. Show a possible running of this algorithm in the network graph
shown in Fig. 6.25.

4. Work out a possible DFS tree rooted at vertex a in the digraph of Fig. 6.26
   by showing the discovery and finish times for each vertex. Show also the tree
   edges, front edges, back edges, and cross edges in the graph.
5. The token-based DFS algorithm is to be executed in the sample graph of
   Fig. 6.27. Work out a possible DFS tree rooted at vertex a. Show the iterations
   of the algorithm in this graph with the contents of the token.
6. Write the pseudocode of the DFS-based cycle detection algorithm that uses
   discovery times and finish times of vertices. Show the step-by-step running of
   this algorithm in the graph of Fig. 6.28.
7. Work out the topological order of vertices in the DAG of Fig. 6.29 using both
   the simple algorithm and the DFS-based algorithm.

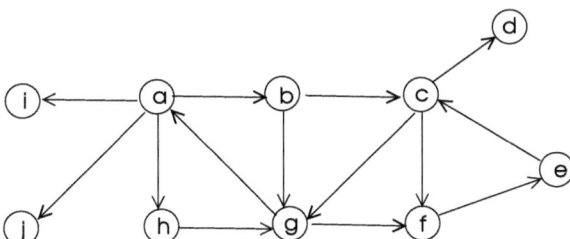

Fig. 6.26 Sample graph for Exercise 4

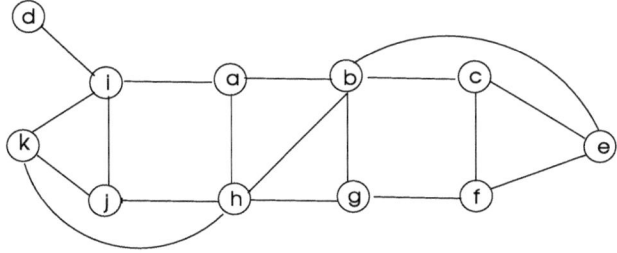

Fig. 6.27 Sample graph for Exercise 5

Fig. 6.28 Sample graph for
Exercise 6

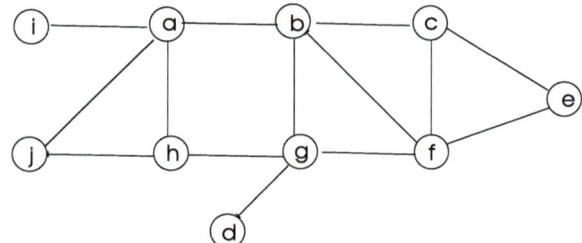

Fig. 6.29 DAG for
Exercise 7

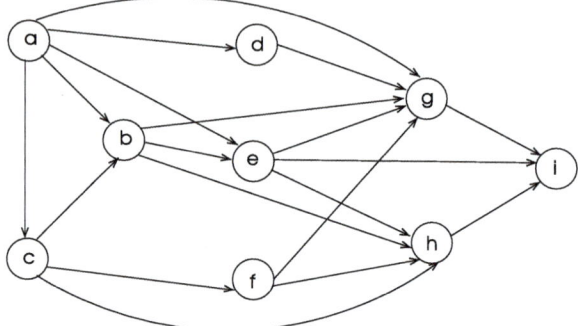

Fig. 6.30 Sample graph for
Exercise 8

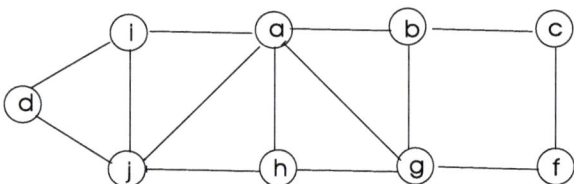

8. Find the BFS tree rooted at vertex g in Fig. 6.30 by showing the levels for
   each vertex.
9. Design the distributed synchronous BFS algorithm of Sect. 6.13 (Algorithm 6.13) with FSMs. Draw the FSM diagram and write the pseudocode
   for this algorithm.
10. Modify Algorithm 6.13 such that termination using a special terminate
    message upcast by leaves is used.

References

1. Awerbuch A (1985) A new distributed depth first search algorithm. Inf Process Lett 20:147150
2. Bader DA, Madduri K (2006) Designing multithreaded algorithms for breadth-first search and
   st-connectivity on the Cray MTA-2. In: Proceedings of the 35th international conference on
   parallel processing (ICPP 2006), pp 523–530

3. Buluc A, Madduri K (2011) Parallel breadth-first search on distributed memory systems. In: International conference for high performance computing, networking, storage and analysis (SC'11), Article 65
4. Cayley A (1857) On the theory of analytical forms called trees. Philos Mag 4(13):172176
5. Cormen TH, Leiserson CE, Rivest RL, Stein C (2009) Introduction to algorithms, 3rd edn, chap 22. MIT Press, Cambridge
6. Erciyes K (2013) Distributed graph algorithms for computer networks, chaps 4 and 5. In: Springer computer communications and networks series. Springer, Berlin (2013). ISBN-10:1447151720
7. Even S, Tarjan RE (1975) Network flow and testing graph connectivity. SIAM J Comput 4(4):507–518
8. Grama A, Gupta A, Karypis G, Kumar V (2003) Introduction to parallel computing, 2nd edn, chap 11. Addison Wesley, Boston
9. Hopcroft JE, Karp RM (1973) An $n^{5/2}$ algorithm for maximum matching in bipartite graphs. SIAM J Comput 2(4):225–231
10. Hopcroft JE, Karp RM (1974) Efficient planarity testing. J ACM 21(4):549–568
11. Peleg D (2000) Distributed computing: a locality-sensitive approach. In: SIAM monographs on discrete mathematics and applications, chap 5
12. Yoo A, Chow E, Henderson K, McLendon W, Hendrickson B, Catalyuurek UV (2005) A scalable distributed parallel breadth-first search algorithm on BlueGene/L. In: Proceedings of the ACM/IEEE conference on high performance computing (SC2005)

Weighted Graphs

7

7.1   Introduction

A paragraph can have weights associated with its edges or its vertices. The weight on an edge typically denotes the cost of traversing that edge and the weight of a vertex commonly shows its capacity to perform some function. Our aim in this chapter is to review algorithms for weighted graphs for two specific tasks; the minimum spanning tree problem and the shortest path problem.

A tree is a connected graph with no cycles and a *spanning tree* of a connected graph is a tree that includes all nodes of the graph as we reviewed in the previous chapter. A *minimum spanning tree* (MST) of a weighted, undirected, and connected graph is the spanning tree with the minimum total cost of edges among all spanning trees of that graph. There can be more than one MST in a graph if edge weights are not distinct. MSTs find a wide range of applications such as connecting a number of cities, components or other objects. In general, our aim in search of an MST of a graph is to use a minimum amount of roads, wires, or any other connecting medium to connect the objects under consideration. MSTs are also used for clustering of large networks consisting of tens of thousands of nodes and hundreds of thousands of edges such as biological networks. Removing a number of heaviest weight edges results in clusters in such networks.

We review sequential, parallel, and distributed algorithms for MST construction in the first part of this chapter. In the second part, we look at algorithms to find the shortest paths between the vertices of a graph. This problem has many practical usages, especially in computer networks when we want to transfer a data packet between the two nodes of such a network with minimum costs. We describe sequential, parallel, and distributed algorithms to find shortest paths between a vertex and all other vertices, and between each pair of vertices in a graph.

K. Erciyes, *Guide to Graph Algorithms*, Texts in Computer Science,
https://doi.org/10.1007/978-3-032-05294-0_7

7.2   Minimum Spanning Trees

In this section, we will first describe the four main sequential algorithms to construct MSTs of weighted graphs. We will then investigate ways of obtaining parallel algorithms from these algorithms followed by the illustration of a distributed algorithm that can be used to find the MST of a computer network. We will also consider conversion between parallel and distributed algorithms for this problem.

7.2.1   Background

Given a weighted, undirected, and connected graph $G = (V, E, w)$, we are searching for the MST $T \subseteq G$ such that $w(T)$ given below is minimized.

$$w(T) = \sum_{(u,v)\in T} w(u, v). \tag{7.1}$$

In search for a solution to this problem, we will consider few seemingly reasonable heuristics. First of all, we do not want heavy-weight edges in the MST and attempt to include as many light edges as possible. We also need to prevent cycles as a tree is required. Lastly, the bridges of a graph are to be included in the MST since excluding these edges leaves the MST disconnected. Two rules defined below will help to form MSTs.

Theorem 3 (cut property) *Consider a weighted, undirected, and connected graph $G = (V, E, w)$. Let A be a subset of edges that is contained in some MST of G. For any cut $(S, V \setminus S)$ of G that has no edges of A crossing the cut, let (u, v) be the least weight edge across this cut such that $u \in S$ and $v \notin S$. Then the edge (u, v) is contained in some MST of G. If the edge weights of G are distinct, then there is a unique MST T^* of G which contains the edge (u, v).*

Proof Consider an MST T that contains the set A and does not contain the edge (u, v). Then there must be a path p that connects the vertex u to the vertex v since the MST T must cover all vertices. Let us combine the edge (u, v) with the path p to form a cycle C in G. The edge (u, v) is across the cut which means there must be at least an edge $(w, z) \in C$ that goes through the cut. Let us now replace (w, z) with (u, v) to form a new tree $T' = T \cup \{(u, v)\} - \{w, z\}$. T' is a spanning tree of G since we added one edge and removed one edge resulting in $n - 1$ edges. Moreover, $w(T') = w(T) + w(u, v) - w(w, z) \leq w(T)$ since $w(u, v) \leq w(w, z)$ which means T' that contains the edge (u, v) is an MST of G. □

The cut property is useful in forming an MST of a graph. Any least weight edge across any cut of the graph can be included in the MST until we have $n - 1$

edges which means the formed tree is an MST. The *cycle property* is also a useful characteristic of an MST.

Theorem 7.1 (cycle property) *Let C be any cycle in G and (u, v) be the maximum weight edge in C. There is no MST of G that contains (u, v).*

Proof Let T be an MST of G and assume the contrary that T contains (u, v). Deleting (u, v) from T results in two subgraphs with vertices V_T and $V - V_T$. The cycle C has another edge $(w, z) \neq (u, v)$ that has exactly one end point in V_T and $w(w, z) < w(u, v)$ since edge (u, v) is the maximum weight edge of C. Form a new tree $T' = T - \{(u, v)\} \cup \{(w, z)\}$. The total weight of T' is less than the total weight of T which is a contradiction. □

Theorem 7.2 *Let $G = (V, E, w)$ be a connected, weighted, undirected graph. If the edge weights of G are distinct, then G has a unique MST.*

Proof Let T be an MST of G. For each edge $(u, v) \in T$, the tree $T' = (V, T \setminus \{(u, v)\}$ has two connected components say P and Q. The edge (u, v) is the only edge of T across the cut between P and Q and it is the least weight unique edge between these two sets by the cut property and because edges of G have distinct weights. Therefore, every MST of G must contain (u, v) and if we consider all edges of T, every MST of G must contain all edges of T. Hence every MST is equal to T. □

7.2.2 Sequential Algorithms

We can have a generic algorithm to build the MST of a graph G as follows. We start with an MST $T = \emptyset$ of G and always add *safe edges* to T that should be in the MST of G. The fundamental algorithms to build the MST of a weighted graph using this method are due to Prim, Kruskal, and Boruvka as we will review next.

7.2.2.1 Prim's Algorithm

This algorithm was initially proposed by Jarnik and then by Prim to find MST of a connected, weighted, directed, or undirected graph $G(V, E, w)$ [16]. The idea of this algorithm is to always select a safe edge using the cut property of Theorem 3 to be included in the set A which has a subset of edges of an MST of G. This algorithm assumes that edges in A form a single tree. It starts from an arbitrary vertex s and includes it in the MST. Then at each step of the algorithm, the minimum weight outgoing edge (MWOE) (u, v) from the current tree fragment T such that $u \in T$ and $v \in G \setminus T$ is found and added to the tree T. If T is an MST fragment of G, $T \cup (u, v)$ is also an MST fragment of G by Theorem 3. Proceeding in this manner, the algorithm finishes when all vertices are included in the final MST as shown in Algorithm 7.1. Figure 7.3 shows the iterations of Prim's algorithm in a graph.

Algorithm 7.1 *Prim_MST*

1: **Input** : $G(V, E, w)$
2: **Output** : MST $T(V, E_T)$ of G
3: $V_T \leftarrow \{s\}$
4: $T \leftarrow \emptyset$
5: **while** $V_T \neq V$ **do**   ▷ continue until all vertices are visited
6:     select the edge (u, v) with minimal weight such that $u \in T$ and $v \in G \setminus T$
7:     $V_T \leftarrow V_T \cup \{v\}$
8:     $E_T \leftarrow E_T \cup \{(u, v)\}$
9: **end while**

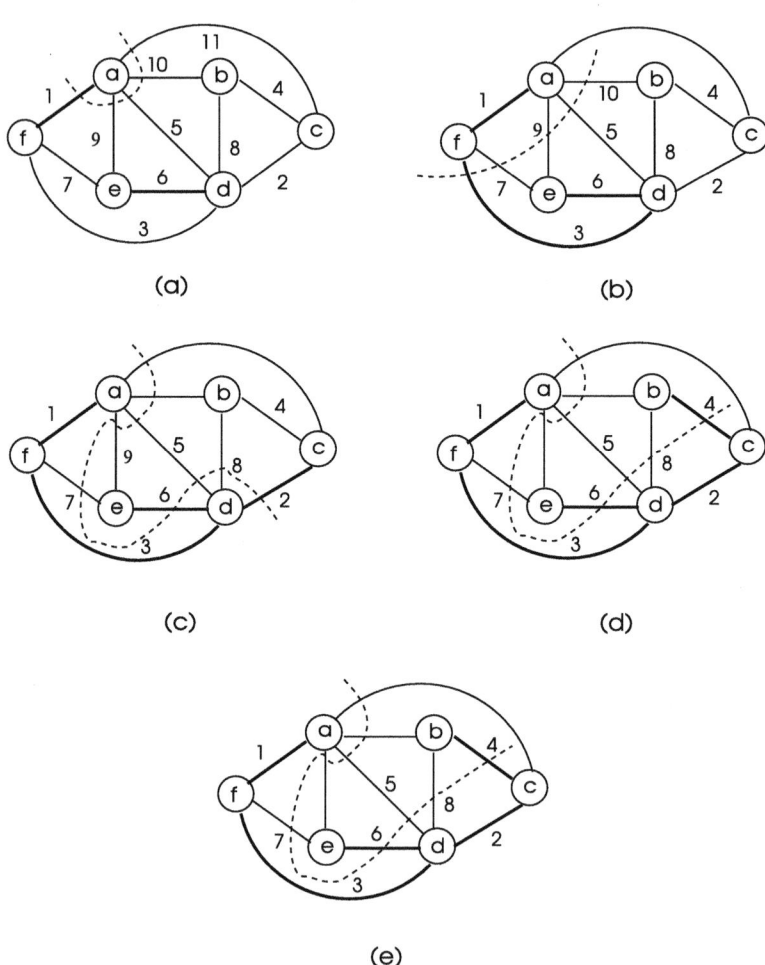

Fig. 7.1 Running of Prim's MST algorithm in a small graph starting from vertex a

Analysis

Theorem 7.3 (correctness) *Prim's algorithm provides an MST of the input graph* $G = (V, E, w)$.

Proof The cut property ensures correctness of this algorithm since we always select the MWOE that is part of the MST by this property. We will show an alternative proof. In each iteration, we add a vertex $v \in V - V_T$ that is connected to a vertex $u \in V_T$ over the lightest edge (u, v) between the two sets. Since edge (u, v) is always between two disjoint sets, it cannot form a cycle with vertices of V_T, hence T is always a tree throughout the algorithm running. Also, since V_T contains all vertices of G in the end as tested in line 5, T is a spanning tree of G.

We now need to check whether T is an MST out of all spanning trees of G and we will do this by induction. Since each vertex must be covered by the MST, the basis of induction is proven. We now want to show that if T_{i-1} is part of the MST, then adding MWOE to it will provide T_i which will also be part of MST. Let us assume T_i is not part of the MST and let T_{i-1} is a partial MST of G and by adding the MWOE (u, v) of T_{i-1} to we obtain T_i according to the rule of Prim's algorithm. We will assume $(u, v) \notin T_{n-1}$ which is the final MST built. In this case, there is another edge (w, z) in the cutset between T_{i-1} and T_i that is part of T_{n-1}. Moreover, (u, v) and (w, z) are edges of a cycle. Deleting (w, z) from G results in another tree T' of G which has a total weight less than T_{n-1} since $w(u, v) \leq w(u, v)$. This means T_i is included in another MST of G which contradicts our initial assumption. □

Theorem 7.4 (complexity) *Prim's algorithm runs in* $O(m \log n)$ *time.*

Proof The main operation performed by this algorithms is the selection of the MWOE at each iteration in line 6 of Algorithm 7.1. An array d can hold the minimum distances to any node in V_T for $\forall v \in V - V_T$, let us call this set V'_T. We can then find the minimum value vertex v of this array to include it in V_T with the corresponding edge (u, v) in E_T. We also need to update the entries in this array as inclusion of the new node v may result in change of values of its neighbors. We need to check $n - 1$ vertices to find the minimum value of d in the first iteration, followed by $n - 2$ iterations in the second step with one less step than the previous one at each step, resulting in $O(n^2)$ steps. Updating of d values requires checking the neighbors of node v at each step, resulting in the sum of degrees of nodes in total which is $2m$. The total time taken therefore is $O(m + n^2)$. We may use a heap data structure to store the values of V'_T as we will show in detail in the next section. Finding the minimum value of V'_T in the heap can be done in $\log n$ time in each step in this case. Updating the d values for neighbors of v requires a further $deg(v) \log n$ at each step. Summing these two operations for n steps results in $O(n \log n + m \log n)$. For a dense graph, we can assume $m \gg n$ and the resulting time is $O(m \log n)$. □

Implementation

Finding the MWOE from the MST fragment is key to the operation of this algorithm. We will use a *min-priority* queue based on a *key* attribute as was described in Sect. 6.2.7. We define $key(v)$ of a vertex v to be the minimum weight of any edge connecting v to a vertex in the tree which is initialized to ∞ since we do not have a tree at start. The queue Q contains all of the vertices of the graph initially and we extract a vertex with the minimum key value from Q at each iteration of the *while* loop. We also assign the parent of each vertex v in $P[v]$ to form the tree structure during iterations. The procedure *ExtractMin(Q)* removes the element with the lowest key from Q. Hence, we invoke this procedure to find MWOE until Q becomes empty. The pseudocode for this algorithm is shown in Algorithm 7.2.

Algorithm 7.2 *Prim_MST2*

1: **Input** : $G(V, E, w)$ undirected, connected, and weighted graph
2:      s source vertex
3: **Output** : MST $T(V, E_T)$ of G
4: **for all** $u \in V$ **do**
5:      $key(u) \leftarrow \infty$
6:      $P[u] \leftarrow \emptyset$
7: **end for**
8: $key(s) \leftarrow 0$
9: $T \leftarrow \emptyset$
10: $Q \leftarrow V$
11: **while** $Q \neq \emptyset$ **do**   ▷ continue until all vertices are visited
12:      $u \leftarrow ExtractMin(Q)$
13:      **for all** $(u, v) \in E$ **do**
14:          **if** $v \in Q$ and $w(u, v) < key(v)$ **then**
15:              $P[v] \leftarrow u$
16:              $key(v) \leftarrow w(u, v)$
17:          **end if**
18:      **end for**
19: **end while**

Analysis of Priority Queue Implementation

We will prove the correctness of this implementation with a priority queue using loop variants as in [5]. We specify a loop variant with three components before each iteration of the *while* loop as follows:

1. The set $V' = \{(v, P[v]) : v \in V - \{s\} - q\}$ is maintained. When we select the lightest edge (u, v) with vertex $u \in q$ we add u to the set $V - Q$ which are the tree vertices. Hence, the edge $(u, P[u])$ is added to MST T.
2. The set $V - Q$ contains vertices that are included in the MST T. This is valid since every time we remove a vertex from Q, we add it to the set of vertices in the MST T.
3. For any vertex $v \in Q$, if $P[v] \neq \emptyset$ then $key(v) < \infty$ and $key(v) = w(u, v)$ with $u \in T$. That is, any vertex v that is outside the partial MST built with a parent

defined has a key value lower than ∞ and is connected to a vertex u in the MST assigned as its parent. We update the key and parent of every neighbor vertex of u without altering any of these values in the tree T and hence this part of the loop variant is maintained.

The time complexity of this implementation depends on how the min-priority queue Q is structured. If binary min-heap described in Sect. 6.2.7 is used, we need to form the heap in $O(n)$ time initially. The *while* loop is run n times with the *ExtractMin* procedure taking $O(\log n)$ at each run resulting in $O(n \log n)$ times to call *ExtractMin*. Testing the *key* values of the neighbors of the vertex u using adjacency list requires $2m$ time in total checking each edge twice. Checking whether $v \in Q$ can be performed in constant time by keeping a boolean variable for each vertex. Assignment of the *key* value at line 16 can be performed by the *DecreaseKey* procedure which requires $O(\log n)$ time. Total time taken therefore is $O(n \log n + m \log n) = O(m \log n)$.

7.2.2.2 Kruskal's Algorithm

Kruskal's MST algorithm takes a different approach by ordering the edges of the graph $G = (V, E, w)$ in nondecreasing weights. Then, starting from the lightest weight edge, edges are included in the partial MST T' as long as they do not form cycles with the edges already contained in T'. This process continues until all edges in the queue are processed as shown in Algorithm 7.3.

Algorithm 7.3 *Kruskal_MST*

1: **Input** : $G = (V, E, w)$
2: **Output** : MST T of G
3: $T \leftarrow \emptyset$
4: $Q \leftarrow$ sorted edges in nondecreasing weights of E
5: **while** $Q \neq \emptyset$ **do**   ▷ continue until all edges are checked
6:    pick an edge (u, v) from Q
7:    **if** (u, v) does not make a cycle with vertices in T **then**
8:       add (u, v) to T
9:    **end if**
10: **end while**

Figure 7.2 shows the iterations of Kruskal's algorithm in a graph. Testing whether addition of an edge creates a cycle is crucial in the operation of this algorithm. When we have a forest of trees, we need to determine whether the endpoints of the edge (u, v) to be considered belongs to the same tree. If they are, including this edge in the MST will create a cycle and therefore we should discard this edge. We also need to merge two trees by the edge under consideration if doing so does not create a cycle. We can use the *Union-Find* data structure for this purpose which has the following defined operations:

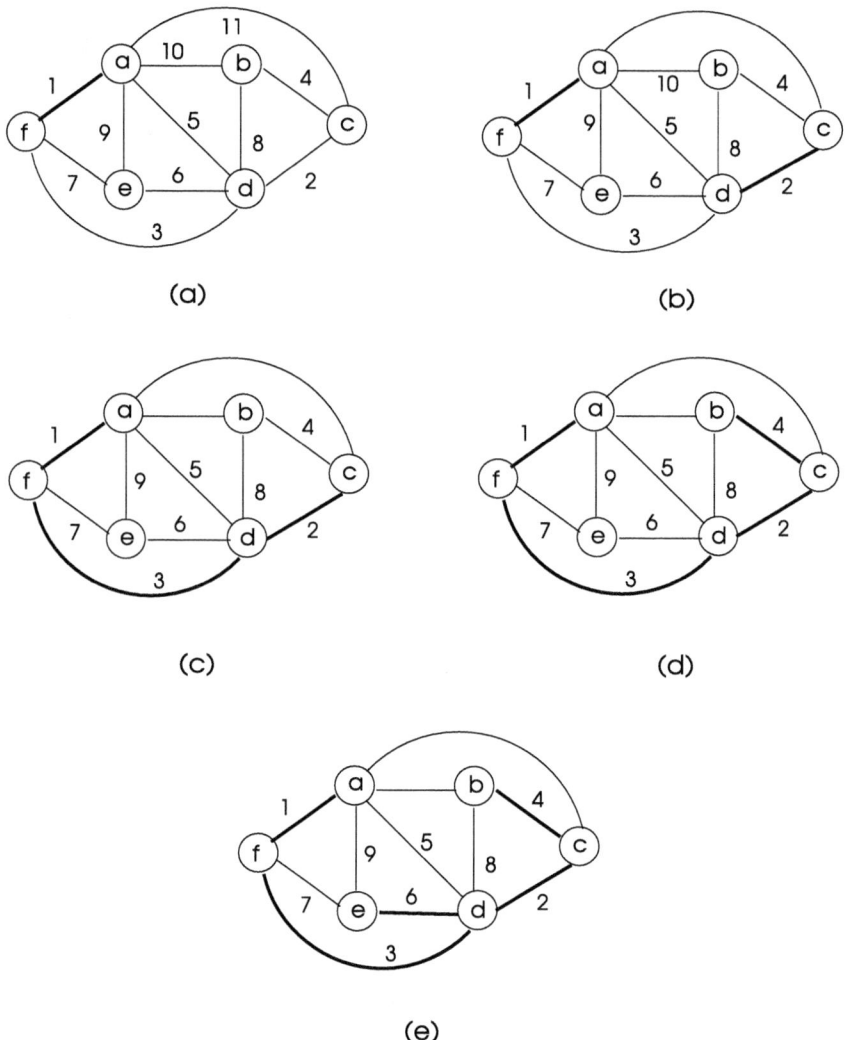

Fig. 7.2 Running of Kruskal's MST algorithm for the same graph of Fig. 7.1. The same MST is obtained as edge weights are distinct

- *MakeSet(x)*: Create a new set consisting only of the element *x*.
- *FindSet(x)*: Find the name (pointer to the representative) of the set containing the element *x*.
- *Union(x, y)*: Merge the sets that contain *x* and *y* to form a new set. The old sets are deleted.

Implementation

Based on these operations, we can restructure Kruskal's algorithm as shown in Algorithm 7.4. All vertices of the graph are components of the forest first. The edges are sorted and inserted into the queue Q, we then test whether the endpoints of an edge (u, v) dequeued from Q are in the same tree. If they are, we know that adding (u, v) to the MST T will create a cycle and we discard this edge. Otherwise, the trees of u and v are merged by the *Union* operation to form a new tree.

Algorithm 7.4 *Kruskal_MST2*

1: **Input** : $G = (V, E, w)$ undirected, weighted graph
2: **Output** : MST $T = (V, E_T)$ of G
3: $T \leftarrow \emptyset$
4: $Q \leftarrow$ sorted edges in nondecreasing weights of E
5: **for all** $v \in V$ **do**
6:     *Make-Set(v)*
7: **end for**
8: **while** $Q \neq \emptyset$ **do**   ▷ continue until all vertices are visited
9:     $(u, v) \leftarrow deque(Q)$
10:     **if** *FindSet(u)* $\neq$ *FindSet(v)* **then**   ▷ check cycles
11:         $E_T \leftarrow E_T \cup \{(u, v)\}$ ▷ include edge in MST
12:         *Union(u, v)*   ▷ merge trees if edge not in cycle
13:     **end if**
14: **end while**

Analysis

Theorem 7.5 (correctness) *Kruskal's algorithm provides an MST of the input graph* $G(V, E, w)$.

Proof At each iteration, we add a vertex $v \in V_T'$ that is connected to a vertex $u \in V_T$ over the cheapest edge (u, v) between the two sets. Since edge (u, v) is always between two disjoint sets, it cannot form a cycle with vertices of V_T, hence T is always a tree throughout the algorithm running. Also, since V_T contains all vertices of G as tested in line 5, T is a spanning tree of G. We now need to prove T is an MST of G. During the ith iteration of the algorithm, let A_i be the subset of edges of the final MST T^*. Note that unlike in Prim's algorithm, A_i may contain a disjoint set of edges. There will not be any edge in A_i that has a greater weight than any edge in $E \setminus A_i$ simply because we include low weight edges in A_i starting from the lowest one. This means if the new edge (u, v) to be added creates a cycle with the existing edges in A_i, it is the highest weight edge in that cycle. By the cycle property, we are rejecting an edge that does not belong to the MST. On the other hand, whenever we accept an edge, it belongs to the MST by the cut property.   □

Initialization by *MakeSet* takes n steps. Since there will be $n - 1$ MST edges, we need to execute the *Union* procedure $n - 1$ times. We also need to test each edge twice for each of its endpoints by the *Find-Set* procedure resulting in $2m$ times invocation. Using linked lists, the *Union* procedure takes $O(n)$ time and *MakeSet* and *FindSet* take $O(1)$ time resulting in $O(n^2)$ time for these procedures. Furthermore, the weights of edges of the graph G can be sorted in $O(m \log m)$ time which is the dominant time for this algorithm. Assuming $m < n^2$, $\log m < 2 \log n$ and hence complexity can be assumed to be $O(m \log n)$.

7.2.2.3 Reverse-Delete Algorithm

As another approach to build an MST of a graph, we can start with all edges of the graph and delete edges that will never be included in the MST until we have a connected graph that has the tree property, that is, it is acyclic or has $n - 1$ edges. We delete edges in the order of decreasing weights as long as removal of a such edge does not disconnect the graph since any bridge of the graph should be contained in the MST. More specifically, the algorithm consists of the following steps:

1. **Input**: An undirected weighted graph $G = (V, E, w)$
2. **Output**: An MST T of G
3. **Sort** edges of G in nondecreasing order into Q
4. Let $T = G$
5. **Repeat**
6. Dequeue the first edge (u, v) from Q
7. Remove (u, v) from T
8. **If** such removal leaves T disconnected **then**
9. Join (u, v) to T
10. **Until** $Q = \emptyset$

Running of this algorithm in a small graph is shown in Fig. 7.3.

Analysis

Theorem 7.6 (correctness) *Reverse edge deletion algorithm provides an MST of the input graph $G = (V, E, w)$.*

Proof This algorithm produces a spanning tree since the resulting structure does not contain any cycles as we delete the heaviest weight edge that lies on a cycle removal of which does not disconnect the graph. We will show the resulting spanning tree T is an MST of G as follows. Let (u, v) be the edge removed during an iteration of the algorithm. Before removal, it must have been on a cycle C as otherwise such removal would disconnect G. Since it is the first edge encountered on C, it is the heaviest weight edge on the cycle C. By the cycle property, the edge (u, v) does not belong to any MST of G. Therefore, this algorithm results in an MST of G since it removes edges that cannot be contained in any MST of the graph G.   □

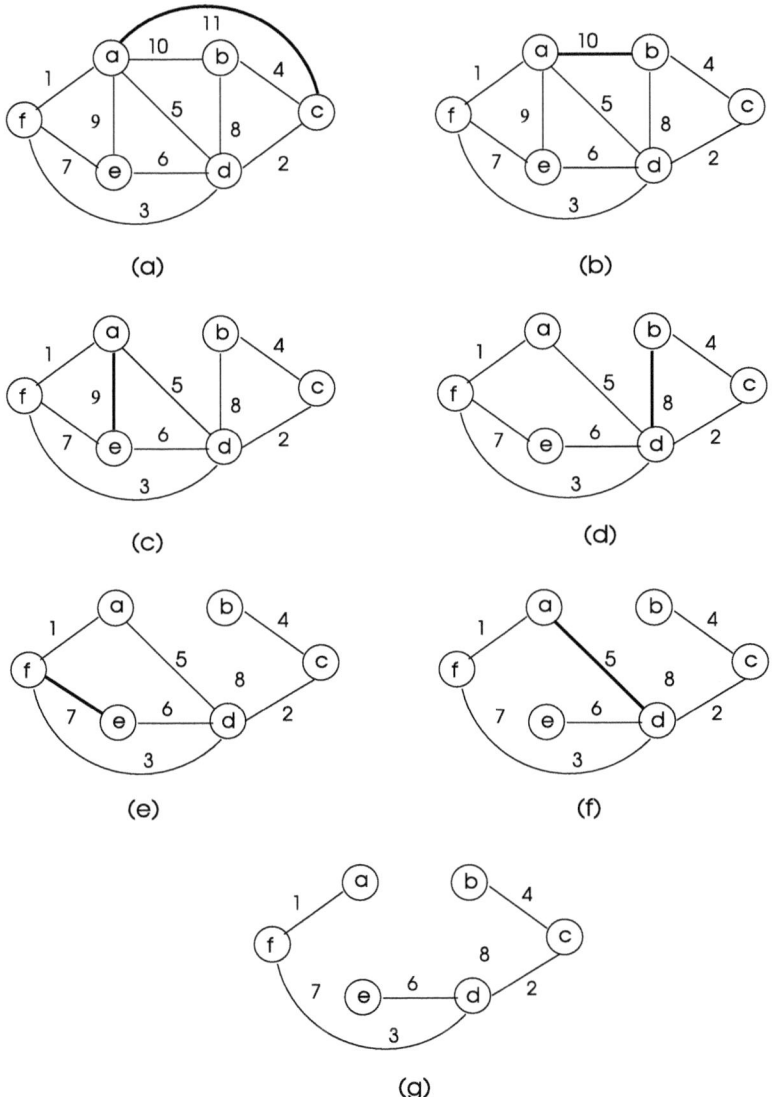

Fig. 7.3 Running of reverse edge deletion algorithm for the same graph of Fig. 7.1. The same MST is obtained as in Prim's and Kruskal's algorithms as edge weights are distinct

The weights of edges of the graph G can be sorted in $O(m \log m)$ time or $O(m \log n)$ time for a dense graph. The main problem with this algorithm is the testing of the connectedness of the graph. This can be performed by the DFS or the BFS algorithm in $O(n + m)$ time after each edge removal resulting in $O(nm + m^2)$ time. Total time taken is then $O(m \log m + nm + m^2)$. It is shown in [18] that removing an edge, checking the connectivity after removal and reinserting the edge if graph is disconnected can be performed in $O(m \log n (\log \log n)^3)$ time per operation.

7.2.2.4 Boruvka's Algorithm

Boruvka's algorithm was the first reported algorithm to construct an MST of a weighted graph $G(V, E, w)$. It was designed to build an electric network of Moravia in Czech Republic in 1926. It starts by finding the lightest weight edge (u, v) incident to each vertex $v \in V$. It then contracts u and v to have a component C which contains u, v and the edge (u, v) in it, and the edge (u, v) is included in the MST. It is possible that the edge (u, v) is the lightest incident edge to the vertex v but another edge, say (u, w), is the lightest edge incident to the vertex u. In such a case, both edges (u, v) and (u, w) are included in the MST, and these edges with all the vertices incident to these edges are placed in the same component. Then, in all of the steps after initialization; two components C_x and C_y are contracted to form a larger component C_z using the lightest weight edge (u, v) between them and the edge (u, v) is included in the MST. If component C_x and/or C_y has lighter edges than (u, v) incident to them, these edges are also included in the MST and these edges with their incident vertices are placed in the new component C_z. This process is repeated until there is only one component which contains all of the vertices providing a spanning tree and the selected edges are the edges of MST as shown in Algorithm 7.5.

Algorithm 7.5 *Boruvka_MST*

1: **Input** : $G(V, E, w)$
2: **Output** : MST $T(V_T, E_T)$ of G
3: Let each vertex $v \in V$ be a component
4: $T \leftarrow \emptyset$
5: **while** there is more than one component **do**
6:     **combine** two neighbor components C_x and C_y using the lightest edge (u, v) between them
7:     $E_T \leftarrow E_T \cup \{(u, v)\}$ ▷ include lightest edge in T
8:     $V_T \leftarrow V_T \cup \{u, v\}$ ▷ include vertices in T
9:     **if** $\exists (p, q) \in C_x : w(p, q) > w(u, v)$ **and/or** $\exists (r, s) \in C_y : w(r, s) > w_{u,v}$ **then**
10:       $E_T \leftarrow E_T \cup (p, q)$ **and/or** $E_T \leftarrow E_T \cup \{(r, s)\}$
11:       $V_T \leftarrow V_T \cup \{p, q\}$ **and/or** $V_T \leftarrow V_T \cup (r, s)$
12:     **end if**
13: **end while**

Figure 7.4 shows two iterations of Boruvka's algorithm in a graph.

Note that we may end up discovering the full MST even with one iteration of this algorithm if lightest edges incident to each vertex are selected as shown in Fig. 7.5.

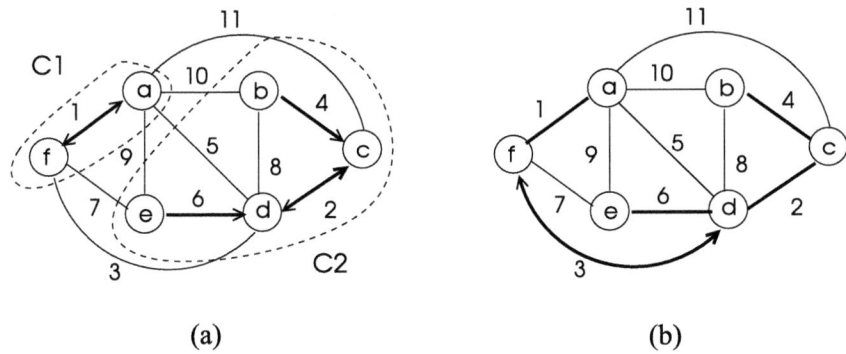

(a)   (b)

Fig. 7.4 Running of Boruvka's MST algorithm in the graph G of Fig. 7.1. The first step in a divides the graph into two components $C_1 = G[a, f]$ and $C_2 = G[b, c, d, e]$ with the lightest weight incident edges included in MST as shown in bold. The lightest edge between these components is (f, d) which is used to merge them and this edge becomes part of the MST as shown in **b**. Arrows show the vertex that the lightest edges are incident. The same MST as in Prim's, Kruskal's algorithms and the reverse-delete algorithms is obtained as edge weights are distinct

Fig. 7.5 Running of Boruvka's MST algorithm when all MST edges are selected in one iteration. MST edges are shown in bold and arrows show the vertex that the lightest edges are incident

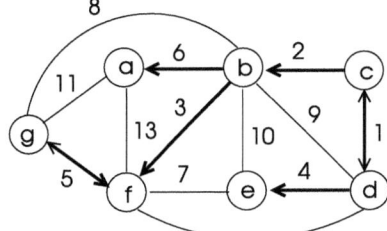

Analysis

Lemma 7.1 *Suppose edge weights of the graph $G = (V, E, w)$ are distinct. Let e_v be the least weight edge incident to a vertex $v \in V$. The MST T of G contains every such edge e_v.*

Proof Let us assume the MST T, which is unique due to distinct edge weights, does not contain some edge $e_v = (u, v)$. Adding the edge (u, v) to T creates a cycle. Let the vertex w be a neighbor of vertex v, then $w(u, v) < w(v, w)$ since (u, v) is the lightest edge incident on v. If we delete (v, w) from T and add (u, v) to T, we obtain $T' = T - \{(v, w)\} \cup \{(u, v)\}$ which is still a tree having $n - 1$ edges and has a total weight smaller than the weight of T resulting in a contradiction.   □

Theorem 7.7 (correctness) *Boruvka's algorithm produces the MST of the input graph $G = (V, E, w)$ that has distinct edge weights.*

Proof The final component T is a tree since we join two tree components by exactly one edge preventing cycles at each iteration. This component T includes all of the vertices of G as we continue until there is one component, it is a spanning tree of G. Edges to be included in the tree T at each iteration are part of the MST of G by Lemma 7.1 hence the resulting tree T is the MST of G. $\square$

Note that we required the edge weights to be distinct to select a unique least weight edge incident to a vertex. This restriction can be relaxed by varying the weights of equal-weight edges slightly to have unique edge weights.

Theorem 7.8 (complexity) *Boruvka's algorithm runs in $O(m \log n)$ time.*

Proof Each step of the algorithm reduces the number of vertices by at least a factor of 2 and therefore the total number of steps is $\log n$. Each step requires $O(m)$ time for contraction resulting in $O(m \log n)$ time. $\square$

7.2.3   Parallel MST Algorithms

Out of the four algorithms we have reviewed, Boruvka's algorithm is most suitable for parallel processing due to relatively independent contraction operations involved. We will however first look at ways of parallelizing Prim's algorithm and then describe briefly how Boruvka's algorithm can run on a distributed memory parallel system.

7.2.3.1   Parallel Prim's Algorithm

Given the input graph $G = (V, E, w)$, we will consider Prim's algorithm to find the MST $T = (V, E_T)$ of G in parallel. We start by considering the modified version of the algorithm with the array $d[n]$ holding the minimum distance values of the nodes that are in $V - V_T$ to V_T for parallel operation. Algorithm 7.6 shows the operation of the sequential algorithm as in [10] where we arbitrarily select a vertex s and first initialize the array values for the neighbors of s. Then at each iteration, we find the minimum value of array d which is the MWOE of the current iteration. The node v having this value with edge (u, v) as the MWOE is included in the MST. We need to update array values as the distance to the new V_T vertex v has now to be considered.

Algorithm 7.6 *Prim_MST3*

1: **Input** : $G = (V, E, w)$
2: **Output** : MST T of G
3: $V_T \leftarrow \{s\}$
4: **for all** $u \in V - V_T$ **do**
5:     **if** $(u, s) \in E$ **then**
6:         $d[u] \leftarrow w(u, s)$
7:     **end if**
8: **end for**
9: $T \leftarrow \emptyset$
10: **while** $V_T \neq V$ **do**   ▷ continue until all vertices are visited
11:     select the minimum element of d with vertex v and edge (u, v)
12:     $V_T \leftarrow V_T \cup \{v\}$
13:     $E_T \leftarrow E_T \cup \{(u, v)\}$
14:     **for all** $u \in N(v)$ **do**   ▷ update distances to V_T considering the new node v
15:         $d[u] \leftarrow min\{d[u], w(u, v)\}$
16:     **end for**
17: **end while**

This algorithm is inherently sequential as we search for the MWOE at each iteration. However, searching for MWOE can be done in parallel within a single iteration. The general idea of the parallel algorithm is to divide the vertices to k processes and have them find the MWOEs in their partitions. The global MWOE can then be found by a special process which broadcasts it to all others for local updates as in lines 14–16 of the sequential algorithm. In the implementation, we have k processes $p_0, \ldots, p_{k-1}$ with p_0 as the supervisor. We divide the vertices into k subsets where each process p_i gets n/k vertices in the set V_i. The array d is 1-D block partitioned to k processes and the weighted adjacency matrix A is also column partitioned to k processes. Each process then finds the minimum value of array d in its partition by using matrix A and the global minimum is computed using the all-to-one reduction at the root process p_0 which then broadcasts it to all processes. The processes update their distances and form their partition of d for the nodes they are responsible. This process continues until array d has no elements left meaning all nodes are in V_T. The pseudocode for this algorithm is depicted in Algorithm 7.6.

Algorithm 7.7 *Parallel_Prim_MST*

1: **Input** : $A[n, n]$: weighted adjacency matrix of $G = (V, E, w)$
2:          $P = \{p_0, p_1, ..., p_{k-1}\}$ ▷ set of k processes
3: **Output** : MST $T = (V_T, E_T)$ of G
4: **if** $p_i = p_0$ **then**   ▷ if I am the root process
5:    **for** $i = 1$ to $k - 1$ **do**
6:        **send** columns $((i - 1)n/k)) + 1$ to in/k of A to p_i
7:    **end for**
8:    **for** *round* $= 1$ to $n - 1$ **do**   ▷ loop for *n*-1 rounds
9:        **all-to-one receive** MWOEs in *values*
10:       $min\_val(u, v) \leftarrow min(values)$ ▷ find the global minimum distance
11:       **one-to-all broadcast** $new\_MWOE(u, v)$ to all processes
12:       $V_T \leftarrow V_T \cup \{v\}$
13:       $E_T \leftarrow E_T \cup \{(u, v)\}$
14:   **end for**
15: **else**   ▷ I am a worker process
16:   *round* $\leftarrow 0$
17:   **receive** my column partition $D[my\_cols]$ from p_0
18:   **while** *round* $< k$ **do**   ▷ get local minimum values from processes
19:       $a_{p_i}(u, v) \leftarrow$ minimum distance between u and v in $d[my\_columns]$
20:       **send** $proc\_min(u, v, d_{p_i})$ to p_0
21:       **receive** $new\_MWOE(u, v)$ from p_0
22:       **for all** $u \in N(v) \wedge u \in d[my\_columns]$ **do**   ▷ update distances to V_T considering the
                  new node v
23:           $d[u] \leftarrow min\{d[u], w(u, v)\}$
24:       **end for**
25:       *round* $\leftarrow$ *round* $+ 1$
26:   **end while**
27: **end if**

We will show the implementation of this parallel algorithm using four processes $p_0, \ldots, p_3$ for the same graph we have used to demonstrate sequential algorithms. Weighted adjacency matrix, sometimes called the distance matrix, A is formed and partitioned as follows:

		p_1			p_2		p_3
	a	b	$\|c$	d	$\|e$	f
a	0	10	$\|11$	5	$\| 9$	1
b	10	0	$\| 4$	8	$\|\infty$	∞
c	11	4	$\| 0$	2	$\|\infty$	∞
d	5	8	$\| 2$	0	$\| 6$	3
e	9	∞	$\|\infty$	6	$\| 7$	7
f	1	∞	$\|\infty$	3	$\| 7$	0

The first entries of the array d would then be as follows:

b	c	d	e	f
10	11	5	9	**1**

The root process gathers the minimum values of 10, 5, and 1 from processes p_1, p_2 and p_3 respectively and determines the global minimum value of 1 between nodes a and f. It then broadcasts this value which is included in V_T, node f is removed from V_T' and all neighbor edges are tested to obtain the new d as below:

b	c	d	e
10	11	**3**	9

This time node d is broadcast to all processes, it is removed from V_T', and d is updated to yield d with (8, 2, 6) for nodes b, c, and e respectively. Three more rounds of the parallel algorithm provide the same MST found by other methods.

Analysis
Each process p_i finds the minimum value and performs the updates in $\Theta(n/k)$ time and the total time is $\Theta(n^2/k)$ for n rounds. It takes $\log k$ time to perform one-to-all communication in each round, resulting in $\Theta(n \log k)$ total time for communication. Total time taken is

$$T_P = \Theta(n^2/k) + \Theta(n \log k). \tag{7.2}$$

Since the sequential time is $\Theta(m \log n)$ for Prim's algorithm, the speedup obtained is

$$S = \frac{\Theta(m \log n)}{\Theta(n^2/k) + \Theta(n \log k)}. \tag{7.3}$$

7.2.3.2 Parallel Kruskal's Algorithm
Kruskal's algorithm grows multiple trees which can be performed in parallel. We implement a similar strategy as in parallel Prim's algorithm by partitioning the adjacency matrix among k processes. The parallel Kruskal algorithm then consists of the following steps [14]:

1. Each process p_i sorts edges in its partition V_i.
2. Each process p_i constructs an MST or a forest of MSTs in its partition using Kruskal's algorithm.
3. MSTs found by process are merged pairwise by a process sending its MST edges to another process. This step can be handled by breaking symmetries using identifiers. The lower identifier process can send its MST edges to an arbitrarily selected higher identifier process. The lower identifier process then should halt.

4. Step 3 is repeated until there is one process that contains the MST.

Merging edges in step 2 can be performed using Kruskal's algorithm. Figure 7.6 shows the operation of this parallel algorithm in a small graph where we partition the adjacency matrix to four processes. Computing the edges of MST in each partition takes $O(n^2/k)$ time and there are $O(\log k)$ merging operations each with a cost of $O(n^2 \log k)$ and each process sends $O(n)$ edges in one merge resulting in a total parallel time of $O(n^2/k) + O(n^2 \log k)$ [14].

7.2.3.3 Parallel Boruvka's Algorithm

Boruvka's algorithm can operate in parallel as we show in the high-level pseudocode of Algorithm 7.8.

Algorithm 7.8 *Par_Boruvka*

1: **Input** : $G(V, E, w)$ undirected, weighted graph
2: **Output** : MST T of G
3: **for all** $v \in V'$ **in parallel do**
4:     Find the lightest edge incident to v
5: **end for**
6: Contract edges
7: Merge adjacency lists
8: Recurse until all edges are processed

Various parallel MST algorithms are based on Boruvka's algorithm. One such approach is reported in [4] where the resulting super vertices after contraction consist of trees of vertices. Neighborhood information is kept in edge lists, one for each vertex. There are at most $n - 1$ elements in each edge list. The steps of the parallel Boruvka's algorithm in this study consists of the following steps:

1. *Choose lightest edge*: The edge list of each vertex is searched to find the lightest weight edge incident to that vertex to form components. Cycles are removed to have a tree for each component.
2. *Find root*: Each vertex finds the root of the tree it belongs using the pointer jumping method.
3. *Rename vertices*: Each process p_i, $1 \le i \le k$ determines the new name of each vertex listed in its edge lists.
4. *Merge*: The edge lists in each component are merged to the root edge list to shrink it into a single super-vertex.
5. *Clean up*: Each process p_i runs the sequential MST algorithm in its edge list.

The parallel running time for this algorithm is given as $\Theta((t_s + t_w)(m \log n/p)$ which results in a speedup comparable to the number of parallel processes but the constant $(s + t_w)$ may be very large [4]. Contraction can be performed using the edge or star contraction methods we have reviewed in Chap. 3 to obtain the parallel Boruvka's algorithm [1].

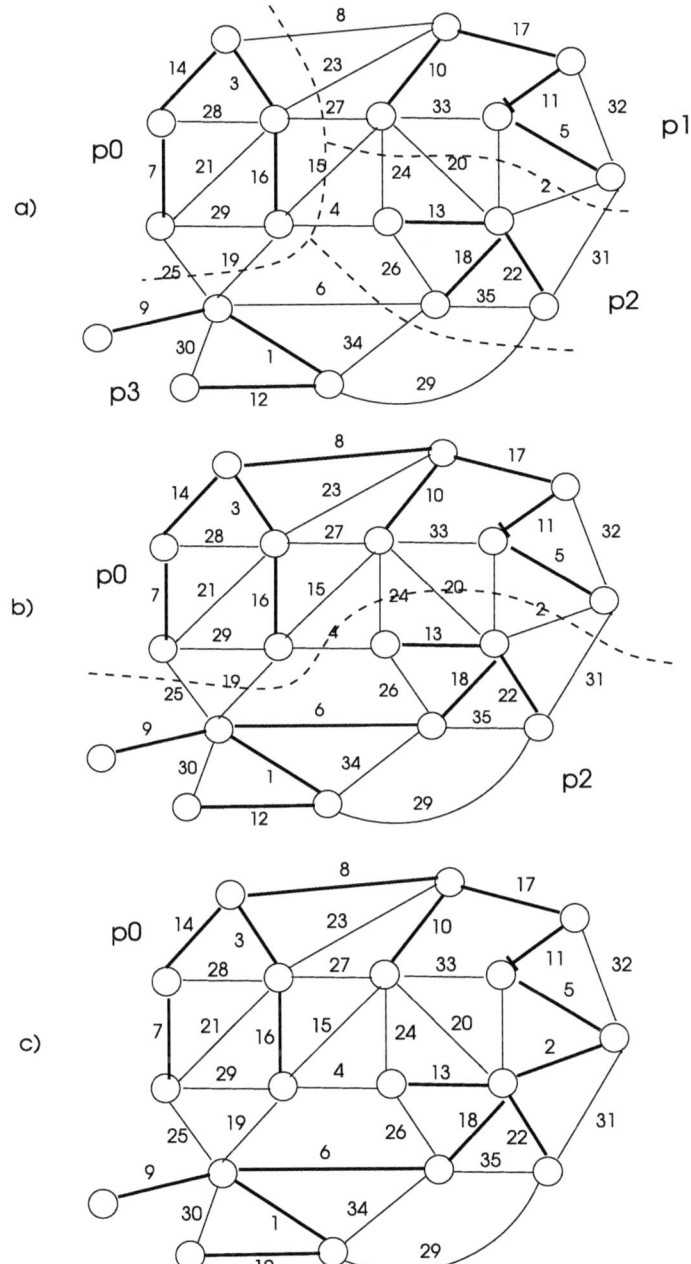

Fig. 7.6 Running of parallel Kruskal's algorithm in a sample graph. There are 4 processes and the adjacency matrix of the graph is partitioned to four processes p_0, p_1, p_2 and p_3 as shown by dashed lines

7.2.4   Distributed Prim's Algorithm

In the distributed version of this problem, we are interested in finding the MST of a network in which every node is involved in the construction. We will consider each of the three sequential algorithms for this purpose. As a first attempt, investigation of Prim's algorithm reveals it is basically sequential in nature. However, the synchronous single initiator (SSI) model of distributed processing may be convenient for this purpose. We will build and use the MST for proper transfer of messages between the root and other nodes in the tree. The processing is performed in synchronous rounds in this model and we have a root process a which initiates each round. In the first round, it includes the lightest edge incident to it in the MST. In each round thereafter, the root solicits the MWOE of each leaf of the partial T which are convergecast to the root. It then finds the smallest of MWOEs received from children and broadcasts this to the members of T which can update their states. In essence, we are processing the graph exactly as in the sequential algorithm but since we do not know the global MWOE beforehand, the special process root has to receive all candidates from each leaf of T and determine the lightest edge. We will describe a possible implementation of this idea similar to [7, 15]. The messages needed are as follows.

- *start*: This is sent by the root to its children in each round. It has a dual purpose; initiation of a new round k and carrying the MWOE (u, v) determined in round $k - 1$ of the partial tree T.
- *reply*: This is the convergecast message from leaves to the root. At each intermediate node, the MWOEs of children are gathered, compared with the own MWOE, and the smallest of them is sent by the *reply* message to the parent.
- *check*: This message is needed to avoid cycles. The newly added node v sends it to neighbors to check the ones already in T and for such a node u, edge (u, v) is marked as *internal* so it will not be considered as MWOE of v in future rounds.
- *status*: This message is returned by a node u as a reply to *check* message from v and contains information about whether $u \in T$ or not.

Algorithm 7.9 shows one way of implementing the procedure we have described. The synchronization is provided by messages only and the root starts the next round only after all convergecast messages from its children are received. It selects the lightest edge (u, v) with v as the new vertex and sends it to the nodes of the partial T in the next round. Any node x in T that has an edge to vertex v marks this edge (x, v) as internal to prevent cycles in T. The vertex v checks whether its neighbors are in T or not. This is again needed to prevent cycles as a neighbor may have become part of T. The leaves start the convergecast process which ends at root with MWOEs received from children.

Algorithm 7.9 *Distributed_Prim_MST*

1: **Input** : $G(V, E, w)$
2: **Output** : MST $T(V_t, E_t)$ of G
3: **if** $a = root$ **then**
4:     let (a, b) be the lightest edge
5:         $T \leftarrow (a, b)$ ▷ a becomes the parent of b
6:     **repeat**   ▷ round k
7:         send $start(v, (u, v), k)$ to children
8:         receive $reply((p, q)_i, we_{(p,q)_i})$ from each child i
9:         **if** $reply = \emptyset$ from all children **then**
10:            send $stop$ to all children
11:        **else**
12:            (u, v) with new node v is the lightest edge of all MWOEs received from children
13:                $T \leftarrow T \cup (u, v)$ ▷ include (u, v) in T
14:        **end if**
15:    **until** $reply = \emptyset$ from all children
16: **else**   ▷ I am node x on T
17:    **while** $stop$ not received **do**
18:        receive $start(v, (u, v), k)$
19:        **if** $(v, x) \in N(x)$ **then**   ▷ check cycles
20:            $state(v, x) \leftarrow internal$
21:        **else if** $x = v$ **then**
22:            **for all** $u \in N(x)$ **do**
23:                send $check$ to u
24:            **end for**
25:            **for all** $u \in N(x)$ **do**
26:                receive $status(res)$ from u
27:                    $state(u, x) \leftarrow res$ ▷ mark neighbor edges as internal or external
28:            **end for**
29:        **end if**
30:        **if** x is a leaf **then**
31:            let (x, z) be my MWOE such that $state(x, z) = external$
32:            $send((x, z), we_{(x,z)}$ to $parent$
33:        **else**   ▷ convergecast data of children
34:            receive $reply((p, q)_i, w_{(p,q)_i})$ from each child i
35:            find the lightest edge $(p, q)_i$ including my MWOE
36:            $send((p, q), we_{(p,q)}$ to $parent$
37:        **end if**
38:    **end while**
39: **end if**

The running of this algorithm in a small network graph is depicted in Fig. 7.7 where the building of the MST is completed in seven rounds.

This algorithm correctly finds the MST of a graph as it mimics the sequential Prim algorithm in a distributed setting. Each step k of the algorithm requires $O(k)$ time and messages. The time and message complexities are therefore both $O(n^2)$.

Looking at other sequential algorithms, Kruskal's algorithm is difficult to be implemented by the nodes in the network as it requires global ordering of the weights of edges. However, Boruvka's algorithms involve independent steps. Since each graph node now is a node in the network, finding the lightest incident edge in

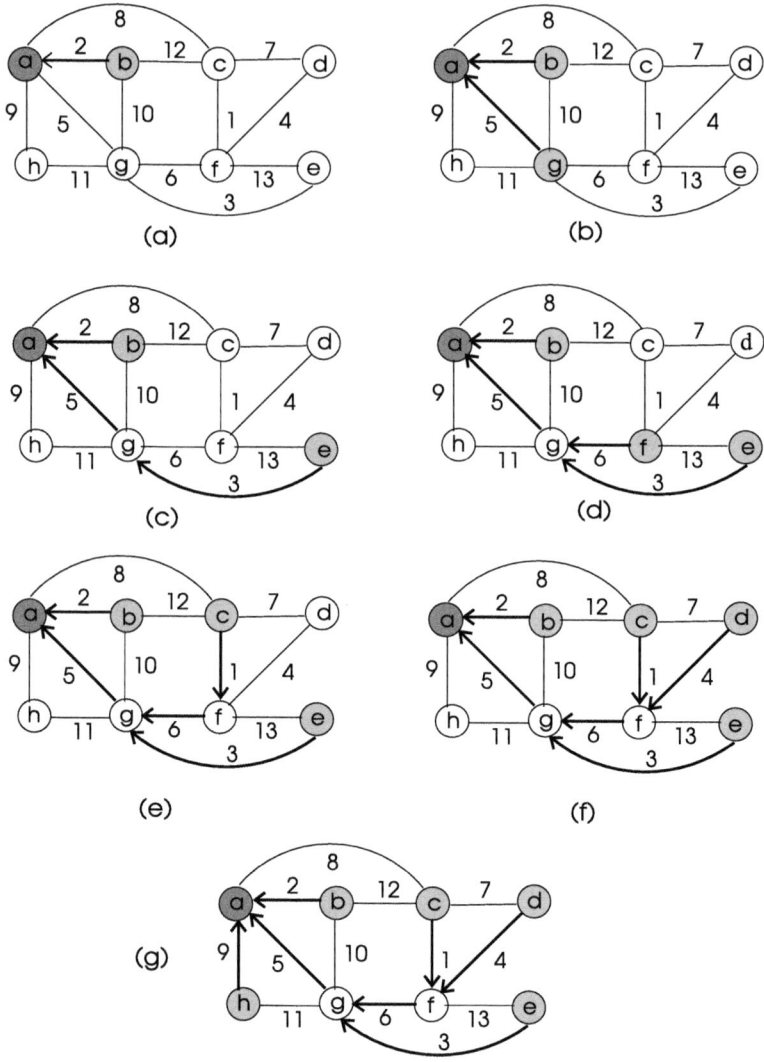

Fig. 7.7 Running of distributed Prim's MST algorithm in a graph from vertex *a* for seven rounds. The gray nodes show the leaves of the tree formed. The downcasting of the *probe* message is in the reverse direction of arrows of bold lines from parent to children and upcasting of *ack* message is in the direction of arrows from children to parent

the initial phase can be done by each node in a single step. We need however to find ways to contract and manage the contracted nodes. A simple yet effective approach is to elect a leader for each contracted component which can find the MWOE of nodes in its component and ask the connected component in the other end of this MWOE for merge operation. The leader may be the lowest identifier node or the newest node in the component. The leaders of each component then communicate with neighboring leaders and decide on the lightest edge between them. A similar method is employed in the algorithm of Gallager, Humblet, and Spira to find the MST of a network [8]. The nature of Boruvka's algorithm provides distributed processing conveniently; however, the choice of the leader has to be performed.

7.3   Shortest Paths

In various real-life situations, we may be interested to find the shortest path, that is, the path with minimum total weight among all other paths between two vertices. For example, the shortest traveling route between two cities may be required. We review algorithms for this problem in this section. In all of these algorithms, we will employ a technique called *relaxation* which can be described as follows. Given a weighted graph $G = (V, E, w)$, we want to find shortest paths from a source vertex $s \in V$ to all other vertices in the graph G. We define a distance value d_v which shows the best estimate of the current distance of v to s and a predecessor vertex u of v which is its parent in the tree formed rooted at the source vertex s. We will be forming a spanning tree T rooted at s at the end of a shortest path algorithm, sometimes referred to as *shortest path tree* in which the sum of weights of a path from s to a vertex v in this tree will be minimum among all possible paths from s to v. Distance of each vertex from the source vertex is set to infinity and its parent is undefined initially. Relaxation then involves checking whether a shorter path of a vertex v through a neighbor vertex u than its current distance is found in which case its distance is updated to go through that neighbor vertex u and its parent is set to u as shown in the following steps performed for each vertex v. We need to add the weight of the edge between these two vertices to the distance of vertex u to get the actual distance.

1. **for all** $u \in N(v)$
2. **if** $d_v > d_u + w(u, v)$
3. $d_v = d_u + w(u, v)$
4. $P_v = u$.

Another issue of concern is whether the graph has negative weights and/or negative cycles.

7.3.1   Single Source Shortest Paths

In the more general case, we can search shortest paths from a single vertex to all other vertices which is called the single source shortest path (SSSP) problem for which we review a fundamental algorithm due to Dijkstra.

7.3.1.1 Dijkstra's Algorithm

Dijkstra proposed an iterative algorithm to find shortest distances from a single source vertex to all other vertices in a weighted, directed, or undirected graph $G = (V, E, w)$ [6]. The general idea of this algorithm is to start from a source vertex s and initially label distance values of all neighbors of s with the weight of edges from s and the rest of the vertices with infinity distance values. and the distance value of s to itself as 0. The neighbor vertex v which has the smallest distance to s is then included in the visited vertices set. Thereafter at each iteration, distance value and predecessor of any neighbor u of the newly included vertex v has been updated if distance through v is smaller than its current distance. This algorithm processes all vertices and eventually forms a spanning tree rooted at source vertex s. We have a vertex set S which shows vertices to be processed, an array D with $D[i]$ showing the current shortest distance of vertex i to the source vertex s and another array P with $P[i]$ which shows the current predecessor of a vertex i along this path as shown in Algorithm 7.3.

Algorithm 7.10 *Dijkstra_SSSP*

1: **Input** : $G(V, E, w)$ ▷ connected, weighted graph G and a source vertex s
2: **Output** : $D[n]$ and $P[n]$ ▷ distances and predecessors of vertices in the tree
3:      tree vertices T
4: **for all** $v \in V \setminus \{s\}$ **do**   ▷ initialize all vertices except source s
5:      $D[v] \leftarrow \infty$
6:      $P[v] \leftarrow \perp$
7: **end for**
8: $D[s] \leftarrow 0$; $P[s] \leftarrow s$
9: $V \leftarrow V'$; $T \leftarrow \emptyset$
10: **while** $V' \neq \emptyset$ **do**
11:      **find** $v \in S$ with minimum distance value
12:      **for all** $(u, v) \in E$ **do**   ▷ update neighbor distances to v
13:          **if** $D[u] > D[v] + w(u, v)$ **then**
14:              $D[u] \leftarrow D[v] + w(u, v)$
15:              $P[u] \leftarrow v$ ▷ update tree structure
16:          **end if**
17:      **end for**
18:      $V' \leftarrow V' \setminus \{v\}$ ▷ remove new vertex from searched
19:      $T \leftarrow T \cup \{v\}$ ▷ add it to tree vertices
20: **end while**

The running of this algorithm is depicted in a digraph in Fig. 7.8. The source vertex is f and the nearest vertex to f is a which is included in the searched vertices. Then all neighbors of a which are b and e are checked whether they have shorter distance to the source vertex f through a. Since these vertices had infinity distances initially, their distances are modified for smaller values through a. Then we find vertex e has the smallest distance value and include it in the searched vertices and update distance values of its neighbors. Note that vertex b has a smaller distance value of 7 through e and therefore its distance is updated and its predecessor becomes e. This process continues until we search all vertices which is performed by removing the shortest distance vertex v from the initial vertex set V' at each iteration.

Analysis

Theorem 7.9 (correctness) *For each vertex $v \in S$ at any time during Dijkstra's shortest path algorithm execution, the path $P_{s,v}$ obtained by the algorithm is the shortest path between the source vertex s and the vertex v.*

We will have proved the correctness of the algorithm by proving this theorem since the set S will contain all of the vertices of the graph at the end of the algorithm.

Proof We will use induction for the proof as in [11]. For the base case, $d(s) = 0$ and $S = \{s\}$ when $|S| = 1$ and hence $P_{s,s}$ is the shortest path.

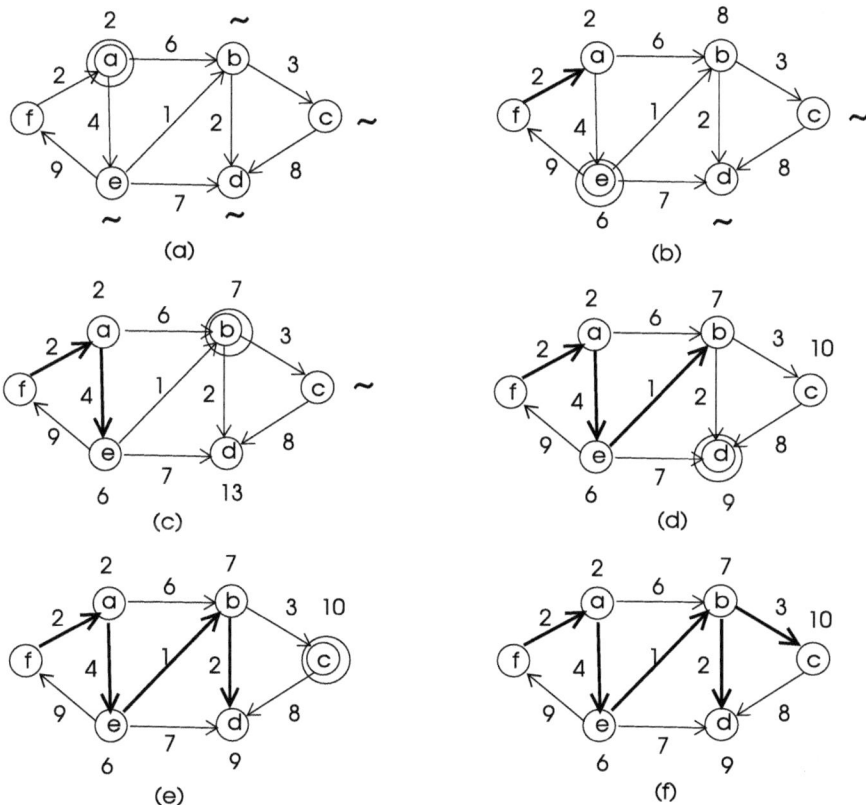

Fig. 7.8 Running of Dijkstra's SSSP algorithm in a sample directed graph from the source vertex f. At each iteration, the vertex with the minimum distance value shown by a large circle is selected

Fig. 7.9 Alternative paths
between two vertices

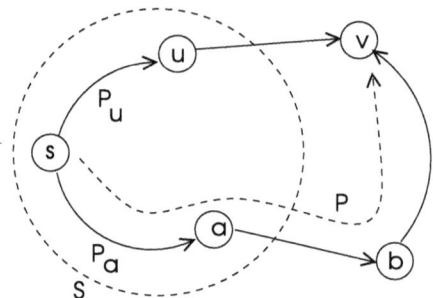

Let us assume adding a vertex $v \notin S$ to S when the size of S is k and $u \in S$ is a neighbor vertex of v on the shortest path $P_{s,v}$ from source vertex s to vertex v. Consider any arbitrary path P from s to v. Our hypothesis is that the total weight of this path is at least as high as the total weight of $P_{s,v}$. Let vertex a be the last vertex on P just before it leaves S and the vertex $b \in \{V \setminus S\}$ be the first vertex that is the neighbor of vertex a on this path as depicted in Fig. 7.9. We know the total weight of path $P_{s,u}$, $w(P_{s,u})$, is the minimum distance to vertex u from the source vertex s by the inductive hypothesis. If $w(a, b) < w(u, v)$ the algorithm would have selected the edge (a, b) rather than the edge (u, v). Therefore, $w(P) \geq w(P_{s,v})$ which means $P_{s,v}$ found during the $k + 1$th iteration of the algorithm is the shortest path from vertex s to vertex v. Note that we have relied on the nonexistence of negative weight edges. □

We need to run the *while* loop of the algorithm $O(n)$ times for n vertices since we process a single vertex at each iteration. We also need to find the smallest distance of unprocessed vertices to the source vertex s in $O(m)$ time since we may need to consider all edges to find the minimum value. Hence the time complexity of this algorithm is $O(nm)$ in this straightforward implementation.

We can improve the performance of this algorithm by using a priority queue. In this case, we will use three priority queue operations; *Insert*, *ExtractMin*, and *DecreaseKey*. We need to insert all vertices in the queue Q by the *Insert* operation, find the minimum value of the queue by the *ExtractMin* operation and *DecreaseKey* operation during relaxation where we update distance values of the neighbors of the selected vertex. When a binary min-heap is used as the priority queue, time to construct the queue takes $O(n)$ time. We need n *ExtractMin* operations for n vertices each with $O(\log n)$ time and $O(m)$ steps of relaxation using *DecreaseKey* during relaxation each with $O(\log n)$ time. Hence, total time taken is $O((n + m) \log n)$. A Fibonacci heap that has an amortized $O(\log n)$ time for *ExtractMin* operation and $O(1)$ amortized time for *DecreaseKey* operation can be used instead of the binary min-heap. In this implementation, the time complexity is reduced to $O(n \log n + m)$.

7.3.1.2 Bellman–Ford Algorithm

It is possible to have negative weights in some graphs and in such cases, Dijkstra's SSSP algorithm fails to provide correct shortest routes from a source vertex as we have counted on nonnegative weights for the correct operation of the algorithm. Dijkstra's algorithm is based on the assumption that a shortest path consists of smaller shortest subpaths. Let $p = \{v_1, v_2, \ldots, v_k, \}$ be a shortest path from a source vertex v_1 to a destination vertex v_k. Then for $2 \leq i < k$, $p = \{v_1, \ldots, v_i, \}$ is also a shortest path for the algorithm to work correctly. Clearly, this assumption is valid only for nonnegative edge weights. Negative weight edges are encountered in some real-life applications such as currency trading and minimum cost flows hence there is a need for a shortest path algorithm in the presence of negative weight edges.

The dynamic algorithm provided by Bellman and Ford works in the presence of edges with negative weights, however, it will only detect negative cycles when there is one [2]. A negative cycle in a graph G is a cycle $\{v_0, v_1, \ldots, v_k, v_0\}$ such that $w(v_0, v_1) + w(v_1, v_2) + \cdots + w(v_k, v_0) < 0$.

The working of this algorithm is simple, it performs relaxation for each vertex progressively that is $1, \ldots, n - 1$ hops away from the source vertex s to allow changes along the longest path which is $n-1$ hops as shown in Algorithm 7.11. We use array D for distance values and array P for identities of predecessor vertices in the tree. Running of this algorithm in a sample undirected graph is shown in Fig. 7.10.

Algorithm 7.11 *BellFord_SSSP*

1: $D[s] \leftarrow 0$
2: **for all** $i \neq s$ **do**   ▷ initialize distances and predecessors
3:     $D[i] \leftarrow \infty$
4:     $P[u] \leftarrow \perp$
5: **end for**
6: **for** $k = 1$ to $n - 1$ **do**
7:     **for all** $\{u, v\} \in E$ **do**   ▷ update distances
8:         **if** $D[u] > D[v] + w(u, v)$ **then**
9:             $D[u] \leftarrow D[v] + w(u, v)$
10:            $P[u] \leftarrow v$
11:        **end if**
12:    **end for**
13: **end for**
14: **for all** $(u, v) \in E$ **do**   ▷ report negative cycle
15:    **if** $D[u] + w(u, v) > D[v]$ **then**
16:        **return** *false*
17:    **end if**
18:    **return** *true*
19: **end for**

Analysis

The following lemma helps to prove correctness of this algorithm [12]:

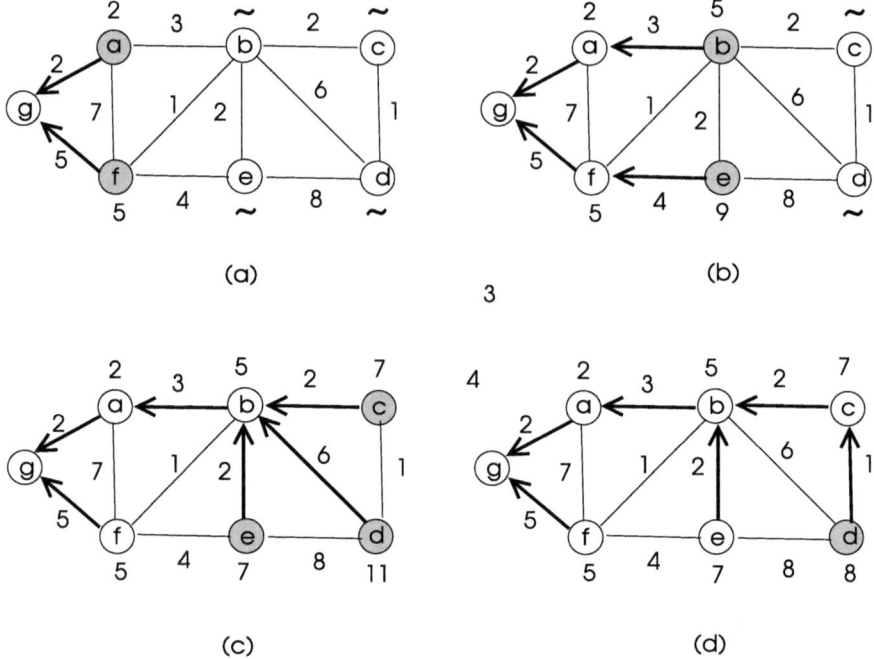

Fig. 7.10 Running of Bellman–Ford algorithm from the source vertex g in 4 iterations. The visited vertices at each iteration are shown in gray and the current distance value of a vertex is shown next to it

Lemma 7.2 (optimality principle) *Let $G = (V, E)$ be a weighted graph with no negative cycles and let u and v be two vertices of G. Let P be a shortest path between u and v with at most k edges and let (w, v) be the final edge in this path. Then, $P' = P \setminus \{w, v\}$ (or $P_{[u,w]}$) is a shortest path from u to w with at most $k - 1$ edges.*

Proof Assume R is a shorter $u - w$ path than P' and $|E(R)| \leq k - 1$. Then, $w(E(R)) + w(w, v) < w(E(P))$. If $v \notin R$, then $R \cup \{(v, w)\}$ is a shorter $u - v$ path than P (Case 1). Otherwise $R_{[v,w]} \cup \{(w, v)\}$ is a nonnegative cycle and thus the cost of $R_{[u,v]}$, $w(E(R_{[u,v]})) = w(E(R_{[u,v]})) + w(w, v) - w(E(R_{[v,w]} + w(w, v)) < w(E(P)) - w(E(R_{v,w})(w + v)) \leq w(E(P))$ (Case 2). In both cases there is a contradiction to the assumption P is a shortest path between vertices u and v with at most k edges. $\quad\square$

Theorem 7.10 (correctness) *Bellman–Ford algorithm correctly computes SSSP paths from a source vertex s to all other vertices of an undirected or directed graph when there are no negative cycles. Let $d(s, v_i)$ be the distance label of vertex v_i after iteration i and $dist(s, v_i)$ be the distance (shortest path) of vertex v_i to source* *vertex s. More specifically, we claim that after k iterations of the algorithm, $d(s, v_i)$ is at most dist(s, v_i).*

Proof We will prove this theorem by induction on the number of iterations.

- *Base case*: The distance labels of every vertex other than the source vertex s have infinite labels when the algorithm starts. After the iteration $i = 1$, only the neighbors of vertex s will have a distance label such that $\forall v \in N(s), d(v, s) = w(s, u)$. Thus, all of these neighbors will have the shortest distance to s when $k = 1$.
- *Inductive case*: Let us consider the kth iteration, and assume theorem holds for all $i < k$. Let P be a shortest $s - v_k$ path with at most k edges and and (v_{k-1}, v_k) be the last edge of P. By Lemma 7.2, the part of path P up to the vertex v_{k-1} $(P_{s,v_{k-1}})$ is a shortest path between s and v_{k-1}; and by the induction hypothesis, $dist(s, v_{k-1}) \le w(E(P_{s,v_{k-1}}))$ after the $(k-1)$th iteration. After the kth iteration, we have $dist(s, v_k) \le dist(s, v_{k-1}) + w(v_{k-1}, v_k) \le w(E(P))$.

The above reasoning is valid when there are no negative cycles in the graph. We now want to prove using contradiction that this algorithm returns *false* when there is a negative cycle. Let us assume graph G contains a negative cycle $C = \{v_0, v_1, \ldots, v_k, v_0\}$ such that $\sum_i^k w(v_i, v_{i+1}) < 0$ with $v_{k+1} = v_0$ and the algorithm returns *true*. There is a path from the source vertex s to v_1 and to all other vertices of C and let $d(v_i)$ be the distance obtained in the first part of the algorithm using relaxation. Since we assumed the algorithm returns *true* without detecting negative cycles, $d(v_{i+1}) \le d(v_i) + w(v_i, v_{i+1})$ for $i = 1, \ldots, k$. When we sum for all vertices in the cycle, we obtain

$$\sum_{i=1}^{k} d(v_{i+1}) \le \sum_{i=1}^{k} (d(v_i) + w(v_i, v_{i+1}))$$

$$\sum_{i=1}^{k} d(v_{i+1}) \le \sum_{i=1}^{k} d(v_i) + \sum_{i=1}^{k} w(v_i, v_{i+1}).$$

Since we sum over the cycle C, $\sum_{i=1}^{k} d(v_{i+1}) = \sum_{i=1}^{k} d(v_i)$ and canceling in the above equation results in the following.

$$0 \le \sum_{i=1}^{k} w(v_i, v_{i+1}).$$

This contradicts our initial assumption and therefore Bellman–Ford algorithm returns *false* when there is a negative cycle in the graph. □

Theorem 7.11 (complexity) *Bellman–Ford algorithm has a time complexity of O(nm).*

Proof We need to have $n - 1$ iterations of the outer *for* loop to consider the longest path in a graph since there may be $n - 1$ changes of the distance of a vertex over this longest path. There may be at most m edge checking at each iteration of the inner loop at line 7 and hence, the total time complexity of this algorithm is $O(nm)$. It is, therefore, a slower algorithm than Dijkstra's SSSP algorithm, however, it allows negative weight edges which may be needed in real-life applications.   □

7.3.1.3 Parallel Dijkstra's Algorithm

We can form a parallel version of Dijkstra's SSSP algorithm in a similar manner to the method of parallel Prim's algorithm. The weighted adjacency matrix A is partitioned columnwise such that each process p_i is assigned n/k consecutive columns of A. Then each p_i computes n/p values of array l. The communication among processes are similar to parallel Prim algorithm and the performance is the same as this algorithm [10].

7.3.1.4 Distributed Algorithms

In a network environment, our aim is to have each node of the network to compute shortest routes from itself to all other nodes in the network.

7.3.1.5 Synchronous Distributed Bellman–Ford Algorithm

We can sketch a synchronous distributed algorithm (*DBF_SSSP*) using the method of Bellman–Ford algorithm in a network environment. There is a special node called the *root* which initiates synchronous round by the *round* message over a spanning tree built prior to the execution of the algorithm. Each node exchanges its distance value to the source node with its neighbors by the *update* message in each round. Any node i that finds it has a shorter path to the source node via a neighbor j makes j its parent and updates its distance to the source by adding the weight of edge (i, j) to the distance of node j. All of this operation is analogous to Bellman–Ford algorithm in a network. There will be $n - 1$ rounds to be initiated by the root as in the sequential case, hence the root should know the number of nodes in the network. Due to the uncertainty in the delivery sequence of messages in a round, an *update* message can reach a node before a *round* message. Therefore, we have included a boolean variable *round_received* which is checked by each node before updating distances. When all of the neighbor messages are received along with a *round* message, updating is performed and another boolean variable *round_over* is set true to enable convergecasting of synchronization messages. A single round of this algorithm is shown in Algorithm 7.12.

Algorithm 7.12 *SDBF_SSSP*

1: **message types** *round, update*
2: **int** *i, j, my_dist, dist*
3: **set of int** *received* ← Ø;
4: **boolean** *round_over* ← *false, round_recvd* ← *false*
5: **while** ¬*round_over* **do**   ▷ A single round executed by each node except the source
6:     **receive** *msg(j)*
7:     **case** *msg(j).type* **of**
8:         $\underline{round(k)}$:   **send** *update(k,my_dist)* to $N(i)$
9:                             *round_recvd* ← *true*
10:       $\underline{update(k, dist)}$: *received* ← *received* ∪ *{j}*
11:                     **if** *received* = $N(i)$ ∧ *round_recvd* **then**
12:                         **for all** $j \in N(i)$ **do**
13:                               **if** *my_dist* > $(dist + w_{ij})$
14:                                   *my_dist* ← $dist + w_{ij}$
15:                                   *parent* ← *j*
16:                       *round_over* ← *true*
17: **end while**

Theorem 7.12 *SDBF_SSSP algorithm correctly finds APSP distances from a source node in $O(n)$ rounds using $O(nm)$ messages.*

Proof Since the distributed algorithm has the same logic as the sequential algorithm, we can conclude each node finds its distance to a source node correctly. We have noted that the root needs to execute $n - 1$ rounds to take the longest path in the network into account, hence time complexity in rounds for this algorithm is $O(n)$. Each edge is in the network is used to send *update* messages in both directions in each round, resulting in a total of $2m$ messages per round. Total number of messages exchanged will, therefore, be $O(nm)$. □

7.3.2 All-Pairs Shortest Paths

In a more general case, we may need to discover shortest paths from all vertices to all other vertices in the graph which is called the *all-pairs shortest paths* (APSP) problem. As a first approach, we can run Dijkstra's SSSP algorithm for each vertex of the graph resulting in $O(n^2 \log n)$ time complexity. When the graph has edges with negative weights, we cannot use Dijkstra's algorithm and using Bellman–Ford algorithm for this purpose yields a time complexity of $O(n^2 m)$ considering running it for n vertices. We will search for algorithms with better performances when dealing with negative weight edges and one such approach is due to Floyd–Warshall described in the next section.

7.3.2.1 Floyd–Warshall Algorithm

Floyd–Warshall Algorithm (*FW_APSP*) solves the APSP problem in linear time using dynamic programming. As often practiced in dynamic programming, the problem is divided into smaller subproblems which are then solved to obtain intermediate results to be used in the overall solution. In this algorithm, negative weight edges are allowed but negative cycles are not. It uses the relaxation method we have seen, this time for distance between all pairs of vertices using each vertex as a pivot in sequence. For convenience, the vertices are labeled with integers $1, \ldots, n$. Let us consider the shortest path p_{ij} with weight $d_{ij}^{(k)}$ between any two vertices $i, j \in V$ with elements taken from $1, 2, \ldots, k$. There are no intermediate vertices when $k = 0$ and therefore $d_{ij}^{(0)} = w_{ij}$. We can define $d_{ij}^{(k)}$ recursively as follows [5]:

$$
d_{ij}^{(k)} = \begin{cases} w_{ij} & \text{if } k = 0 \\ min\{d_{ij}^{(k-1)}, d_{ik}^{(k-1)} + d_{kj}^{(k-1)}\} & \text{if } k \geq 1 \end{cases} \tag{7.4}
$$

We can now implement Floyd–Warshall algorithm as shown in Algorithm 7.6. We have two nested loops to apply the relaxation to the distance between each vertex pair (i, j) to see whether distance between these vertices is shorter through a pivot vertex k than their current distance. Each vertex is assigned as the pivot in sequence and hence we have another outer loop to select pivot vertices resulting in three nested loops as shown in Algorithm 7.13.

We have the distance matrix $D[n, n]$ with elements d_{ij} showing the current distance between vertices i and j which is initialized to infinity for vertices that are not directly connected and to the weight of the edge between them if they are neighbors. The predecessor matrix $P[n, n]$ has entries $p_{i,j}$ which shows the first vertex over the current shortest path between the vertices i and j.

Algorithm 7.13 *FW_APSP*

1: **Input** : $G(V, E, w)$ ▷ connected, weighted directed, or undirected graph G
2: **Output** : $D[n, n]$ and $P[n, n]$ ▷ distances and predecessors of vertices
3:
4: **for** $i = 1$ to n **do**   ▷ initialize
5:     **for** $j = 1$ to n **do**
6:        **if** $(i, j) \in E$ **then**
7:           $D[i, j] \leftarrow w(i, j), P[i, j] \leftarrow j$
8:        **else**
9:           $D[i, j] \leftarrow \infty, P[i, j] \leftarrow \perp$
10:        **end if**
11:     **end for**
12: **end for**
13:
14: **for** $k = 1$ to n **do**   ▷ pivot vertex
15:     **for** $i = 1$ to n **do**
16:        **for** $j = 1$ to n **do**
17:           **if** $D[i, k] + D[k, j] > D[i, j]$ **then**   ▷ relaxation
18:              $D[i, j] \leftarrow D[i, k] + D[k, j], P[j] \leftarrow k$
19:           **end if**
20:        **end for**
21:     **end for**
22: **end for**

Correctness follows from the relaxation rule as we always improve the shortest paths using all possible pivots. We have three nested loops each running n times resulting in a time complexity of $O(n^3)$ for this algorithm with the initialization taking $O(n^2)$ time. A small example graph is depicted in Fig. 7.11.

The contents of the distance matrix initially and for $k = 1$ in sequence are shown below with modified contents displayed in bold figures.

$$D^{(0)} = \begin{matrix} & \begin{matrix} 1 & 2 & 3 & 4 & 5 & 6 \end{matrix} \\ \begin{matrix} 1 \\ 2 \\ 3 \\ 4 \\ 5 \\ 6 \end{matrix} & \begin{pmatrix} 0 & 1 & \infty & \infty & 6 & 2 \\ 1 & 0 & 2 & 1 & 4 & \infty \\ \infty & 2 & 0 & 6 & \infty & \infty \\ \infty & 1 & 6 & 0 & 2 & \infty \\ 6 & 4 & \infty & 2 & 0 & 9 \\ 2 & \infty & \infty & \infty & 9 & 0 \end{pmatrix} \end{matrix} \rightarrow D^{(1)} = \begin{matrix} & \begin{matrix} 1 & 2 & 3 & 4 & 5 & 6 \end{matrix} \\ \begin{matrix} 1 \\ 2 \\ 3 \\ 4 \\ 5 \\ 6 \end{matrix} & \begin{pmatrix} 0 & 1 & \infty & \infty & 6 & 2 \\ 1 & 0 & 2 & 1 & 4 & \mathbf{3} \\ \infty & 2 & 0 & 6 & \infty & \infty \\ \infty & 1 & 6 & 0 & 2 & \infty \\ 6 & 4 & \infty & 2 & 0 & \mathbf{8} \\ 2 & \mathbf{3} & \infty & \infty & \mathbf{8} & 0 \end{pmatrix} \end{matrix}$$

Fig. 7.11 Sample graph for FW_APSP algorithm

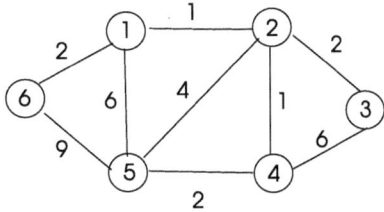

The D matrix is displayed below for $k = 2$. There is no change in distance values when vertex 3 is the pivot and thus we show D values when $k = 4$.

$$
D^{(2)} =
\begin{matrix}
 & \begin{matrix} 1\ 2\ 3\ 4\ 5\ 6 \end{matrix} \\
\begin{matrix} 1 \\ 2 \\ 3 \\ 4 \\ 5 \\ 6 \end{matrix} &
\begin{pmatrix}
0\ 1\ 3\ 2\ 5\ 2 \\
1\ 0\ 2\ 1\ 4\ 3 \\
3\ 2\ 0\ 3\ 6\ 6 \\
2\ 1\ 3\ 0\ 2\ 4 \\
5\ 4\ 6\ 2\ 0\ 7 \\
2\ 3\ 6\ 4\ 7\ 0
\end{pmatrix}
\end{matrix}
\ \rightarrow\
D^{(4)} =
\begin{matrix}
 & \begin{matrix} 1\ 2\ 3\ 4\ 5\ 6 \end{matrix} \\
\begin{matrix} 1 \\ 2 \\ 3 \\ 4 \\ 5 \\ 6 \end{matrix} &
\begin{pmatrix}
0\ 1\ 3\ 2\ 5\ 2 \\
1\ 0\ 2\ 1\ 4\ 4 \\
3\ 2\ 0\ 3\ 6\ 6 \\
2\ 1\ 3\ 0\ 2\ 4 \\
5\ 4\ 6\ 2\ 0\ 6 \\
2\ 4\ 6\ 4\ 6\ 0
\end{pmatrix}
\end{matrix}
\ \swarrow
$$

There are no further changes in D matrix contents when $k = 5$ and $k = 6$. The P matrix is shown below which displays the first vertex on the shortest path from a vertex i to j.

$$
P =
\begin{matrix}
 & \begin{matrix} 1\ 2\ 3\ 4\ 5\ 6 \end{matrix} \\
\begin{matrix} 1 \\ 2 \\ 3 \\ 4 \\ 5 \\ 6 \end{matrix} &
\begin{pmatrix}
0\ 2\ 2\ 2\ 2\ 6 \\
1\ 0\ 3\ 4\ 4\ 1 \\
2\ 2\ 0\ 2\ 2\ 2 \\
2\ 2\ 2\ 0\ 5\ 2 \\
2\ 4\ 4\ 4\ 0\ 4 \\
1\ 1\ 1\ 1\ 1\ 0
\end{pmatrix}
\end{matrix}
$$

7.3.2.2 Parallel APSP Using Dijkstra's Algorithm

We can use Dijkstra's SSSP algorithm to find APSP routes in two different ways as vertex-partitioned and computation-partitioned methods as described in [10].

Vertex-Partitioned SSSP Paths

In the first approach, vertices are partitioned evenly to all processes of the parallel processing system. Each p_i then computes SSSPs for all of the vertices it is responsible. The distance matrix is replicated at each node, and hence there is no interprocess communication. The parallel running time, in this case, is $T_P = \Theta(n^2)$ and since the sequential time is $T_S = \Theta(n^3)$ when Dijkstra's SSSP algorithm of $\Theta(n^2)$ time complexity is run for n vertices, the speedup S obtained and the efficiency E is

$$
S = \frac{\Theta(n^3)}{\Theta(n^2)} = n, \quad E = \Theta(1). \tag{7.5}
$$

If the number of processes k is smaller than the number of nodes, this algorithm has good performance, otherwise, it will scale poorly.

Computation-Partitioned SSSP Paths
We can have the SSSP algorithm running on a number of parallel processes when
$k > n$ as follows. Assuming we have k processes available for parallel computation, we assign k/n processes to each vertex and then run k/n processes in parallel
for each vertex as described in Sect. 7.3.1.3 when parallelizing Dijkstra's SSSP
algorithm. In other words, we have n parallel SSSP computations each of which
is handled by n/k processes.

$$T_P = \Theta(n^3/k) + \Theta(n \log k)$$

$$S = \frac{\Theta(n^3)}{\Theta(n^3/k) + \Theta(n \log k)} = n, \quad E = \frac{1}{1 + \Theta((k \log k)/n^2)} \qquad (7.6)$$

7.3.2.3 Parallel Floyd–Warshall Algorithm

We can form a parallel version of *FW_APSP* algorithm by dividing the task of
matrix D computation among p processes. We will describe a possible partitioning
using 2-D block mapping for this problem as described in [10]. In this approach,
D is divided into blocks of size $(n/\sqrt{p}) \times (n/\sqrt{p})$ with each process assigned a
single block. Processes are positioned in a grid of $\sqrt{p}$ by $\sqrt{p}$ and a process $p_{i,j}$
has a subblock with upper-left corner $((i-1)n/\sqrt{p}+1, ((j-1)n/\sqrt{p}+1$ and a
lower right corner $in/\sqrt{p}, jn/\sqrt{p}$ as shown in Fig. 7.12.

Each process $p_{i,j}$ computes its subblock of D during each iteration but needs
the elements held by processes in its row and columns to be able to perform
this computation. Therefore, we need one-to-all broadcast of D values along rows

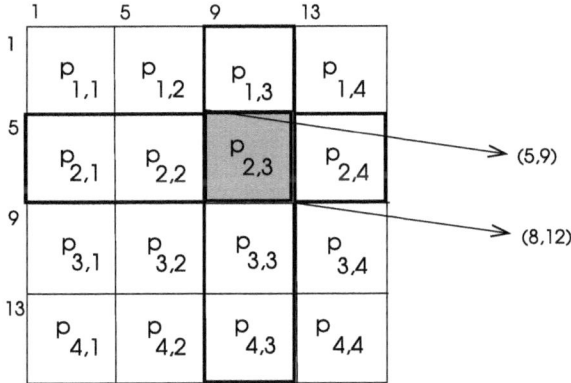

Fig. 7.12 2-D partitioning of 16×16 D matrix for a graph with 16 vertices to 16 processes
in *PFW_APSP* algorithm. The process $p_{2,3}$ for example, has upper left corner coordinates (5,
9) and lower right coordinates (8, 12). This process needs all matrix entries held by processes
$p_{2,1}, p_{2,2}, p_{2,4}$ in its row and by processes $p_{1,3}, p_{3,2}, p_{4,3}$ in its column to be able to compute its
subblock D values for the current iteration

and columns held by processes. Algorithm 7.14 shows one way of realizing this algorithm using 2-D partitioning.

Algorithm 7.14 PFW_APSP

1: **Input** : subblock of D^0 ▷ my postion of the distance matrix
2: **Output** : $D_{i,j}^n$ ▷ shortest path values for my subblock
3:
4: **for** $j = 1$ to n **do**   ▷ update distances and next node
5:      **broadcast** my segment of $D^{(k-1)}$ to all processes in my row
6:      **broadcast** my segment of $D^{(k-1)}$ to all processes in my column
7:      **receive** $D^{(k-1)}$ values from processes in my row and column
8:      **compute** $D^{(k)}$ for my subblock
9: **end for**

Each process $p_{i,j}$ holds $n/\sqrt{p}$ elements of the kth row or column which are broadcast in $\Theta((n\log p)/\sqrt{p})$ time. Synchronization in line 7 requires $\Theta(\log p)$ time and computation of n^2/p values assigned to a process requires $\Theta(n^2/p)$ time resulting a total parallel processing time of

$$T_P = \Theta\left(\frac{n^3}{p}\right) + \Theta\left(\frac{n^2}{\sqrt{p}}\log p\right). \tag{7.7}$$

We know that sequential algorithm has a time complexity of $\Theta(n^3)$, therefore the speedup S can be stated as follows:

$$S = \frac{\Theta(n^3)}{\Theta(n^3/p) + \Theta(n^2\log p/\sqrt{p})} \tag{7.8}$$

Therefore the efficiency E is,

$$E = \frac{1}{1 + \Theta\left(\sqrt{p}\log p/n\right)} \tag{7.9}$$

From the speedup and efficiency equations, we can conclude this algorithm can employ $O(n^2/\log^2 n)$ processes. Synchronization step can be omitted to result in a faster-pipelined version of 2-D algorithm with efficiency $1/(1 + \Theta(p/n^2))$ [10].

7.3.2.4 Distributed Floyd–Warshall Algorithm

In a distributed setting, we need n nodes of the network each of which determines its SSSP to all other nodes at the end of the algorithm. We can have a synchronous distributed version of FW_APSP algorithm with the modification that each node i now holds only a vector $D_i[n]$ which shows its best estimate of its shortest distance to all other nodes in the network. This vector, in fact, corresponds to the row that node i has in the distance matrix D in the sequential algorithm. A local vector $P_i[n]$ held at node i shows the first node along the current shortest path estimate from node i to all other nodes. In order to adapt the sequential algorithm to this network environment in full, we need to have the pivot node k broadcast its local vector $D_k[n]$ so that each node i can compare values in D_k with the values in D_i and update D_i and P_i accordingly as in Algorithm 7.15 where a single round for a node i is shown. We assume the following:

- There is a special node called the *root* which initiates each round.
- A spanning tree is built beforehand to send and receive control messages such as broadcast *round* and convergecast *round_over* messages.
- Nodes have unique integer identifiers in the range $1, \ldots, n$.
- The *root* sends round number r in each round which is interpreted as the parameter k in the sequential algorithm. Any node that finds its identifier equals r will broadcast its D values for all other nodes to compare.

Algorithm 7.15 *DFW_APSP*

1: **set of int** $D[n]$, $P[n]$ ▷ local distance and next node vectors
2:
3: $D[i] \leftarrow 0, P[i] \leftarrow i$ ▷ initialize
4: **for** $j = 1$ to n **do**
5:      **if** $j \in N(i)$ **then**
6:          $D[j] \leftarrow w(i, j), P[j] \leftarrow j$
7:      **else**
8:          $D[j] \leftarrow \infty, P[j] \leftarrow \perp$
9:      **end if**
10: **end for**
11:          ▷ a single round
12: **receive** *round*(r) message
13: **if** $i = r$ **then broadcast** $D_k[n]$ ▷ if I am the pivot, broadcast $D_k[n]$
14: **else**   **receive** $D_k[n]$ from node k ▷ otherwise receive the pivot vector
15: **end if**
16: **for** $j = 1$ to n **do**   ▷ update distances and next node
17:      **if** $D[j] + D[k] < D[j]$ **then**
18:          $D[j] \leftarrow D[j] + D[k]$
19:          $P[j] \leftarrow k$
20:      **end if**
21: **end for**

Figure 7.13 displays the running of this algorithm in a small network. Broadcasting of D_k vector is the main bottleneck in this algorithm. We can have the node r send its D vector to the root which then broadcasts this vector to all nodes over the spanning tree. Toueg provided an asynchronous version of this algorithm by reducing the set of nodes that should receive the D_k values with a time complexity $O(n^2)$ and a message complexity $O(nm)$ [19].

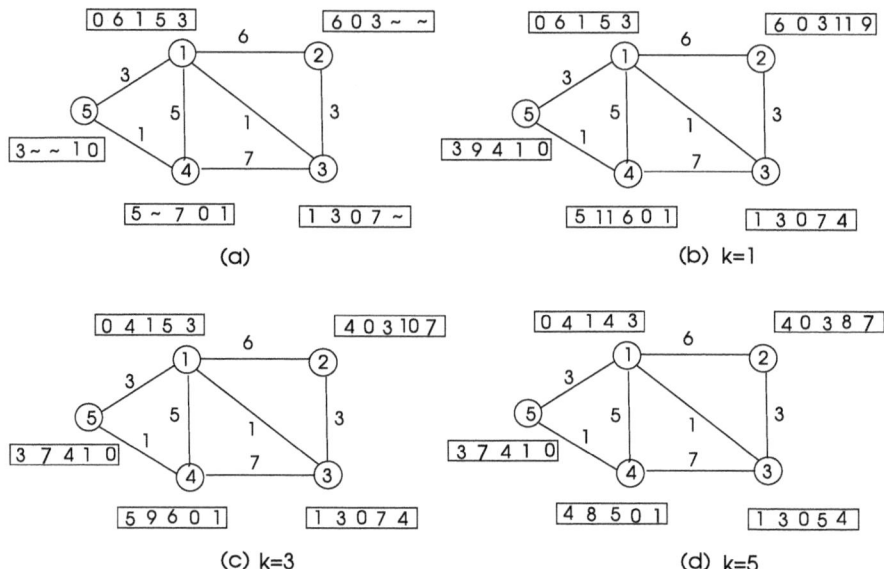

Fig. 7.13 Running of DFW_APSP algorithm in a small network. The current vectors at each node are shown next to them. At iterations $k = 2$ and $k = 4$, there are no changes to previous distance values and these are not shown. After five iterations, all of the shortest paths are determined

7.4 Chapter Notes

We looked at weighted graph algorithms in this chapter where edges in a graph have weights associated with them. We reviewed a fundamental problem in such graphs; finding the MST and described methods for the construction of the MST of a weighted, undirected, and connected graph in this chapter. The MST problem can be solved by greedy sequential algorithms in polynomial time. These algorithms are due to Boruvka, Kruskal, and Prim in chronological order and the Reverse-Delete algorithm, each with a different approach. We showed in detail how to form a parallel version of Prim's algorithm by finding the MWOEs in parallel and also described ways of parallelizing Boruvka's algorithm. Survey of MST algorithms are provided in [9] and in [17].

In a network setting where each node of the graph is a computational node, our first approach is again to consider these sequential algorithms. We described how to obtain a distributed network version of Prim's algorithm and reviewed a method based on Boruvka's algorithm for distributed processing in a network. The contents of the messages and the instants they are sent are important in a network as we discuss. We are interested in both time and message complexities in network algorithms.

As for conversions between the three fundamental methods, we have already performed Seq(Prim) $\rightarrow$ Par(Prim) and Seq(Prim) $\rightarrow$ $Dist$(Prim), and we described ways to achieve Seq(Boruvka) $\rightarrow$ Par(Boruvka) and Seq(Boruvka)

→ *Dist*(Boruvka). One thing to consider is whether we can have conversions such as *Par*(Prim) ↔ *Dist*(Prim). One way of achieving this would be partitioning of the graph into nonoverlapping k partitions and distributing the subgraphs $G_0, \ldots, G_{k-1}$ to processes $p_0, \ldots, p_{k-1}$ of the network. The root process p_0 performs sequential Prim algorithm in its partition until edges that cross the partitions are met. It can then perform the process as in the distributed version of Prim's algorithm by asking for MWOEs from each partition rather than individual nodes. Each node is a partition now and what we have described is a conversion from distributed algorithm to parallel algorithm for this method.

We then considered shortest path problems; these problems are considered as single source shortest path (SSSP) or all pairs shortest paths (APSP) problems. A fundamental algorithm due to Dijkstra solves the SSSP problem but does not work with negative weight edges or negative cycles. Bellman–Ford algorithm works with negative weights and reports negative cycles with increased time complexity. We looked ways of having parallel and distributed versions of these algorithms. Floyd–Warshall algorithm finds APSP paths in a graph with negative-weight edges. We also presented a parallel version and a distributed version of this algorithm. These algorithms provide significant examples of conversion between a sequential, parallel, and distributed versions of the same method.

Exercises

1. Find the MST of the sample graph of Fig. 7.14 using Prim's MST algorithm.
2. Work out the MST of the graph depicted in Fig. 7.15 using both Kruskal's and Boruvka's MST algorithms and show they both result in the same MST of this graph since edge weights are distinct.
3. Work out single source shortest paths from vertex a of the digraph depicted in Fig. 7.16 using *Dijkstra_SSSP* algorithm by showing each iteration.
4. Write the pseudocode of parallel *Dijkstra_SSSP* algorithm and work out its efficiency.
5. Construct single source shortest paths from vertex a of the digraph depicted in Fig. 7.17 using *BF_SSSP* algorithm by showing each iteration.

Fig. 7.14 Sample graph for Exercise 1

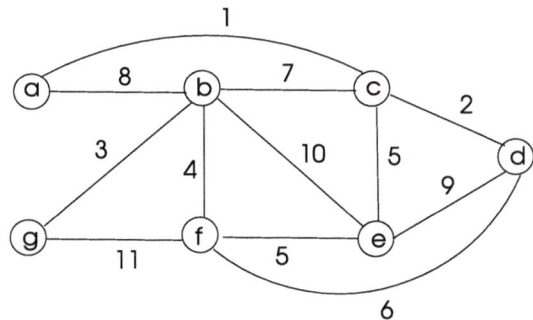

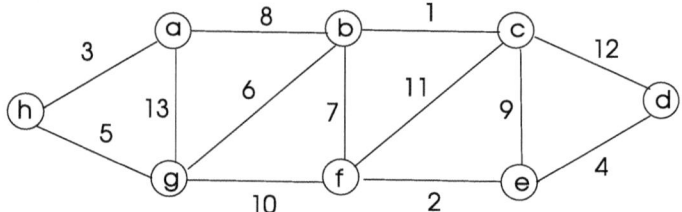

Fig. 7.15 Sample graph for Exercise 2

Fig. 7.16 Sample graph for
Exercise 3

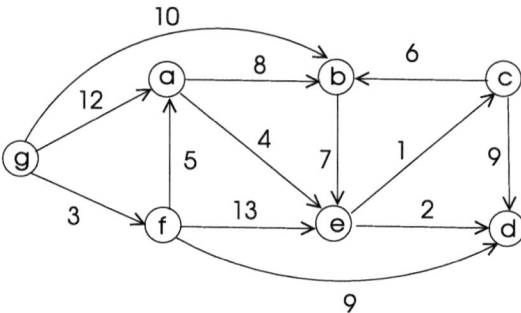

Fig. 7.17 Sample graph for
Exercise 5

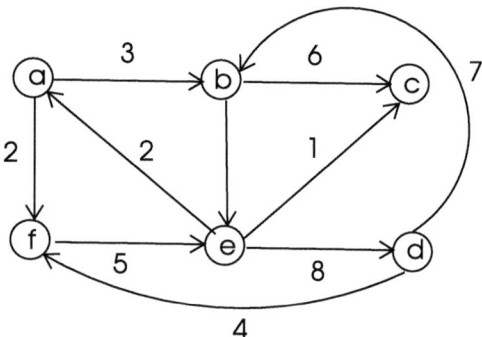

6. Construct all-pairs shortest paths of the digraph depicted in Fig. 7.18 using
   FW_APSP algorithm by showing each iteration.
7. Form the distance matrix D for the graph of Fig. 7.19 and provide a 2-D partitioning of this matrix to 4 processes. Show the data sent by each process during
   parallel running of Floyd–Warshall algorithm for the first two iterations. Work
   out the final D values after k iterations.
8. Modify distributed APSP algorithm *DFW_APSP* pseudocode so that the root
   node may also execute this code by showing starting and ending of each round.

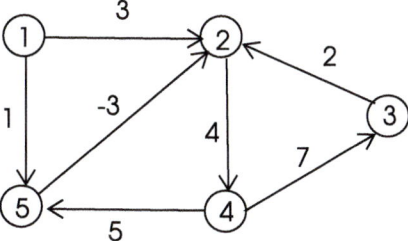

Fig. 7.18 Sample graph for Exercise 6

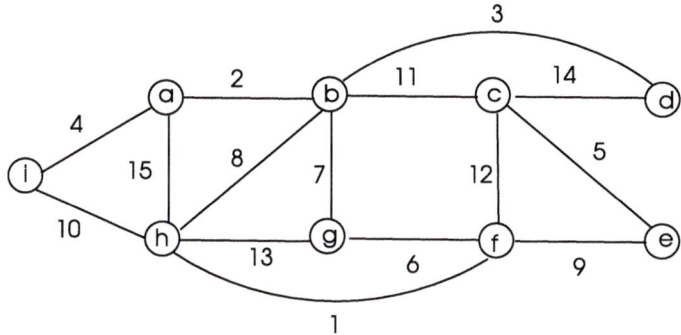

Fig. 7.19 Sample graph for Exercise 7

References

1. Acar UA, Blelloch GE (2017) Algorithm design: parallel and sequential, chap. 18, draft book. Carnegie Mellon University, Department of Computer Science. https://www.parallel-algori thms-book.com/
2. Bellman R (1958) On a routing problem. Q Appl Math 16:87–90
3. Boruvka O (1926) About a certain minimal problem. Prce mor prrodoved spol v Brne III (in Czech, German summary) 3:37–58
4. Chung S, Condon A (1996) Parallel implementation of Boruvka's minimum spanning tree algorithm. Technical report 1297. Computer Sciences Department, University of Wisconsin
5. Cormen TH, Leiserson CE, Rivest RL, Stein C (2009) Introduction to algorithms, chap 23, 3rd edn. MIT Press, Cambridge
6. Dijkstra EW (1959) A note on two problems in connexion with graphs. Numer Math 1:269–271
7. Erciyes K (2013) Distributed graph algorithms for computer networks, chap 6. In: Springer computer communications and networks series. Springer, Berlin
8. Gallager RG, Humblet PA, Spira PM (1983) Distributed algorithms for mininimum-weight spanning trees. ACM Trans Program Lang Syst 5(1):66–77
9. Graham RL, Hell P (1985) On the history of the minimum spanning tree problem. Ann Hist Comput 7(1):4357

10. Grama A, Gupta A, Karypis G, Kumar V (2003) Introduction to parallel computing, chap 10, 2nd edn. Addison Wesley, Boston
11. Kleinberg J, Tardos E (2005) Algorithm design, chap 4, Pearson int. edn. ISBN-13: 978-0321295354, ISBN-10: 0321295358
12. Korte B, Vygen J (2008) Combinatorial optimization: theory and algorithms, chap 7, 4th edn. Springer, Berlin
13. Kruskal JB (1956) On the shortest spanning subtree of a graph and the traveling salesman problem. Proc Am Math Soc 7:4850
14. Loncar V, Skrbic S, Bala A (2013) Parallelization of minimum spanning tree algorithms using distributed memory architectures. In: Transaction on engineering technology. Special volume of the world congress on engineering, pp 543–554
15. Peleg D (1987) Distributed computing: a locality-sensitive approach. In: SIAM monographs on discrete mathematics and applications, chap 5
16. Prim RC (1957) Shortest connection networks and some generalizations. Bell Syst Tech J 36(6):1389–1401
17. Tarjan RE (1987) Data structures and network algorithms. In: SIAM, CBMS-NSF regional conference series in applied mathematics (book 44)
18. Thorup M (2000) Near-optimal fully-dynamic graph connectivity. In: Proceedings of the 32nd ACM symposium on theory of computing, pp 343–350
19. Toueg S (1980) An all-pairs shortest-path distributed algorithm. Technical report RC 8327. IBM TJ Watson Research Center, Yorktown Heights, NY

Connectivity

8

8.1 Introduction

Connectivity is a fundamental concept in graph theory which has both theoretical and practical implications. An undirected graph is connected if there is a path between any pair of its vertices. In a digraph, connectivity implies there is a path between any two of its vertices in both directions. In practice, the study of connectivity is needed for reliable communication networks as connectivity has to be provided in loss of edges (links) or vertices (routers) in these networks. A cut-vertex of a graph G is a special vertex in G removal of which disconnects G. Similarly, removing an edge called bridge of a connected graph G disconnects G. It would be of interest to detect such parts of networks to enhance connectivity around these regions by supplying additional communication devices and links.

We start this chapter by formally defining the parameters of vertex and edge connectivity. We continue by describing algorithms to find cut-vertices and bridges of undirected graphs. Blocks are maximal connected components of a graph without a cut-vertex and we review algorithms to find blocks of graphs. We then review strongly connected components of digraphs along with algorithms to discover them. Connectivity is related to network flows and matching as we will see. Our main goal in a flow network is to find the maximum flow from a source node to a destination node and we show an algorithm to find maximum flow may be used to find how well connected a graph is. A matching of a graph is a set of its disjoint edges and we will see in the next chapter a flow algorithm can be employed to find a maximum matching of a bipartite graph. We also provide parallel and distributed algorithms for most of the topics discussed.

© The Author(s), under exclusive license to Springer Nature Switzerland AG 2026
K. Erciyes, *Guide to Graph Algorithms*, Texts in Computer Science,
https://doi.org/10.1007/978-3-032-05294-0_8

8.2   Theory

We define the basic connectivity parameters in this section.

Definition 8.1 *(connected graph, component)* A graph (directed or undirected) is *connected* if there is a walk between every pair of its vertices. Any graph that does not have this property is *disconnected*. The maximal connected subgraphs of a graph are called *components*.

In other words, a graph $G = (V, E)$ is connected if for every $u, v \in V$, there exists a (u, v) path in G. The vertex-deletion subgraph $F = (V', E')$ of a graph G shown by $G - F$ is obtained by deleting all vertices of V' and their incident edges from G. Similarly, the edge-deletion subgraph $H = (V', E')$ of a graph G shown by $G - H$ is obtained by deleting all edges in E'.

Definition 8.2 *(biconnected graph)* A connected undirected graph $G = (V, E)$ is called *biconnected* if for every vertex $v \in V$, $G - v$ is connected.

That is, a connected and undirected graph is biconnected if it remains connected after removal of any one of its vertices. A cycle, for example, is 2-connected. In practical terms, this means failure of a node in a biconnected computer network will leave it still connected as there will be alternative routes. If a graph is not biconnected, removal of at least one of its vertices will cause it to be disconnected. Such disconnecting vertices are called *cut-vertices* or *articulation points*.

Definition 8.3 *(vertex-cut)* A *vertex-cut* of a connected graph G is a subset V' of its vertices such that $G - V'$ has at least two different components.

When V' consists of a single vertex, this vertex is called the *cut-vertex* (or the *articulation point*) of G. A complete graph K_n of order n does not have a cut-vertex since there is no single vertex removal of which disconnects such a graph. *Edge-cut* of a graph can be defined similarly as follows.

Definition 8.4 *(edge-cut)* An *edge-cut* of a connected graph G is a subset E' of its edges such that $G - E'$ has at least two different components.

When E' consists of a single edge, this edge is called a *cut-edge* or a *bridge*. An edge e is a bridge if and only if it is not included in any cycle of G. If e was part of a cycle, then deleting it from G would not disconnect G. Figure 8.1 displays these concepts.

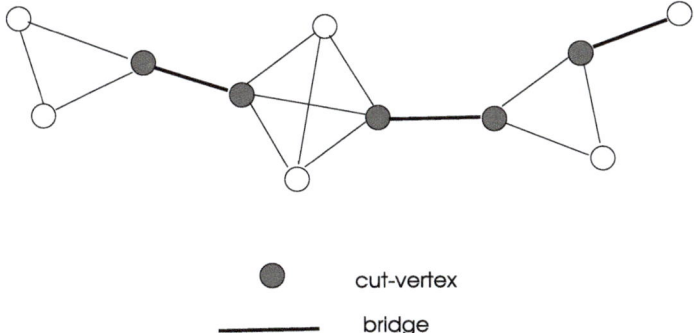

cut-vertex

bridge

Fig. 8.1 Cut-vertices and bridges in a sample graph

8.2.1 Vertex and Edge Connectivity

Definition 8.5 *(vertex connectivity)* The *vertex connectivity* $\kappa(G)$ of a connected graph G is the minimum number of vertices removal of which results in a disconnected or a trivial graph.

It is the cardinality of the minimum vertex-cut of a graph. In a graph G of order n, the maximum degree $\Delta(G)$ will be at most $n - 1$. Hence,

$$0 \le \kappa(G) \le n - 1 \tag{8.1}$$

A graph G is called *k-connected* if $\kappa(G) \ge k$. In other words, a graph is k-connected if removal of k vertices disconnects the graph. Therefore, a *k-connected* graph is also *m-connected* for every integer m with $0 \le m \le k$. For a complete graph K_n, $\kappa(K_n) = n - 1$. We can define edge connectivity similarly as follows.

Definition 8.6 *(edge connectivity)* The *edge connectivity* $\lambda(G)$ (or $\lambda(G)$) of a connected graph G is the minimum number of edges removal of which results in a disconnected graph

A graph G is called *k-edge-connected* if $\lambda(G) \ge k$. Therefore, a *k-edge-connected* graph is also *m-connected* for every integer m with $0 \le m \le k$. For a complete graph K_n, $\lambda(K_n) = n - 1$. For every graph G of order n, we need to remove at most $n - 1$ edges from the highest degree vertex V to make v isolated and hence to have G disconnected. Therefore,

$$0 \le \lambda(G) \le n - 1 \tag{8.2}$$

The vertex connectivity and edge connectivity of a disconnected graph are both 0 since we do not need to remove any vertex or edge to have it disconnected. The vertex connectivity and edge connectivity numbers of the graph in Fig. 8.1 are both unity.

Theorem 8.1 (Whitney [15]) *For every graph G,*

$$\kappa(G) \leq \lambda(G) \leq \Delta(G).$$

Proof A graph G becomes disconnected if all edges incident to a vertex v are removed. The maximum value of edge connectivity will, therefore, be $\Delta(G)$ since edges around the minimum degree vertex v form an edge-cut of G and hence, the inequality at right-hand side holds. In order to prove the left side of the inequality, let us consider, a minimum edge-cut $C \in E$ of G which separates the vertices in G into the subsets S and S'. In the worst case, we would have all vertices in S connected to all vertices in S'. The loose upperbound on connectivity for any graph is that of a complete graph which is $n - 1$. Therefore, $\lambda(G) = |S| \cdot |S'| \leq n - 1$. In the case, when all vertices in S are not connected to all vertices of S', we have at least an edge $(u, v) \notin C$ with $u \in$ and $v \in S'$. □

8.2.2   Blocks

Definition 8.7 *(block)* A *block* or a *biconnected component* of a graph G is a maximal connected subgraph of G without a cut-vertex (articulation point).

Every graph is a union of its blocks. A block B of a graph G may contain a cut-vertices of G although it cannot have a cut-vertex of its own. An edge is a block of a graph G if and only if it is a bridge of G. Therefore, each edge of a tree is its blocks and every isolated vertex of a graph are its blocks. In summary, the blocks of a graph consists of all bi-connected components, all bridges and all isolated vertices. Blocks of a sample disconnected and undirected graph are shown in Fig. 8.2.

Fig. 8.2 Blocks of an undirected graph shown encircled. The bold vertex, for example, is a cut-vertex of the graph but not the cut-vertex of the block it belongs

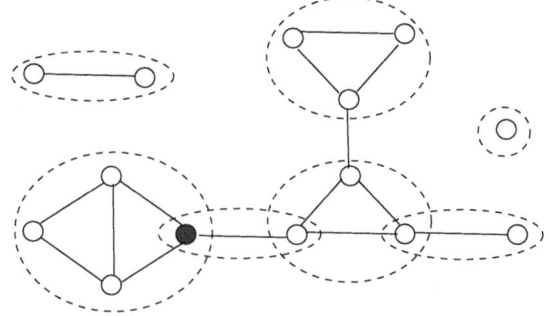

8.2.3 Menger's Theorems

We need to define disjoint paths between two vertices of a graph before stating Menger's theorems for connectivity.

Definition 8.8 *(edge connectivity of two vertices)* Let u and v be two distinct vertices of an undirected graph G. The edge connectivity of vertices u and v, $\lambda(u, v)$, is the least number of edges that are to be deleted from G to have u and v disconnected.

Definition 8.9 *(vertex connectivity of two vertices)* Let u and v be two distinct vertices of an undirected graph G. The vertex connectivity of vertices u and v, $\kappa(u, v)$, is the least number of vertices selected from $V - \{u, v\}$ that are to be deleted from G to have u and v disconnected. We can immediately see that the vertex connectivity of the graph G is the minimum of $\kappa(u, v)$ for each pair of vertices u and v.

Definition 8.10 *(vertex disjoint paths)* Collection of paths between the two vertices u and v of a graph G are called *vertex disjoint* (independent) if they do not share any vertices other than u and v. The greatest number of independent paths between the two vertices u and v is denoted as $\kappa(u, v)$.

Definition 8.11 *(edge-disjoint paths)* Collection of paths between the two vertices u and v of a graph G are called *edge disjoint* (edge-independent) if they do not share any edges. The greatest number of edge-independent paths between the two vertices u and v is denoted as $\lambda(u, v)$.

We will now state Menger's theorems without proving them which provide necessary and sufficient conditions for a graph to be k-connected or k-edge connected.

Theorem 8.2 (Menger's Theorem, vertex version) *Let $\kappa(u, v)$ be the maximum number of vertex disjoint paths between the vertices u and v. A graph is k-connected if and only if each vertex pair in the graph is connected by at least k disjoint paths.*

Theorem 8.3 (Menger's Theorem, edge version) *Let $\lambda(u, v)$ be the maximum number of edge disjoint paths between the vertices u and v. A graph is k-edge-connected if and only if each vertex pair in the graph is connected by at least k edge-disjoint paths.*

8.2.4   Connectivity in Digraphs

Connectivity in digraphs require further specifications as the connection between vertices is not symmetric in such graphs. That is, it may be possible to reach a vertex v from a vertex u but not vice versa. We require that there is a path between every pair of vertices in both directions in a digraph for the connectivity to hold. A strongly connected digraph is defined as follows.

Definition 8.12 *(strongly connected digraph)* A digraph is called *strongly connected* if for every $u - v$ pair of vertices, there is a path from u to v and a path from v to u.

Definition 8.13 *(strongly connected components of a digraph)* A *strongly connected component* (SCC) of a directed graph is a maximal subset of vertices containing a directed path from each vertex to all others in the subset.

The following properties for SCCs in digraphs can be observed.

• Every vertex belongs to exactly one SCC
• Any two SCCs are disjoint
• The SCCs of a graph G form a partition of G.

Definition 8.14 *(weakly connected digraph)* A digraph is *weakly connected* if its underlying undirected graph is connected.

8.3   Sequential Connectivity Algorithms

We will review algorithms to find connected components, articulation points, bridges, SCCs, and blocks of graphs in the next sections.

8.3.1   Finding Connected Components

We can always check connectivity of an undirected graph $G = (V, E)$ by running DFS or BFS from an arbitrary vertex v and recording the visited vertices in a list V' during the search. If visited vertex set $V' = V$, then G is connected. This algorithm works since these searches will always visit every vertex that is connected via a path to the source vertex, forming a spanning tree rooted at the source in the end. Time spent will be $O(n + m)$ as in the DFS or BFS algorithm. We can find the connected components of an undirected graph using the DFS algorithm with a simple modification; run DFS on the graph to get a forest; each tree in the forest formed by a call from the main program is then a connected component as shown in this modified version of DFS in Algorithm 8.1. Every time a return is performed from the DFS procedure, all of the vertices in that component have been visited. We can actually label the vertices with the components they are in by defining an array $label[1 \ldots n]$, where $label[i]$ shows the number of the component vertex i belongs. Time taken to find the components of the graph G using the DFS algorithm is $O(n + m)$.

Algorithm 8.1 *DFS_Component*

1: **Input** : $G(V, E)$, an undirected graph
2: **Output** : $\mathcal{C} = \{C_1, C_2, \ldots, C_k\}$ ▷ components of G
3: **boolean** $visited[1 \ldots n]$
4: $count \leftarrow 0$
5: **for all** $u \in V$ **do**   ▷ initialize
6:     $visited[u] \leftarrow false$
7: **end for**
8: **for all** $u \in V$ **do**
9:     **if** $visited[u] = false$ **then**
10:        $count \leftarrow count + 1$
11:        $DFS(u)$ ▷ call for each connected component
12:    **end if**
13: **end for**
14:
15: **procedure** $DFS(u)$
16:     $visited[u] \leftarrow true$ ▷ first visit
17:     $label[u] \leftarrow count$
18:     $C_{count} \leftarrow C_{count} \cup \{u\}$
19:     **for all** $(u, v) \in E$ **do**   ▷ visit neighbors
20:         **if** $visited[v] = false$ **then**
21:             $DFS(v)$
22:         **end if**
23:     **end for**
24: **end procedure**

8.3.2 Articulation Point Search

An articulation point or a cut-vertex of an undirected graph G is a vertex-cut consisting of a single vertex, hence removing of such vertex will make G disconnected. A *router* is the basic building block of a computer network directing messages coming from its input ports to its output ports. A router that is an articulation point in a computer network is a single point of failure breakdown of which will cause a disconnected and hence, a deficit network. We need to find such articulation points in networks to provide additional links around them to make the network more robust to failures. We will first describe a naive algorithm to find articulation points of an undirected graph and then a DFS-based algorithm with better time complexity.

8.3.2.1 The Naive Algorithm

As a simple approach to find the articulation point of a graph G, we can remove vertices one by one from the graph G and check whether G is connected or not by applying DFS or BFS algorithm after each removal, as shown in Algorithm 8.2. If removal of a vertex v leaves G disconnected, then v is an articulation point. The *for* loop is executed n times and the DFS or BFS traversal takes $O(n + m)$ time, resulting in $O(n(n + m))$ time for this algorithm. We would need to search algorithms for better complexities to be used in large graphs.

Algorithm 8.2 *Naive_AP*

1: **Input** : $G = (V, E)$
2: **Output** : articulation points of G in P
3: **set of vertices** $L \leftarrow \emptyset$
4: **for all** $v \in V$ **do**
5:    $G' \leftarrow G - \{v\}$
6:    **run** $DFS(G', u)$ where u is any vertex in G'
7:    **record** the visited vertices in L
8:    **if** $V' \neq L$ **then**
9:        $P \leftarrow P \cup \{v\}$
10:    **end if**
11: **end for**

8.3.2.2 DFS-Based Algorithm

Tarjan presented an algorithm to find articulation points of graph G using DFS [13]. Before reviewing this algorithm, let us recall the *back edge* property in a DFS tree. A back edge (u, v) of a vertex w in a DFS tree is an edge from any vertex v of the subtree rooted at w to any ancestor vertex u of w in the tree. From this definition, we can see that edge (u, v) is not part of the DFS tree as it forms a loop.

Remark 4 A vertex w with a back edge (u, v) in a DFS tree of a graph G cannot be an articulation point as removal of w does not leave G disconnected.

We can see this is valid as (u, v) still keeps the graph G connected and conversely, removal of a vertex w that does not have a back edge leaves G disconnected and therefore, w is an articulation point. We now have a property to classify vertices; any vertex that does not have a back edge from a vertex in its subtree to one of its ancestors in a DFS tree is an articulation point. The root r of the DFS tree needs special treatment as it has no back edges but it can still be an articulation point if and only if it has more than one child. In such a case, if any two vertices in the subtrees of the children of the root were connected by a non-tree edge, they would be in the same subtree. Therefore, when root r has more than one children, removal of r will leave the graph G disconnected.

Remark 5 The root vertex of a DFS tree of a graph G is an articulation point if and only if it has more than one child.

We can construct an algorithm based on this property only by simply running DFS from each vertex of the graph and checking whether each root has more than one child. This approach requires $O(n(n+m))$ time, however, we can have a better performance by using the back edge property together with this root property as described next.

We need a way to detect back edges and DFS provides this property by the times of visiting the vertices, and hence, the reason for using DFS. We will perform a DFS from any vertex in G and record the discovery times for vertices as they are visited. Let this number be $num(v)$ for a vertex v. We will also record for each vertex v the earliest discovered vertex that is connected to any vertex in the subtree of v and $low(v)$ be the vertex with the lowest number that can be reached from v using 0 or more spanning tree edges and then at most one back edge. We can now see $low(v)$ is the minimum of:

1. $num(v)$ (Rule 1)
2. lowest $num(u)$ among all back edges (v, u) (Rule 2)
3. lowest $low(u)$ among all tree edges (v, u) (Rule 3).

Any vertex v other than the root in the DFS tree is an articulation point if and only if $low(u) \geq num(v)$ for any child u of v meaning there are no back edges from any vertex in the subtree of v to any one of its ancestors. The root is an articulation point if and only if it has more than one child. We can now structure an algorithm based on the foregoing as shown in Algorithm 8.3. The procedure *assign_num* is basically a DFS algorithm which also assigns the *num* values to vertices as they are visited. The second procedure *check_AP* finds the *low* values for vertices by checking the rules above and tests articulation point condition and includes vertices that satisfy this condition in V'.

Algorithm 8.3 *DFS-based_AP*

1: **Input** : connected and undirected graph $G = (V, E)$
2: **Output** : articulation points $V' \subset V$ of G
3: **select** any vertex $v \in V$
4: *counter* $\leftarrow$ 1
5: *assign_num*(v)
6: *check_AP*(v)
7:
8: **procedure** ASSIGN_NUM(vertex v)
9:     $num(v) \leftarrow counter + +$
10:    $visited(v) \leftarrow true$
11:    **for all** $u \in N(v)$ **do**
12:       **if** $visited(u) = false$ **then**
13:          $parent(u) \leftarrow v$
14:          *assign_num*(u)
15:       **end if**
16:    **end for**
17: **end procedure**
18:
19: **procedure** CHECK_AP(vertex v)
20:    $low(v) \leftarrow num(v)$ ▷ Rule 1
21:    **for all** $u \in N(v)$ **do**
22:       **if** $num(u) \geq num(v)$ **then**
23:          *check_ap*(u)
24:          **if** $low(u) \geq num(v)$ **then**
25:             $V' \leftarrow V' \cup \{v\}$ ▷ AP found
26:          **end if**
27:          $low(v) \leftarrow min(low(v), low(u))$ ▷ Rule 3
28:       **else if** $parent(v) \neq u$ **then**
29:          $low(v) \leftarrow min(low(v), num(u))$ ▷ Rule 2
30:       **end if**
31:    **end for**
32: **end procedure**

Running of this algorithm is shown in Fig. 8.3. A simple graph with two artic-ulation points b and d is given in (a). We form a DFS tree shown in (b) for this graph and label every vertex v with $num(v)$ and $low(v)$ as described. For example, vertex g has 7, 1 since it has been discovered last in the DFS and the back edge (g, b) connects it to vertex b which has a num value of 1, therefore, its low value is set to 1. We find d has a descendant vertex e which has a low value of 4 which is equal to the num value of d, therefore, vertex d is an articulation point. Vertex b is an articulation point simply because it has more than one child. Other possible DFS trees rooted at vertices c and e are shown in (c) and (d) of the same figure. We find vertices b and d are again articulation points in both with the same rea-soning as above. The runtime of this algorithm is simply the time it takes for DFS which is $O(n + m)$.

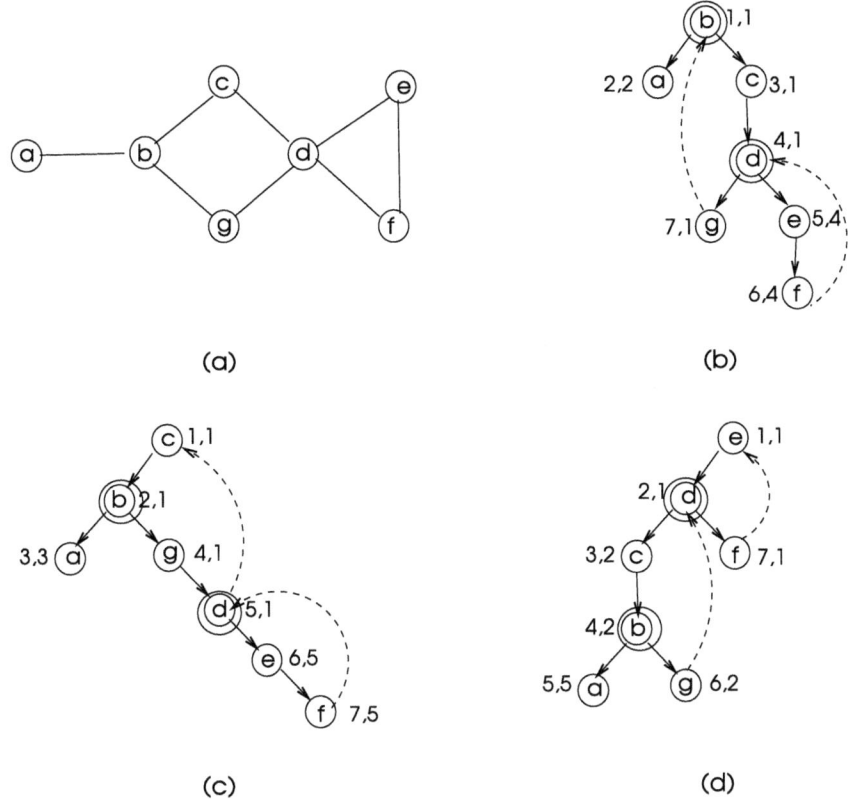

Fig. 8.3 Running of DSF-based AP algorithm in a sample graph for three different DFS trees formed. The articulation points found are vertices b and d in all cases shown in double circles

8.3.3   Block Decomposition

A block or a biconnected component of a graph G is a maximal biconnected subgraph of G. We note that each block of G is connected to one or more blocks by articulation points and an articulation point belongs to more than one block. With this observation, we can structure an algorithm similar to the DFS-based articulation point algorithm as described next.

Hopcroft–Tarjan Algorithm

Hopcroft and Tarjan provided an algorithm to find blocks of a graph using DFS as in the articulation point algorithm [10]. The main idea of this algorithm is the key observation that the blocks are separated by the articulation points of the graph. Note that an articulation point of a graph is not an articulation point of any block it belongs since a block does not contain an articulation point of its own. We can, therefore, discover articulation points in the graph and all vertices between any two articulation points will be a block. This algorithm uses this fact and operates similarly to the DFS-based articulation point finding algorithm with the exception that we push edges visited in a stack until we discover such a cut-vertex and pop all the vertices of edges from the stack into a block data structure when we do. The variables *num* and *low* as in the articulation point algorithm are used and the algorithm consists of the following steps.

1. Start DFS from an arbitrary vertex s of the graph $G = (V, E)$. Set *counter* $\leftarrow$ 1 and $num(s) \leftarrow 1$, $low(s) \leftarrow 1$.
2. Perform DFS as usual and whenever a neighbor vertex v of the vertex u under consideration is encountered, check the edge (u, v).
   a. The vertex v is discovered for the first time and thus (u, v) is a tree edge. Increment counter and set $num(v) = counter$, $low(v) = num(v)$. Push the edge (u, v) onto stack S.
   b. The vertex v has been visited before and $num(v) < num(u)$. Therefore the edge (u, v) is a back edge. Set $low(u) = min\{low(u), num(v)\}$. Push the edge (u, v) onto stack S.
   c. The vertex v has been visited before with $num(v) > num(u)$. Thus, the edge (u, v) is a forward edge. This is valid only when G is a digraph. Since the edge (u, v) has already been processed in this case, we do nothing.
3. Upon backtracking from a vertex v that was searched by using the edge (u, v), set $low(u) = min\{low(u), low(v)\}$. If $low(v) \geq num(u)$, vertex u is an articulation point as in Algorithm 8.3. In this case, pop all edges from the stack S up to and including the edge (u, v). The vertices incident to these edges will form a block of G.
4. When a return from the source vertex s is performed, pop all remaining edges from the stack S and include all incident vertices on these edges in a single block.

Algorithm 8.4 shows a possible pseudocode of this algorithm in more detail. We assume the graph may consist of more than one component.

Algorithm 8.4 *DFS_Block*

1: **Input** : An undirected graph $G = (V, E)$
2: **Output** : Block set $\mathcal{B} = \{B_1, \ldots, B_k\}$
3: *counter* $\leftarrow 1$; *bl_cnt* $\leftarrow 1$ ▷ DFS and block counters
4: **for all** $u \in V$ **do**
5:    *visited*$(u) \leftarrow false$
6: **end for**
7: **select** an arbitrary vertex r to start DFS
8: *num*$(r) \leftarrow 1$; *low*$(r) \leftarrow 1$
9: $DFS\_Block(r)$
10: **pop** all remaining edges on S to B_{bl_cnt}
11: **for all** $w \in V$ **do**   ▷ check other components
12:    **if** $\neg visited(w)$ **then**
13:       $DFS\_Block(w)$
14:       **pop** all remaining edges on S to B_{bl_cnt}
15:    **end if**
16: **end for**
17:
18: **procedure** DFS_BLOCK(u)
19:    *visited*$(u) \leftarrow true$
20:    *counter* $\leftarrow counter + 1$
21:    *num*$(u) \leftarrow counter$; *low*$(u) \leftarrow num(u)$
22:    **for all** $v \in N(u)$ **do**
23:       **if** *visited*$(v) = false$ **then**   ▷ (u, v) is a tree edge
24:          $Push(S, (u, v))$
25:          *parent*$(u) \leftarrow v$
26:          $DFS\_Block(v)$
27:          **if** *low*$(v) \geq num(u)$ **then**   ▷ u is an articulation point
28:             $Form\_Block((u, v))$ ▷ pop edges of the block
29:          **end if**
30:          *low*$(u) = \min\{low(u), low(v)\}$
31:       **else if** *parent*$(u) \neq v$ **then**   ▷ (u, v) is a back edge
32:          $Push(S, (u, v))$
33:          *low*$(u) = \min\{low(u), num(v)\}$ ▷ correct low value
34:       **end if**
35:    **end for**
36: **end procedure**
37:
38: **procedure** FORM_BLOCK$((u, v))$
39:    **repeat**
40:       $(x, y) \leftarrow Pop(S)$
41:       $B_{bl\_cnt} \leftarrow B_{bl\_cnt} \cup \{x, y\}$
42:    **until** $(x, y) = (u, v)$
43:    $\mathcal{B} \leftarrow \mathcal{B} \cup B_{bl\_cnt}$
44:    *bl_cnt* $\leftarrow bl\_cnt + 1$
45: **end procedure**

Operation of this algorithm in a small graph is depicted in Fig. 8.4. DFS is run from vertex a and all the DFS tree edges shown in bold pointing to parents are pushed onto stack S. The edge (d, b) is a back edge since $low(b) < low(d)$. The edge (d, b) is pushed onto S and the low value for vertex d is corrected to 2. Upon return from vertex d to c, the low value of vertex c is corrected and articulation point condition is checked which is false. Next, upon return from vertex c to b, no correction of $low(b)$ is needed, however, vertex b is an articulation point since $low(c) \geq num(b)$ and thus we call the block forming procedure which pops edges from the stack S until and including the edge (b, c). The vertices in the first block B_1 are b, c and d as shown encircled in the graph and bold in the stack. In (b), we continue with the DFS and include edges (b, e) and (e, f) in the DFS tree shown in bold and push these edges onto stack S. The vertices e and f are found to be the articulation points and block B_2 with vertices e and f and block B_3 with vertices b and e are formed. Finally, return from the source vertex a means we remove all the remaining edges which is only the edge (a, b), to form the last block B_4. The time complexity of this algorithm is $O(n + m)$ due to the DFS performed.

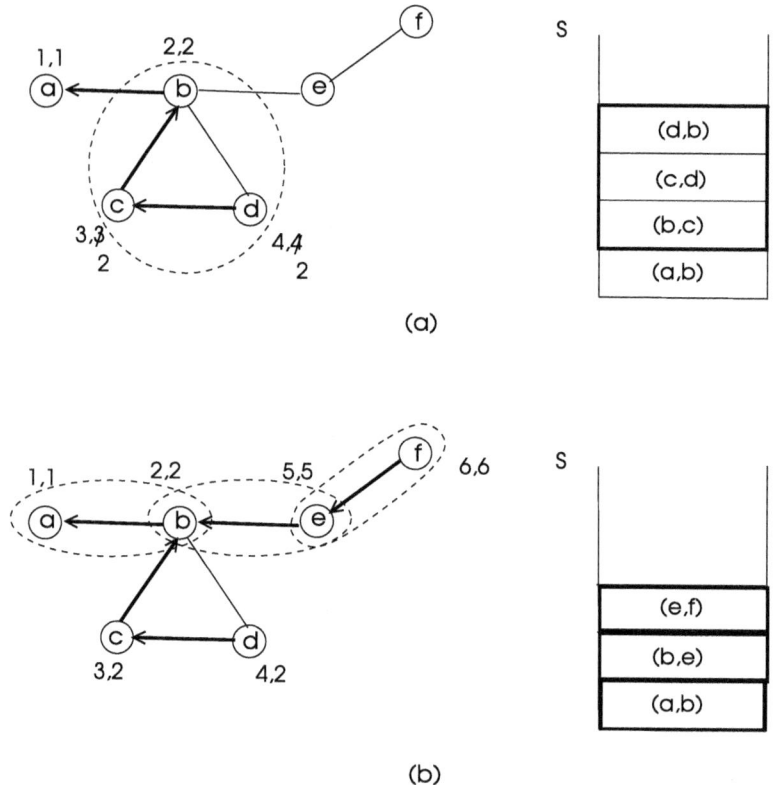

Fig. 8.4 Running of Hopcroft–Tarjan algorithm in a small graph

8.3.4 Finding Bridges

A bridge is an edge of a graph G removal of which increases the number of connected components of G. When a graph G is connected, such removal leaves it
disconnected. A computer network has links between its routers and these links
may fail causing an interruption in network message transfers. We need to find
these deficit links and provide alternative paths around them for a more reliable
network. As another example, a bridge connecting two persons A and B in a connected social network graph shows A and B are friends and if this friendship breaks
up, the social network will have two components and persons in one component
will not have a connection to the other one. Let us review some useful properties
of bridges of a graph.

- Removing an edge that is part of a cycle of a graph G does not disconnect G
  and hence, an edge (u, v) is a bridge if and only if (u, v) is not contained in any
  cycle.
- Consider a bridge (u, v) of a graph G. The vertices u or v are articulation points
  of G if they have a degree greater than 1.

We can apply the same strategy to find bridges of an undirected graph $G = (V, E)$
as we did in finding the cut-vertices; remove each edge one-by-one and check the
connectivity of the graph using DFS or BFS after each removal. If G becomes
disconnected after removing an edge e, this edge e is a bridge (cut-edge) of G.
We need to execute the loop for each edge for a total of m times and checking
connectivity takes $O(n + m)$ time by DFS or BFS resulting in a total time of
$O(m(n + m))$ for this algorithm. Again, this method is not favorable for large
graphs and we look for algorithms with better performances.

Tarjan's Bridge Finding Algorithm

Tarjan provided a linear time algorithm to discover bridges in a graph using the
back edge principle as before [14]. We note that any DFS tree edge (u, v) that has
a back edge from any vertex from the subtree rooted at v to u or any ancestor of u
forms a cycle containing the edge (u, v) and hence (u, v) can not be a bridge. The
algorithm proposed by Tarjan consists of the following steps.

1. Perform a DFS of the graph $G = (V, E)$ from any vertex of G to obtain the
   DFS tree T and label each vertex v with $num(v)$ with respect to its first visit
   time.
2. For each vertex $v \in V$ do the following.
   a. Compute the number of descendants $ND(v)$ of v. This is the number of
      children of v plus 1 as a vertex itself is counted.
   b. Compute $low(v)$ which is the lowest num value reached from v using tree
      edges and at most one back edge.
   c. Compute $high(v)$ which is the highest num value reached from v using tree
      edges and at most one back edge.

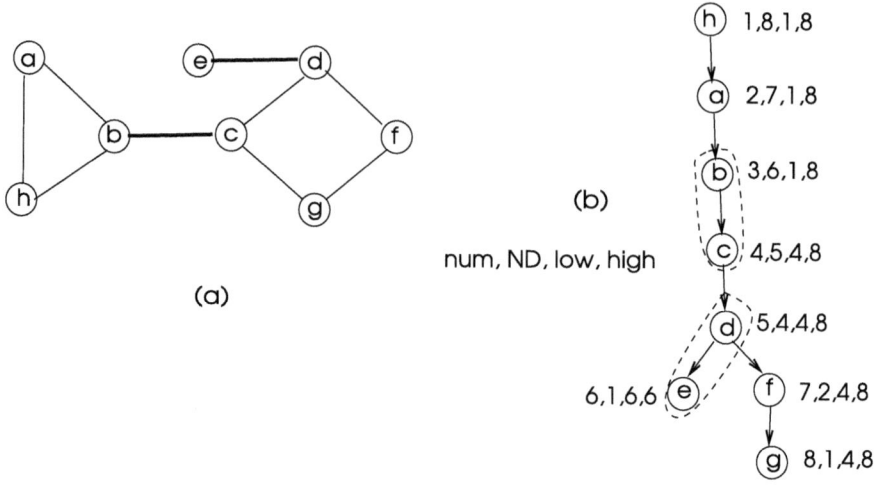

Fig. 8.5 Tarjan's bridge finding algorithm execution in a sample graph

d. *Bridge Condition*: Tarjan showed that for an edge $(u, v) \in T$ with u being parent of v; if $low(v) = num(v)$ and $high(v) < num(v) + ND(v)$ then (u, v) is a bridge [14].

We need to test this condition for every vertex in the DFS tree. In order to do so, we will run the DFS algorithm in the graph G and record the discovery time ($num(v)$) for each vertex v. Then, we compute the values of $ND(v)$, $low(v)$ and $high(v)$ for each vertex and check the bridge condition. Running of this algorithm in a small graph is shown in Fig. 8.5. The edges (b, c) and (d, e) shown in bold are the two bridges of the graph in (a) as can be seen. We run the DFS algorithm and compute the *num, ND, low* and *high* values for each vertex as shown next to each vertex in the DFS tree in (b). Then, we check the bridge condition for every vertex v incident on edge (u, v); $low(v) = num(v)$ and $high(v) < num(v) + ND(v)$. Only vertices c and e satisfy this condition and hence (b, c) and (d, e) are the bridges of this graph. The running time is simply the time for the DFS which is $O(n + m)$.

8.3.5   Strong Connectivity Check

A digraph can be used to model a finite-state machine and strong connectivity in such a digraph implies recovery from a malfunction state as there is always a path from every state to another. We may want to find strongly connected people in a social network who are close friends to analyze such a network. We can check whether a digraph $G = (V, E)$ is strongly connected or not by selecting an arbitrary vertex v, running DFS (or BFS), reversing the direction of edges to obtain the transpose graph G^T and then running DFS (or BFS) from that vertex in G^T again. If the visited vertices in both directions equals V, G is strongly connected.

This method is sufficient as it is possible to get from any vertex u to w via v. The digraph may not be strongly connected, in this case, this algorithm determines the strongly connected component containing the start vertex v. This component called V_c has the common vertices visited during DFS or BFS of G and then G^T as shown in Algorithm 8.5. Since we run DFS or BFS in both directions, the time required for this algorithm is $O(n + m)$.

Algorithm 8.5 *Strong_Conn*

1: **Input** : $G = (V, E)$
2: **Output** : Show whether G is strongly connected
3: **set of vertices** $V_1 \leftarrow \emptyset$, $V_2 \leftarrow \emptyset$, $V_c \leftarrow \emptyset$
4: **pick** any $v \in V$
5: **run** $DFS(G, v)$ (or $BFS(G, v)$)
6: **record** the visited vertices in V_1
7: **reverse** direction of edges to obtain G^T
8: **run** $DFS(G, v)$ (or $BFS(G, v)$)
9: **record** the visited vertices in V_2
10: **if** $V_1 \neq \emptyset \wedge V_2 \neq \emptyset$ **then**
11:    **if** $V_1 = V_2 = V$ **then**
12:        **Output** "G is strongly connected"
13:    **else**
14:        **Output** "G is not strongly connected"
15:    **end if**
16: **end if**

8.3.6   Detecting Strongly Connected Components

Decomposing a digraph into its SCCs is useful in various algorithms as it allows independent runs of the algorithm on each SCC, therefore, allowing parallel processing. There are two fundamental algorithms to detect SCCs in a digraph due to Tarjan [13] and Kosaraju [1]. Both algorithms make use of DFS, Tarjan's algorithm works with a single DFS call while Kosaraju's algorithm requires two DFS calls, however is simpler to implement than Tarjan's algorithm.

8.3.6.1   Tarjan's SCC Algorithm

This algorithm inputs a directed graph $G = (V, E)$ and detects SCCs in this graph [13]. The key idea in this algorithm is the observation that a SCC of a graph is a subtree of a DFS tree. In other words, there will not be a back edge from a subtree rooted at a vertex v from any of its descendants to any of its ascendants if v is the root of a SCC.

The use of a stack appropriately is crucial in the operation of this algorithm. Vertices are pushed on a stack S in the order they are visited. The *invariant property* is that a vertex v is left on the stack S after it is visited if and only if there is a path in G from v to some other vertex already on the stack S. When the call to a vertex

v and its descendants returns, we check whether v has a path to a vertex already on the stack. If there is such a path, the vertex v is left on the stack to maintain the variant. Otherwise, the vertex v is the root of a SCC and thus is popped from the stack together with the SCC that it is assigned as the root.

Each vertex v is assigned $num(v)$ in the order of the DFS first visit and $low(v)$ is the smallest num value to be reached from the vertex v including itself. If $low(v) = num(v)$, vertex v must be removed from the stack S as the root of a SCC. Otherwise, if $low(v) < num(v)$, vertex v must stay on the stack. The pseudocode for this algorithm is shown in Algorithm 8.6.

Algorithm 8.6 *Tarjan_SCC*

1: **Input** : a directed graph $G = (V, E)$
2: **Output** : SCC set $S = \{S_1, \ldots, S_k\}$ of G
3: $i \leftarrow 0; scc\_cnt \leftarrow 1$
4: stack $S \leftarrow \emptyset$
5: **for all** $u \in V$ **do**
6:     $num(u) \leftarrow 0$
7: **end for**
8: **for all** $u \in V$ **do**
9:     **if** $num[u] = 0$ **then**
10:         $SCC(u)$
11:     **end if**
12: **end for**
13:
14: **procedure** SCC(v)
15:     $low(v) \leftarrow num(v) \leftarrow i; i \leftarrow i + 1$ ▷ initialize v
16:     **push** v on S
17:     **for all** $w \in N(v)$ **do**
18:         **if** $num(w) = 0$ **then**   ▷ (v, w) is a tree edge
19:             $SCC(w)$
20:             $low(v) \leftarrow min(low(v), low(w))$
21:         **else if** $num(w) < num(v)$ **then**   ▷ (v, w) is a frond or a cross edge
22:             **if** $w \in S$ **then**
23:                 $low(v) \leftarrow min(low(v), num(w))$
24:             **end if**
25:         **end if**
26:     **end for**
27:     **if** $low(v) = num(v)$ **then**   ▷ v is the root of a component
28:         **while** w is on top of $S \wedge num(w) \geq num(v)$ **do**
29:             **delete** w from S
30:             $S_{scc\_cnt} \leftarrow S_{scc\_cnt} \cup \{w\}$
31:         **end while**
32:         $S \leftarrow S \cup S_{scc\_cnt}$
33:         $scc\_cnt \leftarrow scc\_cnt + 1$
34:     **end if**
35: **end procedure**

This algorithm is a modified DFS procedure that requires $O(n+m)$ time. Testing to find whether a vertex is on the stack can be performed in constant time if a Boolean array is maintained for entries on the stack, as proposed by the author [13].

8.3.6.2   Kosaraju's Algorithm

We can use the transpose of a digraph to find its SCCs based on the observation that the graph G and its transpose G^T has exactly the same SCCs. The algorithm due to Kosaraju is based on the contraction of a digraph defined below.

Definition 8.15 *(contraction of a digraph)* The contraction of a digraph G is another digraph G^{SCC} with SCCs of G as *super vertices* $C_1, \ldots, C_k$ with edges defined as follows. If there is an edge in G from a vertex u in SCC C_x to a vertex v in C_y, then C_x and C_y are connected by an edge in G^{SCC}.

We observe that the contracted digraph G^{SCC}, commonly called *component graph* of G, has no cycles, in other words, it is a directed acyclic graph. If there was a cycle between SCCs, they could be contracted into a larger SCC. Kosaraju's algorithm is based on the idea that same SCCs exist in a graph G and its transpose G^T. We show the high-level description of this algorithm in Algorithm 8.7. It consists of two phases, we first perform a DFS on G to form a DFS forest and place the vertices on a stack with respect to their finish times during DFS. In the second phase, we remove a vertex from stack and perform a DFS on the graph transpose G^T. The second call to DFS is, in fact, to visit the vertices in G^{SCC}. When the search ends, we have all the vertices of a SCC visited. We then continue with the next vertex from the stack until all vertices are visited and placed on the SCCs.

Algorithm 8.7 *Kosaraju_SCC*

1: **Input** : $G = (V, E)$
2: **Output** : SCCs of G
3: **stack** $S \leftarrow \varnothing$
4: **while** $S \neq V$ **do**
5:     **pick** an arbitrary vertex $v \notin S$
6:     **run** $DFS(G, v)$ by putting finished vertices on stack S
7: **end while**
8: **reverse** direction of edges to obtain G^T
9: **while** $S \neq \varnothing$ **do**
10:     $u \leftarrow pop(S)$
11:     **run** $DFS(G, u)$ ▷ form a DFS tree for each vertex on S
12: **end while**
13: vertices in each tree rooted at stack vertices are the SCCs of G

A sample digraph and the operation of the first phase of this algorithm is depicted in Fig. 8.6.

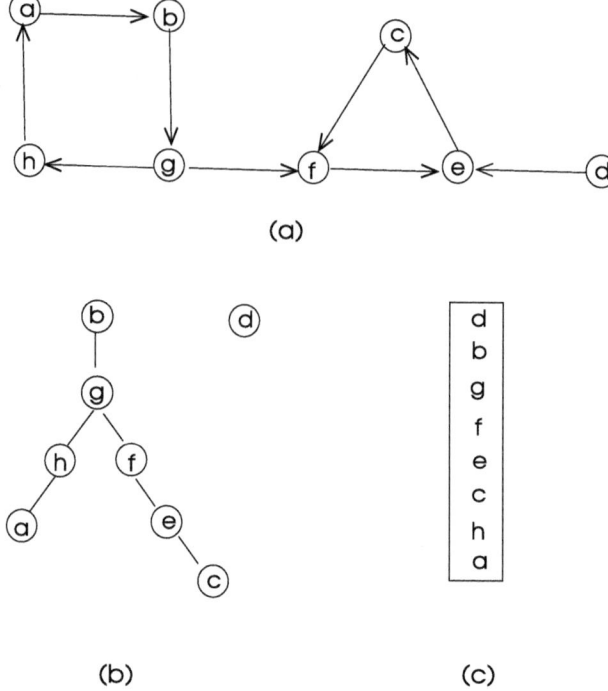

Fig. 8.6 Kosaraju's algorithm first phase; **a** the digraph, **b** DFS tree formed, **c** final contents of the stack

The second phase of the algorithm pops vertices from the stack and performs a DFS on these vertices to obtain SCCs as shown in Fig. 8.7.

Analysis
Given a digraph $G = (V, E)$ with two distinct SCCs C_1 and C_2, consider an edge $(u, v) \in E$ with $u \in C_1$ and $v \in C_2$. We then have the following observation.

Remark 6 It can be shown that $fin(C_1) > fin(C_2)$. Similarly, if (u, v) is an edge in G^T with $u \in C_1$ and $v \in C_2$, then $fin(C_2) > fin(C_1)$ [3].

Theorem 8.4 *Kosaraju's algorithm correctly finds the SCCs of a digraph $G = (V, E)$ in $O(n + m)$ time.*

Proof We will prove the correctness of this algorithm by induction. Let k denote the number of trees formed when DFS is called on G^T. When $k = 0$, the base case holds, and assume the first $k - 1$ trees obtained this way are SCCs of the graph G. Let u be the root of the kth tree and a member of the SCC C_1 of G. For any undiscovered SCC C_x at step k, $fin(C_1) > fin(C_x)$ and all other vertices of C_1 will be descendants of u

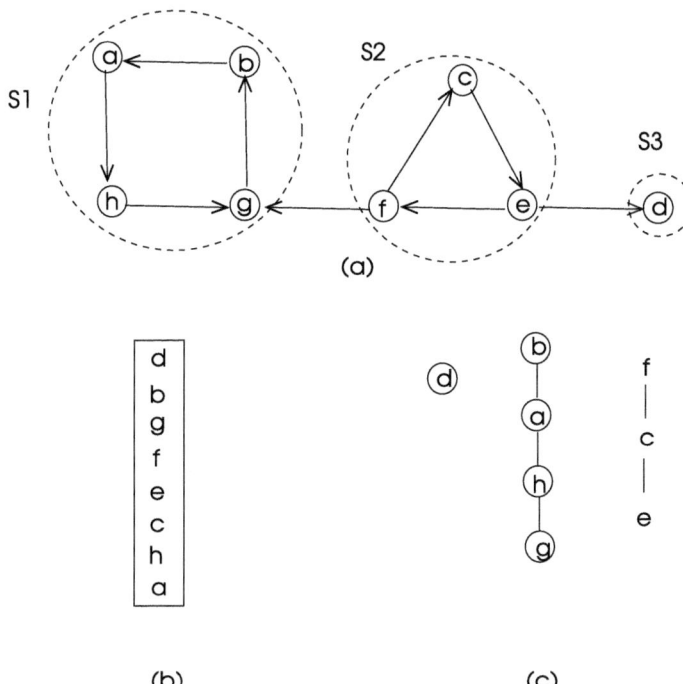

Fig. 8.7 Kosaraju's algorithm second phase; **a** G^T, **b** the stack, **c** SCCs extracted by DFS on stack

in the discovered DFS tree. Any edge that is leaving C_1 in G^T should be directed to SCCs already discovered by the above remark. Therefore, all descendants of u will be only in the SCC C_1 and no other SCCs of G^T [3]. The time complexity of this algorithm is $O(n + m)$ since it involves two DFS calls, first one in G and the second one in G^T.   □

A practical approach to implement Kosaraju's algorithm is depicted in Fig. 8.8. We traverse the graph using DFS and label vertices with their first visit times at numerators shown. Whenever we do a backtrack, the finish times are recorded at denominators as shown. After this first phase is over, we sort the denominators, reverse the direction of the edges of G to obtain G^T and perform a DFS forest starting from the largest denominator vertex. Each discovered component by the DFS is a SCC.

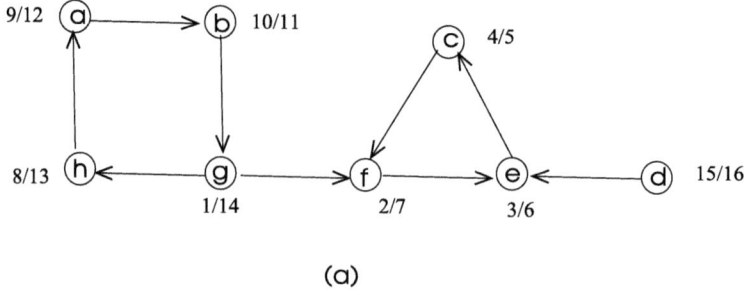

(a)

sorted denomenator: d,g,h,a,b,f,e,c

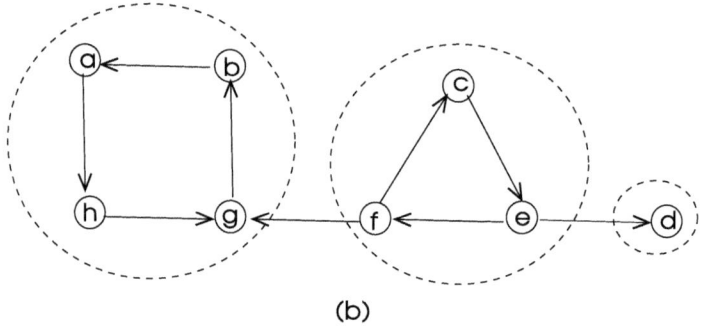

(b)

Fig. 8.8 A practical implementation of Kosaraju's algorithm

8.3.7 Vertex and Edge Connectivity Search

Finding the vertex and edge connectivity numbers of computer networks provide us with the vital information on how reliable they are. Clearly, the larger the connectivity number is, the more robust a network is. We will see efficient ways of finding the vertex and edge connectivity of a graph when we review the network flows in the next section. We review shortly a brute-force algorithm that will find vertex connectivity and then a brute-force algorithm for edge connectivity.

As a first attempt, we can implement the following brute-force strategy. We first find all subsets of the vertices of the network graph; sort these in increasing order and then remove each subset from graph starting from the smaller subsets and check the connectivity of the graph using the BFS (or the DFS) algorithm as shown in the pseudocode of Algorithm 8.8. The *BFS_Conn* algorithm checks the connectivity of the graph using the BFS algorithm and returns true if the graph is connected and false otherwise.

Algorithm 8.8 *Finding k-connectivity value*

1: **Input** : $G(V, E)$, directed or undirected graph
2: **Output** : value of k
3: $\mathcal{P} \leftarrow PowerSet(V)$
4: $\mathcal{P}' \leftarrow Sort(\mathcal{P})$
5: **for all** $s \in \mathcal{P}'$ in increasing order of $|s|$ **do**
6:    $G' \leftarrow G - \{s\}$
7:    $result \leftarrow BFS\_Conn(G')$
8:    **if** $result = false$ **then**
9:       **return** $|s|$
10:   **end if**
11: **end for**

This brute-force method will provide the exact k value for the graph. Although it will work, major problem in practice with this algorithm is its exponential time complexity. For a graph with n vertices, the number of subsets of its power set is 2^n, resulting in 2^n iterations of the *for* loop. The BFS algorithm within each loop iteration also has $O(n + m)$ time complexity resulting in a total time complexity of $O(2^n(n + m))$ which is unacceptable even for moderate size graphs.

The same method can be applied to find the edge connectivity of a graph G, this time by forming the power set of edges of G. The number of the power set is 2^m this time and hence, the total time needed is $O(2^m(n + m))$ in this case.

8.3.8   Transitive Closure

In many cases, we would be interested to find if any two vertices of a graph are connected, that is, there is a path between these two vertices. The below graph derived from the original graph provides this information.

Definition 8.16 *(transitive closure)* The *transitive closure* of a graph $G = (V, E)$ is the graph $G' = (V, E')$ where $(u, v) \in E'$ if there is a path between u and v in the graph G.

The connectivity matrix of the graph defined below provides a suitable representation of a graph that is equivalent to its transitive closure.

Definition 8.17 *(connectivity matrix)* The *connectivity matrix* of a graph $G = (V, E)$ is a matrix C with elements $c_{i,j}$ such that $c_{i,j} = 1$ if there is a path between vertices i and j in G and when $i = j$, and ∞ otherwise.

Thus, finding the transitive closure of a graph is reduced to working out its connectivity matrix. We can set 0 for vertices that are not neighbors instead of ∞ in C to obtain the matrix A and compute the powers of A, say A^k by matrix multiplication using logical *or* and logical *and* instead of scalar multiplication and addition to obtain connectivity of vertices that are $k + 1$ hops away. Since longest path in a graph may be of length $n - 1$ at most, A^{n-1} will be equal to C. Alternatively, running the BFS algorithm for each vertex will provide C or setting weights of edges 1 and running Floyd–Warshall algorithm using the adjacency matrix of the graph with edges having unity weights will also result in the connectivity matrix in $O(n^3)$ time. Warshall's algorithm can also be implemented to find the connectivity matrix C. We have a directed graph $G = (V, E)$ with an adjacency matrix $A[n, n]$, where $A[i, j] = 1$ if $(i, j) \in E$, and compute the matrix C, where $C[i, j] = 1$ if there is a path of length greater than or equal to 1 from i to j as shown in Algorithm 8.9. This algorithm has $\Theta(n^3)$ time complexity due to three nested loops.

Algorithm 8.9 *Warshall's Algorithm*

1: **Input** : $G(V, E)$, directed or undirected graph
2: **Output** : connectivity matrix C
3: **for** $i = 1$ to n **do**   ▷ initialize C
4:     **for** $j = 1$ to n **do**
5:         $C[i, j] \leftarrow A[i, j]$
6:     **end for**
7: **end for**
8: **for** $k = 1$ to n **do**
9:     **for** $i = 1$ to n **do**
10:         **for** $j = 1$ to n **do**
11:             **if** $C[i, j] = 0$ **then**
12:                 $C[i, j] \leftarrow C[i, k] \wedge C[k, j]$
13:             **end if**
14:         **end for**
15:     **end for**
16: **end for**

8.4   Flow-Based Connectivity

Connectivity search that makes use of the network flow algorithms is based on finding the edge or vertex connectivity values between every pair of vertices of a graph. Once we have these values, the edge or vertex connectivity is assigned to the minimum value of the computed values. We will first review network flow method with two basic algorithms to compute flow in a network, and then describe algorithms to find the vertex and edge connectivity using this method.

8.4.1   Network Flows

Let us assume a directed graph $G = (V, E)$ in the usual sense. We will form a *flow network* as follows. Each edge $e \in E$ of G is assigned a nonnegative integer called the *capacity* $c(e)$ and we have a *source* vertex s and a *sink* vertex t. A flow $f(u, v)$ through the edge (u, v) in this network satisfies the following:

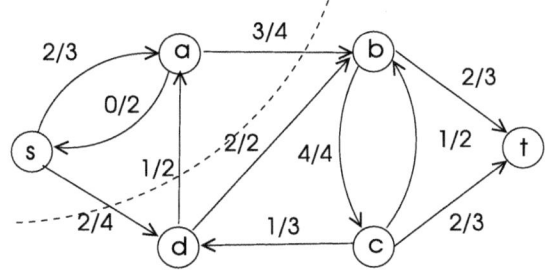

Fig. 8.9 A flow network. A label on an edge is the flow/capacity of the edge

- *Capacity Constraint*: $\forall (u, v) \in E$,

$$0 \le f(u, v) \le c(u, v)$$

which means a flow through an edge may not exceed the capacity assigned to that edge.
- *Flow Conservation*: $\forall u \in \{V - \{s, t\}\}$,

$$\sum_{v \in V} f(v, u) = \sum_{v \in V} f(u, v)$$

In other words, the flow into the vertex u equals the flow out of u for any vertex u except the source vertex s and the sink vertex t.

The value $|f|$ of a flow f is defined as follows.

$$|f| = \sum_{v \in V} f(s, v) - \sum_{v \in V} f(v, s) \tag{8.3}$$

That is, the *flow value* of a flow network is the difference of the sum of flows from the sink vertex s to the sum of flows into s. The *maximum flow problem* is to find a flow with a maximum value in a flow network. An example flow network is depicted in Fig. 8.9.

8.4.1.1 Cuts

Definition 8.18 *(cut)* A cut (S, T) of a flow network divides the network into two sets S and T of disjoint nodes such that $s \in S$ and $t \in T$. The capacity of a cut, $c(S, T)$ is defined as $\sum_{e \in [S,T]} c(e)$.

An edge (u, v) with $u \in S$ and $v \in T$ is called a *forward* edge of the cut (S, T). When $v \in S$ and $u \in T$, the edge (u, v) is said to be a *backward edge*. The flow across a cut (S, T) is the difference between the sum of the flows in forward edges and the sum of the flows in backward edges. The cut shown by a dashed curve in Fig. 8.9 has a flow of $2 + 3 - 1 = 5$ value. Given, a flow network G with any cut (S, T) of G, the following remarks can be made.

Remark 7 The value of flow f in G is equal to the value across the cut (S, T).

Remark 8 The value of flow f across the cut (S, T) does not exceed the capacity of the cut (S, T).

8.4.1.2 Residual Networks

Definition 8.19 *(residual network)* Given a flow f on a graph $G = (V, E)$, the residual network $G_f = (V, E_f)$ has the same set of vertex set as G and edges with positive residual capacities defined as follows.

$$c_f(u, v) = \begin{cases} c(u, v) - f(u, v) & \text{if } (u, v) \text{ is a forward edge} \\ f(u, v) & \text{if } (u, v) \text{ is a backward edge} \end{cases}$$

In other words, the residual capacity of an edge (u, v) is the amount of flow that can be pushed through (u, v) and the residual capacity of the edge (v, u) is the flow that is used. When we are updating a residual network after a flow change, we should always modify these values so that flow conservation at a node of the network is obeyed. For example, if we increase the value of flow through an edge (u, v) by 3 units, then we should increase the value of flow through the edge (v, u) by 3 units to maintain flow network property. Also, flow used by the edge (v, u) may be returned if it will cause a larger network flow by doing so. In summary, G_f has edges that may be utilized to have more flows through them. The residual network of the network of Fig. 8.9 is shown in Fig. 8.10.

Definition 8.20 *(augmenting path)* A path p from the source s to destination t in a residual network G_f is called an *augmenting path* with respect to flow f. We can increase the flow value through an augmenting path P by its residual capacity defined below:

$$c_f(P) = min\{c_f(u, v)\}$$

where the residual capacity $c_f(P)$ of a path P is the minimum residual capacity of its edges. We can push a maximum additional flow through P by the value of $c_f(P)$ as otherwise, we will be violating the capacity constraint in the residual network.

Fig. 8.10 The residual network of the network of Fig. 8.9

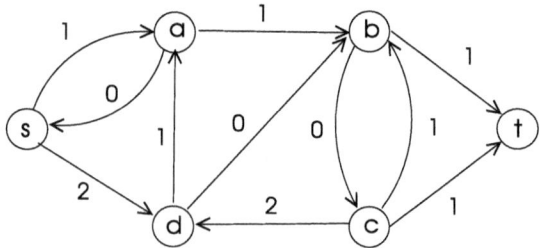

Theorem 8.5 *A flow is maximum if and only if there are no augmenting paths from source s to sink t. The value of this flow $|f| = c(S, T)$ for some cut (S, T) of G.*

Proof If there is an augmenting path from s to t, then we can increase the value of maximum flow f through this path. Therefore, f would not be maximum.

As a consequence, the following theorem can be stated.

Theorem 8.6 *The value of a maximum flow, $|f^*|$, in a flow network G is equal to the capacity of a minimum cut of the network G.*

8.4.1.3 Ford–Fulkerson Algorithm

With this background, we can see that as long as there exists an augmenting path in a residual network, we can increase more flow through that path. Ford–Fulkerson algorithm uses this idea and finds the maximum flow value in a flow network which has a source vertex s and a sink vertex t and positive edge capacities, by gradually incrementing flow in the residual graph. The main idea in this algorithm is to increase the flow through edges of the network by investigating an augmenting path. Whenever such a path is found, flow is augmented by the residual capacity of the path p as shown in Algorithm 8.10.

Algorithm 8.10 *Ford–Fulkerson Algorithm*

1: **Input** : $G(V, E)$, directed or undirected graph
2: **Output** : The value of maximum flow f
3:
4: $f \leftarrow 0$
5: $G_f \leftarrow G$
6: **while** $\exists$ an $(s - t)$ path P of G_f **do**
7:     $c_f(P) \leftarrow \infty$
8:     **for all** $e \in P$ **do**   ▷ compute the residual capacity $c_f(P)$ of path P
9:         **if** $c_f(e) < c_f(P)$ **then**
10:             $c_f(P) \leftarrow c_f(e)$
11:         **end if**
12:     **end for**
13:     **for all** $e \in P$ **do**   ▷ update flows through edges of path P
14:         **if** $c_f(e) > 0$ **then**
15:             $f(e) \leftarrow f(e) + c_f(P)$
16:         **else**
17:             $f(e) \leftarrow f(e) - c_f(P)$
18:         **end if**
19:     **end for**
20: **end while**

Note that when a flow f_x is pushed through an edge (u, v) for the first time, we need to form a new edge (v, u) with label f_x if such an edge does not exist. Figure 8.11 displays the operation of this algorithm in a small network. We start

with 0 flow and the sum of all possible flows to any node u is equal to the sum
of all possible flows from u. We have an arbitrary cut in this network with a value
of 7 as shown and this value is the maximum flow to be attained in this network
as shown by the max-flow min-cut theorem. We then search for augmenting paths
and whenever such a path p is found, flows through all edges of this path are
decreased by the value of the minimum flow along the path p and flow is increased
with this value. We find the edge with the minimum value in the augmenting path
$s - a - b - c - t$ is (s, a) with the value of 3. Hence, flow f is set to 3, the
residual graph is updated to obtain the graph in (c) and proceeding in this manner,
we have the final residual graph in (f) after four iterations which do not have any
augmenting paths and we stop with a final f value of 7 as in the cut.

Analysis
Since flow values are integers, we will be incrementing the flow value $|f^*|$ times
at most where f^* is the maximum flow value, since flow value is incremented by
one at each step in the worst case. Each augmenting path can be found by the DFS
or the BFS algorithm in $O(n + m)$ time resulting in a total time of $O(|f^*|(n + m))$

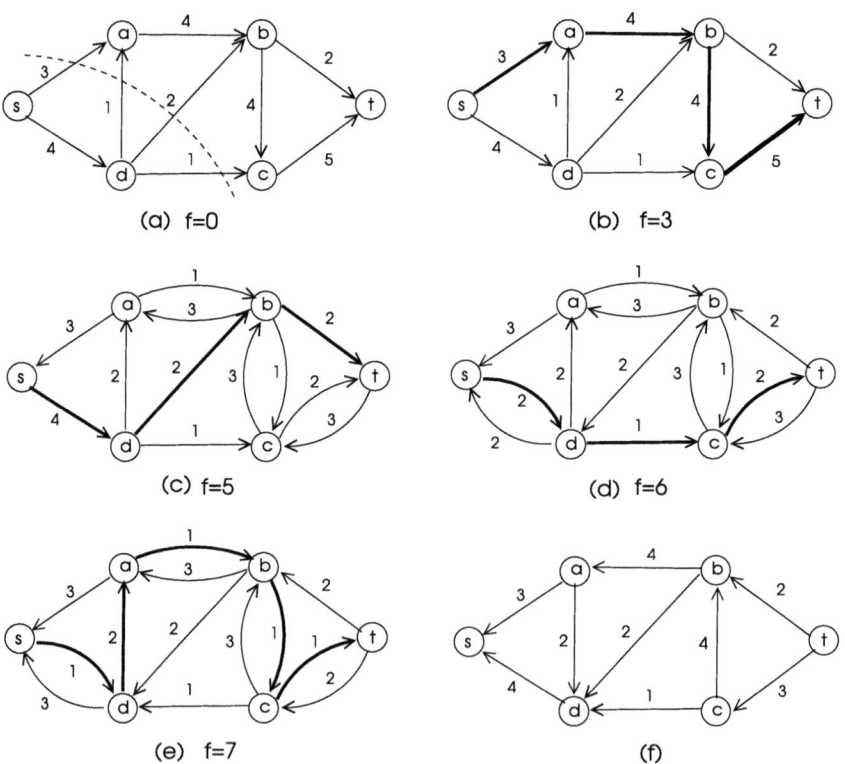

Fig. 8.11 Operation of Ford–Fulkerson algorithm in a small network

time. When the choice of the augmenting paths are done arbitrarily and $|f^*|$ is large, the time complexity of this algorithm may be high.

8.4.1.4 Edmonds–Karp Algorithm

The Edmonds–Karp algorithm works similar to Ford–Fulkerson algorithm with the exception of selecting the augmenting path with the minimal length among all possible augmenting paths. Selection of the shortest augmenting paths is done by the BFS algorithm.

Algorithm 8.11 *Edmond–Karp Algorithm*

1: **Input** : $G(V, E)$, directed or undirected graph
2: **Output** : The value of maximum flow f
3:
4: $f \leftarrow 0$
5: $G_f \leftarrow G$
6: **while** G_f contains an $s - t$ path P **do**
7:     $P \leftarrow$ an $s - t$ path in G_f with minimum length
8:         **augment** f using P
9:         **update** G_f
10: **end while**

The minimum length path p can be determined in $O(n + m)$ time using BFS. Having found the shortest path p, we can augment f in $O(n)$ and update of G_f takes also $O(n)$ time resulting in $\approx O(m)$ time for one iteration of the *while* loop. There are $O(n)$ iterations resulting in a total time of $O(m^2 n)$ [2].

8.4.2 Edge Connectivity Search

The edge connectivity $\lambda(u, v)$ of two vertices u and v of a simple graph G is the least number of edges deletion of which makes u and v disconnected. In an undirected graph, $\lambda(u, v) = \lambda(v, u)$ and in the case of a directed graph this equality may not hold. We can see that when an undirected graph G is not trivial, the edge connectivity of G, $\lambda(G)$, is the minimum value of $\lambda(u, v)$ for each pair of unordered vertices u and v. For a digraph G', $\lambda(G')$ is the minimum value of $\lambda(u, v)$ for each pair of ordered vertices u and v.

With this background, we can compute the value of $\lambda(G)$ for an undirected or a directed graph if we have a method to find the connectivity values for each pair of vertices. This method is in fact based on the maximum flow algorithm we have reviewed above. Even provided an algorithm based on the maximum flow to compute $\lambda(u, v)$ for each pair of vertices and the graph edge connectivity is simply the minimum of all the values computed [6]. The algorithm to find $\lambda(u, v)$ consists of the following steps [4, 6].

1. **Input**: An undirected or a directed graph $G = (V, E)$ and a pair of vertices u and v.
2. **Output**: The value of $\lambda(u, v)$.
3. Form the network $G' = (V', E')$ by implementing the following.
   a. Replace each edge $(x, y) \in E$ with arcs (x, y) and (y, x).
   b. Designate u as the source and v as the sink vertex.
   c. Assign a capacity of 1 to each arc.
4. Find the maximum flow function f in G'.
5. $\lambda(u, v) \leftarrow f$

We have $n(n - 1)/2$ unordered pairs in an undirected graph and $n(n - 1)$ ordered pairs in a digraph to check. Hence, we need to call the above procedure that many times in these graphs. It was shown in [6] that the time complexity of this algorithm is $O(nm)$.

Let us consider the graph G of Fig. 8.12. The edge-cut C shown in dashed line separates the vertices into two subsets of G_1 and G_2. If C is minimum edge-cut and we select a single vertex $a \in G_1$ and check connectivity $\forall v \in G_2$, it can be seen that $\kappa(G)$ is the minimum of these values.

We can, therefore, have an algorithm consisting of the following steps [5]:

1. **Input**: An undirected or a directed graph $G = (V, E)$.
2. **Output**: The value of $\lambda(G)$.
3. Find the minimum edge-cut C of G that separates G into G_1 and G_2.
4. Select an arbitrary vertex $u \in G_1$.
5. Compute $\lambda(u, v), \forall v \in G_2$.
6. $\lambda(G) \leftarrow min\{\lambda(u, v) \mid v \in G_2\}$.

This algorithm requires $n(n - 1)/2$ edge connectivity computations instead of $n(n - 1)$ of the first algorithm, however, a problem with this algorithm is to determine the minimum edge separator C. Even and Tarjan noticed that computation of the minimum edge-cut is not necessary, and hence, we can compute $\lambda(u, v)$

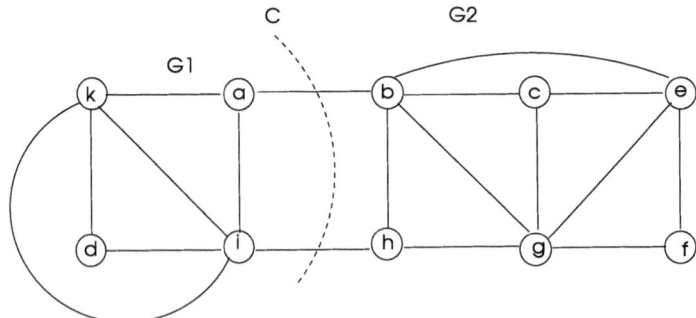

Fig. 8.12 A sample graph

for all vertices $v \neq u$ of the graph G and implement the algorithm above without computing the minimum edge-cut [7]. A set Y that contains vertices both from G_1 and G_2 such that $Y\hat{G}_1 \neq \emptyset$ and $Y\hat{G}_2 \neq \emptyset$ is called a λ-covering of graph G. Selecting vertices from such a small set in line 3 of the above algorithm results in an algorithm with better performance.

8.4.2.1 A Spanning Tree-Based Algorithm

Esfahanian and Hakimi proposed an algorithm to form a spanning tree with a large number of leaves and few intermediate vertices [5]. Consider a spanning tree $T = (V, E(T))$ of a graph G with $L \in V$ as non-leaf vertices. Then L is a λ-covering of G which means both partitions G_1 and G_2 contain at least one vertex that is an intermediate vertex of T. Using this reasoning, selecting a spanning tree of a graph G with as few intermediate vertices as possible results in an algorithm that has to select from only a few such vertices to compute edge connectivity values. The pseudocode of the spanning tree forming algorithm is shown in Algorithm 8.12 where $I(u)$ refers to the set of all edges incident at a vertex u, and $A(u)$ to refers to the set of all vertices adjacent to u.

Algorithm 8.12 *Spanning Tree Algorithm*

1: **Input** : $G(V, E)$
2: **Output** : A spanning tree T of G
3: $V(T) \leftarrow \emptyset; E(T) \leftarrow \emptyset$
4: **select** $u \in V(G)$
5: $V(T) \leftarrow \{u\} \cup A(u)$
6: $E(T) \leftarrow E(T) \cup I(u)$
7: **while** $|E(T)| < |V(T)| - 1$ **do**
8:     select a leaf vertex $w \in T$ such that $|A(w)\hat{(}V(G) - V(T))| \geq |A(x)\hat{(}V(G) - V(T)|, \forall x \in T$
       that is a leaf vertex
9:     **for all** $v \in A(w) \cup ((V(G) - V(T))$ **do**
10:        $V(T) \leftarrow V(T) \cup \{v\}$
11:        $E(T) \leftarrow E(T) \cup \{(w, v)\}$
12:     **end for**
13: **end while**

Once a spanning tree is constructed using Algorithm 8.12, the graph edge connectivity can be computed by the following algorithm.

1. **Input**: An undirected or a directed graph $G = (V, E)$.
2. **Output**: The value of $\lambda(G)$.
3. Construct a spanning tree T of G using Algorithm 8.12.
4. $P \leftarrow$ the minimum of the order of the leaves or the inner vertices of T.
5. Select an arbitrary vertex $u \in P$.
6. Compute $\lambda(u, v), \forall v \in P \setminus \{u\}$.
7. $c \leftarrow min\{\lambda(u, v)\}$.
8. $\lambda(G) \leftarrow min\{c, \delta(G)\}$.

8.4.2.2   A Dominating Set-Based Algorithm

A dominating set of a graph $G = (V, E)$ is the set $V' \subset V$ of its vertices such that any vertex $v \in V$ is either an element of V' or a neighbor of a vertex in V'. Edge connectivity of a graph G can be computed by using a dominating set of G as shown by Matula [11]. A dominating set D of a graph is first formed and a vertex $u \in D$ is selected. The edge connectivity values for u and v, $\forall v \in D$ are then computed using the maximum flow algorithm. The edge connectivity of the graph G, $\lambda(G)$, is then the minimum of the minimum λ value and the smallest degree of the graph as in the above algorithm. Clearly, a smaller dominating set of G results in better performance. Finding the minimum order dominating set of a graph is NP-hard, however, finding a minimal dominating set which is not contained in any other dominating set of G can be found using a greedy approach. Algorithm 8.13 shows how to compute a small dominating set. Matula showed that using this dominating set algorithm $\lambda(G)$ can be computed in $O(nm)$ time [11].

Algorithm 8.13 *Minimal Dominating Set Algorithm*

1: **Input** : $G(V, E)$
2: **Output** : A minimal dominating set D of G
3:
4: **select** $u \in V$
5: $D \leftarrow \{u\}$
6: $G' \leftarrow G - u$
7: **while** $G' \neq \emptyset$ **do**
8:     select a vertex v in G'
9:     $D \leftarrow D \cup \{v\} \cup N(v)$
10:    $G' \leftarrow G - D$
11: **end while**

8.4.3   Vertex Connectivity Search

The vertex connectivity $\kappa(u, v)$ of two vertices u and v of a simple graph $G = (V, E)$ is the least number of vertices deletion of which makes u and v disconnected. When $(u, v) \in E$, $\kappa(u, v) = n - 1$. The method employed to find vertex connectivity is similar to the edge connectivity computation. We can find the edge connectivity values of all vertex pairs in G using maximum flow method and assign the minimum of these values as the edge connectivity of G as shown in [6]. This algorithm we will call Even's algorithm consist of the following steps.

1. **Input**: An undirected or a directed simple graph $G = (V, E)$ and a pair of vertices u and v.
2. **Output**: The value of $\kappa(G)$.
3. Form the network $G' = (V', E')$ by implementing the following.

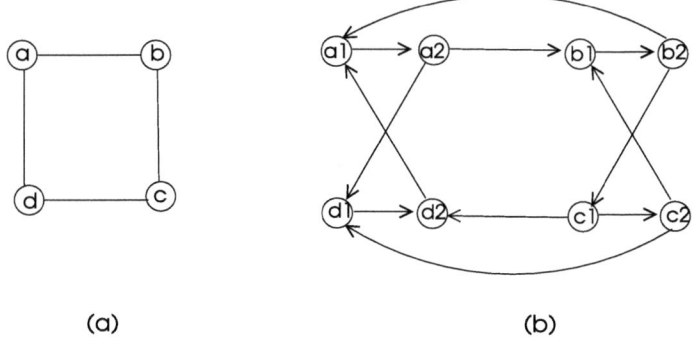

(a)   (b)

Fig. 8.13 Obtaining the expanded directed graph G' in **b** from the sample graph G in **a**

    a. Replace each vertex a with two vertices $a_1, a_2 \in E'$ which are connected by an edge $e_{a_1 a_2}$. Each edge $(a, b) \in E$ is then replaced by two edges $e_1 = (a_2, b_1)$ and $e_2 = (b_2, a_1)$.

    b. Designate u as the source and v as the sink vertex.

    c. Assign a capacity of 1 to each arc in G'.

4. Find the maximum flow function f in G'.

5. $\kappa(u, v) \leftarrow f$.

The G' graph obtained this way will have $2n$ vertices and $2n + m$ edges. Figure 8.13 displays the directed graph G' obtained from an undirected graph G using this procedure. The arcs in G' are labeled with unity values and the maximum flow in this network from vertex u_2 to v_1 is computed. It was shown in [6] that the time complexity of the above algorithm is $O(mn^{2/3})$.

Even and Tarjan showed that we do not need to find $\kappa(u, v)$ for each pair of vertices u and v and we need only compute values for the set $V' \subset V$ of vertices with $|V'| = \kappa + 1$ and update the minimum value of $\kappa(u, v)$ as we do as shown in Algorithm 8.14 [7].

Algorithm 8.14 *Even–Tarjan Algorithm*

1: **Input** : $G(V, E)$, directed or undirected graph
2: **Output** : The value of $\kappa(G)$
3:
4: $min\_conn \leftarrow n - 1$
5: $i \leftarrow 1$
6: **while** $i \leq min\_conn$ **do**
7:     **for** $j = i + 1$ to n **do**
8:         **if** $i > min\_conn$ **then**
9:             **break**
10:         **else if** v_i and v_j are not adjacent **then**
11:             compute $\kappa(v_i, v_j)$ using Even's algorithm
12:             $min\_conn \leftarrow \min\{min\_conn, \kappa(v_i, v_j)\}$
13:         **end if**
14:     **end for**
15: **end while**
16: $\kappa(G) \leftarrow min\_conn$

8.5   Parallel Connectivity Search

We may need to find whether a graph representing a network is connected and its connectivity parameters. We present parallel algorithms for this purpose in this section.

8.5.1   Computing the Connectivity Matrix

The $n \times n$ connectivity matrix C of a graph G of order n is defined as follows.

$$C[i, j] = \begin{cases} 1 & \text{if there is a path of length 0 or more from } v_i \text{ to } v_j \\ 0 & \text{otherwise} \end{cases}$$

We can obtain the connectivity matrix C from the adjacency matrix A of a graph G by multiplying A $n - 1$ times by itself, in other words, taking the $n - 1$th power of A as noted. However, we need to perform the required addition and multiplication in the usual matrix multiplication as Boolean addition (logical or operation) and Boolean multiplication (logical and operation). Before performing the Boolean multiplication, we need to generate matrix B which differs from the adjacency matrix A with all diagonal elements as 1's instead of 0's. This matrix B now has 1's for all paths in G that have 0 or 1 length. Multiplying B by itself provides B^2 which shows paths of length 2 or less and in general, B^k contains 1's for paths of length k or less between any two vertices.

The maximum path length in a graph G with n vertices can be $n - 1$ and hence, we need to find B^{n-1}. The required number of Boolean multiplications is then $\lceil \log(n - 1) \rceil$. For example, to find B^8, we need to find $B \times B$ to yield B^2; then

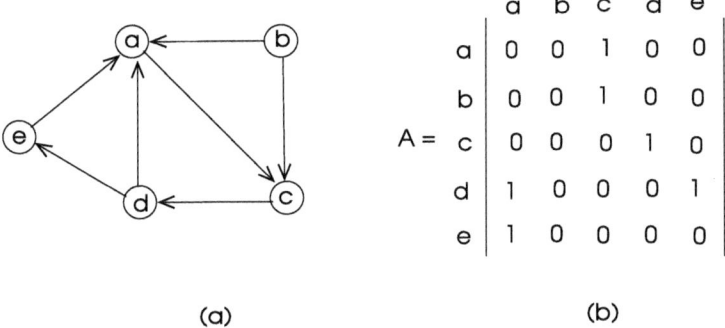

a | b | c | d | e

$$A = \begin{array}{c|ccccc} & a & b & c & d & e \\ \hline a & 0 & 0 & 1 & 0 & 0 \\ b & 0 & 0 & 1 & 0 & 0 \\ c & 0 & 0 & 0 & 1 & 0 \\ d & 1 & 0 & 0 & 0 & 1 \\ e & 1 & 0 & 0 & 0 & 0 \end{array}$$

(a)   (b)

Fig. 8.14 A sample undirected graph and its adjacency matrix A

$B^2 \times B^2$ to give B^4 and finally $B^4 \times B^4$ to obtain B^8 for a total of $\lceil \log 7 \rceil = 3$ multiplications. $C = B^m$ where $m = 2^{\lceil \log (n-1) \rceil}$ when $n - 1$ is not a power of 2 since $B^m = B^{n-1}$ when $m > n - 1$ [2]. A sample graph and its adjacency matrix is shown in Fig. 8.14.

B^2 and B^4 for this sample graph are as follows.

$$B = \begin{bmatrix} 1 & 0 & 1 & 0 & 0 \\ 0 & 1 & 1 & 0 & 0 \\ 0 & 0 & 1 & 1 & 0 \\ 1 & 0 & 0 & 1 & 1 \\ 1 & 0 & 0 & 0 & 1 \end{bmatrix}, B^2 = \begin{bmatrix} 1 & 0 & 1 & 1 & 0 \\ 0 & 1 & 1 & 1 & 0 \\ 1 & 0 & 1 & 1 & 1 \\ 1 & 0 & 1 & 1 & 1 \\ 1 & 0 & 1 & 0 & 1 \end{bmatrix}, C = B^4 = \begin{bmatrix} 1 & 0 & 1 & 1 & 1 \\ 1 & 1 & 1 & 1 & 1 \\ 1 & 0 & 1 & 1 & 1 \\ 1 & 0 & 1 & 1 & 1 \\ 1 & 0 & 1 & 1 & 1 \end{bmatrix}$$

Now, checking the entry $C[i, j]$ shows whether there is a path from vertex i to j. We can see that the vertex b can reach all other vertices whereas it cannot be reached by any other vertex as evident from the graph. In order to parallelize this algorithm, we can use any of the parallel matrix multiplication procedures such as the one described in Sect. 4.7.1 using Boolean multiplication and addition instead of multiplication and addition of real numbers.

8.5.2 Finding Connected Components in Parallel

We have seen how to find connected components of an undirected graph in Sect. 8.3.1. We will now describe two algorithms to find the components of an undirected graph in parallel.

8.5.2.1 Using the Connectivity Matrix

We know how to work out the connectivity matrix C in parallel from the previous section. Let us now define a matrix D which has entries as follows [2].

$$D[i,j] = \begin{cases} v_j & \text{if } C[i,j] = 1 \\ 0 & \text{otherwise} \end{cases}$$

The row i of D has the names of the vertices vertex i is connected which are in fact in the same component as i since there is a path between each of them and vertex i. We can now assign a vertex v_i to component k if k is the smallest index for which $D[i,k] \neq 0$. The parallel formation of this algorithm has three steps as follows.

1. Form the matrix C in parallel.
2. Construct matrix D from C in parallel.
3. Assign a component to each vertex in D.

8.5.2.2 Using the Adjacency Matrix

We have reviewed how to find the connected components of an undirected graph sequentially using the DFS algorithm. Every time we called the DFS procedure from the main program, a new component was processed since each call would process every vertex in a component. We will now attempt to construct a parallel algorithm based on the sequential algorithm. Let A be the adjacency matrix of the undirected graph that we are investigating. We will row-wise partition A to k parallel processes such that each process p_i gets $\lfloor n/k \rfloor$ rows. Each process then finds DFS spanning forests for the subgraphs that include the edges they are responsible. The final step involves merging of the spanning forests pairwise until one spanning forest remains. Merging of the forests can be done efficiently using the *find* and *union* operations [9]. Figure 8.15 displays an undirected graph with two components. The adjacency matrix has 8 rows which can be partitioned to two processes p_0 (rows 1–4) and p_1 (rows 5–8).

Process p_0 has the subgraphs shown in Fig. 8.16a and p_1 has the subgraphs displayed in Fig. 8.16c. Each process then constructs the DFS spanning trees for their subgraphs shown in (c) and (d) of the same figure.

These two spanning forests are then merged using the find and union procedures to find the components $C_1 = \{a, b, c, d, e\}$ and $C_1 = \{f, g, h\}$ as shown in Fig. 8.17.

Analysis

When we use 1-D block mapping with k processes, computing the spanning forest at each process p_i needs $\Theta(n^2/k)$ time since $n/k \times k$ block of the adjacency matrix is assigned to each process. Merging of the spanning forests take $\Theta(n \log k)$ time since there are $\log k$ merging each taking $\Theta(n)$ time. The communication cost of sending spanning forests is also $\Theta(n \log k)$. The parallel runtime of this algorithm is then,

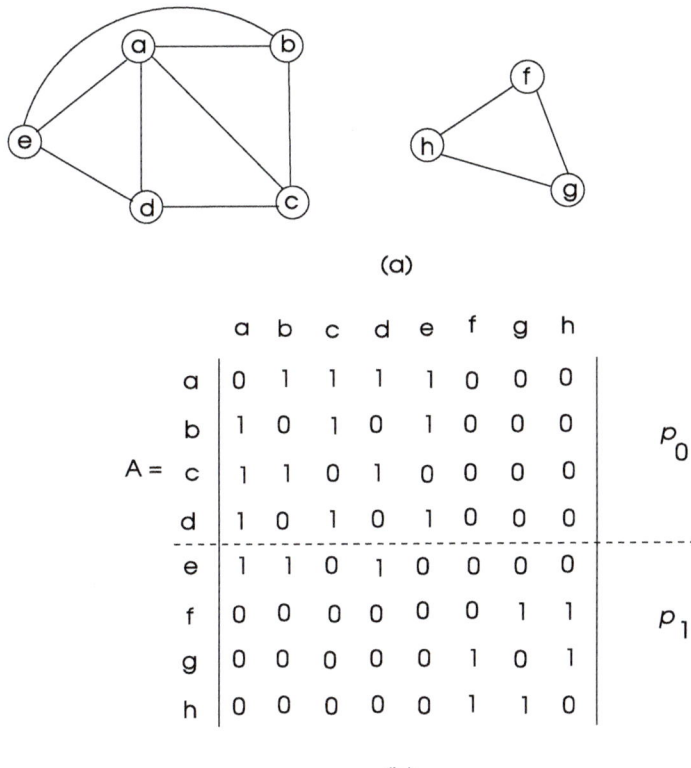

$$\begin{array}{c|cccccccc} & a & b & c & d & e & f & g & h \\ \hline a & 0 & 1 & 1 & 1 & 1 & 0 & 0 & 0 \\ b & 1 & 0 & 1 & 0 & 1 & 0 & 0 & 0 \\ c & 1 & 1 & 0 & 1 & 0 & 0 & 0 & 0 \\ d & 1 & 0 & 1 & 0 & 1 & 0 & 0 & 0 \\ e & 1 & 1 & 0 & 1 & 0 & 0 & 0 & 0 \\ f & 0 & 0 & 0 & 0 & 0 & 0 & 1 & 1 \\ g & 0 & 0 & 0 & 0 & 0 & 1 & 0 & 1 \\ h & 0 & 0 & 0 & 0 & 0 & 1 & 1 & 0 \end{array}$$

(b)

Fig. 8.15 A sample undirected graph with two components

$$T_P = \Theta\left(\frac{n^2}{k}\right) + \Theta(n \log k) \tag{8.4}$$

The speedup S and efficiency E considering the sequential complexity of $\Theta(n^2)$ for this algorithm are then as follows.

$$S = \frac{\Theta(n^2)}{\Theta(n^2/k) + \Theta(n \log k)} \tag{8.5}$$

$$E = \frac{1}{1 + \Theta((k \log k)/n)} \tag{8.6}$$

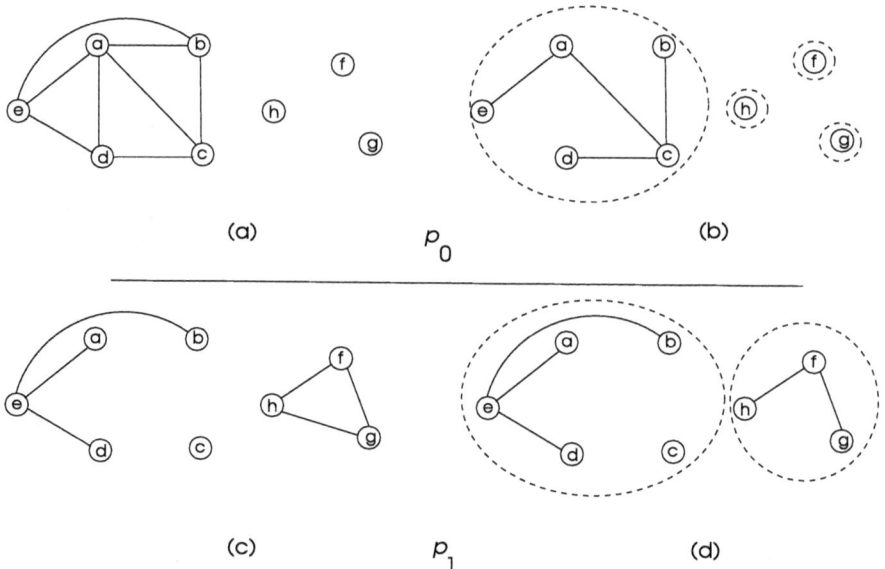

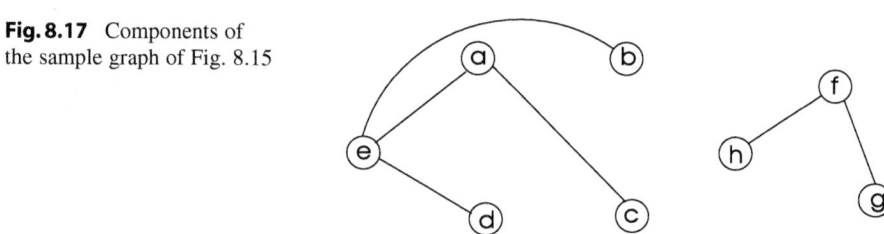

Fig. 8.16 Partitioning of the sample graph of Fig. 8.15 to two processes

Fig. 8.17 Components of
the sample graph of Fig. 8.15

8.5.3 A Survey of Parallel SCC Algorithms

The classical algorithms of Tarjan and later Kosaraju to find SCCs of a digraph are
difficult to parallelize due to the inherently sequential operation of the DFS algorithm employed in both. A parallel algorithm called divide and conquer strong
components (DCSC) algorithm by Fleischer et al. [8] uses a different approach
by partitioning the digraph into three disjoint subgraphs and processing these subgraphs recursively in parallel. We will briefly describe this parallel algorithm as it
can be used in practice with some modifications.

Given a digraph $G = (V, E)$, the descendants $Desc(G, v)$ of a vertex v are
the vertices in G that are reachable from v including itself. The predecessors
$Pred(G, v)$ of a vertex v can be defined similarly as the set of vertices from
which the vertex v is reachable. The remaining vertices in graph G are called

the *remainder* shown by $Rem(G, v) = V \setminus \{Desc(G, v) \cup Pred(G, v)\}$. It is shown that,

$$SCC(G, v) = Desc(G, v) \cup Pred(G, v)$$

and any SCC of G is a subset of $Desc(G, v)$, $Pred(G, v)$ or $Rem(G, v)$. The designed algorithm makes use of this property by first selecting a random vertex v, finding its predecessor and descendant sets and then finding the SCC that contains this vertex. It then recurses on the remaining vertices in parallel as shown in Algorithm 8.15. It was shown in [8] this algorithm has an expected time complexity of $O(n \log n)$ in the serial case. Later on, McLendon et al. extended this algorithm by a simple modification to improve performance [12].

Algorithm 8.15 *Finding SCCs in parallel*

1: **procedure** DCSC(G)
2:     **if** $G = \emptyset$ **then**
3:         **return**
4:     **end if**
5:     **select** a vertex $v \in V$
6:     $SCC(G, v) = Desc(G, v) \cup Pred(G, v)$
7:     **output** $SCC(G, v)$
8:     **do in parallel**
9:         $DCSC(Pred(G, v) \setminus SCC(G, v))$
10:         $DCSC(Desc(G, v) \setminus SCC(G, v))$
11:         $DCSC(Rem(G, v))$
12:     **end do**
13: **end procedure**

8.6   Distributed Connectivity Algorithms

In a network environment, our aim is to have each node of the network find out the connectivity values of the graph that represents the network.

8.6.1   A Distributed *k*-Connectivity Algorithm

We know that the lowest degree $\delta(G)$ of a graph G is an upper bound on the value of the connectivity $\kappa(G)$ since we can isolate this vertex by removing all edges incident to it to have G disconnected. This concept can be used in a distributed setting by nodes exchanging their degrees to estimate $\delta(G)$. Three localized distributed algorithms to determine the value of $\kappa(G)$, say k, that works with neighbor knowledge only are proposed by Jorgic et al. [16, 1]. In the first algorithm called local neighbor discovery (LND), each node discovers its degree d_i by first sending *hello* messages to neighbors and counting the responds from them. Each node then exchanges degree information with their neighbors and send this data to neighbors.

Repeating this process p times results in degrees of nodes transferred to all nodes within p hops from them. Nodes can then simply sort the degrees they received in total and denote the lowest value as the value of k. The pseudocode of a possible implementation is shown in Algorithm 8.16 and although the original algorithm uses the *time-to-live* field of the message which is initialized to p, we provide a SSI algorithm version that works in p rounds to achieve the same function.

Algorithm 8.16 *Local Neighborhood Detection*

1: **boolean** $round\_over$
2: **message type** deg
3: **set of int** $degs, all\_degs$
4: $degs \leftarrow d_i; all\_degs \leftarrow d_i$
5: **for** $i = 1$ to p **do**
6:     $round\_over \leftarrow false$
7:     **while** $\neg round\_over$ **do**
8:         **send** $degs$ to $N(i)$
9:         **receive** $deg(j)$
10:         $degs \leftarrow deg(j).d_j$
11:         $all\_degs \leftarrow all\_degs \cup degs$
12:         $received \leftarrow received \cup \{j\}$
13:         **if** $received = N(i)$ **then**
14:             $round\_over \leftarrow true$
15:         **end if**
16:     **end while**
17:     $degs \leftarrow \{\emptyset\}$
18: **end for**
19: $k \leftarrow min\{d_j \in all\_degs \cup d_i\}$

All of the edges of the graph will be traversed in both directions in each round, so there will be $O(pm)$ messages in total. Although this linear time may look favorable, a high value of p is needed to estimate k more correctly. Moreover, $\delta(G)$ is an upper bound on the value of k so the actual value may be much lower. In the second algorithm called *local subgraph connectivity detection* (LSCD) proposed by the same authors, a further test is made to find a subgraph of p-hop neighbors of a given node is k-connected. A node v determines that the graph is k-connected when both of the following conditions are satisfied:

- All of the p-hop neighbors of v have at least a degree of k
- The union of v and subgraph of p-hop neighbors of v is k-connected.

The third algorithm searches for critical nodes removal of which will disconnect graph.

8.7   Chapter Notes

Connectivity is a fundamental concept in graph theory that has immediate applications in the computer networks. A digraph is strongly connected if there is a path in both directions between any vertex pair u, v in such a digraph. Finding strongly connected components which are subgraphs of a digraph that are strongly connected has various applications such as detecting highly connected regions in biological and social networks. A one-way road system in a city should also be strongly connected. We reviewed two linear time algorithms due to Tarjan and Kosaraju to detect such SCCs of digraphs.

Removal of an articulation point or a bridge from a connected undirected graph G leaves G disconnected. We need to find these vertices and edges of a computer network to reinforce additional routers and links in these areas to have a more robust network. We described a linear time DFS-based algorithm that makes use of the simple property that any vertex on a DFS tree of a graph that does not have a back edge from its subtree to its ancestors is an articulation point.

A block of a graph is a maximal connected subgraph without any articulation points. We reviewed two linear time algorithms to identify blocks in a graph. We can find connectivity and strong connectivity of a graph in parallel as demonstrated by two algorithms. Finally, we described a heuristic distributed algorithm that estimates the vertex connectivity of a network.

Exercises

1. Show the articulation points, bridges, and the blocks in the undirected graph of Fig. 8.18.
2. Write the pseudocode of the DFS-based articulation point search algorithm as one main procedure. Identify articulation points in the sample undirected graph of Fig. 8.19 using the DFS-based algorithm. Show the *low* and *num* values of vertices at each iteration.
3. Find the bridges of the same graph of Fig. 8.19 using Tarjan's algorithm.

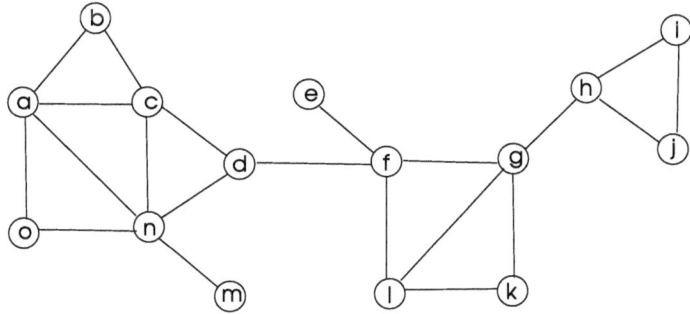

Fig. 8.18 Sample graph for Exercise 1

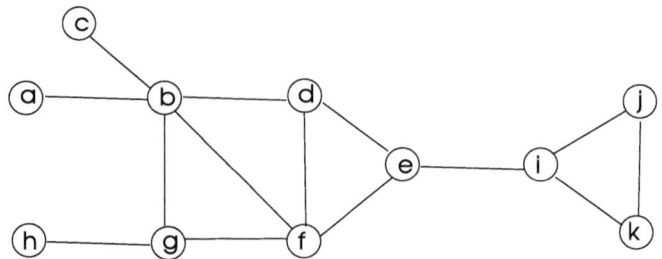

Fig. 8.19 Sample graph for Exercises 2 and 3

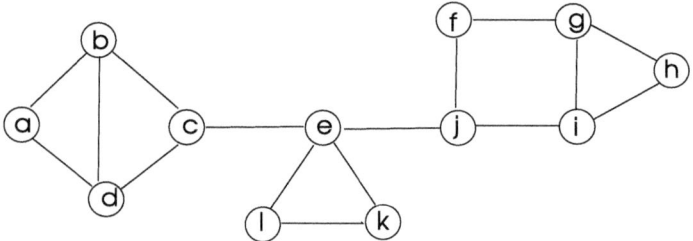

Fig. 8.20 Sample graph for Exercise 4

Fig. 8.21 Sample graph for
Exercise 5

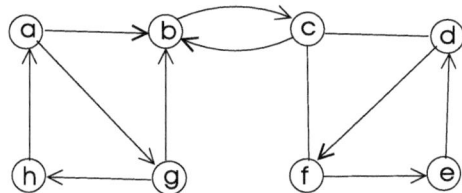

4. Work out the blocks of the sample graph depicted in Fig. 8.20 using the
   Hopcroft–Karp algorithm by showing the iterations of the algorithm.
5. Find the SCCs of the digraph in Fig. 8.21 using Kojarasu's algorithm. Show
   the contents of the stack and the DFS trees formed.
6. Find the maximum flow in the network of Fig. 8.22 using the Ford–Fulkerson
   algorithm by showing all iterations of the algorithm.

Fig. 8.22 Sample network for Exercise 6

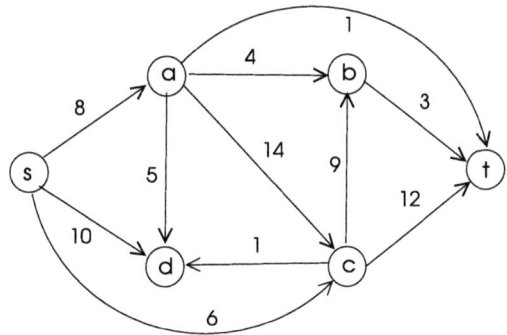

References

1. Aho AV, Hopcroft JE, Ullman JD (1983) Data structures and algorithms. Addison-Wesley
2. Akl SG (1989) The design and analysis of parallel algorithms. Prentice Hall, Englewood Cliffs, p 07632
3. Cormen TH, Leiserson CE, Rivest RL, Stein C (2009) Introduction to algorithms, 3rd edn. MIT Press, Cambridge
4. Esfahanian AH (1988) On the evolution of connectivity algorithms. In: Wilson R, Beineke L (eds) Selected topics in graph theory. Cambridge University Press, Cambridge
5. Esfahanian AH, Hakimi SL (1984) On computing the connectivities of graphs and digraphs. Networks 355–366
6. Even S (1979) Graph algorithms. In: Computer software engineering series. Computer Science Press. ISBN-10: 0914894218, ISBN-13: 978-0914894216
7. Even S, Tarjan RE (1975) Network flow and testing graph connectivity. SIAM J Comput 4:507–518
8. Fleischer L, Hendrickson B, Pinar A (2000) On identifying strongly connected components in parallel. In: Parallel and distributed processing, pp 505–511
9. Grama A, Karypis G, Kumar V, Gupta A (2003) Introduction to parallel computing, 2nd edn. Addison-Wesley, New York
10. Hopcroft J, Tarjan R (1973) Algorithm 447: efficient algorithms for graph manipulation. Commun ACM 16(6):372–378
11. Matula DW (1987) Determining edge connectivity in $O(mn)$. In: Proceedings, 28th symposium on foundations of computer science, pp 249–251
12. McLendon W III, Hendrickson B, Plimpton SJ, Rauchwerger L (2005) Finding strongly connected components in distributed graphs. J Parallel Distrib Comput 65(8):901–910
13. Tarjan RE (1972) Depth first search and linear graph algorithms. SIAM J Comput 1(2):146–160
14. Tarjan RE (1974) A note on finding the bridges of a graph. Inf Process Lett 2(6):160–161
15. Whitney H (1932) Congruent graphs and the connectivity of graphs. Am J Math 54:150–168

Matching

9

9.1 Introduction

A *matching* M of a graph $G = (V, E)$ is a subset of edges of G that do not share any endpoints. In other words, each vertex of G is incident to at most one edge of M. We can also view a matching M as a set of independent edges in G. Matchings can be used in a variety of applications such as channel frequency assignment in radio networks, alignment of protein interaction networks when similarity between two or more such networks is investigated [8]. It is also used in multilevel graph partitioning algorithms [15].

Maximal matching is the set M of edges that cannot be enlarged any further by the addition of new edges. In an unweighted graph, *maximum matching* of a graph is the set of edges that has the maximum cardinality among all matchings in that graph. In an edge-weighted graph, our aim is to find a matching with the maximum (or minimum) total weight. Finding a maximum (weighted) matching in an unweighted or weighted graph is one of the rare graph problems that can be solved in polynomial time. However, there are various approximation algorithms to improve the runtime of matching. Also, approximation algorithms turn out to be easier to implement with significantly less lines of code than exact algorithms. Unweighted or weighted matching in bipartite graphs can be treated separately than general graph matching as the structure of bipartite graphs can be exploited for designing conceptually different algorithms than the general case.

In this chapter, we review the matching problem in general graphs and bipartite graphs for both unweighted and weighted cases. We describe sequential, parallel, and distributed algorithms for these graphs.

K. Erciyes, *Guide to Graph Algorithms*, Texts in Computer Science,
https://doi.org/10.1007/978-3-032-05294-0_9

9.2   Theory

We will review theory, parameters and notation used in matching in unweighted or weighted general graphs and bipartite graphs starting with unweighted matching in this section.

9.2.1   Unweighted Matching

A *matching* of a graph $G = (V, E)$ is a set of edges $M \subseteq E$ which do not pairwise share endpoints. A vertex is *matched* or *saturated* if it is incident to an edge of matching, otherwise, it is a *free* or an *unmatched* or an *unsaturated* vertex. An edge that is part of a matching is called *matched* and *unmatched* otherwise. A *maximal matching* (MM) of a graph G cannot be made larger by the addition of a new edge meaning it is not contained in any larger matching. A *maximum matching* (MaxM) of a graph G has the maximum size among all matchings of G. In *perfect matching*, every vertex of the graph is incident to an edge that is included in the matching. In other words, all vertices of a graph are saturated in a perfect matching. MM and MaxM of a sample graph are displayed in Fig. 9.1.

We need few more definitions before reviewing sequential, parallel, and distributed matching algorithms.

Definition 9.1 (*alternating path*) An *M- alternating* path of a matching M is a path that alternates between edges in M and edges in $E - M$ (edges not in M).

Definition 9.2 (*augmenting path*) An *M- augmenting* path of a matching M is a path that is an alternating path that starts and end in unsaturated vertices.

Augmenting and alternating paths are displayed in Fig. 9.1. An augmenting path of a matching M that contains k edges contains exactly $k + 1$ edges that are not in M for a total of $2k + 1$ edges. We try to find augmenting paths since we can increase the size of a matching that has an augmenting path by switching the

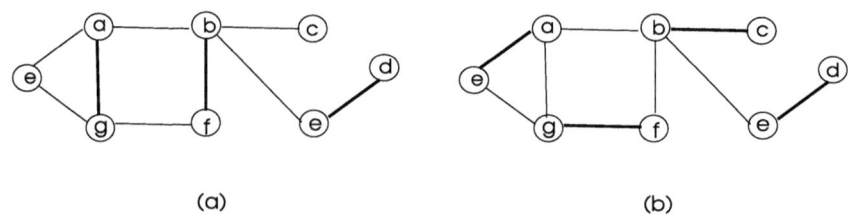

(a)   (b)

Fig. 9.1 **a** A MM of size 3, **b** A MaxMM with size 4 of a sample graph. Path (e, a, g, f, b, c) is an augmenting path in (**a**) and path (g, f, b, c) is an alternating path in (**b**). There is no augmenting path in (**b**) since the matching is maximum. We can also see that matching in (**b**) is perfect as each vertex is saturated

edges of matching in the path with edges that do not belong to matching in the path.

Definition 9.3 (*alternating tree*) An *alternating tree* is rooted at a free vertex and each path of this tree is an alternating path.

For example, the tree rooted at vertex e in Fig. 9.1b is an alternating tree with branches e, a, b, c, e, g, f and e, a, b, e, d.

Definition 9.4 (*graph factor*) *spanning subgraph* or a *factor* of a graph G contains all vertices of G. A k-regular spanning subgraph of G is called its k-factor. A 1-factor of G is a perfect matching of G that saturates all vertices of G.

Definition 9.5 (*symmetric difference of two graphs*) Symmetric difference of two graphs G and G' shown as $G \oplus G'$ (or $G \Delta G'$) is a graph G'' that is induced on the edge set that contains edges in either G or G' but not in both which is equal to $(G - G') \cup (G' - G)$.

This means we need to find edges that are present in only one of the input graphs and include vertices incident on those edges. Two graphs and their symmetric difference is shown in Fig. 9.2.

Lemma 9.1 *Given a graph G with a matching M and an augmenting path P of M, $M \oplus P$ is again a matching with a cardinality one more than that of M.*

Proof The symmetric difference of a matching M with an augmenting path P with respect to M, that is an augmenting path that is incident with edges of M, results in a new matching M' which has edges that were not part of matching in P as matched edges and the matched edges in P are now unmatched. In other words, we switch the edges of matching in P with edges that are not in matching. The new matching M' is in fact formed by taking the symmetric difference of an augmenting path P of a matching M with M, to result in a matching that has a size one unit larger than M

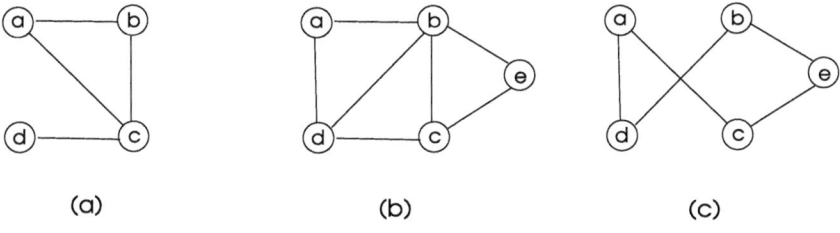

      (a)   (b)   (c)

Fig. 9.2 Symmetric difference of the two graphs in **a** and **b** is shown in **c**

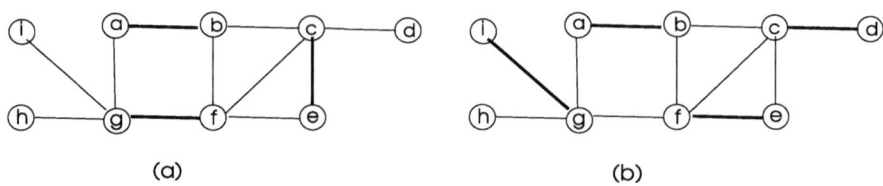

Fig. 9.3 **a** A MM M of size 3 is shown in sample graph in **a**. However, there is an augmenting path $P = \{i, g, f, e, c, d\}$ in this matching and the new matching M' obtained by taking symmetric difference of M with P results in a matching that has a size of 4 as shown in **b**

as shown below.

$$M' = M \oplus P = (M - P) \cup (P - M).$$

Now to prove this lemma, we see that P has odd number of edges and its edges alternate between as edges in M and edges not in M. Edges in the complement of the path P have the same set of neighbors in M as in M' and vertices in P have exactly one neighbor in M', therefore M' is also a matching of G.   □

This process is called *augmenting the matching M* and can be used in a number of maximum matching algorithms. The matching in augmenting path in Fig. 9.3a is augmented to have a matching with an incremented size in (b).

We can now state an important theorem by Berge which forms the basis of few fundamental matching algorithms [3].

Theorem 9.1 (Berge) *A matching M is maximum if and only if there are no augmenting paths with respect to M.*

Proof We know by Lemma 9.1 that if matching M has an augmenting path, it is not of maximum cardinality. Now we need to prove the reverse direction of the statement; if there are no augmenting paths of a matching M, then M is maximum. Let us assume M is not maximum when there are no augmenting paths. This means there exists a matching M' that is maximum such that $|M'| > |M|$. Let us form $H = M \oplus M'$. Each vertex v of H is incident to at most one M edge and one M' edge, therefore each vertex of H has at most a degree of 2. Also, each path of H alternates with edges in M and edges in M'. Since each path is alternating, each cycle must be even. As the size of M' is greater than the size of M, H has a path P that has more maximum matching edges than M edges. This path P begins and ends with a maximum matching edge, therefore it is an augmenting path of M, meaning M can be enlarged using P. This contradicts the first assumption that M has no augmenting paths.   □

This gives us a simple algorithm template to find maximum matching M of a graph consisting of following steps:

1. $M \leftarrow \emptyset$
2. **while** $\exists$ an augmenting path P with respect to M
3. $M \leftarrow M \oplus P$
4. **end while**
5. **return** M

We need a method to find the augmenting path P and we will see it is more convenient to have different procedures for bipartite graphs and general graphs in the next sections.

9.2.2 Weighted Matching

In the matching of a weighted graph, our aim is to find the matching with the maximum or minimum total weight. A *maximum weighted matching* (MaxWM) of a weighted graph $G = (V, E, w)$ with $w : E \rightarrow \mathbb{R}^{+}$ has the maximum total weight among all weighted maximal matchings of G where the weight of a matching M is defined as $w(M) = \sum_{e \in M} w(e)$. Similarly, the *minimum weighted matching* (MinWM) of G has the total least weight among all weighted maximal matchings of G. By a maximal weighted matching (MWM) of a weighted graph G, we mean a weighted matching of G which cannot be enlarged by a new edge. We will see both MaxWM and MinWM have practical applications. Figure 9.4 displays MWM nd MaxWM of a sample graph.

9.3 Unweighted Bipartite Graph Matching

A matching M in a bipartite graph $G = (A \cup B, E)$ has the same property, that is, no vertex $v \in A$ or $v \in B$ is incident to more than one edge in M. The unweighted and weighted maximal and maximum matchings in bipartite graphs are defined similar to general graphs. Bipartite matchings in two sample graphs are shown in Fig. 9.5. A vertex cover of a graph $G = (V, E)$ is a subset V' of its vertex set such that every edge $e \in E$ is incident to at least one vertex in V'.

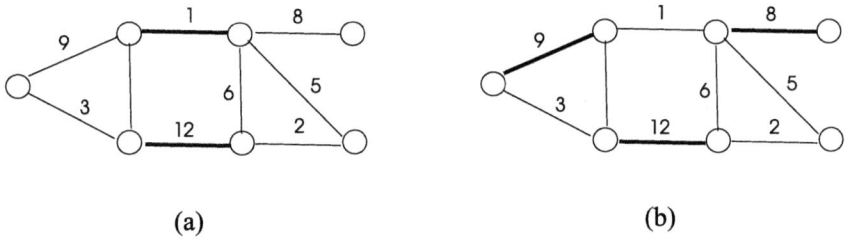

(a)   (b)

Fig. 9.4 **a** A MWM of total weight 13, **b** A MaxMM with total weight 29 of a sample graph

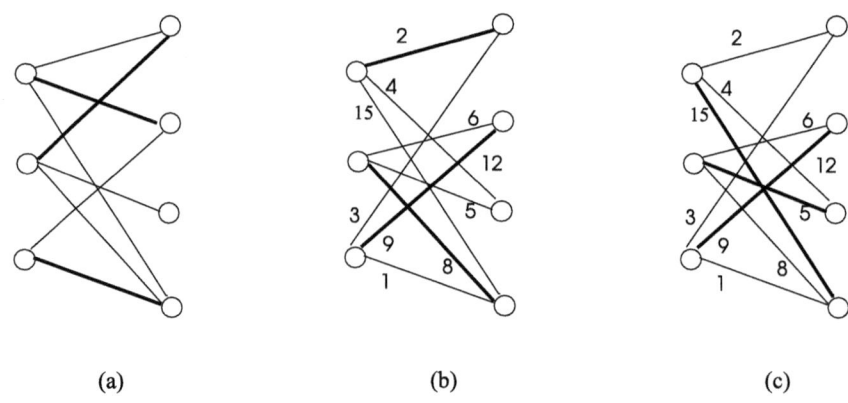

Fig. 9.5 **a** A MM of an unweighted bipartite graph, **b** A maximal weighted matching of magnitude 19 of a weighted bipartite graph, **c** the maximum weighted matching of magnitude 29 of of the same weighted bipartite graph of **b**. The matching edges are shown in bold

Given a bipartite graph $G = (A \cup B, E)$, the neighborhood $N(S)$ of a set of vertices $S \subset A$ is a set of vertices in B such that $\forall u \in S$ and $(u, v) \in E$, $v \in N(S)$. In other words, $N(S)$ is the union of the neighbors of the vertices in S.

Theorem 9.2 (Hall) *A bipartite graph* $G = (A \cup B, E)$ *contains a matching that saturates every vertex in* A *if and only if,*

$$|N(S)| \geq |S|, \text{ for all } S \subset A. \tag{9.1}$$

Proof In the forward direction of the proof, every vertex in $S \subset A$ is matched to a vertex in $N(S)$ under the matching M and since all vertices in A are saturated with any two distinct vertices in S being matched to two distinct vertices in $N(S)$, it follows $|N(S)| \geq |S|$.

In the reverse direction, we need to prove if $|N(S)| \geq |S|$, then G has a matching that saturates every vertex of A. For this condition, assume M^* is a maximum matching in G and $u \in A$ is not a saturated vertex by M^*. Let us consider the set W which contains vertices that can be reached from u by an alternating path. Let $S = W \cap A$ and $T = W \cap B$. We can see that every vertex in $S \setminus \{u\}$ is saturated and every vertex in T is also saturated which means $|T| = |S| - 1$. We can then write $|N(S)| = |T|$ which means $|N(S)| < |S|$ and therefore a contradiction. We can now conclude that no such vertex exists and hence every vertex in S is saturated. $\square$

Theorem 9.3 (König 1931) *For any bipartite graph* $G = (A \cup B, E)$, *the maximum size of a matching* $\alpha(G)$ *is equal to the minimum size of a vertex cover* $\beta(G)$ [16].

We will show a constructive proof of this theorem that results in an efficient algorithm to find the minimum vertex cover of a bipartite graph when we review vertex cover algorithms in Chap. 10.

9.3.1 A Sequential Algorithm Using Augmenting Paths

We will use the approach of enlarging the matching with augmenting paths to find a maximum matching in an unweighted bipartite graph. In order to do so, let us consider a bipartite graph $G = (A \cup B, E)$ and construct a digraph $G' = (A \cup B, E')$ where each edge $e \in E'$ is directed from A to B if $e \notin M$ and it is directed from B to A if $e \in M$. We can see that there is an augmenting path with respect to matching M in G if and only if there is a directed path from an unmatched vertex in A to a unmatched vertex in B in the graph G'. Let us now rewrite the generic matching algorithm formally as shown in Algorithm 9.1 with the procedure *Find_AP* to find the augmenting paths.

Algorithm 9.1 *MaxM_UBG1*

1: **Input:** An undirected bipartite graph $G = (A \cup B, E)$
2: **Output:** Maximum matching M of G
3:
4: $M \leftarrow$ any edge $e \in E$
5: **repeat**
6:    $P \leftarrow Find\_AP(G, M)$
7:    **if** $P \neq \emptyset$ **then**
8:       $M \leftarrow M \oplus P$
9:    **end if**
10: **until** $P = \emptyset$
11:
12: **procedure** FIND_AP($G(A \cup B), M$)
13:    $A' \leftarrow$ a set of free vertices in A
14:    $B' \leftarrow$ a set of free vertices in B
15:    **direct** unmatched edges from A to B, matched edges from B to A to construct G'
16:    **find** the shortest path P from A' to B' in G'
17:    **if** P not found **then**
18:       return $\emptyset$
19:    **else**
20:       return P
21:    **end if**
22: **end procedure**

Correctness is evident since the procedure *Find_AP* returns an augmenting path P if such a path exists in G' as it starts from a free vertex and ends at a free vertex using matched edges.

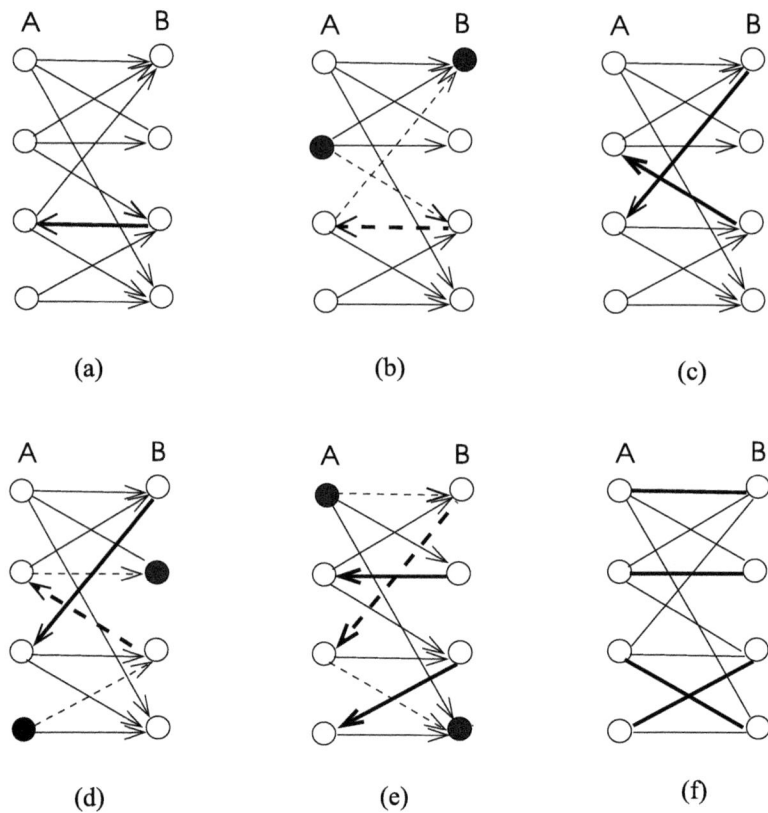

Fig. 9.6 Running of *MaxM_BPG1* in a small bipartite graph $G = (A \cup B, E)$. The initial arbitrarily selected matching M in **a** is shown in bold. Using this matching an augmenting path shown in dashed lines starting from a bold vertex in A and ending at a bold vertex in B is shown in **b** which is XORed with M to obtain the matching in **c**. We find an augmenting path now with this matching shown in dashed lines shown in **d** to form matching in **e** in which another augmenting path shown in dashed lines is discovered. The final matching obtained by XORing this path with current matching does not have any augmenting paths as all vertices are now saturated after 3 steps

Theorem 9.4 *Algorithm MaxM _UBG1 has O(nm) complexity.*

Proof The upperbound on the size of the matching is $n / 2$ and at each step we can only extend the current matching by 1 resulting in $O(n)$ time for the loop that calls *Find_AP*. An augmenting path can be found in $O(m)$ time searching all of the edges in the worst case. The total time needed is therefore $O(nm)$.   □

Running of this algorithm in a sample graph is depicted in Fig. 9.6.

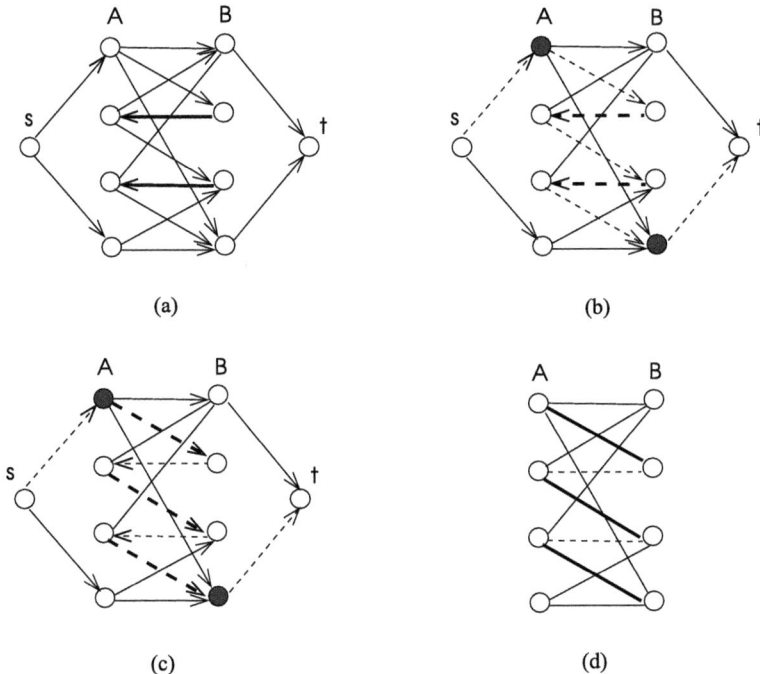

Fig. 9.7 Running of *FindBFS_AP* in the graph of Fig. 9.6. The two existing matched edges are made directed from *B* to *A* and all other edges are directed from *A* to *B*. A vertex *s* is added to the left with directed edges to all unmatched vertices in *A* and directed edges from unmatched vertices in B to the new vertex *t*. We can now run BFS from *s* to *t* to find the shortest path shown as dashed lines in **b** of the figure. We know augment this path to obtain the augmenting path shown in **c** and the final maximum matching is shown in **d** which is different than the maximum matching found in Fig. 9.6 but has the same size

Finding Augmenting Paths Using BFS

We still have not shown how to search for an augmenting path in G'. One way of achieving this is by adding a source vertex *s* to the left of the bipartite graph and connecting it by directed edges to all free vertices of *A*. A sink vertex *t* is also added to all of the free vertices in *B* by directing all edges from vertices of *B* to vertex *t* as shown in Fig. 9.7. We then run BFS from *s* and return the shortest path as shown in procedure *FindBFS_AP* in Algorithm 9.2. The shortest path starting from *s* and ending in *t* will be an augmenting path as it goes through free vertices in *A* and *B* and it has to go through some matched edges to be able to return to *B*. The running time for BFS is $O(n + m) \approx O(m)$ in a dense graph and hence the total time of Algorithm 9.1 using BFS-based approach is $O(nm)$ since the size of the matching can be at most $n / 2$.

Algorithm 9.2 *MaxM_UBG2*

1: **procedure** FINDBFS_AP($G(A \cup B)$, M)
2:     $A' \leftarrow$ a set of free vertices in A
3:     $B' \leftarrow$ a set of free vertices in B
4:     **direct** unmatched edges from A to B, matched edges from B to A to construct G'
5:     **add** s, t to G'
6:     **run** BFS from s of G' to obtain shortest path P
7:     **if** P not found **then**
8:         return $\emptyset$
9:     **else**
10:        return $P \setminus \{s, t\}$
11:    **end if**
12: **end procedure**

The operation of this algorithm is shown in Fig. 9.7 on the same bipartite graph of Fig. 9.6 with different initial matching.

9.3.2   A Flow-Based Algorithm

Finding maximum matching in an   unweighted bipartite graph $G = (A \cup B, E)$ can be performed using maximum flow as follows. We add a source vertex s and sink vertex t, connecting s to all vertices of A using directed edges and all vertices of B to vertex t again using directed edges. We also direct edges from A to B and give a capacity of unity to all edges to obtain graph G' as shown in Fig. 9.8. We now solve the maximum network flow problem in this graph using Ford–Fulkerson algorithm. The edges that are used in the maximum network flow correspond to the edges of the maximum matching of G. We will have a flow through an edge or not since capacities are all 1.

The capacities of the edges of G' are all 1, therefore flow through an edge is either 0 or 1. Each $u \in A$ has exactly one incoming edge of flow from vertex s, then there can be at most one edge (u, v) with $v \in B$ that can have a flow of 1 by the flow conservation law. Each vertex $v \in B$ has exactly one outgoing edge that can pass flow to vertex t and hence, at most one of its incoming edges can carry the maximum flow, again by the flow conservation law. This is to say the edges of the maximum flow are disjoint resulting in a matching. Therefore, each $u \in A$ will be matched with at most one vertex of B forming a maximum matching in G. When we have a flow of value k, this corresponds to a matching of size k with the same set of edges of G. Since we attempt to maximize the flow, we are maximizing the matching.

Time complexity of this algorithm is the same of Ford–Fulkerson which is $O(kC)$ where k is the number of edges that the flow runs through and C is the maximum flow in the graph which is $|A| = n$. The number of edges in the newly formed graph G' is $2n + m$ since we added $2n$ new edges to the vertices s and t. The complexity of this algorithm is therefore $O(n^2 + nm)$.

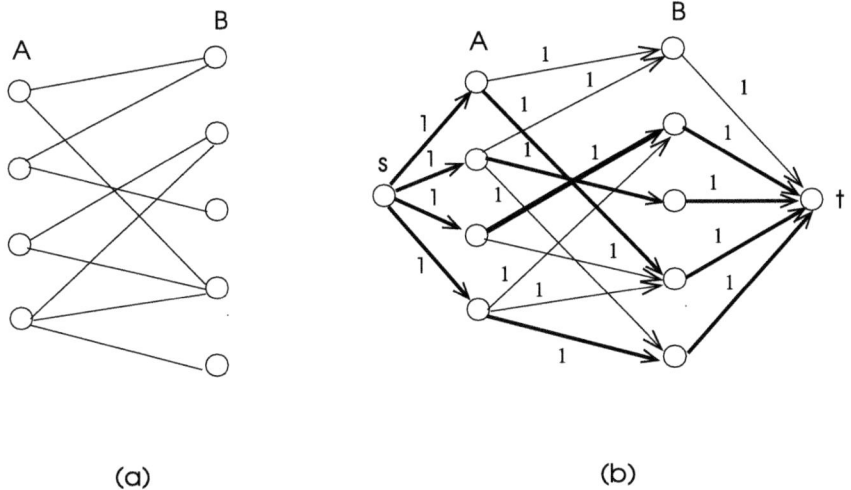

Fig. 9.8 The flow-based re-construction of an example bipartite graph in **a** is shown in **b**. The maximum flow edges which correspond to maximum matching edges are shown in bold

9.3.3   Hopcroft–Karp Algorithm

The Hopcroft–Karp algorithm also makes use of augmenting paths while find-ing the maximum matching in a bipartite graph. This algorithm however searches many paths simultaneously rather than one by one as in the previous algorithm and brings down the time complexity to $O(\sqrt{n}m)$ [14]. The working principle of this algorithm is based on the following lemma.

Lemma 9.2 *Given a bipartite graph $G = (A \cup B, E)$ with a matching M and a maximum matching M^* for this graph, let $|M^*| - |M| = k$ for some integer k. Let P be the symmetric difference $M \oplus M^*$. Then P contains at least k disjoint augmenting paths.*

Proof All of the edges in the set E' obtained by $M \oplus M^*$ have a maximum degree of 2 meaning the connected components of the subgraph G' induced by E' are simple paths and cycles. Let us consider paths and cycles separately in G'. Each cycle has the same number of edges in M^* as in M, however, each M-augmenting path has exactly one less edge in M as in M^*. In $M \oplus M^*$, we have exactly k more edges in M^* than edges in M. Therefore, G' contains k vertex disjoint augmenting paths of M.   □

The algorithm consists of a number of phases and all possible vertex disjoint augmenting paths are searched in each phase. The symmetric difference of the union of all of these paths with the existing matching is computed to yield the new matching as shown in the high-level description of the algorithm in Algorithm 9.3.

Algorithm 9.3 *HK1_UBM*

1: **Input**: An undirected bipartite graph $G = (A \cup B, E)$
2: **Output**: Maximum matching M of G
3: $M \leftarrow \emptyset$
4: **while** M is not maximum **do**
5:     find $\mathcal{P} = \{P_1, ..., P_k\}$ a maximal set of vertex disjoint shortest M-augmenting paths
6:     $M \leftarrow M \oplus \mathcal{P}$
7: **end while**

Finding the disjoint augmenting paths of a bipartite graph $G = (A \cup B, E)$ in each phase can be done by a modified BFS algorithms as follows. The BFS algorithm is run for each unmatched vertex $v \in A$ to form layers starting at v using alternating paths of unmatched and matched edges to form an alternating edge tree rooted at v. The BFS algorithm stops when one or more unmatched vertices in B are reached since we are looking for shortest augmenting paths. A path reaching an unmatched vertex in B will be an augmenting path since it started from an unmatched vertex and traversed alternating edges. All of the unmatched vertices reached in B are stored in the set F. After this first part of the phase is over, a modified DFS algorithm is run for each vertex in F until an augmenting path ending at a free vertex in A is found. The modified DFS algorithm should run through alternating edges to discover an augmenting path. Each discovered path P_x is added to the set $\mathcal{P}$, vertices in P_x are removed from the BFS tree with the orphan vertices and this procedure is repeated for other free vertices in B. At the end of each phase, the new matching is formed by XORing the set $\mathcal{P}$ with the existing matching M. The detailed version of this algorithm is depicted in Algorithm 9.4.

Algorithm 9.4 *HK2_UBM*

1: **Input**: An undirected bipartite graph $G = (A \cup B, E)$
2: **Output**: Maximum matching M of G
3: $M \leftarrow \emptyset, \mathcal{P} \leftarrow \emptyset$
4: **while** $\mathcal{P} \neq \emptyset$ **do**   ▷ continue until no augmenting paths found
5:     **for all** $u \in A$ **do,**
6:         $BFS\_Mod(u)$ ▷ run modified BFS
7:     **end for**
8:     $F \leftarrow$ all reached free vertices in B
9:     **for all** $v \in F$ **do,**
10:        $P_v \leftarrow DFS\_Mod(v)$ ▷ run modified DFS
11:        **remove** P_v and orphan vertices from BFS graph
12:        $\mathcal{P} \leftarrow \mathcal{P} \cup P_v$
13:     **end for**
14:     $M \leftarrow M \oplus \mathcal{P}$
15:     $\mathcal{P} \leftarrow \emptyset$
16: **end while**

Correctness of the algorithm is evident since the BFS and DFS algorithms discover augmenting paths based on their operations. Also, since we delete the augmenting path found during DFS from the BFS trees, the paths discovered are disjoint making it possible to include the union of them to matching at once. Running of this algorithm in a bipartite graph $G = (A \cup B, E)$ is depicted in Fig. 9.9 with $A = \{a, b, c, d, e\}$ and $B = \{1, 2, 3, 4, 5\}$. The first iteration of the algorithm starts with $M = \{\emptyset\}$ and all of the vertices are unmatched. The BFS from all of vertices in A ends in all of the free vertices in B which results in a BFS tree same as the original graph which is not shown. Therefore, the free vertex set F has $\{1, 2, 3, 4\}$ at the end of BFS and we stop BFS in the first layer since we have reached free vertices in B. We now run DFS from each of the vertices in F to find paths to be included in $\mathcal{P}$. We should delete the edges and vertices found in the path along with any remaining orphan vertices before searching the next path. We have selected the matching shown in (a) by always opting for the first free vertex in A from left while running DFS. In the second phase, we run the BFS from the free vertices d and e in A to obtain the BFS trees shown in (a) rooted at vertices d and e which end at free vertices 4 and 5 in B. Note that vertex 3 is not a free vertex and we need not run DFS from there but we had to stop at layer 3 since we reached free vertices in B. Running DFS from vertices 4 from first tree and 5 from the second one results in the final maximum matching of the graph with size 5 in two phases which is a perfect matching since each vertex is matched as shown in (b). If we had selected the augmenting path from vertex 5 in the BFS tree on the left, we would have edges $(5, c)$ and $(2, d)$ matched and would need a third phase that would select the augmenting path $(e, 3), (3, b), (b, 5)(5, c), (c, 4)$, however, we would arrive at the same maximum matching.

Analysis

We will first state a lemma to aid the analysis of the complexity of this algorithm.

Lemma 9.3 *If the shortest augmenting path with respect to a matching M in a graph G has l edges, then the size of the maximum matching in G has a maximum size of,*

$$|M| + \frac{|V|}{l+1}.$$

Proof Let M^* be the maximum matching of G. Then $M^* \oplus M$ contains $|M^*| - |M|$ vertex disjoint augmenting paths by Lemma 9.2. Every such augmenting path has at least l edges, hence at least $l+1$ vertices. Therefore, we can have at most $|V|/(l+1)$ of such paths meaning M can be increased at most that much. □

We can now state the time complexity of this algorithm.

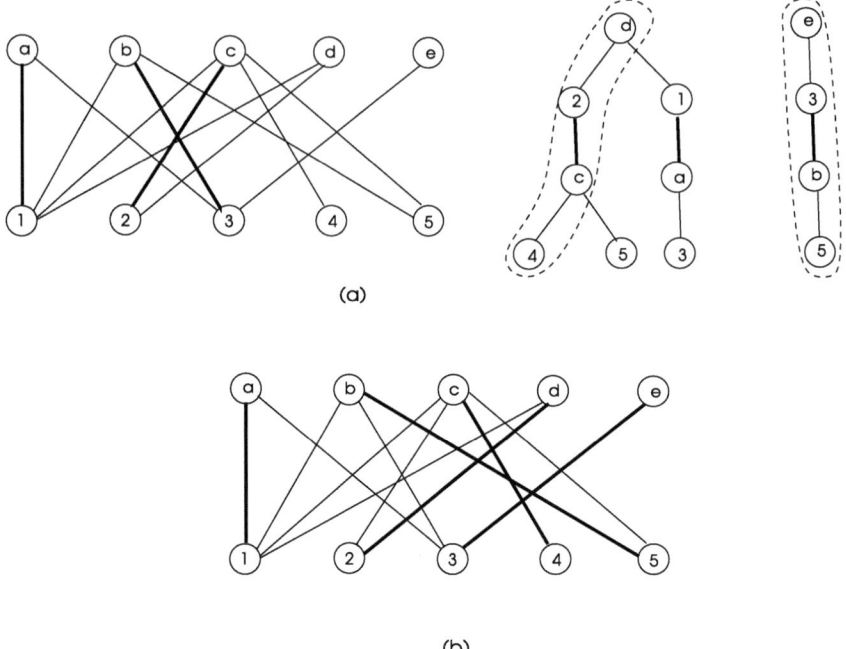

(a)

(b)

Fig. 9.9 Running of Hopcroft–Karp algorithm in a small bipartite graph. Augmenting paths in BFS trees are enclosed in dashed lines

Theorem 9.5 *Hopcroft–Karp algorithm has a running time of* $O(\sqrt{n}m)$.

Proof Each phase of the algorithm increases the length of the shortest augmenting path by at least one. Therefore, the length of the shortest augmenting path after $\lfloor\sqrt{n}\rfloor$ iterations will be at least $\lfloor\sqrt{n}\rfloor + 1$. There will be at most $|n|/(\sqrt{n}+1) \le \sqrt{n}$ augmenting paths left and hence, the algorithm will run for another $\sqrt{n}$ iterations at most. The total number of loop execution will therefore be $2\sqrt{n}$ times. Each iteration of the while loop requires $O(m)$ time due to BFS and DFS algorithms making the time complexity of this algorithm $O(\sqrt{n}m)$.   □

Parallel Hopcroft–Karp Algorithm

The disjoint BFS operations during the first part of each phase and disjoint DFS operations in the second part of Hopcroft–Karp algorithm make it suitable for parallel processing. The BFS-based graph construction can be performed in parallel by initiating the modified BFS algorithm simultaneously from free vertices in the left set of the bipartite graph. This approach among other methods such as lookahead DFS algorithm is experimented in [2] using multithreads.

9.4   Unweighted Matching in General Graphs

We will review sequential, parallel, and distributed matching algorithms in unweighted general graphs in this section.

9.4.1   Sequential Algorithms

The first sequential algorithm is a greedy one that selects legal edges iteratively and the second algorithm finds MaxM in linear time.

9.4.1.1   Greedy Algorithm
For an unweighted graph G, a greedy algorithm to find MM M can be designed so as to pick an edge (u, v) randomly, include it in M and remove all edges that are incident to u or v from G as shown in Algorithm 9.5.

Algorithm 9.5 *Seq_MM*

1: **Input** : $G = (V, E)$
2: **Output** : MM M of G
3: $M \leftarrow \emptyset$
4: $E' \leftarrow E$
5: **while** $E' \neq \emptyset$ **do**
6:     **select** any $(u, v) \in E'$
7:     $M \leftarrow M \cup \{(u, v)\}$
8:     $E' \leftarrow E' \setminus \{\{(u, v)\} \cup$ all $(u, x) \in E' \cup$ all $(v, y) \in E'\}$
9: **end while**

This process is repeated until there are no edges left. Operation of this algorithm is depicted in Fig. 9.10 The greedy algorithm is correct since we never select any adjacent edges to be included in M (matching rule) as these are deleted from graph and we continue until graph becomes empty meaning there can be no more edges added to M (MM rule). The number of iterations of the *while* loop has an upper bound as the number of edges and hence the time complexity of this algorithm is $O(m)$.

9.4.1.2   Edmond's Blossom Algorithm
We have seen the main method of obtaining a maximum matching in unweighted graphs; start with some initial matching M, find an augmenting path P with respect to M and form $M \oplus P$ to get a matching M' which has a size one unit larger than M; and repeat this procedure until no more augmenting paths are found. We described methods to find augmenting paths in bipartite graphs, however, finding augmenting paths in a general graph is more difficult due to odd alternating cycles which do not exist in bipartite graphs. Let us take a look at the graph in Fig. 9.11a where the current matching is shown with bold edges and let us try to find an augmenting path from the free vertex a. The augmenting path search may traverse the 5-cycle

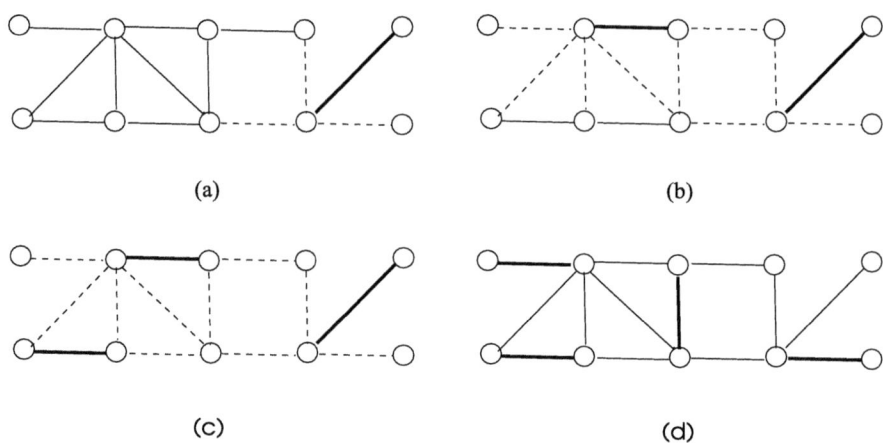

(a)   (b)

(c)   (d)

Fig. 9.10 Three iterations of the greedy matching algorithm in a sample graph results in MM of cardinality 3 as shown in **a**, **b** and **c**. Matching edges are shown in bold and the deleted edges are shown as dashed. A MaxM of the same graph is displayed in **d**

and fail to find the augmenting path that has a and j as endpoints by ending with edge (i, e) or edge (f, e) instead. We need a way to find augmenting paths in the presence of such odd alternating cycles.

Edmond presented a linear time algorithm to overcome this difficulty in unweighted general graphs [7]. This algorithm improves the current matching by finding augmenting paths in the graph as in other matching algorithms but by also taking care of odd alternating cycles. The general idea of this algorithm is to detect such cycles and remove them by shrinking them to super nodes and then carry on search of augmenting paths. A *blossom* in a graph G is an odd cycle consisting of $2k + 1$ edges with exactly k edges belonging to the current matching M as shown in Fig. 9.11a where the blossom consists of vertices e, f, g, h, i and the *stem* is an even-length alternating path of vertices a, b, c, d and e, starting from a free vertex and ending at the *base* (or the *tip*) of the blossom. The base vertex of the blossom is connected to the stem and is both part of the stem and the blossom. The stem and the blossom form the *flower*. The essence of this algorithm relies on the following theorem which we state without proof.

Theorem 9.6 *Let $G = (V, E)$ be an unweighted general graph with a matching M. Let B a blossom discovered in this graph and $G' = G \setminus B$ with a matching M' of the graph obtained after contracting G using B as a single vertex. M' is a maximum matching in G if and only if M is a maximum matching in G.*

We can therefore investigate augmenting paths in the contracted graph G' by contracting blossoms as we perform the search and whenever an augmenting path P' is discovered in G', we uncontract blossoms to get G and mark the corresponding augmenting path P in G. Last step of the iteration is to update current matching

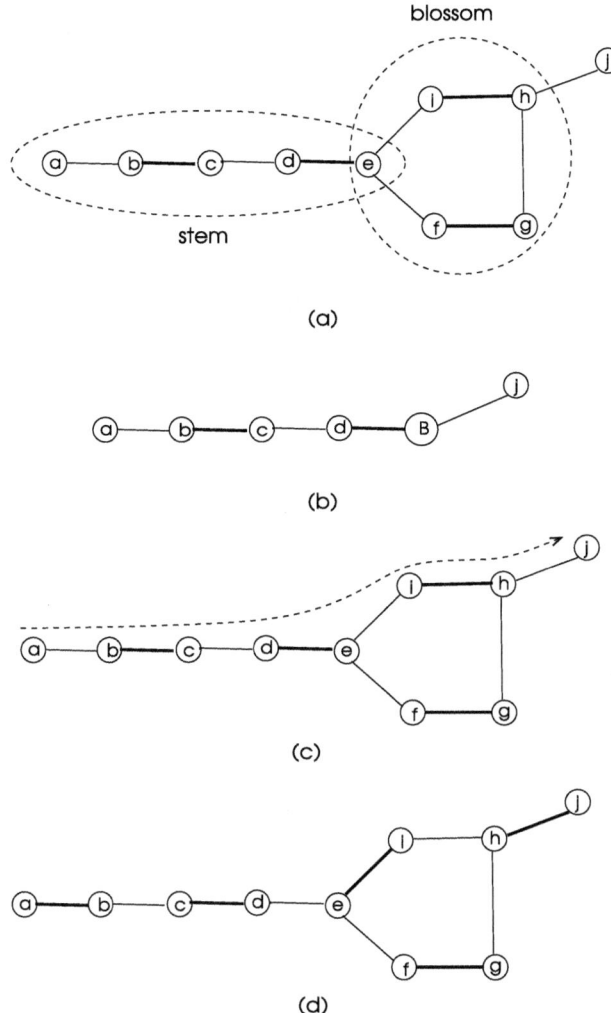

Fig. 9.11 The flower, blossom and stem of a sample graph and contracting and uncontracting of the blossom

to obtain $M \leftarrow M \oplus P$. An M'-augmenting path in G' exists if and only if there is an M-augmenting path in G. If we find an M'-augmenting path P' in G', we can form an augmenting path P in G after uncontracting the blossom B with a base vertex b_B as follows:

1. If P' starting from a free vertex u in G' goes through b_B and ends at a free vertex v in G', then P' is replaced by a path $u \rightarrow (x \rightarrow \ldots \rightarrow y)$ such that the edges in the blossom included in P are alternating.

2. If P' starting from a free vertex u in G' ends at b_B, the path $u \to v_B$ is replaced by the path $u \to (x \to \ldots \to y)$ such that path $P = u \to y$ is alternating and y is a free vertex.

The contraction of the blossom of the graph G in Fig. 9.11a to get G' is depicted in (b) where we have an augmenting path and no more blossoms, therefore we can uncontract the blossom in (c) to mark the augmenting path shown by dashed lines. As this path runs through the blossom B, we select alternating edges in B to complete the augmenting path that ends at vertex j. Finally, we form the new matching $M \leftarrow M \oplus P$ with size 5 which in fact is maximum for this graph as there are no blossoms or augmenting paths left. Another example when the alternating path ends in a blossom is shown in Fig. 9.12. We apply the same strategy, shrink the blossom B to obtain G' first in (b), search for an augmenting path in G' and when such a path P finishing at B is found as shown in (c), unshrink B and mark the edges of the augmenting path inside the blossom accordingly (c). Finally, perform $M \leftarrow M \oplus P$ to obtain the matching M of size 5 in (d) which is maximum as there are no other blossoms or augmenting paths.

As we have seen in these examples, there are three possibilities while searching for an augmenting path in the graph G;

1. No augmenting path found: In this case, the current matching M is maximum and algorithm terminates
2. An augmenting path P is found: $M \leftarrow M \oplus P$; continue
3. A blossom B is found: The blossom B is contracted by replacing it with its base vertex.

Therefore, this algorithm will either find an augmenting path or a blossom or conclude that these do not exist in the graph in which case, the matching is maximum and the algorithm stops. We show a high-level pseudocode of Edmond's algorithm in Algorithm 9.6. The algorithm starts to build a BFS tree from an exposed vertex and labels edges at even levels as *outer* (O) and at odd level as *inner* (I). Whenever we find two outer vertices that are neighbors, a blossom with an odd cycle is encountered. This blossom is contracted to a single vertex and the search continues. When we find an augmenting path P in the contracted graph, then all of the blossoms found so far are uncontracted and the new matching $M \leftarrow M \oplus P$ is computed.

Algorithm 9.6 *Edmond_Blossom*

1: **Input** : $G = (V, E)$
2: **Output** : MaxM M of G
3: $M \leftarrow \emptyset$
4: $S \leftarrow$ an arbitrary free vertex in V
5: **for all** $v \in V$ such that v is saturated **do**
6:     **search** for simple paths starting from v
7:     **shrink** any blossoms found
8:     **if** any found path P ends at a saturated vertex **then**
9:         $M \leftarrow M \oplus P$ ▷ P is an augmenting path
10:     **else if** no augmenting paths found **then**
11:         **discard** v in future searches
12:     **end if**
13: **end for**

A more detailed example with two nested blossoms is shown in Fig. 9.13. We start a BFS from a free vertex a and label vertices as inner and outer corresponding to odd and even levels respectively. Vertices c and e are both outer vertices, therefore a blossom is detected and this is contracted to vertex B_1 in the new graph G'. We find another blossom (B_2) in G' and this is contracted to give the new graph G''. Note that in between G' and G'' formation, we have not encountered an augmenting path, otherwise we would have uncontracted B_1 in G' to get a new matching. We find an augmenting path in G'' and therefore uncontract blossoms and mark the augmenting path P through them so as to alternate. Finally the new matching M is formed by $M \leftarrow M \oplus P$.

Analysis

The algorithm is based on Berge's theorem, it attempts to find an augmenting path in a general graph and when such a path is found, it enlarges it. It only remains to show contracting and uncontracting of blossoms do not disturb the augmenting paths found.

Theorem 9.7 *Edmond's algorithm has a time complexity of $O(n^2 m)$.*

Proof There will be at most n augmentations and there will be at most $n / 2$ times of blossom shrinking between any two augmentations. The alternating tree can be constructed in $O(m)$ time, therefore the total time taken is $O(n^2 m)$. $\square$

An improvement to the running time of this algorithm to $O(\sqrt{n} m)$ was provided by Micali and Vazirani [19] and a complete proof was given in [25].

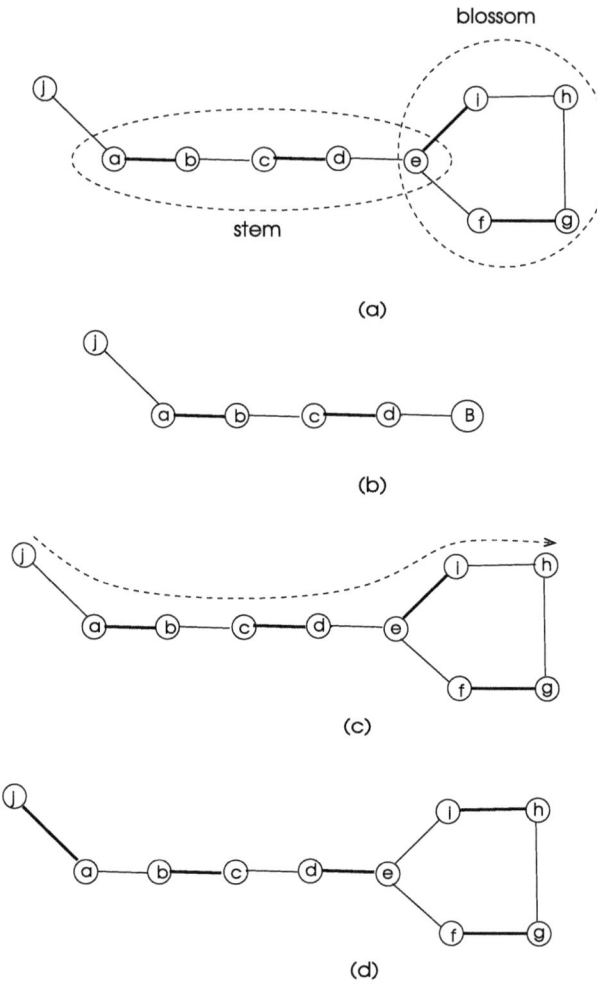

Fig. 9.12 Searching an augmenting path when path ends in a blossom

9.4.2   A Greedy Distributed Algorithm

In a distributed environment, we want to find MM of a network such that each node of the network should decide whether it is saturated or adjacent to a node that is saturated in the end. We will describe a distributed algorithm that uses *edge coloring*. An edge coloring of a graph G is assignment of colors in the form of integers to each edge of the graph such that no two adjacent edge has the same color as we will review in more detail in Chap. 11. Edges of the same color constitute an edge *color class* and we can see instantly that any edge coloring of a graph G provides a matching of G since edges in a color class cannot be adjacent.

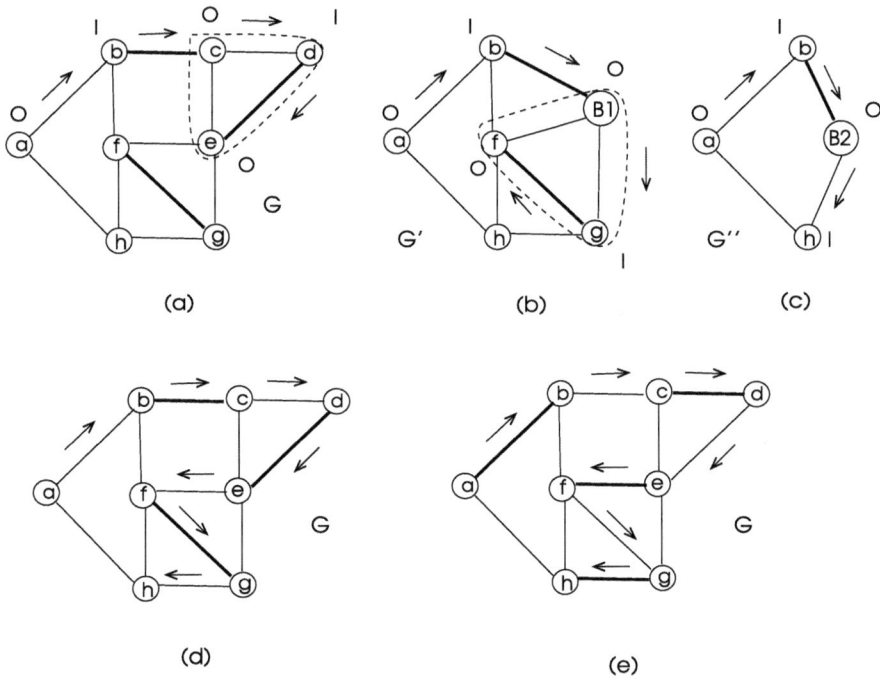

Fig. 9.13 Working of Edmond's algorithm in a simple graph

However, when attempting to find a maximal matching using this method, we need to be careful since the union of edge color classes clearly contains adjacent edges. However, we can start with color class 1 for example, include all edges of this class in matching since these are not adjacent by the definition of edge coloring and then continue with color class 2. We should include an edge in this class only if it is not adjacent to any previously matched edge.

A distributed algorithm based on this observation is proposed in [12] to find a maximal matching in a network which is already edge colored with k colors. It is an SSI algorithm working in rounds under the control of a *root* node. There are k rounds starting with round 1 and at round r, any node that has an incident edge (u, v) colored with r checks wether it can include (u, v) in matching legally. That is, there are no other edges adjacent to (u, v) that are included in the matching in the previous rounds. We will sketch a possible implementation of this algorithm as in [9] but by using a FSM. There are three states of a node as follows, also as shown in Fig. 9.14.

- UNMATCHED: Initially, all nodes are in UNMATCHED state which means they can compete to be a matched node.
- MATCHED: Any node that has an incident edge incident to it which is determined to be a matching edge enters this state.

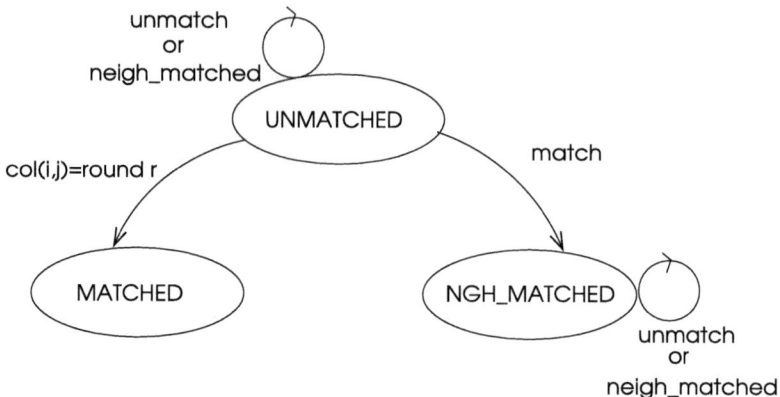

Fig. 9.14 The FSM of the matching algorithm for node i with a neighbor node j

- NEIGH_MATCHED: When a node has a neighbor that is MATCHED, it is assigned to this state.

We have the following message types:

- *round(r)*: Sent by the root i to initiate round r.
- *match(r)*: Sent by a node i that is matched to its neighbors.
- *unmatch*: Sent by a node i that does not have an incident edge with the same color of the round number r. This is needed for synchronization.
- *neigh_match(r)*: Sent by a node i that does not have an incident edge with the same color of the round number r.

The pseudocode for a single round of this distributed algorithm for a node i is shown in Algorithm 9.7.

Algorithm 9.7 *Edgecol_MM*

1: **int** *i,j*   ▷ *i* is this node, *j* is the sender of a message
2: **message types** *round, match,unmatch*
3: **states** UNMATCHED, MATCHED, NGH_MATCHED
4: **boolean** *round_over, round_recvd ← false*
5: *curr_neighs ← N(i); received, neighs_removed ← {∅}*
6: *currstate ←* UNMATCHED
7: **for** *r =* 1 to *k* **do**
8:          { round *r* for all nodes}
9:          **while** ¬*round_over* **do**
10:              **receive** *msg(j)*
11:                 **case** *msg(j).type* **of**
12:                     *round(r)*:   **if** *currstate ≠* MATCHED **then**
13:   **if** (∃*j ∈ curr_neighs* such that *col(i, j) = r*) **then**
14:   *currstate ←* MATCHED
15:   **send** *match* to *j*
16:   **send** *neigh_match* to *curr_neighs \ {j}*
17:   **else send** *unmatch* to *curr_neighs*
18:   *round_recvd ← true*
19:                     *match(r)*:   *currstate ←* NEIGH_MATCHED
20:   *received ← received ∪ {j}*
21:                     *neigh_match(r)*:   *received ← received ∪ {j}*
22:   *neighs_removed ← neighs_removed \ {j}*
23:                     *unmatch(r)*:   *received ← received ∪ {j}*
24:
25:              **if** *round_recvd ∧ (received = curr_neighs)* **then**
26:                  *curr_neighs ← {curr_neighs \ neighs_removed}; received ← ∅*
27:                  *round_recvd ← false ; round_over ← true;*
28:              **end if**
29:          **end while**
30: **end for**

The operation of this algorithm in a small sample network is depicted in Fig. 9.15. This algorithm correctly finds a maximal matching in a network since we obey the matching rule in each round by not considering adjacent edges of the matched edges and also, we continue until each color class is considered and thus the matching is maximal. There will be a total of *k* rounds for a *k- edge-colored* network and each edge will be traversed at most once by the *match, unmatch* or *neigh_match* messages and thus the total number of messages transferred is $O(km)$.

9.5   Weighted Bipartite Graph Matching

Edges of a weighted bipartite graph $G = (A \cup B, E, w)$, $w : E \to \mathbb{R}$ have weights associated with them. Our aim is to search for a total maximum or minimum weight maximal matching in such graphs.

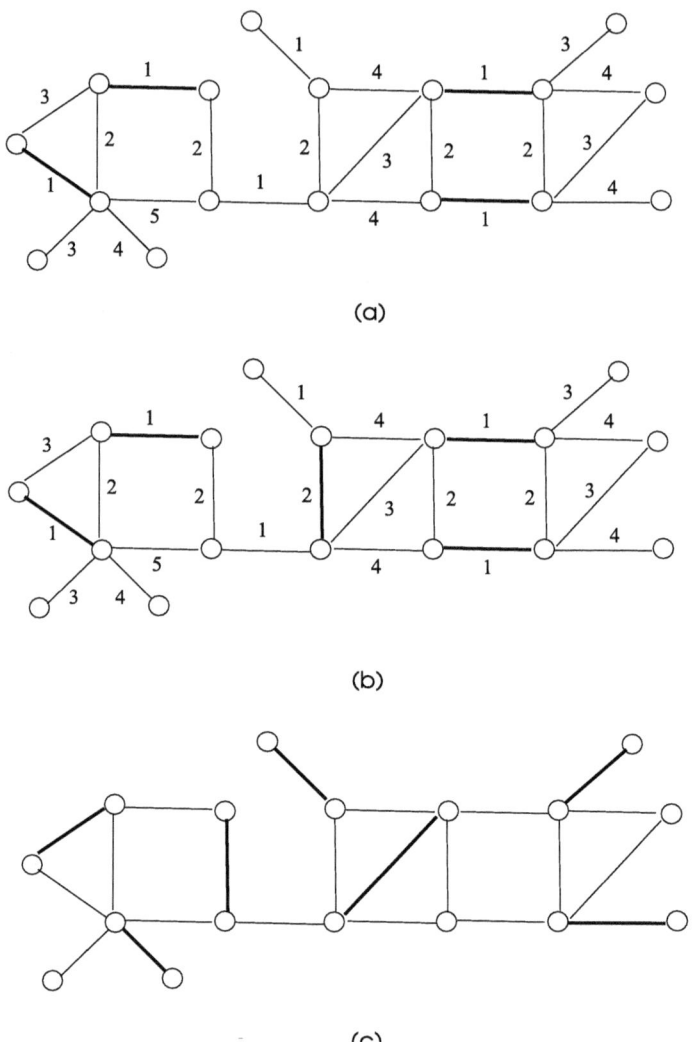

(a)

(b)

(c)

Fig. 9.15 Running of Algorithm 9.7 in a sample network. The first and second rounds are shown in **a** and **b** respectively. In only two rounds a maximal matching of size 5 is obtained. The maximum matching of size 7 for this network is shown in **c**

9.5.1 Greedy Algorithm

We can implement a greedy strategy in which we always select the greatest weight available edge from all available edges. We need to sort the weights of edges initially and then check availability. The running time of this algorithm is dominated by the sorting operation and hence we need $O(m \log m)$ and the approximation factor is 1/2 [22].

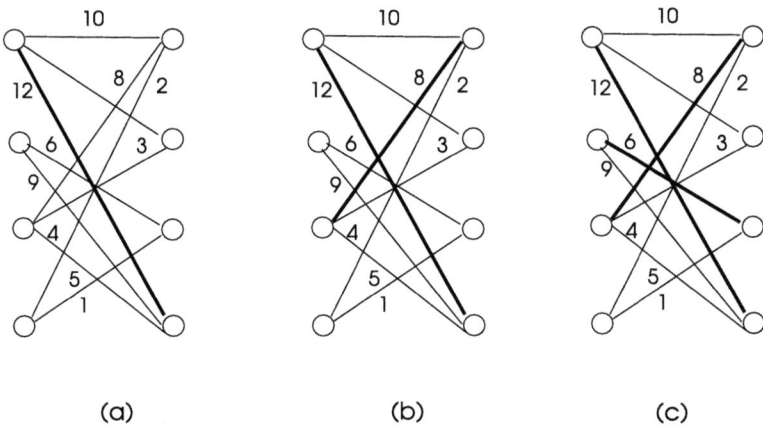

(a)   (b)   (c)

Fig. 9.16 Running of the greedy algorithm to find MWM in a weighted bipartite graph with weights shown next to edges. The largest weight available edge is selected in each step to obtain the final matching of total weight 26 shown with bold lines in **c** in 3 iterations

---

Algorithm 9.8 *MWM_BPG*

1: **Input**: An undirected weighted bipartite graph $G = (A \cup B, E)$
2: **Output**: Maximal weighted matching M of G
3:
4: $Q \leftarrow$ sorted edges of G in the order of decreasing weights
5: **while** $Q \neq \emptyset$ **do**
6:      $(u, v) \leftarrow Q.front$
7:      $M \leftarrow M \cup (u, v)$
8:      **remove** all edges incident to u or v from Q
9: **end while**

Running of this algorithm in a small bipartite graph is shown in Fig. 9.16.

9.5.2 The Hungarian Method

The Hungarian method, so-called by its developer Kuhn as it relies on the earlier ideas of two Hungarian mathematicians König and Egervary, finds the maximum matching in a weighted complete bipartite graph with the same order of bipartite vertex sets in linear time [17]. This method solves the *assignment problem* which aims to assign objects such as machines, people, processors to tasks by finding minimum or maximum weighted matching in such a graph. Let us assume we are given a set of people and a set of tasks these people can perform which are the vertices of the bipartite graph consecutively. The weight on an edge (u, v) shows the time required by person u to perform task v and our aim is to have the minimum amount of time to get all tasks done. We could have processors of a multiprocessing system instead of people and tasks would be software modules running on these machines in this case. We will describe this method in two equivalent approaches; using the cost matrix and as a graph-theoretic method (Kuhn-Munkres algorithm). These two approaches have the same time complexity.

9.5.2.1 Matrix Interpretation

We can consider the assignment problem as a weighted bipartite graph $G = (A \cup B, E)$ where we need to assign vertices in A to the vertices in B to result in optimal mapping. The graph can be represented by the cost matrix C with elements c_{ij} denoting the cost of assigning $a_i \in A$ to $b_j \in B$. The order of A and B should be equal and if this is not provided, we can simply add dummy rows or columns with 0 entries to make them equal. This algorithm relies on the following two observations.

- If a number is added to or subtracted from all of the entries of any one row or column of a cost matrix C_i to get a cost matrix C_{i+1}, then on optimal assignment for the cost matrix C_{i+1} is also an optimal assignment for the cost matrix C_i.
- We have an optimal assignment of a_i to b_j if $c_{ij} = 0$. In other words, if we can reduce the cost of assigning an element of A to an element of B to zero, this assignment is optimal.

The approach of the matrix interpretation of the Hungarian method is then to transform the original cost matrix C_1 to a matrix C_k to provide zero assignment in each row and column using add and subtract operations. We can have the following steps of the algorithm to achieve this goal.

1. *Reduce rows*: Subtract the least value of each row from all of the entries in its row.
2. *Reduce columns*: Subtract the least value of each column from all of the entries in its column.
3. *Cover zeroes*: Cover the zero entry rows and columns using minimum number of lines.
4. **if** number of lines is n goto 6.
5.   **else** Find the smallest uncovered element x. Subtract x from all of the uncovered elements of C and add x to elements that are at the intersection of the covering lines in 3. Goto 1.
6. *Assignment*: Select a row or column with only one zero and assign. If not found, select arbitrarily. Select other assignments so that no two tasks are assigned to same persons.

Let us see the operation of this algorithm through an example shown in Fig. 9.17a.

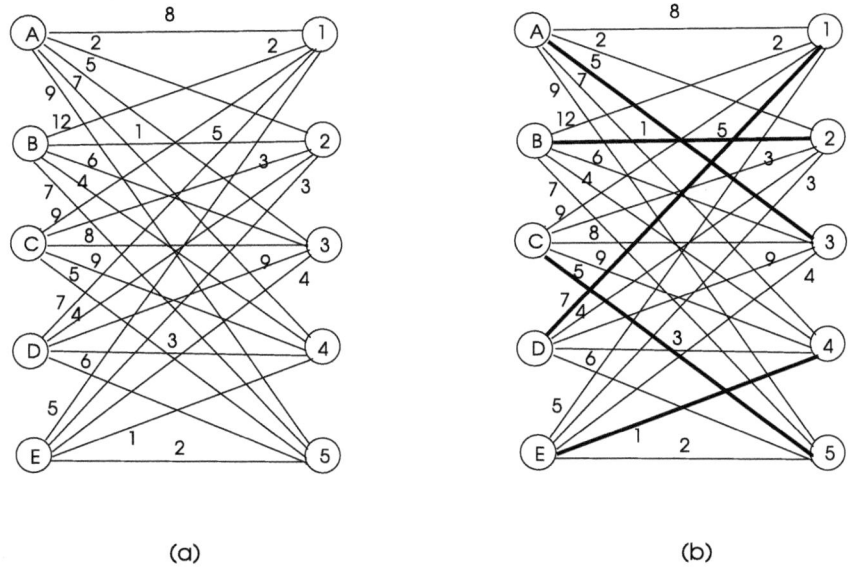

Fig. 9.17 A sample weighted fully connected bipartite graph to test the Hungarian algorithm

The cost matrix C for this graph is given as below and the first two steps of the algorithm which reduces rows and then columns results in the matrices shown.

$$
\begin{array}{c}
\begin{array}{cc}
 & 1\,2\,3\,4\,5 \\
A & \begin{pmatrix} 8\ 2\,5\,7\,9 \\ 12\ 1\ 6\,4\,7 \\ 9\ 3\,8\,9\,5 \\ 7\ 4\,9\,3\,6 \\ 5\ 3\,4\,1\,2 \end{pmatrix}
\end{array}
\begin{pmatrix} 2 \\ 1 \\ 3 \\ 3 \\ 1 \end{pmatrix}
\end{array}
\rightarrow
\begin{array}{cc}
 & 1\,2\,3\,4\,5 \\
\begin{array}{c} A \\ B \\ C \\ D \\ E \end{array} & \begin{pmatrix} 6\ 0\,3\,5\,7 \\ 11\ 0\,5\,3\,6 \\ 6\ 0\,5\,6\,2 \\ 4\ 1\,6\,0\,3 \\ 4\ 2\,3\,0\,1 \end{pmatrix}
\end{array}
\rightarrow
\begin{array}{cc}
 & 1\,2\,3\,4\,5 \\
\begin{array}{c} A \\ B \\ C \\ D \\ E \end{array} & \begin{pmatrix} 2\,0\,0\,5\,7 \\ 7\,0\,2\,3\,6 \\ 2\,0\,2\,6\,2 \\ 2\,1\,3\,0\,3 \\ 2\,2\,0\,0\,1 \end{pmatrix}
\end{array}
$$

$$(4 \quad 0 \quad 3 \quad 0 \quad 1)$$

Covering rows and columns with zeroes results in the first left matrix C below with covered entries shown in bold. Since the number of covered lines is 4 which is less than 5, we need to continue with the algorithm. We select the lowest uncovered value which is 2 and subtract 2 from all of the uncovered values and add it to the entries at the intersection of the covered entries to obtain the second matrix and cover this matrix this time again with covered rows and columns shown in bold figures.

$$
\begin{array}{c}
\begin{array}{ccccc} 1 & 2 & 3 & 4 & 5 \end{array}\\
\begin{array}{c} A \\ B \\ C \\ D \\ E \end{array}
\begin{pmatrix}
2 & 0 & 0 & 5 & 7 \\
7 & 0 & 2 & 3 & 6 \\
2 & 0 & 2 & 6 & 2 \\
2 & 1 & 3 & 0 & 3 \\
2 & 2 & 0 & 0 & 1
\end{pmatrix}
\end{array}
\rightarrow
\begin{array}{c}
\begin{array}{ccccc} 1 & 2 & 3 & 4 & 5 \end{array}\\
\begin{array}{c} A \\ B \\ C \\ D \\ \end{array}
\begin{pmatrix}
2 & 2 & 0 & 7 & 7 \\
5 & 0 & 0 & 3 & 4 \\
0 & 0 & 0 & 6 & 0 \\
0 & 1 & 1 & 0 & 3 \\
2 & 4 & 0 & 0 & 1
\end{pmatrix}
\end{array}
$$

We find that the number of covered rows and columns is 5 which is the number of vertices of the bipartite graph, therefore we stop and move on to the assignment step. We search for single 0 rows first as this means that person can only do the task that has 0 in its column. Person A has such a property and we assign task 3 to her and delete task 3 column as this task cannot be assigned to another person. Person F can do tasks 3 and 4 but since task 3 is already assigned, we have to assign task 4 to her and delete column 4 from the matrix. Similarly, person C is assigned task 2 and person E can only be assigned to task 1 which leaves person D only with task 5 although she is capable of performing tasks 1, 2, 3 and 5. The assignments in 0 locations are shown in bold in the final cost matrix in below left and the actual assignment in the original cost matrix using these values is shown in below right. For the total time taken, we calculate this as $5 + 1 + 5 + 7 + 1 = 19$ units from the original cost matrix. This matching is depicted in the bipartite graph of Fig. 9.17b.

$$
\begin{array}{c}
\begin{array}{ccccc} 1 & 2 & 3 & 4 & 5 \end{array}\\
\begin{array}{c} A \\ B \\ C \\ D \\ E \end{array}
\begin{pmatrix}
2 & 2 & 0 & 7 & 7 \\
5 & 0 & 0 & 3 & 4 \\
0 & 0 & 0 & 6 & 0 \\
0 & 1 & 1 & 0 & 3 \\
2 & 4 & 0 & 0 & 1
\end{pmatrix}
\end{array}
\leftrightarrow
\begin{array}{c}
\begin{array}{ccccc} 1 & 2 & 3 & 4 & 5 \end{array}\\
\begin{array}{c} A \\ B \\ C \\ D \\ E \end{array}
\begin{pmatrix}
8 & 2 & 5 & 7 & 9 \\
12 & 1 & 6 & 4 & 7 \\
9 & 3 & 8 & 9 & 5 \\
7 & 4 & 9 & 3 & 6 \\
5 & 3 & 4 & 1 & 2
\end{pmatrix}
\end{array}
$$

9.5.2.2 Kuhn-Munkres Algorithm

The Kuhn-Munkres algorithm was first proposed by Kuhn [17] and later analyzed by Munkres [20] to solve the assignment problem efficiently. Before describing the operation of this algorithm, we need to make few definitions.

Definition 9.6 (*vertex labeling*) A vertex labeling of a graph $G = (V, E)$ is a function $l : V \rightarrow \mathbb{R}$. A legal labeling allows labeling two vertices u and v of a bipartite graph G such that,

$$ l(u) + l(v) \geq w(u, v), \forall(u, v) \in E \text{ (labeling rule)}. \tag{9.2} $$

Definition 9.7 (*equality graph*) The *equality graph* of a graph $G = (V, E)$ with respect to a labeling function l is a graph $G_l = (V, E_l)$ such that

$$E_l = \{l(u, v) : w(u, v) = l(u) + l(v)\} \tag{9.3}$$

We can see immediately that G_l is a spanning subgraph of G since it contains all vertices of G and $E_l \subseteq E$ as we select edges of G in G_l that only provide equivalence. A feasible labeling of a bipartite graph $G = (A \cup B, E)$ can be achieved by assigning each vertex of the set B the maximum of the weights of all edges incident to it and 0 to each vertex of the set A. That is,

$$\forall u \in A, l(u) = 0, \text{ and } \forall v \in B, l(v) = max\{w(u, v)\}. \tag{9.4}$$

This way, we make sure labeling rule is obeyed and an initial equality graph is obtained. A bipartite graph that is labeled accordingly and its equality graph are depicted in Fig. 9.18. Note that when the bipartite graph is not fully connected, we need to append edges with 0 weights.

The following theorem due to Kuhn and later Munkres provides the basis for this graph-theoretic assignment algorithm.

Theorem 9.8 (Kuhn-Munkres) *Given a weighted bipartite graph $G = (A \cup B, E)$ and its equality graph $G_l = (A \cup B, E_l)$, M is a perfect matching in G_l if and only if M is a maximum weight matching in G.*

Proof Let us assume G_l contains a perfect matching M. The matching M is also a perfect matching in G since all edges of G_l are contained in the edge set E of G which means,

$$w(M) = \sum_{e \in M} w(e) = \sum_{v \in V(G)} l(v) \tag{9.5}$$

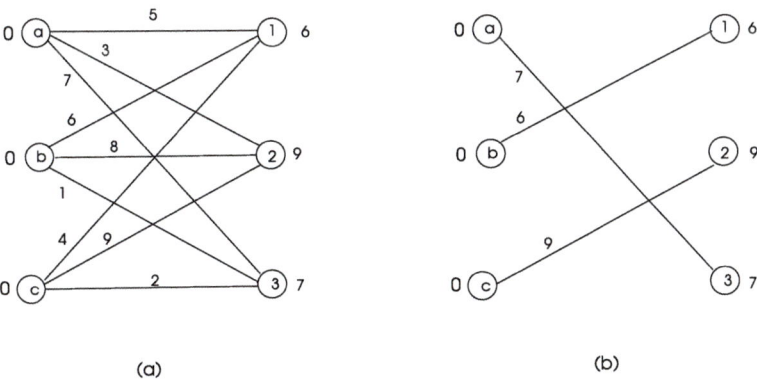

(a)   (b)

Fig. 9.18 **a** A legally labeled weighted bipartite graph, **b** Its equality graph

Let us assume M' is any matching in G. Then, the labeling rule implies,

$$w(M') = \sum_{e \in M'} w(e) \leq \sum_{v \in V(G)} l(v). \qquad (9.6)$$

Therefore $w(M') \leq w(M)$ which means M is a maximum matching of the graph G. □

Thus, finding the maximum weight matching in the original graph G is reduced to finding a perfect matching of the equality graph G_l. We can now form the steps of the algorithm based on this theorem.

1. l is a legal labeling of G, M is an initial matching of G_l
2. **while** M is not perfect matching in G_l
3.   **find** an augmenting path P in G_l
4. $M \leftarrow M \oplus P$
5.   **if** $P = \emptyset$
6. $l' \leftarrow l$ such that $E_l \subset E_{l'}$.

Finding new labeling l' is crucial in the operation of this algorithm. For a legal labeling of the graph G, let us first define the neighborhood relations of a vertex in G_l and the set S,

$$N_l(u) = \{v : (u, v) \in E_l\}, N_l(S) = \bigcup_{u \in S} N_l(u) \qquad (9.7)$$

For $S \subseteq A$ and $T = N_l(S) \neq B$, let us define parameter a_l

$$a_l = min\{u \in S, v \notin T\}\{l(u) + l(v) - w(u, v)\}. \qquad (9.8)$$

Now, the improved labeling l' for any vertex of G can be specified in terms of the previous labeling l using a_l as follows.

$$l'(x) = \begin{cases} l(x) - a_l, & \text{if } x \in S \\ l(x) + a_l, & \text{if } x \in T \\ l(x) & \text{otherwise} \end{cases} \qquad (9.9)$$

We can now write the pseudocode of the Kuhn-Munkres algorithm as shown in Algorithm 9.1.

Algorithm 9.9 *Kuhn-Munkres_WBM*

1: **Input**: An undirected weighted complete bipartite graph $G = (A \cup B, E)$
2: **Output**: Maximum weighted matching M of G
3:
4: $E_l \leftarrow \{(u, v) \in E(G) : w(u, v) = l(u) + l(v)\}$ ▷ labeling of G according to Eqn. 9.4
5: $M \leftarrow$ some initial matching of G
6: **while** A is not M-saturated in G_l **do**
7:    **select** an unmatched vertex $x \in A$
8:    $S \leftarrow \{x\}, T \leftarrow \emptyset$
9:    **repeat**
10:      **if** $N_{G_l}(S) = T$ **then**
11:        $a_l \leftarrow min_{u \in S, v \notin T}\{l(u) + l(v) - w(u, v)\}$
12:        **for all** $u \in S$ **do**
13:          $l(u) \leftarrow l(u) - a_l$
14:        **end for**
15:        **for all** $v \in T$ **do**
16:          $l(v) \leftarrow l(v) + a_l$
17:        **end for**
18:        **update** G_l
19:      **end if**
20:      **select** $v \in \{N_{G_l}(S) - T\}$
21:      **while** v is M-saturated and $N_{G_l}(S) \neq T$ **do**
22:        $u \leftarrow$ a vertex in A matched to v in M
23:        $S \leftarrow S \cup \{u\}, T \leftarrow T \cup \{v\}$
24:        **if** $N_{G_l}(S) \neq T$ **then**
25:          **select** $v \in \{N_{G_l}(S) - T\}$
26:        **end if**
27:      **end while**
28:    **until** v is M-unsaturated
29:    $P \leftarrow M$-augmenting path from x to v
30:    $M \leftarrow M \oplus P$
31: **end while**

An example operation of this algorithm in a small weighted bipartite graph is depicted in Fig. 9.19.

Analysis

There are n phases of the algorithm and at each phase the size of the matching is incremented by 1. Initial slack calculation takes $O(n^2)$ time. When a vertex moves from S to T, we compute slacks for all $y \in T$. In each phase, at most n vertices go from S to T, so n slack re-calculations, each at $O(n)$ time, for a total of $O(n^2)$. The algorithm takes $O(n^3)$ time in total.

9.5.3   The Auction Algorithm

Matching in weighted bipartite graphs problem can be solved   efficiently using the auction method which is based on game theory. Auctions in everyday life involves an auctioneer opening bidding and bidders submitting bids and the object

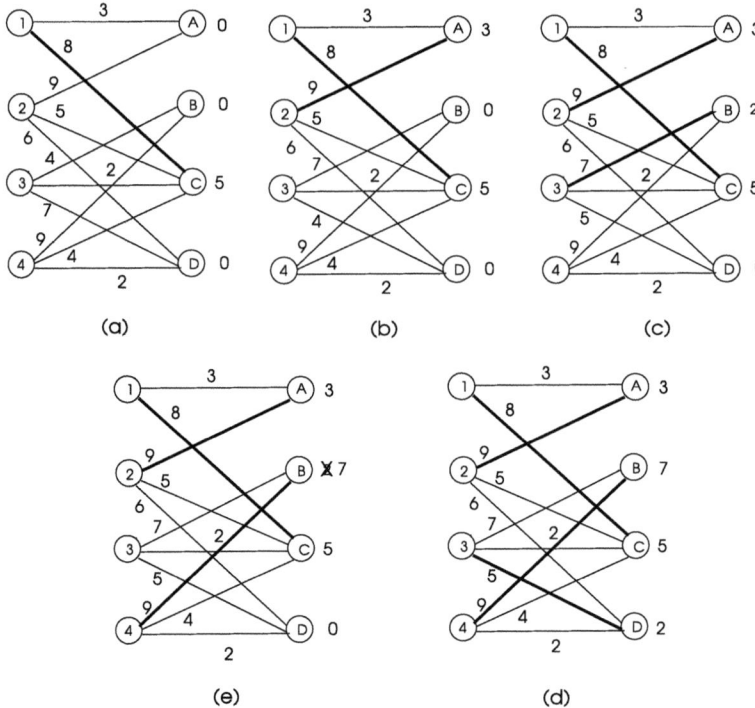

(a)   (b)   (c)

(e)   (d)

Fig. 9.19 Running of Kuhn-Munkres algorithm in a small weighted bipartite graph. The edges of the matching obtained at each iteration are shown in bold

under consideration is acquired by the bidder that offers the highest price. Auction algorithms are based on this principle in which a bipartite graph $G = (A \cup B, E)$ is considered with vertex set A as buyers and B as objects [4]. Each object i has a price p_i associated with it and the weight of an edge between a bidder i and an object j, $w(i, j)$ shows the amount that bidder i values object j, in other words, it is the cost of the object as seen by buyer i. The algorithm consists of the bidding phase and the assignment phase. Each object can be sold to only one person and each person can buy only one object.

For each object, a buyer has a *benefit* and a *price* to be the owner of that object. The *profit* for an object by a buyer is the difference between the benefit and the price for an object. Algorithm 9.10 displays the pseudocode for the sequential auction algorithm as adapted from [24]. At each iteration, the first element of the buyers from the set B is selected, then an object with the maximum profit for that buyer is found. The second highest profit yielding object is also computed and the bid is computed as the difference of the first two best profits. The object that provides this bid is then assigned to the buyer in the assignment phase. The new price for the object is then increased by the bid and a small value designated as the ε which may be initialized to $\delta \leftarrow 1/(n+1)$. Iterations continue until each buyer is assigned to an object.

Algorithm 9.10 *Auction_Alg*

1: **Input** : $G(A \cup B, E, w)$ ▷ undirected weighted bipartite graph
2: **Output** : Matching M
3: $B = \{1, ..., n_1\}$: set of buyers
4: $M \leftarrow \emptyset$
5: **initialize** ε
6: **for** $j=1$ to n_2 **do**   ▷ initialize prices for objects to all zeroes
7:     $p_j \leftarrow 0$
8: **end for**
9: **while** $B \neq \emptyset$ **do**   ▷ start auction
10:     **select** $i \in B$ ▷ select an available buyer
11:     $j_i \leftarrow max_j\{w_{ij} - p_j\}$ ▷ find the best object for this buyer
12:     $s_i \leftarrow w_{ij_i} - p_{j_i}$ ▷ profit for the best object
13:     **if** $s_i > 0$ **then**
14:         $t_i \leftarrow max_{j \neq j_i}\{w_{ij} - p_j\}$ ▷ second best profit
15:         $p_{j_i} \leftarrow p_{j_i} + s_i - t_i + \varepsilon$ ▷ update the bid for object
16:         $M \leftarrow M \cup (i, j_i); B \leftarrow B \setminus \{i\}$ ▷ assign buyer to object
17:         $M \leftarrow M \setminus (k, j_i); B \leftarrow B \cup \{k\}$ ▷ release previous owner k
18:         **update** ε
19:     **else**
20:         $B \leftarrow B \setminus \{i\}$ ▷ no object with profit found for buyer i found,
21:     **end if**
22: **end while**

We will show the operation of this algorithm using the weighted bipartite graph of Fig. 9.19 as an example in Fig. 9.20. We have four buyers 1, 2, 3 and 4 and four objects A, B, C and D with initial bids all set to zeroes and $\varepsilon = 0$. We start with the lowest index buyer 1 who has the highest profit at object C with cost 8 and the second highest profit is object A with profit 3. The bid is therefore $8 - 5 = 3$ for object C as shown. Buyer 1 is assigned to this object and we start the second iteration with buyer 2. Similarly, buyer 2 is assigned to object A with bid 3 which is the difference of its two best profits as shown in (b) and buyer 3 is assigned to the object B with the bid 2 as depicted in (c). We have a different situation in (d) where buyer 4 can bid 7, which is the difference between its two best profits, for object B. This bid is higher than the current bid of 2 for object B and therefore we release buyer 3 from object B and assign buyer 4 to this object. Finally, buyer 3 is re-assigned this time to object D as shown in (d) which is the maximum weighted matching for this bipartite graph.

Recently, it was shown in [21] that the expected time complexity of the auction algorithm for random bipartite graphs where each edge is independently selected with probability $p \geq \frac{c \log n}{n}$ with $c > 1$ is $O(\frac{n \log^2 n}{\log np})$. Also in this study, the expected time complexity of this algorithm in a shared memory parallel system with $O(\log n)$ processors is shown to be $O(n \log n)$ (Fig. 9.20).

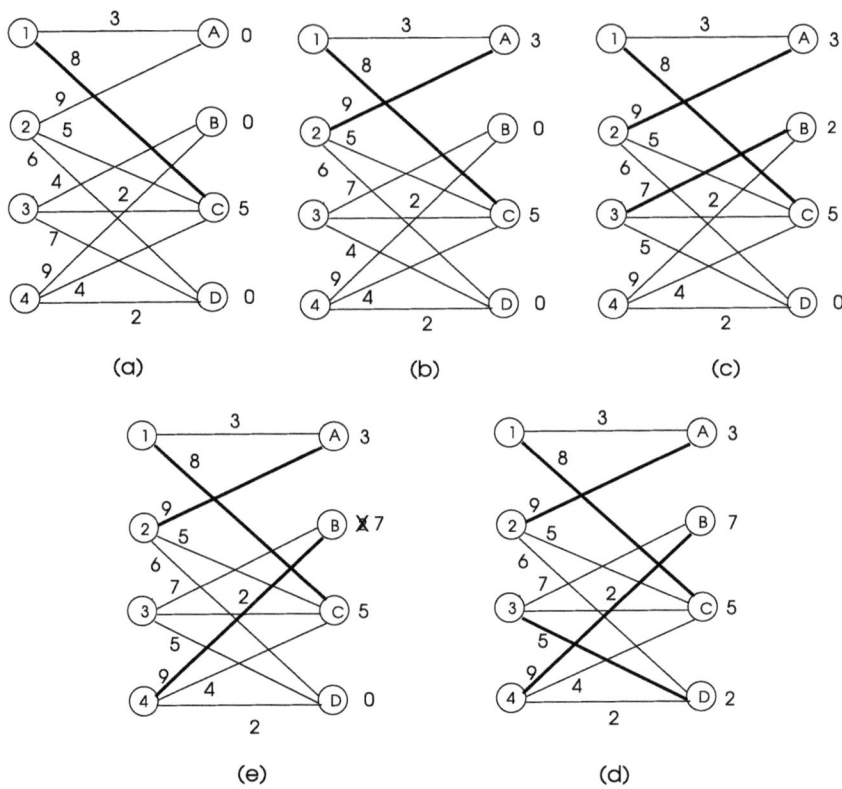

Fig. 9.20 A sample weighted bipartite graph to test the Auction algorithm. We have the same maximum matching as in Fig. 9.19

Parallel Auction Algorithms

The bidding and the assignment phases of the auction algorithm are both convenient for parallel processing. When each of these steps is performed simultaneously, a synchronous parallel algorithm is formed. Also, a buyer can make bids at arbitrary times in asynchronous operation. The asynchronous parallel mode of operation along with shared memory synchronous procedure are described and analyzed in [5]. Parallelization by distributing the vertices of the weighted bipartite graph to a set of parallel processes is presented in [24]. Each process performs the bidding phase of the algorithm for free buyers in its set and then these bids are exchanged with other processes to determine the largest global bid. Therefore, the bidding step is performed independently by each process with synchronization in the end. The implementation was carried in a distributed memory computer using MPI. Other implementations of parallel auction algorithm were performed on distributed memory computers in [6, 23].

9.6   Weighted Matching in General Graphs

As a first approach to design an approximation algorithm to find the maximal weight matching in a weighted graph, we can use the same strategy as in the unweighted graph case. This time, we need to change line 6 of Algorithm 9.5 to select the globally heaviest weight edge. This algorithm requires sorting of the edges with respect to their weights and hence the dominant time taken is this step resulting in a time complexity of $O(m \log m)$. This method results in the same MWM for the sample graph in Fig. 9.4. This algorithm has an approximation factor of 1/2 [22].

9.6.1   Preis Algorithm

Preis came up with a greedy weighted matching algorithm that has better performance than the global greedy algorithm [22]. The idea of this algorithm is to select the *locally heaviest* edges rather than a globally maximum weight one. A locally heaviest edge is the edge with largest weight among all of its adjacent edges. Selection of a locally heaviest edge is done arbitrarily choosing an edge (u, v) but if an adjacent edge that has a larger weight is found, then that edge is selected. The operation of this algorithm is depicted in Algorithm 9.11. We can see the local operations are independent and for this reason, this approach is suitable for distributed and also parallel matching.

Algorithm 9.11 Preis_MWM

1: **Input** : $G = (V, E)$
2: **Output** : MM M of G
3: $M \leftarrow \emptyset$
4: $E' \leftarrow E$
5: **while** $E' \neq \emptyset$ **do**
6:     select some locally heaviest weight edge $e \in E$
7:     $M \leftarrow M \cup \{e\}$
8:     $E' \leftarrow E' \setminus \{e$ and its adjacent edges$\}$
9: **end while**

The iterations of this algorithm are illustrated in Fig. 9.21. The time complexity of this algorithm is $O(m \log n)$ with an approximation ratio of 2 [22].

9.6.2   Hoepman's Distributed Matching Algorithm

In a distributed setting, we aim to have a maximal matching where nodes of a computer network are actively involved in the matching process. Hopeman modified the sequential algorithm of Preis so that nodes cooperate to find the locally heaviest weight edge [13]. The main idea is that if nodes u and v at the ends of

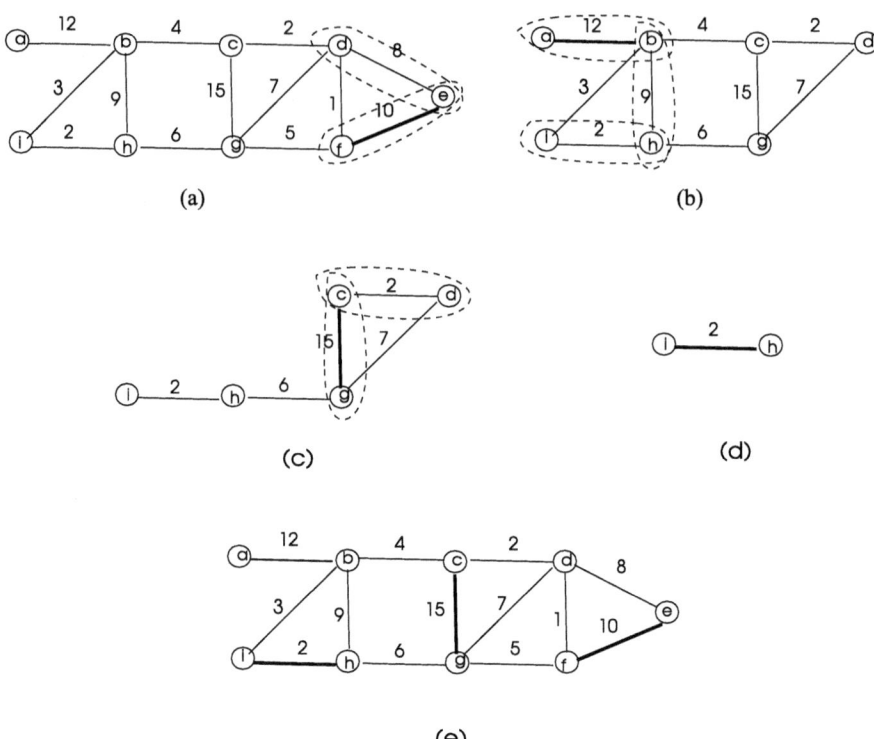

Fig. 9.21 The operation of Preis' algorithm in a sample graph. The first selected edge is (d, e) but an adjacent edge (e, f) has a greater weight so (e, f) is checked and found to be locally heaviest and included in the matching M in **a**. All adjacent edges to (e, f) are removed from the graph to obtain the subgraph in **b**. This time edges (i, h), then (h, b) and then (b, a) are selected in sequence to find the locally heaviest edge (b, a) which is added to M in **b**. The third iteration selects (c, d) and (c, g) in turn to add (c, g) to M. The last edge to add is (i, h) as shown in **d** and the final matching with a total weight 39 is shown in **e**

an edge (u, v) decide (u, v) is the heaviest edge incident to both of them, there cannot be a heavier edge adjacent to this edge and hence it can be included in the MWM. There are two message types, *request* and *drop*; a node u that finds (u, v) is the heaviest edge incident to it sends *request* to neighbor node v. If this node finds (u, v) is the heaviest weight edge incident to it, it replies by a *request* message and (u, v) is included in the MWM as shown in Algorithm 9.12.

Algorithm 9.12 *Hoepman_MWM*

```
 1: set of int R, S
 2: message types req,drop
 3: R ← ∅
 4: N ← N(i)
 5: c ← candidate(i, S)
 6: if c ≠⊥ then
 7:    send request to c
 8: end if
 9: while S ≠ ∅ do
10:    receive msg(j)
11:    case msg(j).type of
12:          req:   R ← R⋃{u}
13:          drop:  S ← S \ {u}
14:                    if u = c then c ← candidate(i, S)
15:                          if c ≠ ⊥ then send req to c
16:       if c ≠ ⊥ ∧ c ∈ R then
17:          for all ∈ N \ {c} do
18:             send drop to w
19:          end for
20:          S ← ∅
21:       end if
22: end while
```

Analysis

An edge of the network graph may be traversed by at most two messages, either by *req* from two nodes at its endpoints or a *req* and a *drop* message. Therefore, total number of messages exchanged will be 2 *m*. Since this algorithm imitates the sequential Preis algorithm, the output is the same matching produced is the same of the global heaviest matching algorithm with the same approximation ratio of 1/ 2 [13].

9.6.3   Parallel Algorithm Methods

In search of a parallel algorithm for the matching problem, we can partition the graph and distribute the vertices to processors. Each process then performs the following for its partition of the graph.

1. **while** there are edges left
2. A pointer from each vertex is set to point to the neighbor heaviest vertex.
3. **if** two vertices *u* and *v* point to each other, edge (*u*, *v*) is included in the matching.

We need to be careful while considering the border vertices in the partitions. This can be handled by the introduction of *ghost vertices* which are the non-member vertices that are connected to the border vertices of a partition. In this case, when a border vertex v is matched in a partition i, the process p_i responsible for the partition i should inform processes p_j which holds ghost vertices that are neighbors of v of the matching.

Parallelizing Hoepman's Algorithm

Manne et al. described the similarity between the Hoepman algorithm and the Luby's parallel algorithm to build an independent set of a graph we will describe in Chap. 10. A sequential version of Hoepman algorithm is first developed which in fact is similar to Preis algorithm using the notation of Algorithm 9.12. This algorithm searches for dominating edges in the graph. A parallel version of this algorithm is then formed which allocates a block of vertices to each process p_i of the parallel processing system. The graph under consideration is partitioned by replicating the border vertices and each process runs the sequential algorithm in its partition to result in a global maximal matching. The authors show that the designed parallel algorithm is efficient up to 32 processors [18].

9.7   Chapter Notes

We reviewed matching and related problems in this chapter. Finding maximum matching of an unweighted or weighted general graph can be performed in polynomial time as we have seen, in fact, this problem is one of the very few problems related to graphs that can be performed in polynomial time. However, there are various approximation algorithms to reduce the linear time. It was shown in [11] that any unweighted greedy matching is a 1/2 approximation to the maximum matching. Moreover, any greedy weighted matching that selects legal edges with maximum weights has the 1/2 approximation to the maximum weight matching [1]. The sequential algorithms we have reviewed in this chapter are shown in Table 9.1. The greedy algorithm and the algorithm due to Preis are approximation algorithms and all other listed algorithms are exact. Edmonds provided the first polynomial time algorithm for weighted matching with $O(n^2m)$ time complexity [7].

It should be noted there are various improvements to these basic algorithms. For example, Gabow improved the running time for weighted matching to $O(nm + n^2 \log n)$ [10]. There are various parallel algorithms for unweighted or weighted,

Table 9.1 Sequential matching algorithms

	Bipartite graphs	General graphs
Unweighted matching	Augmenting-path algorithm: $O(nm)$	Greedy algorithm: $O(nm)$
	Hopcroft–Karp algorithm: $O(n^2)$	Edmond's algorithm: $O(n^2m)$
Weighted matching	Hungarian algorithm: $O(n^3)$	Preis' algorithm: $O(nm)$
	Flow-based algorithm $O(m^2 + mn)$	

general or bipartite graphs. We saw how Hopcroft–Karp algorithm can be conveniently parallelized due to disjoint BFS and DFS processing in each phase. Auction algorithm for weighted bipartite graph matching can be modified to have parallel processing in the bidding and assignment phases. Also, we can partition a general graph and distribute partitions to parallel processes which perform matching and the results can then be gathered at a root process which merges them to find the global matching. A recent survey of parallel algorithms for maximum matching in bipartite graphs is provided in [2].

In many cases, approximation matching algorithms turn out to be faster at the expense of returning an approximate solution rather than an exact one. For very large graphs, they may be preferable as times involved may be very high. We have also described how a series of conversions from one type of algorithm can lead to efficient solutions. The algorithm that sorts edges and then includes legal edges to matching has $O(m \log m)$ complexity due to sorting process and has an approximation ratio of 1/2. Preis came up with the idea of selecting local heaviest edges which are independent of each other to result in a better time complexity of $O(m)$. Hoepman later on provided a distributed version of this algorithm with the same approximation ratio as we reviewed. Finally Manne et al. presented a parallel approximation matching algorithm based on Hopeman's work. We can see the sequence of development here are a sequential algorithm; an improved sequential algorithm; a distributed algorithm from the improved sequential algorithm and a distributed memory parallel algorithm that builds upon the distributed algorithm. This path, although usually in less steps, is commonly followed in various graph problems as we saw.

Distributed matching algorithms need careful consideration as matching of an edge incident to a node in a network requires notification to two-hop neighbors since they will be affected. Matching has numerous applications and hence there is need for parallel and distributed algorithms with better performances.

Exercises

1. Given the graph of Fig. 9.22 with initial matching shown in bold, find augmenting paths iteratively to obtain a maximum matching for this graph.
2. Work out the maximum matching in the bipartite graph of Fig. 9.23 using the augmenting path algorithm.
3. Find the maximum matching in the bipartite graph of Fig. 9.24 using the Hopcroft–Karp algorithm showing the BFS trees constructed in all iterations.

Fig. 9.22 Sample graph for Exercise 1

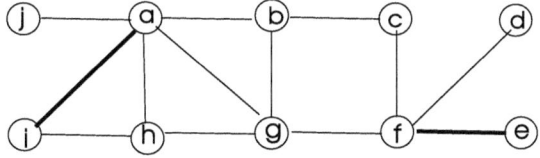

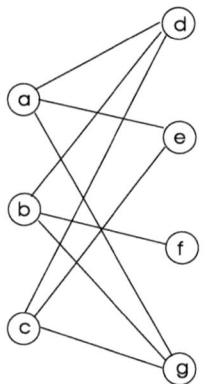

Fig. 9.23 Sample graph for Exercise 2

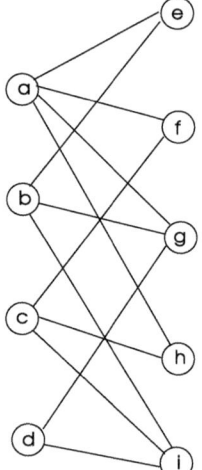

Fig. 9.24 Sample graph for Exercises 3 and 4

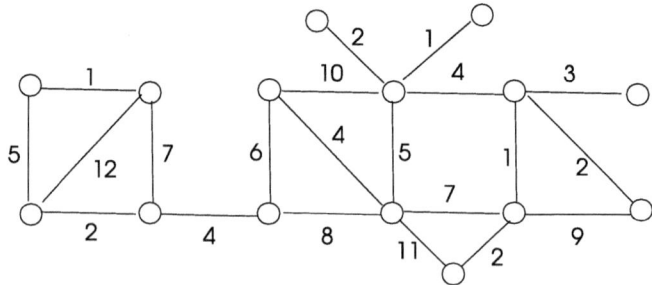

Fig. 9.25 Sample graph for Exercise 6

4. Determine the maximum matching in the graph of Fig. 9.24 this time using the maximum flow method of Ford–Fulkerson algorithm.
5. A multiprocessor system has 5 computers $P_1, \ldots, P_5$ that should finish 5 tasks $1, \ldots, 5$. The time to finish tasks for each processor is given in the below cost matrix. Work out the minimum time to finish all tasks by these 5 processors using the Hungarian algorithm. Show each step of the algorithm.

$$C = \begin{matrix} & \begin{matrix} 1 & 2 & 3 & 4 & 5 \end{matrix} \\ \begin{matrix} P_1 \\ P_2 \\ P_3 \\ P_4 \\ P_5 \end{matrix} & \begin{pmatrix} 8 & 1 & 4 & 3 & 2 \\ 2 & 5 & 9 & 6 & 4 \\ 6 & 2 & 3 & 4 & 5 \\ 1 & 4 & 7 & 9 & 3 \\ 5 & 0 & 8 & 1 & 2 \end{pmatrix} \end{matrix} \tag{9.10}$$

6. Design a distributed synchronous weighted matching algorithm that finds the minimal weighted matching in a network of computers. A node v that has the least degree among its neighbors has the privilege to propose to its neighbor u if (u, v) is the least weight edge incident to it. Work out the time and message complexities of this algorithm and show its operation in the network depicted in Fig. 9.25.
7. Provide an extension to the pseudocode of the edge coloring based maximal matching algorithm (Algorithm 9.7) so that when there are no more nodes that are in UNMATCHED state, the algorithm at a node is stopped without waiting to finish k rounds. The root is also informed of this condition.

References

1. Avis D (1983) A survey of heuristics for the weighted matching problem. Networks 13:475–493
2. Azad A, Halappanavar M, Rajamanickam S, Boman EG, Khan AM, Pothen A (2012) Multi-threaded algorithms for maxmum matching in bipartite graphs. IPDPS 2012:860–872
3. Berge C (1957) Two theorems in graph theory. Proc Natl Acad Sci USA 43:842–844
4. Bertsekas DP, Tsitsiklis JN (1989) Parallel and distributed computation: numerical methods. Prentice-Hall, Englewood Cliffs
5. Bertsekas DP, Castanon DA (1991) Parallel synchronous and asynchronous implementations of the auction algorithm. Parallel Comput 17:707–732
6. Bus L, Tvrdik P (2009) Towards auction algorithms for large dense assignment problems. Comput Optim Appl 43(3):411–436
7. Edmonds J (1965) Paths, trees and flowers. Can J Math 17:449–467
8. Erciyes K (2015) Distributed and sequential algorithms for bioinformatics. Springer computational biology series. Springer, Cham
9. Erciyes K (2015) Distributed graph algorithms for computer networks. Springer computer and communications series. Springer, London
10. Gabow HN (1976) An efficient implementation of Edmonds' algorithm for maximum matching on graphs. J Assoc Comput Mach 23:221–234

11. Hausmann D, Korte B (1978) K-greedy algorithms for independence systems. Z Oper Res 22(1):219–228
12. Hirvonen J, Suomela J (2012) Distributed maximal matching: greedy is optimal. In: Kowalski D, Panconesi A (eds) PODC12. Proceedings of 2012 ACM symposium on principles of distributed computing, Madeira, Portugal, 161–8 July 2012
13. Hoepman JH (2004) Simple distributed weighted matchings. Technical report, Nijmegen institute for computing and information sciences (NIII)
14. Hopcraft J, Karp RM (1973) An $O(n^{2.5})$ algorithm for maximum matching in bipartite graphs. SIAM J Comput 2:225–231
15. Karypis G, Kumar V (1998) A parallel algorithm for multilevel graph partitioning and sparse matrix ordering. J Parallel Distrib Comput 48(1):71–95
16. König D (1931) Graphen und matrizen. Math. Lapok 38:116–119
17. Kuhn HW (1955) The Hungarian method for the assignment problem. Nav Res Logist Q 2:83–97
18. Manne F, Bisseling RH (2007) A parallel approximation algorithm for the weighted maximum matching problem. In: Wyrzykowski R, Karczewski K, Dongarra J, Wasniewski J (eds) Proceedings of seventh international conference on parallel processing and applied mathematics (PPAM 2007). LNCS, vol 4967. Springer, Berlin, pp 708–717
19. Micali S, Vazirani V (1980) An O(sqrt(| V |) | E |) algorithm for finding maximum matching in general graphs. In: Proceedings of 21st annual symposium on on foundations of computer science, IEEE, pp 17–27
20. Munkres J (1957) Algorithms for the assignment and transportation problems. J Soc Ind Appl Math 5(1):32–38
21. Naparstek O, Leshem A (2014) Expected time complexity of the auction algorithm and the push relabel algorithm for maximal bipartite matching on random graphs. Random Struct Algorithms 48:384–395
22. Preis R (1999) Linear time 1/2-approximation algorithm for maximum weighted matching in general graphs. In: Meinel C, Tison S (eds) Symposium on theoretical aspects of computer science (STACS) 1999. LNCS, vol 1563, Springer, Berlin, 259–269
23. Riedy J (2010) Making static pivoting scalable and dependable. Ph.D. thesis, EECS Department, University of California, Berkeley
24. Sathe M (2012) Parallel graph algorithms for finding weighted matchings and subgraphs in computational science. Ph.D. thesis, University of Basel
25. Vazirani VV (1994) A theory of alternating paths and blossoms for proving correctness of the $O(\sqrt{V}E)$ general graph maximum matching algorithm. Combinatorica 14(1):71–109

Independence, Domination, and Vertex Cover

10

10.1   Introduction

Detecting subgraphs of a graph with special properties may be useful in various implementations. In this chapter, we study detection of three such special subgraphs: *independent sets*, *dominating sets*, and *vertex cover*. We will see that all of these problems are equivalent in terms of complexity and finding solution to one of them yields the solution to the other ones.

An independent set of a graph is a subset of its vertices such that no two vertices in this set are neighbors. Finding a maximum independent set of a graph G which is the maximum order independent set among all independent sets of G is NP-hard. An independent set of a graph G is a clique in the complement of G. A dominating set of a graph is a subset of vertices such that each vertex of the graph is either in this set or a neighbor to a vertex in this set. Finding a minimum dominating set of a graph is again NP-hard. The last problem we study in this chapter is the vertex cover of a graph which consists of vertices such that every edge of the graph is incident to at least one vertex in this set. Independent sets, dominating sets, and vertex covers may be used for various real-life complex network applications such as detecting clusters in biological networks and routing in computer networks. These special subgraphs may also serve as building blocks of more complex graph algorithms. In this chapter, we investigate theoretical properties of these vertex sets in graphs, show how they are related, and study sequential, parallel, and distributed algorithms to find them.

10.2   Independent Sets

An *independent set* of a graph is a subset of its vertices such that no vertex in this set is adjacent to any other vertex contained in this set. We can formally define the independent set as follows.

© The Author(s), under exclusive license to Springer Nature Switzerland AG 2026
K. Erciyes, *Guide to Graph Algorithms*, Texts in Computer Science,
https://doi.org/10.1007/978-3-032-05294-0_10

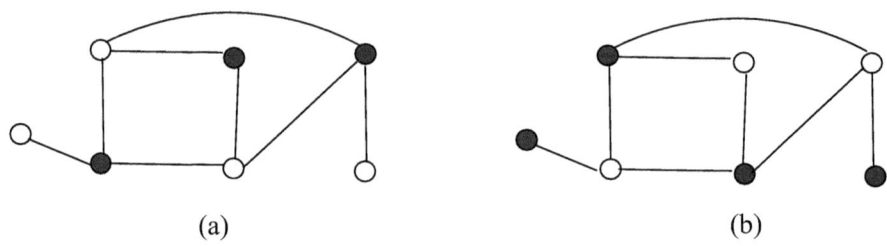

Fig. 10.1 **a** An MIS with order 3. **b** A MaxIS with order 4 of a sample graph. The vertices in the independent sets are shown in bold

Definition 10.1 (*Independent set*) An independent set of a graph $G = (V, E)$ is a subset I of its vertices such that $\forall u, v \in I, (u, v) \notin E$.

An independent set is *maximal* if it cannot be enlarged further by additional vertices, that is, a maximal independent set (MIS) is not properly contained in any other independent set of G. A *maximum independent set* (MaxIS) of a graph G is the largest order independent set of G among all its independent sets. The number of vertices in an MaxIS is called the *independence number*, $\alpha(G)$, of the graph G. Figure 10.1 displays maximal and maximum independent sets of a sample graph. Finding a MaxIS of a graph is NP-hard and the decision version of this problem which is determining whether a graph has a maximum independent set of order k or more is NP-complete [2]. However, we can find the MIS of a graph in linear time as we will see. When each vertex in the graph has an associated weight, our aim is to find the independent set with maximum total weight (MaxWIS). This problem is again NP-hard and is also very hard to approximate.

10.2.1 Reduction to Clique

We can reduce the problem of finding an independent set in a graph to finding a clique. A clique of a graph $G = (V, E)$ is a subset C of its vertices such that every vertex in C is connected to all other vertices in C. In an independent set I of G, which is not necessarily maximal, no two vertices are adjacent. Therefore, when we take the complement of a graph to obtain graph $\overline{G}$, the set I will be a clique since the complement will exhibit all nonexisting edges in G. Figure 10.2 displays a graph with an independent set and the complement of this graph has the independent set as a clique.

This duality shows that the clique problem is at least as hard as the independent set problem and also the independent set problem is at least as hard as the clique problem which means they can be reduced to each other in polynomial time stated as follows:

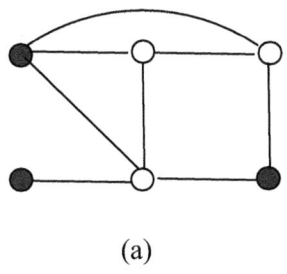
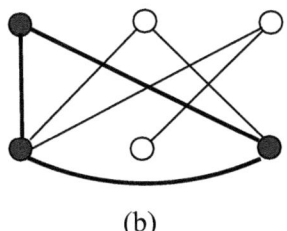

(a)   (b)

Fig. 10.2 **a** An MIS with order 3 in a sample graph. **b** The same vertices form a clique in the complement of the sample graph

$$\text{Independent Set Problem (IND)} \leq_p \text{Clique Problem (CLIQUE)}$$

$$\text{Clique Problem (CLIQUE)} \leq_p \text{Independent Set Problem (IND)}$$

10.2.2 Sequential Algorithms

We will review four sequential algorithms to find the MIS of a graph, starting with a random greedy one. The second algorithm uses a heuristic and the third one considers labels of vertices while selecting members of MIS while the fourth algorithm is a general method that searches an independent set at each step.

Algorithm 10.1 *Seq_MIS1*

1: **Input** : $G = (V, E)$ an undirected unweighted graph
2: **Output** : MIS I of G
3: $I \leftarrow \emptyset$
4: $V' \leftarrow V$
5: **while** $V' \neq \emptyset$ **do**
6:    **select** any $v \in V'$
7:      $I \leftarrow I \cup \{v\}$
8:      $V' \leftarrow V' \setminus \{\{v\} \cup N(v)\}$
9: **end while**

10.2.2.1 The Random Greedy Algorithm

As a first attempt, we will adopt a greedy strategy to form the MIS. Intuitively, we can arbitrarily pick a vertex v of the graph G, include this vertex in the MIS, and remove v and all of its neighbors $N(v)$ together with incident edges on these vertices from the graph, simply because its neighbors cannot be included in the MIS. We proceed in this manner until there are no more vertices left. Algorithm 10.1 displays the code for this procedure.

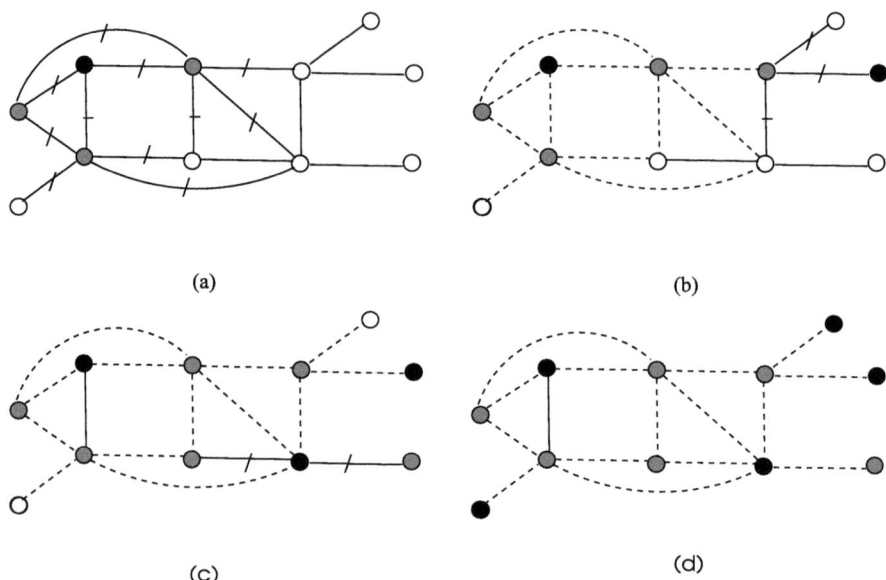

Fig. 10.3 Running of *Seq_MIS1* in a sample graph. The first three iterations are shown in **a–c**; and the last two iterations are shown in **d**. The independent set vertices are shown in bold and the deleted neighbor vertices in gray with the deleted edges marked with dashed lines

The output set *I* is an independent set since we never include any neighbor of the selected vertex *v* in the graph to obey the IS property, that is, no two vertices in this set should be adjacent. It is also maximal since we proceed until there are no vertices left and hence cannot enlarge *I* any further. This algorithm requires $O(n)$ steps as we may end up selecting a vertex and its single neighbor repeatedly as in the case of a linear network. The running of this algorithm in a sample graph is shown in Fig. 10.3.

10.2.2.2 The Lowest Degree First Algorithm

As another approach to form the MIS of a graph, we can select the vertex with the lowest degree at line 5 of Algorithm 10.1 instead of a random vertex. We call this algorithm the Lowest Degree First algorithm (LDFA). This heuristic is reasonable since our aim is to have an independent set as large as possible which means we want to remove as few neighbors of a selected vertex as possible. Hence, we always select vertices with the least number of neighbors. The running of this algorithm is depicted in Fig. 10.4. Note that we had eight steps instead of four steps of the random greedy algorithm but we obtained an MIS of size 6 which is maximum for this graph.

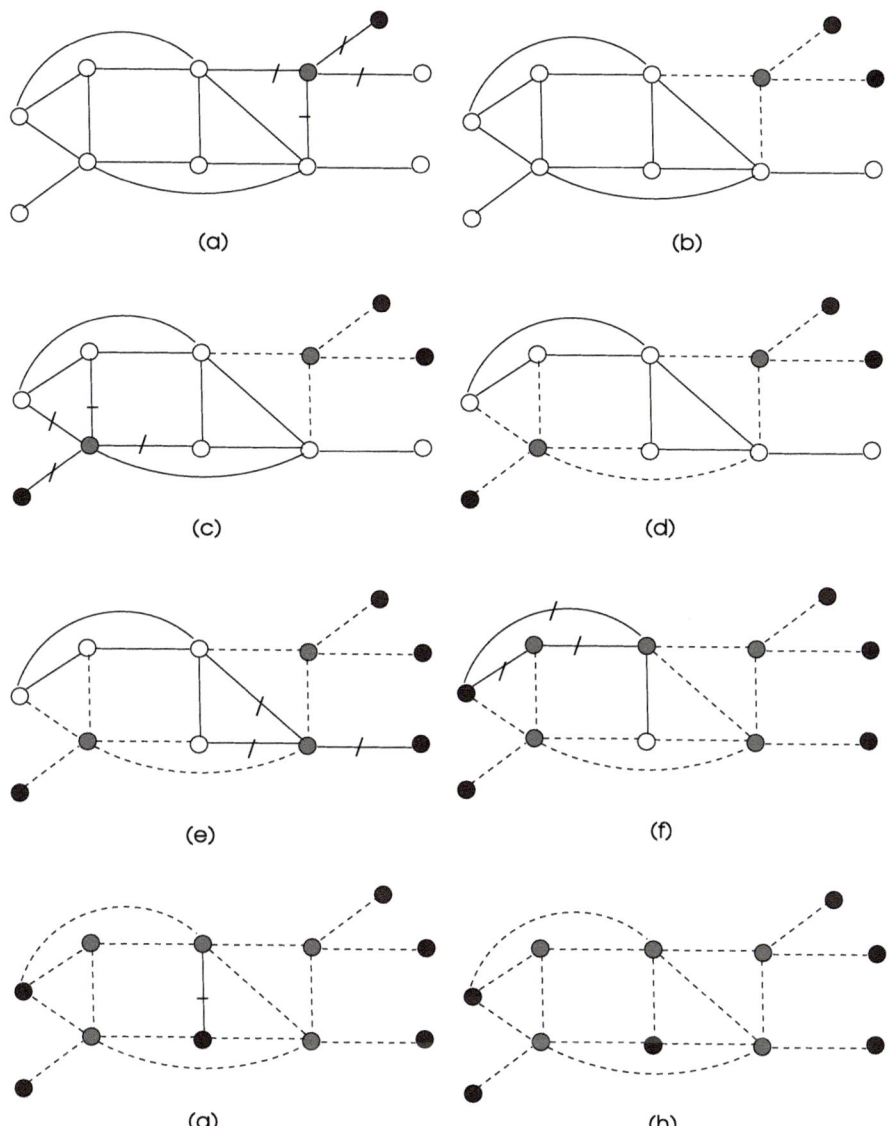

Fig. 10.4   Running of LDFA in the same graph of Fig. 10.3. The final MIS is shown in **h**

Theorem 10.1 LDFA provides an MIS I such that $|I| \geq n/(\Delta(G)+1)$ where $\Delta(G)$ is the maximum degree of the graph.

Proof A vertex u is included in $V \setminus I$ when a vertex $v \in N(u)$ is selected to be in the set I. If we mark such a vertex u each time one of its neighbors is selected to be in I, it can be marked at most $\Delta(G)$ time. Therefore, $|V \setminus I| \leq \Delta(G)|I|$. Also, a vertex can either be in the MIS or a neighbor of a vertex in the MIS. Therefore, $|I| + |V \setminus I| = n$. Hence, $(\Delta(G)+1)|I| \geq n$ which means $|I| \geq n/(\Delta(G)+1)$. $\square$

Corollary 10.1 *LDFA is an* $1/(\Delta(G)+1)$ *approximation algorithm for the maximum independent set problem.*

10.2.2.3 The Lexicographically First Algorithm

In a slightly different manner but using the greedy approach again, we can label n vertices of a graph with order n as $v_1, \ldots, v_n$ and find the MIS in sequence using these labels as shown in Algorithm 10.2. In this case, we know which vertex to select at each iteration and this algorithm is called Lexicographically First MIS algorithm (LFA). However, this algorithm does not improve the runtime of the previous one as it has a similar greedy approach as the first one and has a time complexity of $O(n + m)$ time since we check neighbors of each vertex.

Algorithm 10.2 *Seq_MIS3*

1: **Input** : $G = (V, E)$ an undirected unweighted graph
2: **Output** : MIS I of G
3: $I \leftarrow \varnothing$
4: **for** $i = 1$ to n **do**
5:    **if** $I \cap N(v_i) = \varnothing$ **then**
6:       $I \leftarrow I \cup \{v_i\}$
7:    **end if**
8: **end for**

10.2.2.4 The Incremental Algorithm

Yet another approach to find MIS is to gradually enlarge the current independent set I by new independent sets formed from vertices not in the set I. Instead of selecting a single vertex in line 4 of Algorithm 10.1, we select an independent set of the graph G this time. The selected independent set I' of the input graph G is then added to the MIS I and all vertices of I', their neighbors with their incident edges are removed from the graph since these neighbors cannot be considered further. The IS I' need not be an MIS of G. The algorithm terminates when there are no more vertices left to consider as shown in Algorithm 10.3.

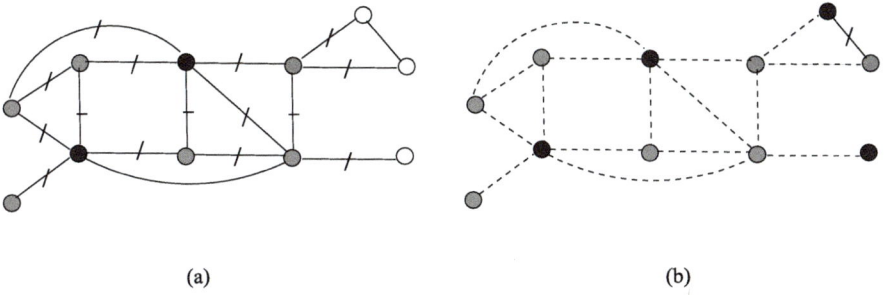

Fig. 10.5 Finding MIS of a sample graph by selecting an IS at each step. The independent set vertices are shown in bold and the deleted neighbor vertices in gray with the deleted edges marked with dashed lines

---

Algorithm 10.3 *Seq_MIS4*

1: **Input** : $G = (V, E)$ an undirected unweighted graph
2: **Output** : MIS I of G
3: $I \leftarrow \emptyset$
4: $G' \leftarrow G$
5: **while** $G' \neq \emptyset$ **do**
6:     **select** any independent set I' of G'
7:        $I \leftarrow I \cup \{I'\}$
8:        $G' \leftarrow G' - I'$
9: **end while**

Clearly, the choice of the independent set at line 6 determines the performance of this algorithm. We will see that randomization in this selection provides algorithms with good performances. An example operation of this algorithm is shown in Fig. 10.5.

10.2.2.5 MIS Construction Using Vertex Coloring

A *vertex coloring* of a graph is assigning colors to its vertices in the form of integers $1, \ldots, k$ such that no two adjacent vertices receive the same color as we will review in the next chapter. We can make use of vertex coloring to find the MIS of a graph.

A color class $V' \in V$ in a vertex colored graph $G = (V, E)$ is a set of vertices of G that have the same color. Since no two adjacent vertices of a graph have the same color, each color class is an independent set of G. We can exploit this property to find the MIS of a graph G by first including all vertices of color 1 in the MIS then iteratively checking vertices of increasing colors to be included in the MIS or not as shown in Algorithm 10.4. If a vertex u has a neighbor v that is already in the MIS, we cannot include vertex u in the MIS. We will assume that the graph is colored with k colors before running this algorithm.

Algorithm 10.4 *Seq_MIS5*

1: **Input** : $G = (V, E)$ an undirected unweighted graph
2: **Output** : MIS I of G
3: $I \leftarrow$ vertices of color 1
4: **for** $i = 2$ to k **do**
5:      $I \leftarrow I \cup$ all allowed vertices of color i if MIS property is not distorted
6: **end for**

Running of this algorithm in a sample graph is depicted in Fig. 10.6 for a sample graph colored with four colors. Time complexity of this algorithm is $O(km)$ since we need to run the for loop $O(k)$ times and we may need to check all of the edges at each run. We also need the time C to color vertices of the graph G and thus total time is $C + O(km)$.

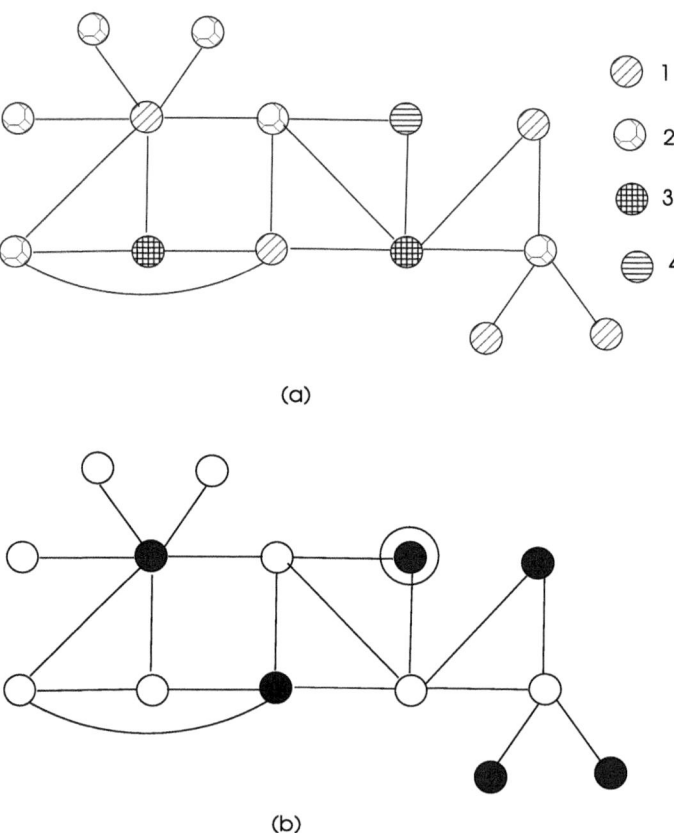

(a)

(b)

Fig. 10.6 Running of vertex coloring-based MIS algorithm in a sample graph which is colored with four colors. The MIS shown by bold vertices is almost formed by including all vertices of color 1 in the MIS in the first step. The only vertex added to the MIS after the first step when k = 4 is shown inside a large circle

---

Algorithm 10.5 *Luby_MIS*

1: **Input** : $G = (V, E)$
2: **Output** : MIS I of G
3: $I \leftarrow \emptyset$
4: $G' = (V', E') \leftarrow G = (V, E)$
5: **while** $V' \neq \emptyset$ **do in parallel do**
6:   **choose** a random set of vertices $V' \in V(G')$ by choosing each vertex with probability $1/(2d(v))$
7:   **for all** $(u, v) \in E(G')$ **do in parallel do**
8:     **if** $(u \in V' \wedge v \in V')$ **then**
9:       remove the endpoint with lower degree to obtain S from V'
10:     **end if**
11:   **end for**
12:   $I \leftarrow I \cup V'$
13:   $G' \leftarrow G' \setminus \{S \cup N(S)\}$
14: **end while**

10.2.3 Luby's Parallel MIS Algorithm

We can implement selection of an independent set in Algorithm 10.3 using various approaches. An efficient randomized parallel Monte Carlo algorithm was proposed by Luby in 1986 to find MIS of a graph which proceeds as follows [7]. Each vertex v is marked with probability $1/(2d(v))$ in parallel, where $d(v)$ is the degree of v, to be included in the independent set or not. This marking may produce edges with both endpoints marked to be in the MIS since assignment is done in parallel independently and hence corrections are needed. The next step identifies such edges and for each such edge (u, v) with both u and v covered in the MIS, the vertex with the higher degree is selected and in the case of a tie, vertex identifiers are used to select only one of such vertices. The selected vertices and their neighbors are then deleted from the graph and this process is repeated until the graph becomes empty as shown in Algorithm 10.5.

Algorithm 10.6 *SMLuby_MIS*

1: **Input** : $G = (V, E)$
2: **Output** : $MIS\ I[n]$ of G
3: **boolean** $I[n], C[n], R[n]$
4: **for** $j = ((i-1)*n/k) + 1$ to $i*n/k$ **do**   ▷ initialize my partition
5:    $C[i] \leftarrow 1$
6:    $I[i] \leftarrow 0$
7: **end for**
8: **repeat**
9:    **generate** a random number for each available vertex i ($C[i] \neq 0$) in my partition w.r.t
      $1/(2d(v))$
10:    **check** the random numbers for neighbors of my vertices in R
11:    **for** any vertex i in my partition that has highest random number than its neighbors **do**
12:       $C[i] \leftarrow 0$
13:       $I[i] \leftarrow 1$
14:       **for all** $j \in N(i)$ **do**
15:          $C[j] \leftarrow 0$
16:       **end for**
17:    **end for**
18: **until** $C[i] = 0, \forall 1 \leq i \leq n$

The selection of the nodes to be included in the independent set can be implemented on a EREW-PRAM using $O(m)$ processors with each execution taking $O(\log n)$ time. It can be shown that the execution of the *while* loop is $O(\log n)$ times [7] resulting in a total time of $O(\log^2 n)$ for this algorithm. We will use this algorithm as the basis of a distributed algorithm as described in the next section.

We can form a shared memory parallel version of Luby's algorithm as described in [3]. We have three vectors of n elements for n vertices; C, I, and R as shown in Algorithm 10.6. $I[i]$ shows whether vertex i is included in the MIS or not, $C[i]$ displays whether vertex i is a candidate to be included with 0 meaning it is either in MIS or a neighbor of a vertex in the MIS and therefore cannot be included, and finally $R[i]$ holds the generated random number for vertex i at each iteration. The vector C is initialized to all 1 s meaning all vertices are candidates and I is initialized to all 0 s since no MIS member is determined. The vectors are 1-D partitioned among k processes and each process performs the following steps at each iteration until MIS is found which is determined by all entries of vector C becoming 0 as shown in Algorithm 10.6 for parallel process i for a total number of k processes. This algorithm requires synchronization at each step, otherwise its performance is similar to Algorithm 10.5 when synchronization is not considered.

10.2.4 Distributed Algorithms

We can use various heuristics to design distributed algorithms for the MIS problem in a network setting. We describe three distributed MIS algorithms in this section; the first algorithm uses identifiers of nodes to break symmetries and the second algorithm is a distributed version of Luby's parallel algorithm with the third one being another randomized distributed algorithm with a better performance. Although the last two algorithms have similar structures, we show two common ways of implementation, using finite-state machines in the first one and a more straightforward approach in the second one by showing control messages explicitly.

Algorithm 10.7 $Dist\_MIS1$

1: **Input**: unweighted undirected graph $G = (V, E)$
2: **Output**: MIS I of G
3: **message types**: in_mis, $neigh\_mis$
4: **states**: {IDLE, INMIS,NONMIS}   ▷ states of a node
5: $state \leftarrow$ IDLE   ▷ initialize
6: **while** $state = IDLE$ **do**   ▷ not in MIS or a neighbor to a MIS vertex
7:   **if** $id > ids$ of all current neighbors **then**
8:     $state \leftarrow$ INMIS
9:     **send** $in\_mis(i)$ to $N(i)$ ▷ inform neighbors of INMIS state
10:   **else if** $in\_mis(j)$ is received from neighbor j **then**
11:     $state \leftarrow$ NONMIS
12:     **send** $neigh\_mis(i)$ to $N(i)$
13:   **else if** $neigh\_mis(j)$ is received from neighbor j **then**
14:     $N(i) \leftarrow N(i) \setminus \{j\}$ ▷ remove neighbor j from neighbor list
15:   **end if**
16: **end while**

10.2.4.1 Greedy Distributed Algorithm

In our first attempt to find a distributed MIS algorithm, we will modify the greedy MIS algorithm (*Seq_MIS1*), this time, decisions to join the MIS should be made locally based on some criteria. We will assume each node has a unique identifier and the node with the largest identifier among neighbors joins the MIS as shown in Algorithm 10.7 executed by an active node i in each synchronous round. We assume each node knows the identifiers of its neighbors initially. Each node of the network can be in one of the following states.

- IDLE: This is the initial state for a node.
- INMIS: A node assigned to the MIS enters this state and informs neighbors by the *in_mis* message.
- NONMIS: A node in this state has given up being in MIS because one of its neighbors has become a member of MIS. It informs its neighbors by the *neigh_mis* message.

A node that is included in the MIS should not participate in the algorithm in further rounds. This is accomplished by changing its state and informing its neighbors by *in_mis* message so they should as well remain inactive in further rounds. Any node that is adjacent to a neighboring node of an MIS node is informed by the

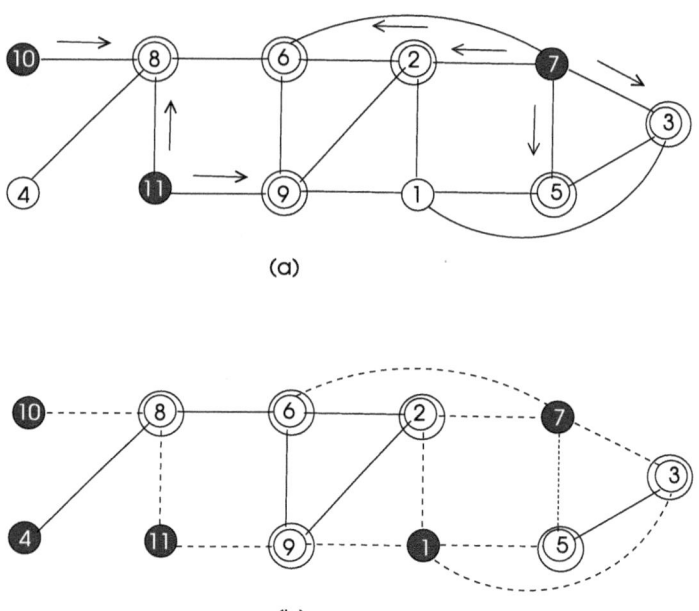

(a)

(b)

Fig. 10.7 Execution of *Dist_MIS1* in a sample graph with unique node identifiers. The in_mis messages are shown by arrows, the MIS nodes are in black, and the neighbor nodes that enter NONMIS state are shown by double circles. In only two rounds shown in **a** and **b**, the MIS is formed

neighbor by the *neigh_mis* message so that it is dropped from the active neighbors list. Operation of this algorithm in a sample network is shown in Fig. 10.7.

The time complexity of this algorithm is $O(n)$ as there can be purely sequential operation as in a linear network with increasing/decreasing identifiers or a general network with neighbors that have increasing or decreasing identifiers in a sequence. The number of messages transmitted is proportional to m. This algorithm may therefore turn out to be slow.

10.2.4.2 The First Randomized Distributed Algorithm
In search of an algorithm for better performance, we will attempt to convert Luby's algorithm to a synchronous distributed algorithm which we will call *Dist_Luby* that works in rounds. Each round is made up of two phases; in the first phase, each active node i marks itself with probability $1/2d(i)$ to be in the MIS. In this MARKED/UNMARKED state, it exchanges status with neighbors to conclude the phase. In the second phase, if node i is marked and there is another MARKED neighbor node, the higher degree one is included in the MIS with the other one being dropped from active nodes again by exchanging status messages. The pseudocode for node i is shown in Algorithm 10.8. This algorithm completes

in expected number of $O(\log n)$ rounds [12] and since the number of messages is proportional to the number of edges, total number of messages is $O(m \log n)$.

Algorithm 10.8 $Dist\_MIS2$

1: **Input**: unweighted undirected graph $G = (V, E)$
2: **Output**: MIS I of G
3: **message types**: $info$ ▷ sent by nodes to inform their states to neighbors
4: **states**= {IDLE, INMIS, NONMIS} ▷ states of a node as idle, in, or neighbor to a node in MIS
5: **init_states**= {MARKED, UNMARKED}   ▷ initial states at each round
6: $I \leftarrow \emptyset$; $state \leftarrow$ IDLE   ▷ initialize vertex cover C
7: **while** $state = IDLE$ **do**   ▷ not in MIS or a neighbor to a MIS vertex
8:      **mark** $init\_state$ MARKED with probability $1/(2d(i))$ ▷ Phase 1
9:      **send** $info(init\_state)$ to $N(i)$
10:      **receive** $info(init\_state_j)$ from $\forall j \in N(i)$
11:      **if** $init\_state$=MARKED **then**   ▷ Phase 2
12:         **if** $\nexists j \in N(i)$ with $state_j$=MARKED **then**
13:           $state \leftarrow$ INMIS
14:         **else**
15:           $state \leftarrow$ NONMIS
16:         **end if**
17:         **send** $info(state)$ to $N(i)$
18:         **receive** $info(state_j)$ from $\forall j \in N(i)$
19:      **end if**
20:      **if** $\exists j \in N(i)$ with $state_j$=INMIS **then**
21:         $state \leftarrow$ NONMIS
22:      **end if**
23:      **if** $\exists j \in N(i)$ with $state_j$=NONMIS **then**
24:         $state \leftarrow$ NONMIS
25:         $N(i) \leftarrow N(i) \setminus \{j\}$ ▷ remove neighbor j from neighbor list
26:      **end if**
27: **end while**

10.2.4.3 The Second Randomized Distributed Algorithm

Our last distributed algorithm is a more recent randomized one that has better performance than the distributed version of Luby's algorithm in which each active node i picks a random number r between 0 and 1 in each round. If node i has the largest r among its neighbors, it assigns itself to MIS and informs its neighbors of its decision so they will not participate in MIS selection in next rounds. The detailed ready to be coded pseudocode for a node is shown in Algorithm 10.9 where we show explicitly the logic for terminating a round [1]. This algorithm has the same time and message complexity as the first one since the logic is similar.

Algorithm 10.9 $Dist\_MIS3$

1: **set of int** $curr\_neighs \leftarrow N(i); received, values \leftarrow \varnothing$
2: **message types** $round, info$
3: **states** INMIS, NONMIS
4: **boolean** $inflag, outflag, round\_recvd, round\_over \leftarrow false$
5: **while** $\neg round\_over$ **do**
6:    **receive** $msg(j)$
7:    **case** $msg(j).type$ **of**
8:                    $\underline{round(k)}$:   **if** $inflag$ **then** $state \leftarrow$ INMIS
9: $inflag \leftarrow false$
10:                       **else if** $outflag$ **then** $state \leftarrow$ NONMIS
11:                                   $outflag \leftarrow false$
12:                       **else draw** $rval \in [0, 1]$
13:                       **send** $info(k,rval, state)$ to $curr\_neighs$
14:                       $round\_recvd \leftarrow true$
15:           $\underline{info(k, r, sta)}$: $received \leftarrow received \bigcup \{j\}$
16:                       $values \leftarrow values \bigcup \{j\}$
17:                       **if** $sta =$ INMIS **then** $outflag \leftarrow true$
18:                       **if** $sta =$ INMIS/NONMIS
19:                         $lost\_neighs \leftarrow lost\_neighs \bigcup \{j\}$
20:    **if** $round\_recvd \wedge (received = curr\_neighs)$ **then**
21:      **if** $\forall x \in curr\_neighs : rval > r_x \in values$ **then**
22:        $inflag \leftarrow true$
23:      **end if**
24:      $round\_over \leftarrow true; curr\_neighs \leftarrow curr\_neighs \setminus lost\_neighs$
25:      $round\_recvd \leftarrow false; received, values, lost\_neighs \leftarrow \{\varnothing\}$
26:    **end if**
27: **end while**

10.3   Dominating Sets

A *dominating set* of a graph is a subset of its vertices such that every vertex is either in this set or adjacent to a vertex in it. We can formally define this set as follows:

Definition 10.2 (*Dominating set*) A *dominating set* of a graph $G = (V, E)$ is a subset D of its vertices such that $\forall v \in V$, either $v \in D$ or $v \in N(u)$ where $u \in D$. Equally, a set $D \in V$ is a dominating set if $\bigcup_{v \in D} N[v] = V$. In other words, the union of closed neighborhoods of each vertex of the graph should be equal to the vertex set of the graph.

There are various applications of dominating sets in communication networks and surveillance systems. A dominating set can be used effectively as a backbone to transfer messages in a wireless network as we will see in Chap. 14.

Every maximal independent set is a dominating set as every vertex of the graph will be either in this set or adjacent to a vertex in this set. However, not every

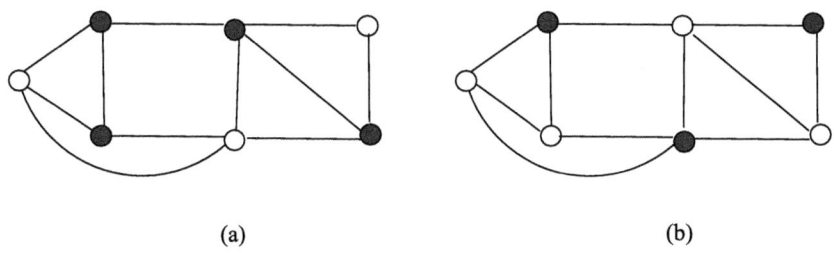

Fig. 10.8 **a** A connected DS with order 4. **b** An unconnected MDS with order 3 of a sample graph. The vertices in the dominating sets are shown in black

dominating set is an independent set since members of a dominating set can be neighbors. A *minimum dominating set* (MinDS) of a graph is the set with the minimum order among all dominating sets of that graph. The cardinality of MinDS of a graph G is called the *domination number* ($\gamma(G)$) of G. A minimal dominating set (MDS) of a graph does not contain any other dominating sets of that graph as a proper subset as shown in Fig. 10.8. In other words, removing a vertex from such a set will destroy the dominating set property of this set. In a *connected dominating set* (CDS), there is a path between each pair of vertices in the dominating set, consisting only of dominating set vertices. Formally, $\forall u \in D \wedge v \in D$, there is a path $u, x_1, \ldots, x_k, v$ such that $u, x_1, \ldots, x_k, v \in D$.

A *k-dominating set* D of a graph $G = (V, E)$ consists of vertices such that every $v \in V - D$ is adjacent to at least k elements of D. A *k-distance dominating set* which is sometimes confused with the *k-dominating set* concepts is a set of vertices that have a distance of at least k to at least one of the vertices of the dominating set. Note that the latter definition loosens the general dominating set definition.

Finding MinDS of a graph is NP-hard [2] and we are mostly interested in finding minimal dominating sets of graphs when we review sequential, parallel, and distributed algorithms for this purpose in this section.

10.3.1   A Greedy Sequential Algorithm

For the design of a greedy algorithm, we will use coloring of vertices such that vertices in MDS will be shown in *black*, their dominated neighbors are colored *gray* and any other vertex in the graph is *white* with all vertices initialized to *white*. The *span* of a vertex v is the number of *white* neighbors it has, including itself if it is *white*. The heuristic we will use is to always select a *white* or a *gray* vertex with the highest span in the graph. Since our aim is to find a MDS, we are trying to cover as many *white* vertices as possible at each step with this heuristic. The pseudocode for this algorithm called *Span_MDS* is depicted in Algorithm 10.10.

Algorithm 10.10 *Span_MDS*

1: **Input** : An undirected unweighted graph $G = (V, E)$
2: **for all** $v \in V$ **do**   ▷ all nodes are white initially
3:    $color[v] \leftarrow white$
4: **end for**
5: $V' \leftarrow V, D \leftarrow \emptyset$
6: **while** $\exists u \in V', color[u] = white$ **do**
7:    **select** $v \in V'$ with the highest span
8:    $color[v] \leftarrow black$ ▷ select a MDS vertex
9:    $V' \leftarrow V' \setminus \{v\}$
10:    **for all** $w \in (V' \cap N(v))$ **do**   ▷ color its *white* neighbors *grey*
11:       **if** $color[w] = white$ **then**
12:          $color[w] \leftarrow grey$
13:       **end if**
14:    **end for**
15: **end while**

Figure 10.9 displays operation of this algorithm in a sample graph where the vertex with the highest span is colored *black* at each iteration. We can see that after three iterations, there are no *white* vertices left and the algorithm terminates. If we always select a *gray* vertex with the highest span at each iteration after the first one, we have a connected DS as shown in Fig. 10.9d.

The *Span_MDS* algorithm provides a DS of a graph G since there are no white vertices left when this algorithm terminates. This means all of the vertices of G end up in *gray* or *black* color as dominator or dominated vertices. The DS is minimal (MDS) since at each step, one or more *white* vertices are colored by coloring a

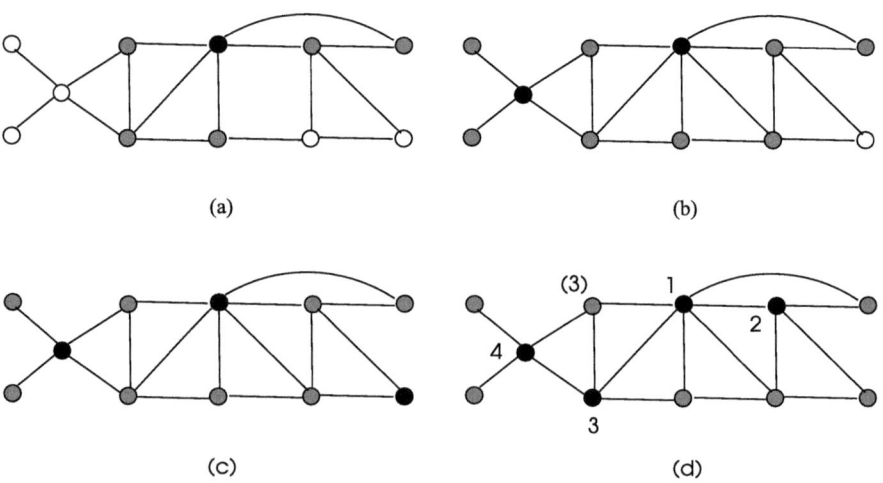

(a)   (b)

(c)   (d)

Fig. 10.9 Running of *MDS_Span* in a sample graph. The three iterations are shown in **a–c** with the dominating set vertices in black and the dominated vertices in gray. A connected DS is shown in **d** for the same graph with the iteration step numbers displayed next to included vertices

single vertex *black* that has the highest span. Removing this vertex from graph will leave one or more *white* vertices and hence the DS obtained is minimal being not contained in a larger DS. The time complexity of this algorithm is $O(n)$ since we may need to have n iterations as in the case of a linear network. The approximation ratio of this algorithm is $\ln \Delta$ as shown in [12].

In certain graphs, this algorithm may yield a dominating set which may be far from optimal. For example, in a graph where two vertices u and v are connected by n vertices using separate paths, this algorithm may start by coloring u or v black and then attempt to color all of the intermediate vertices black, resulting in $n - 1$ steps where coloring of u and v black suffices to form a DS in two steps.

10.3.1.1 Guha–Khuller Algorithms

Guha–Khuller algorithms provide improvements to *Span_MDS* so that such cases are handled more efficiently [4]. The first algorithm presented by these authors starts by coloring all vertices white. It then colors the highest degree vertex of the graph *black* and all of its neighbors *gray*. Thereafter at each step, the white neighbors of *gray* and *white* nodes are found and depending on the number of white neighbors, either a gray vertex or a *gray–white* vertex pair is colored *black* to be able to color the most number of possible *white* vertices *gray*. This process continues until there are no more *white* vertices left. This algorithm provides a MCDS with an approximation ratio of $2(1 + H(\Delta))$ where H is the harmonic function [4]. Running of this algorithm is depicted in Fig. 10.10.

The second algorithm also colors all vertices *white* initially. A *piece* is defined as a *white* vertex or a connected set of *black* vertices. This algorithm has two main phases. In the first phase, the vertex u *black* coloring of which will yield the greatest decrease in the number of pieces is selected; vertex u is colored *black* and its neighbors are colored *gray*. This phase continues until there are no *white* vertices left. Since the MDS obtained may not be connected, vertices of the MDS are connected using a Steiner tree based algorithm in the second phase. The output MCDS has an approximation ratio of $3 + \ln \Delta$ [4].

10.3.1.2 MIS-based Algorithms

An alternative way of constructing a MCDS is to first form an MIS of the graph and then connect the vertices in this MIS to get a MCDS in the second step. We can use any algorithm to find MIS such as the lowest degree first greedy algorithm. Connecting the MIS vertices can be performed using various heuristics such as selecting the vertex with the highest degree between the MIS vertices or using a Steiner tree based algorithm as in Guha–Khuller second algorithm.

10.3.2 A Distributed Algorithm to Find MDS

A simple distributed algorithm based on the span of a node in its two-hop neighborhood in the network can be formed as follows. Each node exchanges its span with all of its two-hop neighbors and if it has the highest span among all of these

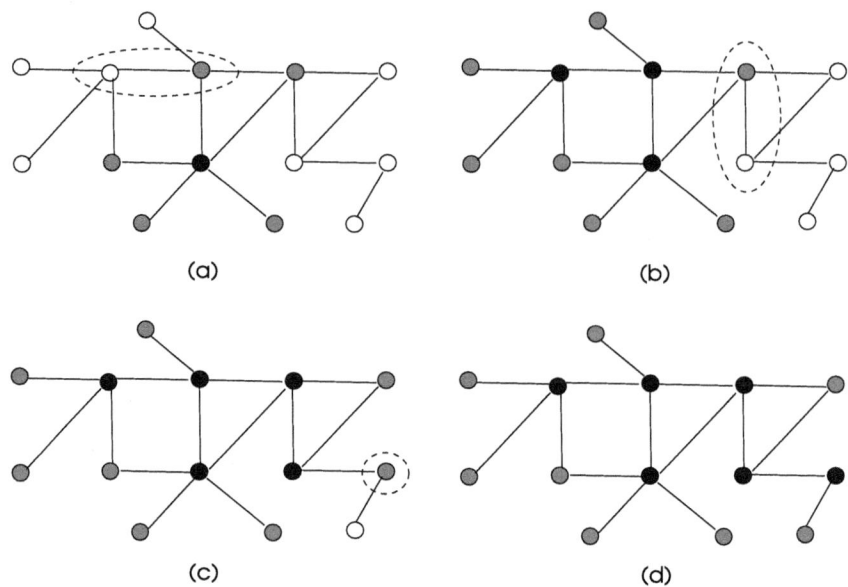

Fig. 10.10 Running of Guha–Khuller's first algorithm in a sample graph. Selected MCDS vertices are colored in black and their neighbors are shown in gray. The selected vertex or vertex pairs are shown inside dashed regions. The MCDS is formed after four iterations of the algorithm

neighbors, it enters the MDS. The algorithm is executed by a node i until it has no white neighbors which are neither MDS or dominated nodes as shown in Algorithm 10.11. It can be shown that this algorithm computes a MDS with $\ln \Delta + 2$ approximation ratio [12]. The number of rounds needed is $O(n)$ since there will be at least one new node entering the dominating set at each round.

Algorithm 10.11 *Dist_CDS*

1: Input $G = (V, E)$
2: $S \leftarrow V, MIS \leftarrow \emptyset$
3: **while** $\exists j \in N(i) : color[j] = white \wedge state \neq$ INCDS **do**   ▷
4:    $s \leftarrow span(i)$
5:    **send** s to nodes at distance of at most 2
6:    **receive** spans of nodes at distance 2
7:    **if** $s >$ all spans **then**
8:       $state \leftarrow$ INCDS
9:    **end if**
10: **end while**

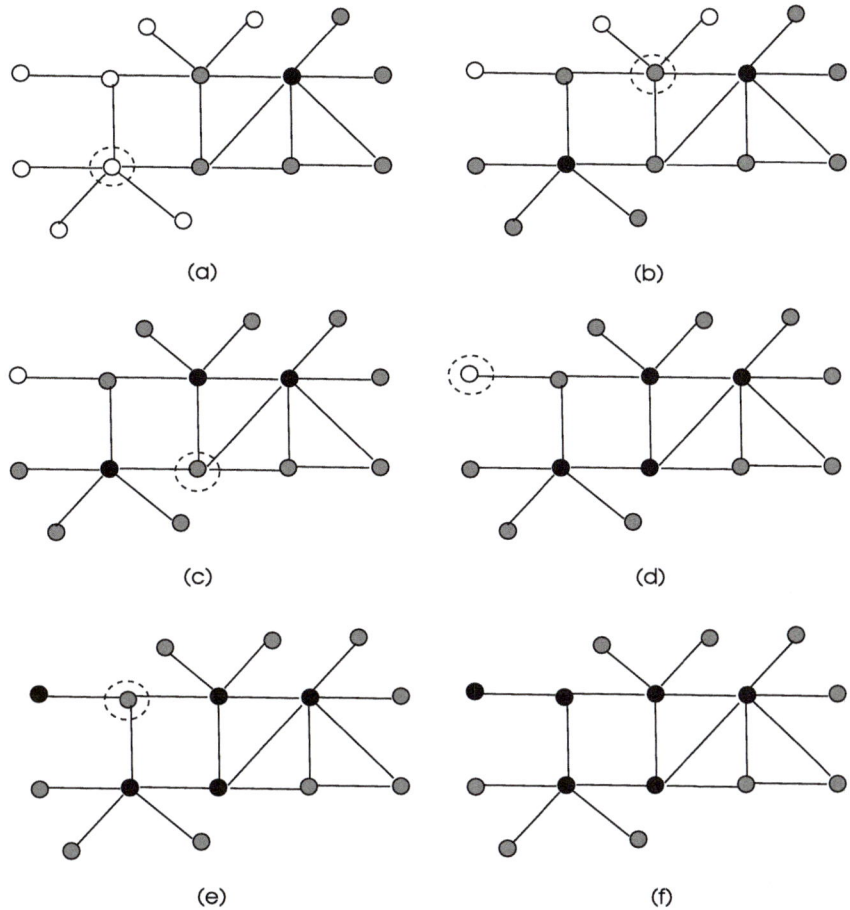

Fig. 10.11 Running of Guha–Khuller's second algorithm in a sample graph. The selected vertex at each iteration is shown inside a dashed circle to be shown in *black* in the next iteration. Note that we could have opted to include the *white* vertex in MDS in **c** since it doing so also results in one less piece. Also, we could have selected the *gray* vertex next to the *white* one in **d** to result in one less step. The unconnected MDS in **e** is connected in **f**. The MCDS is formed after six iterations of the algorithm

10.4 Vertex Cover

A *vertex cover* or a *cover* of a graph is a subset of its vertices such that every edge is incident to at least one vertex in this subset. Vertex cover has numerous applications, such as placing stores in a region so that every road leads to at least one store, in bioinformatics [10] and in chemistry [9]. We can define this set property formally as follows:

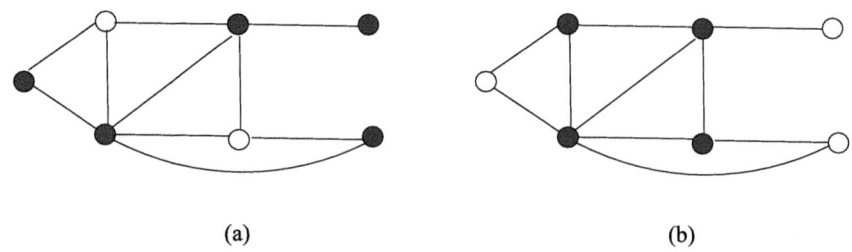

Fig. 10.12 a An unconnected MVC with order 5. **b** A connected MinVC with order 4 of a sample graph. The vertices in the vertex covers are shown in gray. The vertex cover in **b** is connected

Definition 10.3 (*Vertex cover*) A vertex cover of a graph $G = (V, E)$ is a subset V' of its vertices such that $\forall(u, v) \in E$, either u, or v or both are in V'.

A *minimum vertex cover* (MinVC) of a graph is the set with the minimum order vertex cover among all vertex covers of that graph. A *minimal vertex cover* (MVC) of a graph does not contain any other vertex covers of that graph as depicted in Fig. 10.12. In other words, removing a vertex from a MVC will destroy the vertex cover property of this set. Finding MinVC of a graph is NP-hard [2] and commonly, our goal is to find a minimal vertex cover of a graph.

We had stated the following conclusion in Chap. 3 as an example of a reduction and it would be appropriate to restate it here. A set V' is a vertex cover of a graph $G = (V, E)$ if and only if $V \setminus V'$ is an independent set of G. Since V' is a vertex cover, every edge (u, v) has at least one endpoint in V'. If both of vertices u and v are in $V \setminus V'$, then $(u, v) \notin E$ as otherwise V' will not be a vertex cover since it does not have a vertex incident to the edge (u, v). Therefore, $V \setminus V'$ is an independent set of G. We saw that I is an independent set of a graph G if and only if I is a clique in $\overline{G}$. Thus, all of the three problems of independent set, clique, and vertex cover are equivalent.

Vertices of a graph may have weights associated with them depicting some physical property such as the capacity of a router in a computer network. In such a case, our aim is to find a vertex cover with the minimal total weight. The minimum weight vertex cover (MinWVC) is the vertex cover with minimum total weight among all weighted vertex covers of a graph. Viewed from another perspective, we may require to find a minimal connected vertex cover (MCVC) which is to say that there is a path between each pair of vertices in the cover consisting of a subset of vertices in this set only. We will review sequential, parallel, and distributed algorithms to find unweighted and weighted minimal vertex covers in this section.

10.4.1   Unweighted Vertex Cover

Unweighted vertex cover algorithms assume the vertices of the graph have no weights (or sometimes unity weights) assigned to them.

10.4.1.1 Unweighted General Graph Vertex Cover

As a first natural approach, we may select the vertex with the highest degree first to be included in the vertex cover and continue by always selecting the highest degree vertices. Instead of selecting the vertices with initial static degrees, we select the current highest degree at each iteration since deleting a vertex and its incident edges will cause a decrease in the degrees of its active neighbors. However, this approach does not yield a fixed approximation ratio. In fact, the approximation ratio is $\Theta(\log n)$ which is not favorable as it depends on the number of vertices.

We have already reviewed how to compute the vertex cover of a graph from a matching which yielded a constant approximation ratio of 2 as an example of an approximation algorithm (See Sect. 3.8.2). In this algorithm, we find a maximal matching of a graph by selecting an arbitrary legal edge and including both ends of the selected edge in the MVC. The selected edge and its adjacent edges are deleted from the graph and the process is repeated until there are no edges left. As for a parallel vertex cover algorithm, we can always use a parallel maximal matching algorithm by including both endpoints of matching edges in the vertex cover consequently.

10.4.1.2 Unweighted Bipartite Graph Vertex Cover

In any bipartite graph, the number of edges in a maximum matching equals the number of vertices in a minimum vertex cover by König's theorem [5]. We know how to find a maximum matching of a bipartite graph by the augmenting path algorithm in $O(nm)$ time or by the Hopcroft–Karp algorithm in $O(\sqrt{n}m)$ time (See Sect. 9.3). Once we have a maximum matching M^* of a bipartite graph $G = (A \cup B, E)$, we know the size of the minimum vertex cover, $|V^*|$, equals $|M^*|$ but we need to find the elements of the set VC. We will first provide a method to do so and then prove its correctness. But first of all, we note $\forall (u, v) \in M^*$, we need to include either vertex u, or vertex v but not both in V^* simply because one endpoint of the edge (u, v) is sufficient to cover it. Our procedure to find the vertices contained in the minimum vertex cover is based on this observation. If we mark all possible vertices that can be reached using alternating paths from unmatched vertices in A in sets S and T for those in A and B respectively, the minimum vertex cover set V^* is $(A \setminus S) \cup (B \cap T)$ as shown in Algorithm 10.12.

Algorithm 10.12 *Bipartite_MinVC*

1: **Input**: an undirected, unweighted bipartite graph $G = (A \cup B, E)$
2:        a maximum matching M^* of G
3: **Output**: minimum vertex cover V^* of G
4:
5: **form** G' by directing matched edges from B to A and unmatched edges from A to B
6: **for all** $u \in A$ of G' that is free **do**
7:    run $DFS(u)$ and insert all vertices visited in A in S, and visited in B in T
8: **end for**
9: $V^* \leftarrow (A \setminus S) \cup T$

An example bipartite graph with a maximum matching shown in bold lines is depicted in Fig. 10.13a. We first form G' by orienting the edges. The only unmatched vertex in A is d, we therefore run the DFS algorithm from this vertex visiting vertices $\{r, a, p, b, s, d\}$ shown in gray. The minimum vertex cover is then $(\{a, b, c, d, e\} \setminus \{a, b, d, e\}) \cup (\{p, q, r, s, t\} \cap \{p, r, s\} = \{c, e, p, r, s\}$ with order 4 which is the size of the maximum matching as shown in Fig. 10.13b. We can see only one end of matched edges are included in the minimum vertex cover.

Theorem 10.2 Algorithm 10.12 correctly constructs a minimum vertex cover of a bipartite graph from its maximum matching in $O(n + m)$ time.

Proof We will first prove V^* is a vertex cover, then it is a minimum one. Let us assume V^* is not a vertex cover. Then $\exists (u, v) \in E$ with $u \in A$ and $v \in B$ such that $u \notin V^*$ and $v \notin V^*$. Since $V^* = ((A \setminus S) \cup T)$, we must have $u \in S$ and $v \in B \setminus T$. We have two possibilities in such a case:

1. Assume edge $(u, v) \in M^*$. Since $u \in S$, we must have visited it by DFS. However, we can only reach u over a matched edge meaning $v \in T$ resulting in a contradiction.
2. Assume edge $(u, v) \notin M^*$. In this case, since $u \in S$, v is included in T which is a contradiction again since we assumed $b \notin T$.

We will now show $|V^*| \geq M^*$ and since $|M^*| \geq V^*$, we will have proven $|M^*| = |V^*|$. The following observations can be stated [11].

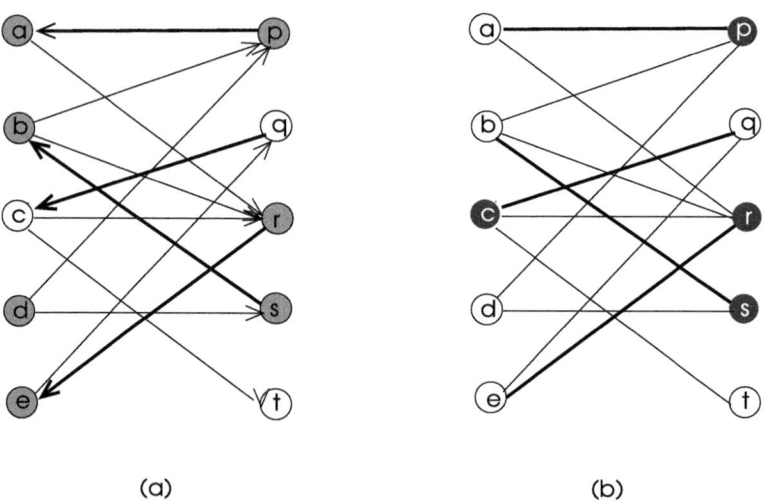

(a)   (b)

Fig. 10.13 A sample bipartite graph to test minimum vertex cover algorithm from maximum matching. The shaded vertices in **a** are visited by DFS from vertex d and the dark vertices in **b** are contained in the minimum vertex cover

1. $\nexists u \in (A \setminus S)$ that is unmatched since we start each DFS from an unmatched vertex in A which is included in S.
2. $\nexists u \in T$ that is unmatched since this would mean there exists an augmenting path of G with respect to M^* and therefore M^* would not be maximum.
3. $\nexists (u, v) \in M^*$ such that $u \in (A \setminus S)$ and $B \in T$. If such an edge existed, u would be in S when b was found to be in T which results in a contradiction since $u \in (A \setminus S)$.

We can conclude that each vertex of the set V^* is incident to a matched edge by the first two observations and two endpoints of a matched edge are not both included in V^* by the last observation.   □

10.4.1.3 Distributed Algorithms

We can use any of the distributed matching algorithms described in Chap. 9 to find matching edges. Any node that has an incident matching edge can then mark itself as being a member of the MVC in a distributed setting. This approach will provide a MVC with an approximation ratio of 2 as in the sequential case since we are merely imitating the sequential operation in a network environment. We will describe two synchronous distributed algorithms; the first one is the distributed version of the greedy algorithm that favors high-degree nodes and the second one has fixed number of rounds providing a constant approximation ratio.

Greedy Algorithm

The *highest degree vertex first* heuristic provided an approximation ratio that depended on the order of the graph. We will however describe a distributed version of this algorithm to give another example of how to convert a sequential algorithm to a distributed one.

The distributed MVC algorithm using this heuristic works in synchronous rounds. Each active node compares its current degree with its active neighbors and if it has the highest degree, it marks itself to be in the MVC and informs its neighbors of its decision. The execution of this algorithm in a sample graph is depicted in Fig. 10.14.

Parnas–Ron Algorithm

Parnas and Ron provided a distributed synchronous unweighted vertex cover algorithm that works for a constant number of rounds dependent on the highest degree Δ in a graph [8]. Each node in the network checks whether its degree is greater than $\Delta(G)/2^i$ in each round i. If this check returns a true value, it becomes part of the vertex cover.

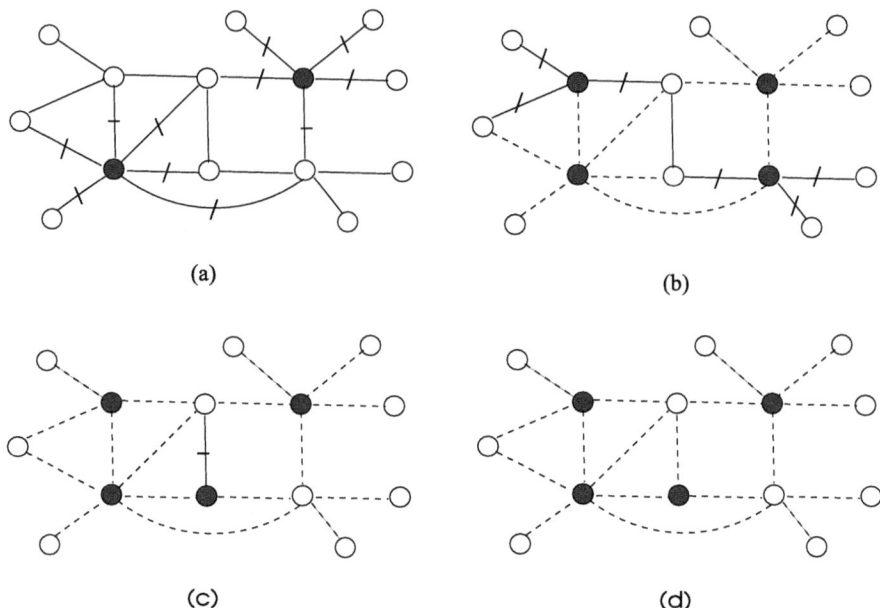

Fig. 10.14 Execution of the greedy distributed vertex cover algorithm in a sample undirected graph. Nodes included in the cover are shown in bold with deleted edges as dashed at each iteration. The final vertex cover is shown as black nodes in d

---

Algorithm 10.13 *Parnas_MVC*

1: **Input**: unweighted undirected graph $G = (V, E)$, $\Delta(G)$
2: **Output**: vertex cover C of G
3: **message types**: *in_cover*   ▷ sent by a node entering VC
4: **states**: {INVC, NONVC}   ▷ states of a node as in or not in the vertex cover
5: $C \leftarrow \emptyset$; *state* $\leftarrow$ NONVC   ▷ initialize vertex cover C
6: **for** $i = 1$ to $\log \Delta(G)$ **do**
7:     **if** $d(i) \geq \Delta(G)/2^i$ **then**
8:         *state* $\leftarrow$ INVC
9:         **send** *in_cover*(i) to $N(i)$
10:    **end if**
11:    **if** *in_cover*(j) is received from neighbor j **then**
12:        $N(i) \leftarrow N(i) \setminus \{j\}$ ▷ remove neighbor j and the edge (i, j) from graph
13:    **end if**
14: **end for**

A major drawback with this algorithm is that $\Delta(G)$ parameter must be broadcast to all nodes before the execution of the algorithm. The operation of this algorithm in the same sample graph of Fig. 10.14 is shown in Fig. 10.15.

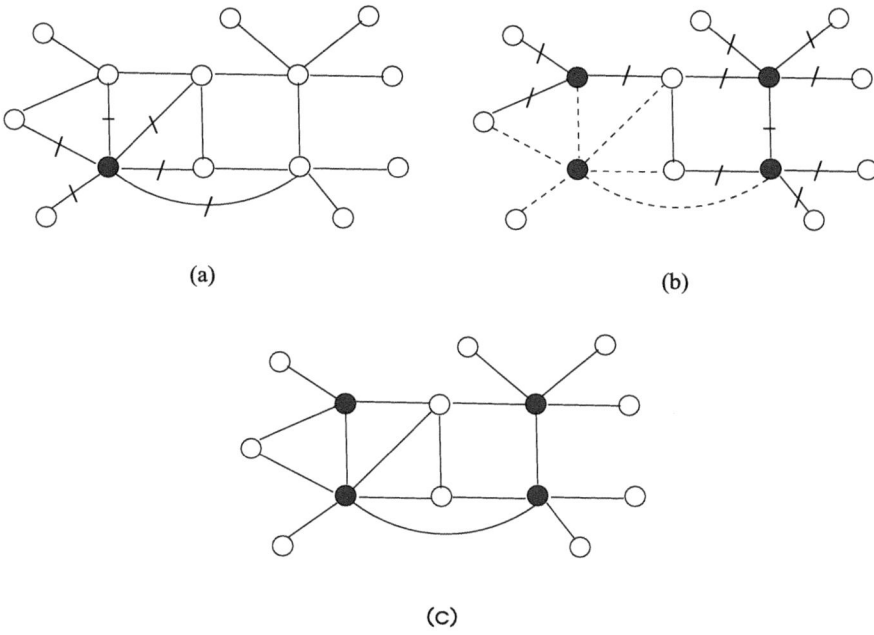

(a)

(b)

(c)

Fig. 10.15 Running of Parnas–Ron algorithm in the same sample graph of Fig. 10.14

Theorem 10.3 *Parnas_MVC* algorithm correctly constructs a minimal vertex cover VC such that $|V^*| \leq |VC| \leq (2 \log \Delta(G) + 1) \cdot |V^*|$ where V^* is the minimum vertex cover.

Proof In the last iteration of the algorithm, all vertices that have degree ≥ 1 will be included in the vertex cover, therefore all of the edges of the graph will be removed with at least one of their incident vertices included in the vertex cover, hence the algorithm correctly constructs a vertex cover.

  The number of iterations is at most $\log \Delta(G)$ since after $\log \Delta(G)$ iterations, the remaining vertices in the graph will have 0 degree. At each iteration, there are at most $2|V^*|$ new vertices that are added from $G - V^*$ to VC. At the beginning of the ith iteration, the degree of each vertex is at most $d/2^{i-1}$. Therefore, the number of edges between V^* and $G - V^*$ is at most $|V^*| \cdot \Delta(G)/2^{i-1}$. Let us assign x_i to the number vertices in $G - V^*$ of degree at least $d/2^i$ at the beginning of the ith iteration. Hence, $x_i \cdot d/2^i \leq |V^*| \cdot d/2^{i-1}$; therefore, $x_i \leq 2|V^*|$. Since we have at most $\log \Delta(G)$ iterations, it follows that the total number of vertices included in VC is at most $2|V^*| \cdot \log \Delta(G)$ [8].   □

10.4.2 Weighted Vertex Cover

When vertices have weights, we search for a minimal weighted vertex cover. Finding the vertex cover with the minimum total weight is NP-hard as most of the problems we have studied. The Pricing algorithm is a sequential approximation algorithm for MWVC problem as described next together with parallel and distributed algorithms for the MWVC problem.

10.4.2.1 Pricing Algorithm
The main idea of this algorithm is to cover an edge with the minimum weight vertex it is incident. An edge e pays a price $p_e \geq 0$ to be covered by vertex u it is incident and the sum of prices assigned to edges that are incident to a vertex u should not exceed the weight w_u of a vertex. Formally,

- an edge $e \in E$ pays a price $p_e \geq 0$ to be covered by a vertex that it is incident.
- $\forall u \in V$, $\Sigma_{u,v}\, p_e \leq w_u$.

When the sum of the prices of edges incident to a vertex equals its weight, the vertex is said to be *tight*. A possible implementation of this algorithm is shown in Algorithm 10.14 where each vertex $v \in V$ has a capacity c_i which is initialized to its weight and the active edge set S is initialized to E. The algorithm inspects each edge $e_{uv} \in S$ and if u or v has a remaining capacity, it charges the edge e with the lower of the capacities. When the capacity of u or v becomes 0, it is labeled as a *tight* node and included in the $MWVCV'$. The algorithm stops when each vertex has a tight vertex at least incident to one of its endpoints, meaning all of the edges are covered by tight vertices on one or both ends. The execution of this algorithm is shown in Fig. 10.16.

Algorithm 10.14 $Pricing\_MWVC$

1: Input $G(V, E)$
2: $S \leftarrow E, V' \leftarrow \emptyset$
3: **while** $S \neq \emptyset$ **do**
4:   pick any $e_{uv} \in S$
5:   **if** $c_u \neq 0 \vee c_v \neq 0$ **then**
6:     $q \leftarrow$ node with $\min\{c_u, c_v\}$
7:     $p_e \leftarrow c_q, q \leftarrow tight$
8:     $V' \leftarrow V' \cup \{q\}; S \leftarrow S \setminus \{e\} \cup$ any other edge incident at q
9:   **end if**
10: **end while**

Theorem 10.4 *Pricing_MWVC* algorithm correctly constructs a MWVC of a graph in $O(n)$ iterations with an approximation ratio of 2.

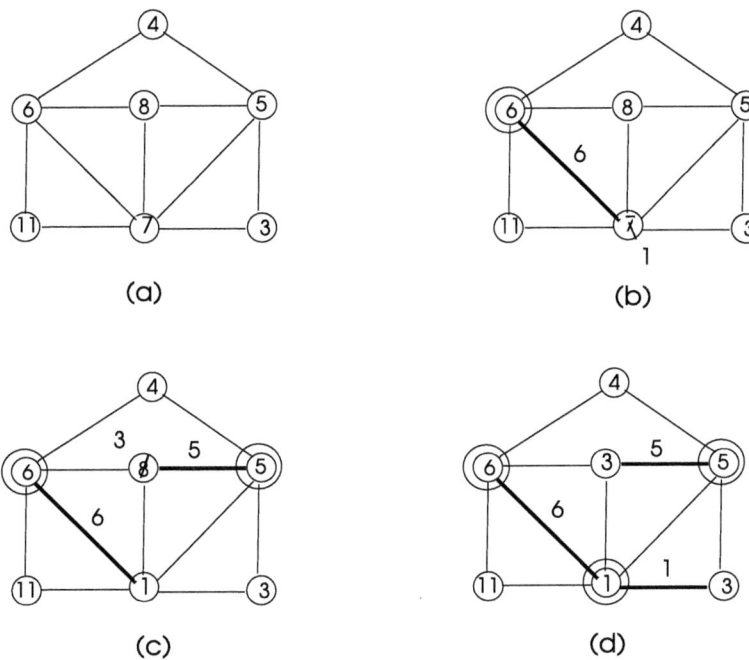

Fig. 10.16 Running of *Pricing_MWVC* algorithm in a sample graph. The weights of vertices are shown inside them and tight vertices are shown with double circles. After five iterations, MWVC is formed as shown in **f** with a total weight of 18

Proof This algorithm correctly constructs a vertex cover since we continue exploring all edges until there are no edges left with a tight vertex on one or both of its endpoints, therefore all of the edges are covered. There will be at least one tight vertex at each iteration, resulting in $O(n)$ time complexity

Let V' be the set of all tight vertices at the end of the algorithm and V^* the minimum vertex cover vertices. We need to show $w(V') \le 2w(V^*)$. Since all vertices in V' are tight, we can write the following equation.

$$w(V') = \sum_{v \in V'} w_v = \sum_{v \in V'} \sum_{e=(u,v)} p_e \le \sum_{v \in V} \sum_{e=(u,v)} p_e. \tag{10.1}$$

Since each edge is counted twice, we can write the above as

$$= 2 \sum_{e \in E} \le 2w(V^*). \tag{10.2}$$

□

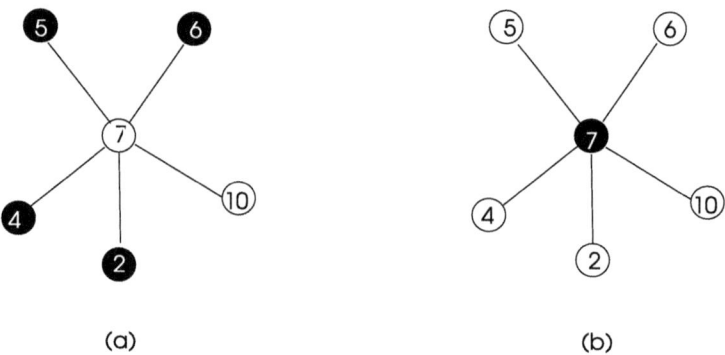

Fig. 10.17   a A star graph with MWVC vertices shown in black formed by the greedy distributed MWVC algorithm that has a total weight of 17 for the cover **b** the optimal MWVC for the same graph

Fig. 10.18 Handling of ghost vertices

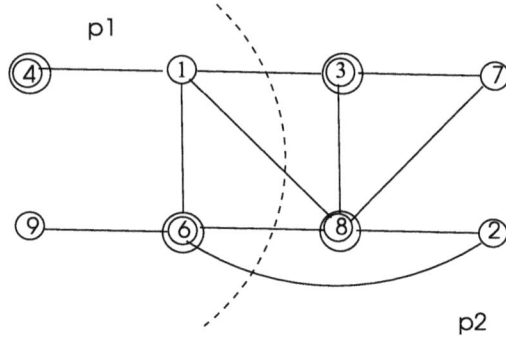

10.4.2.2 Distributed Algorithms

We can again use any of the distributed weighted matching algorithms with the simple modification of including both endpoints of the matched edges in the minimal vertex cover. As another simple approach, we can use the greedy method in a synchronous distributed algorithm where each node that has the smallest weight in its neighborhood enters the vertex cover set. The algorithm designed this way is basically very similar to the greedy distributed algorithm that used degrees of nodes and the locally highest degree node is assigned to the vertex cover in the unweighted vertex cover case. However, in certain network topologies, it may yield a solution that is far from optimal as shown in the star configuration of Fig. 10.17.

10.4.3   Parallel Algorithms

Parallel algorithms whether shared memory or distributed memory for MVC in unweighted or weighted case are scarce. A basic approach to perform parallel finding of vertex cover can be achieved by partitioning the graph into a number of approximately same order of subgraphs and have each process find vertex covers in its subgraphs. Each process can work independently to form its vertex covers, however, care must be taken when dealing with vertices that appear on the borders of partitions. One way of dealing with this problem is to use the *ghost vertex* concept in which border vertices are replicated at each process and the choice of a ghost vertex to be included in which partition can be handled by breaking symmetries using unique vertex identifiers as we have briefly described in Chap. 4.

This concept is illustrated in Fig. 10.18 where a simple graph with 8 vertices is partitioned to two parallel processes p_1 and p_2. Each vertex has a unique integer identifier and the border vertices between partitions are 1, 3, 6, 8, and 2. Each process is responsible for the higher identifier border vertex it is assigned. In this respect, edges (1, 8) and (1, 3) have to be covered by process p_2 and (6, 2) is covered by process p_1. Vertices in cover are shown by double circles.

Algorithm 10.15 displays a possible pseudocode of a parallel algorithm to perform parallel vertex cover in which a special process, the *root*, partitions the graph using a suitable algorithm, sends each partition to processes, and gathers the partial minimal vertex cover results to form the final minimal vertex cover of the graph. Each process finds the minimal vertex cover in its partition by including higher identifier neighbors of border vertices in its partition and sends the partial minimal vertex cover to the *root*.

Algorithm 10.15 *Par_MVC*

1: **Input**: unweighted undirected graph $G = (V, E)$, $\Delta(G)$
2: **Output**: minimal vertex cover MVC of G
3: **if** $i = root$ **then**
4:     **partition** G into k subgraphs $G(1), ..., G(k)$
5:     **for** $i = 1$ to k **do**
6:         **send** $G(i)$ to p_i with neighbors of border vertices
7:     **end for**
8:     **for** $i = 1$ to k **do**
9:         **receive** $VC(i)$ from p_i
10:         $MVC \leftarrow MVC \cup VC(i)$
11:     **end for**
12: **else**
13:     **receive** $G(i)$ with neighbors of border vertices
14:     **for all** $u \in$ border vertices **do**
15:         **if** $\exists v \in$ neighbors of border vertices such that $(u, v) \in E$ **then**
16:             **if** $id(u) > id(v)$ **then**
17:                 $G(i) \leftarrow G(i) \cup \{v\}$
18:             **else**
19:                 $G(i) \leftarrow G(i) \setminus \{u\}$
20:             **end if**
21:         **end if**
22:     **end for**
23:     **find** vertex cover $VC(i)$ of $G(i)$
24:     **send** $VC(i)$ to *root*
25: **end if**

10.5   Chapter Notes

We have reviewed three problems in graphs; the maximum independent set, the minimum dominating set, and the minimum vertex cover problems. All of these problems are NP-hard and the algorithms proposed in literature are approximation or heuristic algorithms that find maximal or minimal solutions rather than maximum or minimum order subgraphs. We can employ greedy algorithms commonly by the use of some heuristic but algorithms with better performances are frequently sought. As we have noted in the highest degree first algorithm to find a minimal vertex cover, a seemingly natural heuristic may not provide a constant approximation ratio.

We saw the independent set, clique, and vertex cover problems are computationally equivalent; a vertex set V' of a graph $G = (V, E)$ is a vertex cover of G if and only if $V \setminus V'$ is an independent set of G. Also, the set $V \setminus V'$ is a clique in $\overline{G}$ if and only if V' is an independent set of G. A dominating set may be connected and an independent set may be used as the first step of forming a connected dominating set. A vertex cover may also be connected and also vertices may have weights associated with them depicting some physical parameter attributed to the nodes of the system that the graph represents. The weighted versions of these problems commonly require different considerations than unweighted ones.

The parallel algorithms for these problems are scarce and with the recent advancements resulting in the availability of data for very large real networks, there is an increasing need for parallel algorithms. In some cases, fast, efficient, deterministic parallel algorithms have been developed for these problems but these algorithms may be quite complicated. For parallel computation of MDS, we can use a similar algorithm as the one used for vertex cover by partitioning the graph to a set of processes each of which runs an MDS algorithm in its partition.

Distributed network algorithms are at a more fairly investigated level for these problems as we have noted. In many cases, these algorithms are derived from sequential ones, however, there is still need for algorithms with better performances.

Exercises

1. Propose a heuristic to find the IS in Algorithm 10.2 and implement this method with Algorithm 10.2 to find the MIS of Fig. 10.19.
2. In order to find the MaxIS of a rooted tree, we can include all leaves of the tree in the MaxIS and move upwards in the tree by not including one level in the MaxIS and next level in the MaxIS.
   a. Write the pseudocode for this sequential algorithm.
   b. Prove that this algorithm finds the MaxIS for a rooted tree.
   c. Show how the sequential code can be converted to a distributed algorithm.
   d. Propose a method to find the MaxIS of a tree in parallel using this method.
3. Implement Guha–Khuller first algorithm to find the MDS in the example graph of Fig. 10.20.

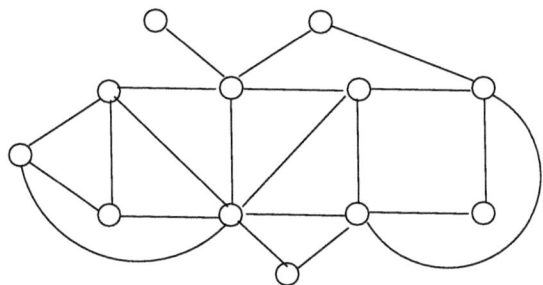

Fig. 10.19   Sample graph for Exercise 1

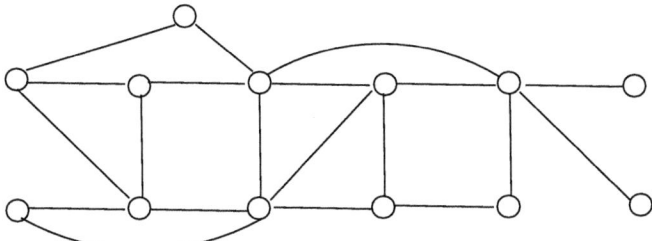

Fig. 10.20   Sample graph for Exercise 3

4. Implement the distributed span algorithm to find the MDS in the example graph of Fig. 10.21.
5. Find the MVC of the example bipartite graph of Fig. 10.22 using matching. Show all iterations of the algorithm.
6. Work out the MWVC in the sample graph of Fig. 10.23 using Pricing algorithm.
7. Implement the greedy distributed weighted vertex cover algorithm to find the MWVC in the graph shown in Fig. 10.24, where weights of vertices are shown inside them.

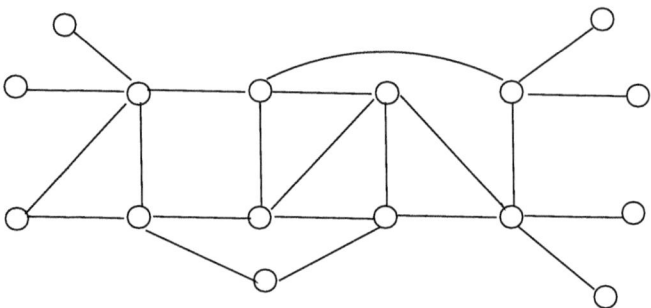

Fig. 10.21 Sample graph for Exercise 4

Fig. 10.22 Sample graph for
Exercise 5

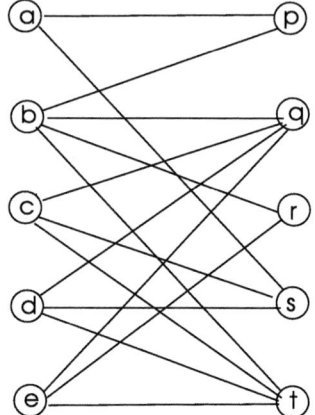

Fig. 10.23 Sample graph for
Exercise 6

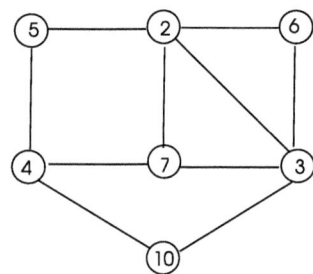

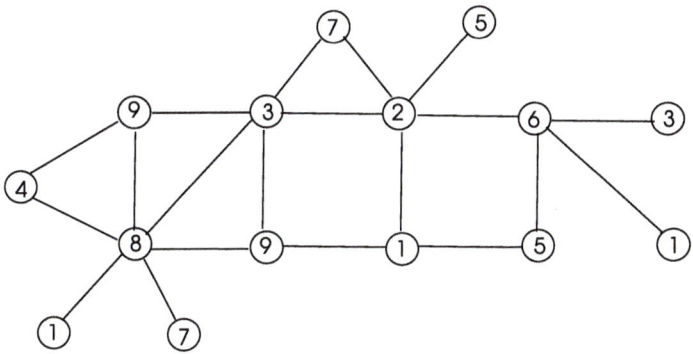

Fig. 10.24 Sample graph for Exercise 7

References

1. Erciyes K (2013) Distributed graph algorithms for computer networks (Chap. 10). Computer communications and networks series, Springer, Berlin. ISBN 978-1-4471-5172-2
2. Garey MR, Johnson DS (1978) Computers and intractability: a guide to the theory of NP-completeness. Freeman, New York
3. Grama A, Gupta A, Karypis G, Kumar V (2003) Introduction to parallel computing (Chapter 10), 2nd edn. Addison Wesley, Reading
4. Guha S, Khuller S (1998) Approximation algorithms for connected dominating sets. Algorithmica 20(4):374–387
5. König D (1931) Graphen und Matrizen. Math Lapok 38:116119
6. Koufogiannakis C, Young N (2009) Distributed and parallel algorithms for weighted vertex cover and other covering problems. In: The 28th ACM SIGACT-SIGOPS symposium on principles of distributed computing (PODC 2009)
7. Luby M (1986) A simple parallel algorithm for the maximal independent set problem. SIAM J Comput 15(4):1036–1053
8. Parnas M, Ron D (2007) Approximating the minimum vertex cover in sublinear time and a connection to distributed algorithms. Theor Comput Sci 381(1):183–196
9. Rhodes N, Willett P, Calvet A, Dunbar JB, Humblet C (2003) Clip: similarity searching of 3d databases using clique detection. J Chem Inf Comput Sci 43(2):443448
10. Samudrala R, Moult J (2006) A graph-theoretic algorithm for comparative modeling of protein structure. J Mol Biol 279(1):287302
11. Stein C (2012) IEOR 8100. Lecture notes. Columbia University
12. Wattenhofer R (2016) Principles of distributed computing (Chapter 7). Class notes. ETH Zurich

Coloring

11

11.1   Introduction

Coloring in a graph refers either to *vertex coloring*, *edge coloring* or both in which case it is called *total coloring*. Each vertex is assigned a color from a set of colors such that no two adjacent vertices have the same color in vertex coloring. This method has many applications including channel frequency assignment and scheduling of jobs. Assignment of frequency channels to radio stations may be modeled by coloring of a graph with the vertices representing radio stations and an edge connects two stations if they are within interference distance to each other. Different colors in this case correspond to different broadcast frequencies. As a scheduling example, we may need to assign final exams in a university so that no student takes two exams at the same time. We can represent each exam by a vertex in a graph and an edge connects two vertices a and b if a student is taking both final exams a and b. If the colors of vertices represent time slots for final exams, our aim is to color each vertex of the graph such that two adjacent vertices receive a different color, meaning a student taking both exams will attend them in different time slots. Edge coloring is the process of assigning colors to the edges of a graph such that no two edges incident to the same vertex are assigned the same color. Edge coloring may be used in planning a timetable for teachers to teach courses in a school to achieve the minimum amount of course time. A bipartite graph with teacher and course partitions of vertices can be formed, and we search a minimal edge-coloring of this graph. Then, we find a maximum time value that any teacher is involved in teaching, which is the maximum time spent in teaching all of the courses. Time division multiple access network communication protocols for sensor networks may be realized using edge coloring, representing each time slot with a color [7].

The main goal of any coloring method is to use a minimum possible number of colors. Since this is an NP-hard problem for vertex and edge coloring [10], various heuristics are commonly employed. Parallel vertex coloring algorithms attempt to

K. Erciyes, *Guide to Graph Algorithms*, Texts in Computer Science,
https://doi.org/10.1007/978-3-032-05294-0_11

concurrently color different regions of the graph under consideration by a number of processes to achieve speedup. A distributed graph coloring algorithm on the other hand, is executed by each node of the network graph so that each node determines its color in the end.

We can have parallel and distributed edge coloring algorithms as in vertex coloring. Total graph coloring can be achieved by both coloring vertices and edges of a graph. We start with the vertex coloring problem in this chapter by describing sequential, parallel, and distributed algorithms for this task and continue with algorithms for edge coloring.

11.2   Vertex Coloring

A vertex coloring of a graph is the assignment of colors to its vertices such that no two adjacent vertices have the same color. It can be formally defined as follows.

Definition 11.1 (*Vertex coloring*) A *vertex coloring* or coloring of a graph $G = (V, E)$ is an assignment function $\phi : V \rightarrow C$ such that $\forall (u, v) \in E, \phi(u) \neq \phi(v)$, where C is a set of colors elements of which are commonly the elements of $\mathbb{N}^+$.

For a graph with n vertices, the set C with n elements will provide its coloring, however, our aim is to find the minimum number of colors in the optimization version of the *vertex coloring problem*. The decision version of this problem seeks to find an answer to the question. "can we color the vertices of a graph with at most k colors?." We will consider connected and simple graphs for this problem. A *k-coloring* of a graph G is the coloring of G using k colors. In a *proper vertex coloring* of a graph, adjacent vertices are colored with distinct colors. The vertices of the same color in a graph form a *color class*.

Definition 11.2 (*Chromatic number*) The *chromatic number* $\chi(G)$ of a graph G is the minimum number of colors required to color its vertices properly.

Finding $\chi(G)$ of G is an NP-hard problem [10]; however, we can specify an upper bound on the value of this parameter as we will see.

Remark 11.1 Any subgraph H of a graph G can be colored with less colors than G, that is, $\chi(H) \leq \chi(G)$.

While coloring G we will have colored all of its subgraphs and it is probable that we use less colors to color its subgraphs.

Remark 11.2 The chromatic number $\chi(G)$ of a graph G with n vertices is n if and only if $G = K_n$. That is, $\chi(K_n) = n$.

This is valid since all vertices are connected to all other vertices in K_n and hence, we need n distinct colors to color such a graph.

Remark 11.3 The chromatic number $\chi(G)$ of a star graph S_n with n vertices is 2 since we can color all of the vertices connected to the center with the same color and the center with another color.

There are some interesting properties of vertex coloring as follows.

- A bipartite graph has no odd length cycles and hence can be colored with 2 colors. We can color a bipartite graph by running the BFS algorithm of Chap. 7 and color vertices at odd levels with color 1 and the vertices at even levels with color 2.
- Since a tree is a bipartite graph, we can color a tree with two colors.
- A cycle graph with an even number of vertices is a bipartite graph and therefore, we can color such a graph with two colors as in Fig. 11.1a.
- A cycle graph with an odd number of vertices is not a bipartite graph. We can color such a graph with n vertices using a total of three colors; two colors for $n - 1$ vertices and a third color for the nth vertex as shown in Fig. 11.1b.

Theorem 11.1 (Brook's Theorem [4]) *For a connected graph G that is not fully connected or does not have an odd cycle,*

$$\chi(G) \leq \Delta(G) + 1. \tag{11.1}$$

Lovasz gave a short and simple algorithmic proof of this theorem by considering three cases [15].

Remark 11.4 The equality holds in only two cases.

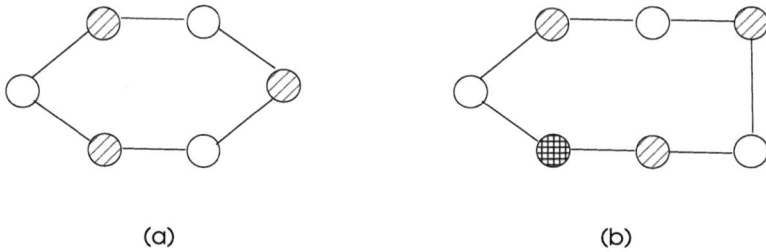

(a)   (b)

Fig. 11.1 Coloring of even and odd-cycle graphs

- In a complete graph K_n with $n \geq 3$, $\Delta(K_n) = n - 1$ and $\chi(K_n) = n$. Therefore,

$$\Delta(K_n) + 1 = n = \chi(K_n)$$

- For an odd-cycle graph G, $\chi(G) = 3$ and since $\Delta(G) = 2$, $\chi(G) = \Delta(G) + 1$.

We will see the greedy algorithm presented in the next section also has a complexity as this upper bound. Brook showed that the equality holds only for odd-cycle graphs and complete graphs. However, this upper bound on a chromatic number of a graph may turn out to be very far from the real value as in the star graph S_n with n vertices; $\Delta(S_n) = n - 1$ and $\chi(S_n) = 2$ in such a graph.

11.2.1 Relation to Independent Sets and Cliques

A subset I of vertices of a graph $G = (V, E)$ is called an *independent set* if no two vertices in I are neighbors and the maximum independent set (MaxIS) is an independent set with the maximum order as we described in Sect. 10.2. A maximal independent set (MIS) on the other hand is an independent set which cannot be enlarged by the addition of vertices. We have reviewed a method to obtain an MIS from a k-colored graph in Chap. 10 and the reverse operation of coloring vertices of a graph using independent sets is also possible. We observe that vertices in an independent set can be colored with the same color since they are not adjacent. In fact, we will use this property in designing vertex coloring algorithms as we will see. Given a k-chromatic graph G, we can partition the vertices of G into k disjoint independent sets $I_1, \ldots, I_k$ which are called *color classes*. Each vertex in color class I_i can then be colored by i. In other words, we can find k disjoint independents sets of a graph G such that $\bigcup_{i=1}^{k} I_i = V$ which are not necessarily maximal, we can color all elements of each set with a new color and the chromatic number for this graph is k.

Remark 11.5 If a graph G can be partitioned into k disjoint independent sets but not less, then $\chi(G) = k$.

Each coloring class I_i of G is an independent set. Therefore,

$$|I_i| \leq \alpha(G)$$

where $\alpha(G)$ is the maximum independence number of G. Since

$$|I_1| + |I_2|, \ldots, |I_k| \leq k\alpha(G) = \chi(G)\alpha(G)$$

we can conclude,

$$\chi(G) \geq \lfloor \frac{n}{\alpha(G)} \rfloor. \tag{11.2}$$

A *clique* of a graph G is a complete subgraph of G. The *clique number* $\omega(G)$ of a graph G is the order of its largest clique. There is a relation between a clique and an independent set of a graph G as we have noted in Sect. 10.2.1, a subset V' of vertices of $G = (V, E)$ is a clique if and only if V' is a maximal independent set in $\overline{G}$.

Theorem 11.2 Let $\omega(G)$ *be the clique number of graph G, that is, it is the order of the largest clique of G. Then,*

$$\chi(G) \geq \omega(G). \tag{11.3}$$

Proof Since all of the vertices of a clique are adjacent to each other, each such vertex must be colored with a different color. Therefore, the order of the maximum clique in a graph G sets a lower bound on the chromatic index of G. ☐

11.2.2 Sequential Algorithms

Since coloring the vertices of a graph with its chromatic number of colors is an NP-hard problem, various heuristics are proposed in the literature that approximate the number of colors to $\chi(G)$. In this section, we first present a greedy coloring algorithm template and then review four algorithms using different heuristics which are the random algorithm, the first-fit algorithm, the largest-degree-first algorithm and the saturation-based-ordering algorithm. All of these algorithms may be classified as greedy approaches as they select the vertex that best meets the required criteria at each iteration.

11.2.2.1 Algorithm Template

We need to color every vertex of the graph with colors obeying the coloring principle, that is, each vertex is assigned a color that does not conflict with the already assigned colors of the neighbors. We will form an algorithm template without specifying the selection criteria as shown in Algorithm 11.1 and then discuss various heuristics for the selection of the vertex to be colored. In the simplest form, a vertex is selected randomly and the smallest available color is assigned to this vertex in *random selection*. The operation of this algorithm in a simple graph is depicted in Fig. 11.2a.

Algorithm 11.1 *Coloring_Template*
1: **Input** : $G = (V, E)$
2: **Output** : $\phi : V \rightarrow C$ where $C = \{1, 2, ..., n\}$
3: $V' \leftarrow V$
4: **while** $V' \neq \varnothing$ **do**
5:     **select** a vertex $v \in V'$ according to some heuristic
6:     $\phi(v) \leftarrow$ the smallest legal color from C
7:     $V' \leftarrow V' \setminus \{v\}$
8: **end while**

Selection of vertex v in line 5 of the algorithm can be performed using various heuristics as follows.

- *Identifier-Based Algorithm*: In this case, vertices of the graph are numbered from 1 to n to yield a vertex set $V = \{v_1, \ldots, v_n\}$ and the vertices are colored in sequence obeying the rules of coloring; that is, coloring each vertex with the minimum possible color that does not conflict with neighbors. This algorithm is also called the *first-fit* algorithm and uses at most $2\chi(G)$ colors on the average [11]. It is simple and fast in general sense but can yield an approximation ratio of $n / 4$ in some special graphs [12] requiring $O(m)$ running time.
- *Largest-Degree-First (LDF) Algorithm*: It makes sense to color the large degree vertices first to have a less number of colors since the low-degree vertices can usually be colored in a more flexible way as proposed in [20]. The operation of this algorithm is shown in the graph of Fig. 11.2b and we can see it results in one less color than the greedy approach. This algorithm can be implemented to have $O(m)$ time complexity (Fig. 11.3).
- *Saturation-Degree-Ordering (SDO) Algorithm*: A further refinement to the LDF algorithm can be provided as follows [3]. The *saturation degree* $s(v)$ of a vertex

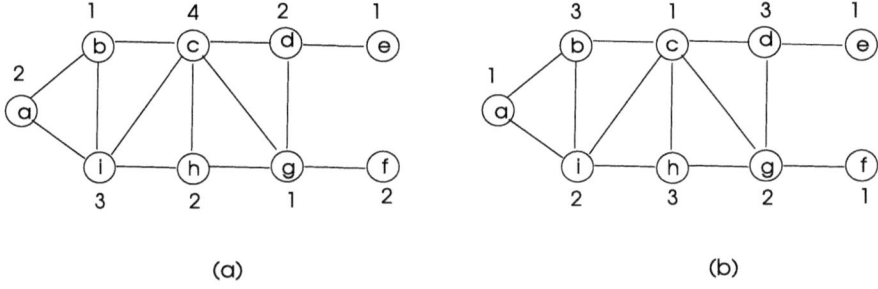

(a)   (b)

Fig. 11.2 Random selection; **a** A random coloring heuristic by Algorithm 11.1 selects vertices *b-g-e-a-d-h-i-c* in sequence and uses four colors shown next to vertices as integers; **b** The largest-degree-first heuristic uses three colors for this graph, irrespective of the choice of vertices when degrees are the same

Fig. 11.3 Saturation-based
ordering. The order of
vertices selected are
c-g-i-d-b-a-e-f

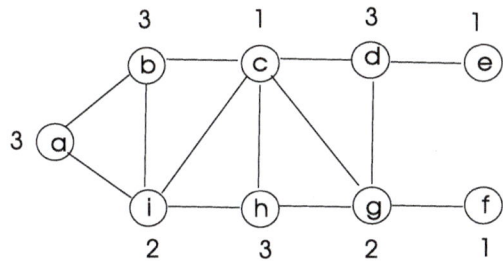

v is defined as the number of distinct colors currently assigned to its neighbors. This parameter is dynamic and a greedy algorithm based on the saturation
degrees can be designed to always select the vertex with the highest value of
this parameter. In case of ties, the vertex with the highest degree is selected
which means we are searching the largest value of the pair $(s(v), deg(v))$ of
all vertices $v \in V$ that are not assigned a color. Note that such an algorithm
will start by the largest degree vertex v of the graph and assign the minimum
color to v, and will continue with the largest degree neighbor vertex of v. The
operation of SDO algorithm is shown in Fig. 11.3. This algorithm has $O(n^2)$
time complexity [3].

- *Incident-Degree-Ordering Algorithm*: This heuristic is a modified form of the
  SDO. The incident degree of a vertex is the number of its colored neighbors.
  Note that the colors of neighbors need not be distinct as in the saturation-based
  heuristic. The vertex that has the highest incident degree is selected at each
  iteration of the algorithm [6]. Vertex identifiers are used in case of ties as in the
  saturation degree algorithm. It is a linear time algorithm running in $O(m)$ time.

11.2.2.2 Independent Set-Based Algorithms

The main idea of independents set-based algorithms is that the vertices in an independent set can be assigned the same color as they are nonadjacent. We can,
therefore, find (maximal) independent sets of a graph G and color all of the vertices of this set with a new color, remove vertices from graph G, and continue until
all vertices are colored as depicted in Algorithm 11.2. Clearly, the performance of
this generic algorithm is influenced by the method used in finding the independent
sets.

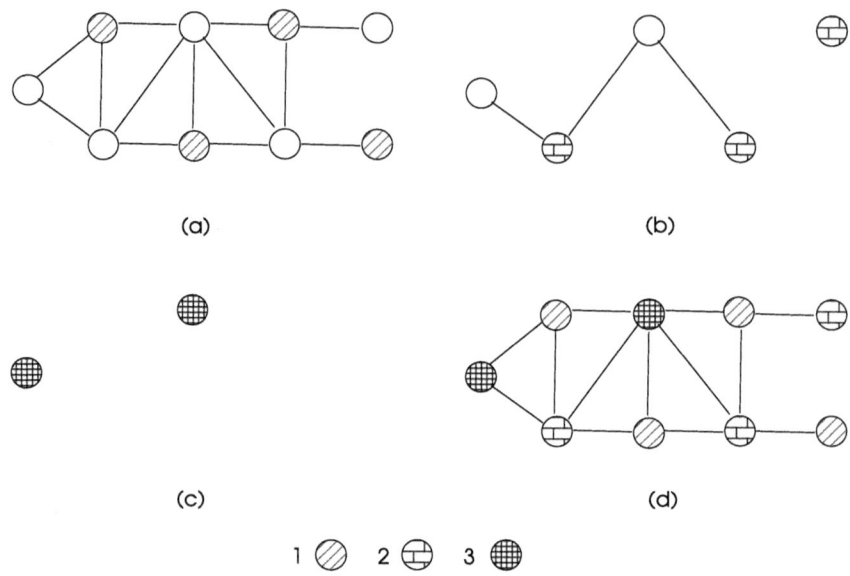

1 ⊘   2 ⊕   3 ⊕

Fig. 11.4 Maximal independent set-based vertex coloring of the same graph of Fig. 11.2; **a, b**, and **c** displays the iterations of the algorithm, where a new color is assigned to each new independent set as shown by different patterns; **d** shows the final coloring of the graph

---

Algorithm 11.2 *IS_Vcolor*

1: **Input** : $G = (V, E)$
2: **Output** : $\phi : V \rightarrow C$ where $C = \{1, 2, ..., k\}$
3: $V' \leftarrow V$
4: **while** $G' \neq \varnothing$ **do**
5:     $I \leftarrow Find\_MIS(G')$
6:     **color** the vertices in I with a new color
7:     $V' \leftarrow V' \setminus I$
8:     $G' \leftarrow$ graph induced by V'
9: **end while**

11.2.3   Parallel Algorithms

There are only few algorithm for coloring vertices of a graph in parallel. We describe independent-set-based algorithms and the identifier-based algorithm for this purpose next.

11.2.3.1 Parallel Independent Set-Based Algorithms

We can use the independent set-based coloring algorithm for parallel coloring vertices of a graph since we can perform finding a maximal independent set in parallel. Algorithm 11.3 displays the code for parallel independent-set-based vertex coloring algorithm. The code within the *while* loop is executed in parallel synchronously.

Algorithm 11.3 *ParIS_Vcolor*

1: **Input** : $G = (V, E)$
2: **Output** : $\phi : V \to C$ where $C = \{1, 2, ..., k\}$
3: $G' \leftarrow G$
4: **while** $G' \neq \varnothing$ **in parallel do**
5:     $I \leftarrow Find\_MIS(G')$
6:     **color** the vertices in I with a new color
7:     $V' \leftarrow V' \setminus I$
8:     $G' \leftarrow$ graph induced by V'
9: **end while**

We have seen how an independent set can be constructed in parallel using Luby's Monte Carlo method in Sect. 10.2.3. In this algorithm, vertices were assigned random permutations $1, \ldots, n$ at each iteration and a vertex with a local minimum value was colored these values were used to break symmetries and decide the colors of vertices. We can simply implement this algorithm for this purpose and then color the vertices of the graph accordingly.

Jones–Plassmann Algorithm

In another and more recent approach, Jones and Plassmann presented independent set-based parallel graph coloring algorithm (*JP_Color*) [13]. Their approach is different than Luby's algorithm as the random numbers are assigned only once at the beginning of the algorithm and do not change. Also, assigning of the colors to the independent set vertices is assigned individually for each vertex, that is, each vertex in the set is colored with the minimum color that does not exist in its neighbors as shown in Algorithm 11.4. Each vertex v is assigned a random number which is its weight $w(v)$ initially. If the weight of a vertex is greater than all of the weights assigned to its neighbors, then v is assigned to the independent set I with ties broken by unique vertex identifiers. This step is performed in parallel and the elements of the set I are also colored in parallel with legal colors. Different than Algorithm 11.3, the independent set I formed at each step need not be MIS as in Luby' method and the vertices of I may be colored with different colors. Jones and Plassmann showed that the expected runtime of this algorithm in bounded degree graphs using PRAM model is $O(\log n / \log \log n)$ [13].

Algorithm 11.4 *JP_Vcolor*

1: **Input** : $G = (V, E)$
2: **Output** : $\phi : V \rightarrow C$ where $C = \{1, 2, ..., k\}$
3:
4: $G' \leftarrow G$
5: **while** $G' \neq \varnothing$ **do**
6:    **for all** $v \in V'$ **in parallel do**
7:       $I = \{v\}$ such that $w(v) > w(u), \forall u \in N(v)$
8:       **for all** $u \in I$ **in parallel do**
9:          **assign** u the minimum color not used by $N(u)$
10:       **end for**
11:    **end for**
12:    $V' \leftarrow V' \setminus I$
13:    $G' \leftarrow$ graph induced by V'
14: **end while**

The parallel Largest-Degree-First (PLDF) algorithm has a similar structure to *JP_Alg* with the difference that the weights that are assigned are the degrees of the vertices and the ties are broken by selecting a random number. A parallel vertex coloring method using graph partitioning is described in [9]. An experimental study reported in [1] compares parallel independent set, Jones and Plassmann and LDF algorithms for parallel vertex coloring.

11.2.3.2 Cole–Vishkin Algorithm

Cole and Vishkin proposed a method to reduce the colors used in a graph in parallel [5], however, the same technique can be implemented in a distributed algorithm to color the nodes of a network. Basically, given a proper k-coloring of a graph G with a possible large k, it aims to find a coloring of G with a smaller k value, hence, it is a color reduction technique. The general idea of this algorithm is to initially have nodes assigned unique labels of $\log n$ bits. Then, new node labels that are much smaller than the previous ones are computed at each iteration of the algorithm.

Each node in the graph has at most one successor meaning this algorithm can be used in directed paths, directed cycles. The k-coloring is reduced to $\log k$ coloring in one step and 6-coloring of the graph is determined after few iterations. Each node v in parallel sends its color to its successor. A node v receiving the color c_u of its predecessor u, compares it with its color c_v and finds the rightmost bit that is different. It then sets its new color as the concatenation of the index with the value in the index as shown in Algorithm 11.5. The iterations continue until we have $k = 6$, that is, colors in the range $0, \ldots, 5$. We need other methods if further reduction is needed. An example reduction is as follows:

$$
\begin{array}{llll}
0001 & 0001 & & \\
0100 & 1001 & 0111 & \\
0110 & 1001 & 1011 & 100
\end{array}
$$

Algorithm 11.5 *CV_Vcolor*

1: **Input** : $G = (V, E)$
2: **Output** : $\phi : V \to C$ where $C = \{0, 1, 2, 3, 4, 5\}$
3: $G' \leftarrow G$
4: **for all** $v \in V$ **do**
5:     $color(v) \leftarrow v_i$
6: **end for**
7: **while** $\exists v_i : color(v_i) > 5$ **in parallel do**
8:     assume $color(v_i)$ and $color(v_{i-1})$ as little-endian bit strings
9:     let j be the smallest bit index x they differ
10:     $c \leftarrow j \cup x$
11:     $color(v_i) \leftarrow c$
12: **end while**

Lemma 11.1 *Algorithm 11.5 correctly finds a legal coloring of a graph G.*

Proof The initial coloring of vertices is legal. We need to show the newly formed colors are also legal. Let us assume two cases.

- *Case 1*: Successor of vertex u, vertex v, chooses the same index. Since bit values are different, they have different colors assigned.
- *Case 2*: When they have different indexes, they have different colors. □

The $\log^* n$ is an extremely slowly growing function of n. It is defined as follows.

Definition 11.3 Starting with n, $\log^* n$ is the number of times logarithm on base 2 is applied until reaching a number smaller than or equal to 2. That is, $\log^* n = \min\{i : \log^i n \leq 2\}$. In other words, $\forall n \leq 2, \log^* n = 1$ and $\forall n > 2, \log^* n = 1 + \log^* (\log n)$. For example, $\log^* 2^{32} = 1 + \log^* 32 = 2 + \log^* 16 = 3 + \log^* 4 = 4$.

Theorem 11.3 *Algorithm 11.5 computes a 6-coloring of a graph in $\log^* n$ time.*

Proof Let n_j be the maximum number of bits used by the color c_v of vertex v after iteration j and let $n_0 = \lceil \log n \rceil$ be the number of bits used for initial coloring of nodes. We can state $n_{j+1} \leq \lceil \log n_j \rceil + 1 \leq \log n_j + 2$. We can continue to find $n_1 \leq \log n_0 + 2$ and $n_2 \leq \log (\log n_0 + 2) + 2 \leq \log \log n_0 + 3$ when $\log n_0 \geq 2$. We can see for $j = 1, 2, \ldots$ with $\log^{(j)} n_0 \geq 3$, $n_j \leq \log^{(j)} n_0 + 3$. Hence, when the number of iterations $j = \log^* n_0$, $n_j \leq 5$. Since $n_0 = \lceil \log n \rceil$, the number of bits for c_v is at most 5. The number of bits after two more iterations is reduced to 3 and hence, the number of possible colors becomes 8. Another iteration makes the palette size 6 as the first part of the color has 3 possible values [2]. □

Note that this algorithm can only reduce the colors to 6 colors. For example, let us assume node u is the predecessor of vertex v and $\phi(v) = 101_B$ and $\phi(u) = 011_B$, the index is 001_B and the node v will set 011_B which will be the same as the color value of u.

11.2.4 Distributed Vertex Coloring

In a distributed setting, our aim is to have each node in the network assigned a legal color. We will first present a simple color reduction algorithm using identifiers of nodes as initial colors and then a synchronous algorithm that breaks symmetries using the identifiers of nodes and an algorithm to color the nodes of a tree in this section.

11.2.4.1 A Synchronous Reduction Algorithm

We will use the SSI model of distributed computing with a single initiator that starts synchronous rounds. In this algorithm, we assume each node of the network has a unique identifier i and is colored with that identifier initially. Our aim is to reduce the coloring of the nodes in synchronous rounds and use the round number for this purpose. The rounds are numbered from $\Delta + 1$ to n since we know a legal coloring requires a chromatic number less than $\Delta + 1$ by Brook's theorem and hence, we want to reduce any initial coloring number larger than this value. Any node i that finds it has a color equal to the round number changes its color to the smallest color not used by its neighbors and informs neighbors of this choice as shown in Algorithm 11.6. Note that unique identifiers are used to select the running node and hence, there will be only one node executing the code at any time. This algorithm is simple requiring no symmetry breaking technique but its operation is inherently sequential.

Figure 11.5 displays the execution of this simple algorithm in a small network. Note that the operation is sequential and the number of colors used is exactly $\Delta + 1$ for $n > \Delta + 1$ in this algorithm.

Algorithm 11.6 $Dist\_Vcol1$

1: **Input**: neighbor list $N(i)$
2: **Output**: color my_color of node i in the network
3: **for** $r = \Delta + 1$ to n **do**
4: **if** $my\_color = r$ **then**
5:    $my\_color \leftarrow$ first free color not used by neighbors
6:    inform neighbors of my color by the *status* message
7: **else if** *status* message received from a neighbor **then**
8:    update available color list
9: **end if**
10: **end for**

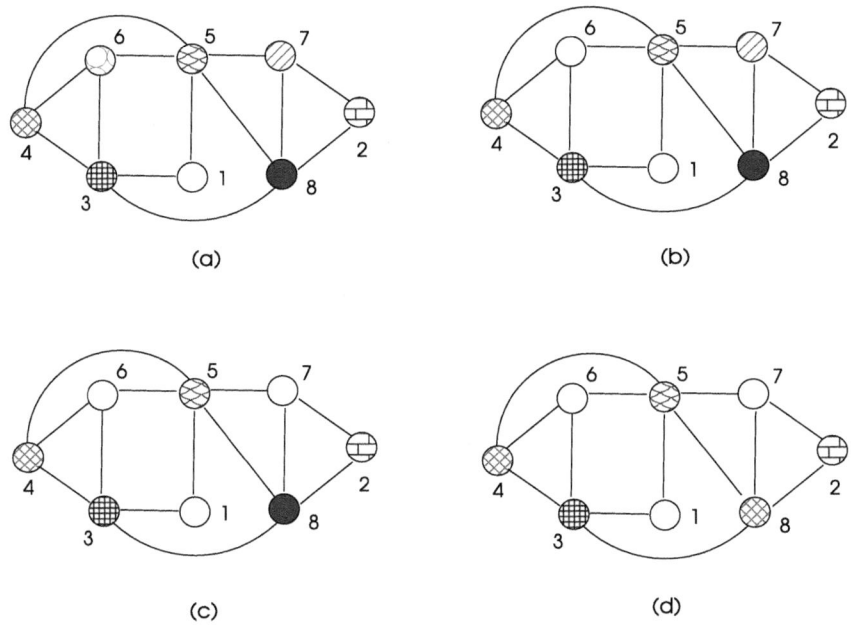

Fig. 11.5 Distributed coloring of a sample graph using Algorithm 11.6. Rounds 6, 7, and 8 provide a legal coloring with 5 ($\Delta + 1$) colors

Theorem 11.4 *Algorithm 11.6 provides legal coloring of a network with $\Delta + 1$ colors in $n - \Delta + 1$ time using $O(\Delta n)$ messages.*

Proof Since the initial coloring is legal and the color changing nodes perform legal coloring, that is; selecting a color not used by neighbors, the final coloring is legal and uses up to $\Delta + 1$ colors for $n > \Delta + 1$ in $n - \Delta + 1$ rounds. The only data messages sent in a round is by the node i that has an identifier equalling the round number and therefore, this would be $O(\Delta)$ messages. The total number of messages exchanged will then be $O(\Delta n)$. $\qquad\Box$

11.2.4.2 A Synchronous Rank-Based Algorithm

We use the SSI model again for its simplicity in this algorithm. The unique identifier of nodes are used to assign priorities this time to break symmetries. The general idea of this algorithm is to give priority to nodes that have the highest (or lowest) identifier. In each round, an uncolored node that finds it has the highest identifier among its uncolored neighbors assigns a legal color to itself and informs its decision to all of its neighbors. One way of implementing this algorithm for a node i at each round is shown in detail with needed data structures in Algorithm 11.7.

Algorithm 11.7 $Dist\_Vcol2$

1: **Input**: neighbor list $N(i)$
2: **Output**: color my_color of node i in the network
3: **message types**: $decided(x, col)$, $undecided(x)$ ▷ sent by node x coloring itself by col and a
    message by undecided node
4: **states**: {COLORED, UNCOLORED}   ▷ states of a node as colored or uncolored
5: $neigh\_colors \leftarrow \varnothing$
6: $my\_state \leftarrow$ UNCOLORED
7: **send** my_id to $N(i)$
8: **receive** ids of neighbors
9: **while** my_state = UNCOLORED **do**
10:     **if** $my\_id > id$s of all active neighbors **then**
11:         $color \leftarrow$ first free color $c \notin neigh\_colors$
12:         $state \leftarrow colored$
13:         **send** $decided(i, c)$ to $N(i)$
14:     **else if** $decided(j, col)$ received from neighbor j **then**
15:         $N(i) \leftarrow N(i) \setminus \{j\}$ ▷ remove neighbor j from active neighbor list
16:         $neigh\_colors \leftarrow neigh\_colors \cup \{col\}$
17:     **end if**
18: **end while**

The operation of this algorithm is depicted in Fig. 11.6 for the same sample graph of Fig. 11.5. This graph is colored with three legal colors in four rounds. We can apply a different criteria such as degrees of vertices or a random number picked between 0 and 1 to break symmetries instead of vertex identifiers resulting basically in a very similarly structured algorithm.

Theorem 11.5 *Algorithm 11.7 correctly colors the nodes of a network in $O(n)$ time with $O(\Delta + 1)$ colors.*

Proof Correction is evident since coloring rule is applied at each step. As in other rank-based greedy distributed algorithms we have reviewed, number of rounds required may be as high as the number of vertices n as in the case of a network with sorted identifier neighbors. The maximum number of colors used will be $\Delta + 1$ as there will always be a free color in that range.   □

11.2.4.3 Tree Coloring

A network that is configured as a rooted tree can be colored using two colors only. The root starts the algorithm by coloring itself with color 0 and sending its color to its children which color themselves with color 1. Continuing in this manner, two colors suffice to color the whole tree. The steps of this algorithm for node i are as follows.

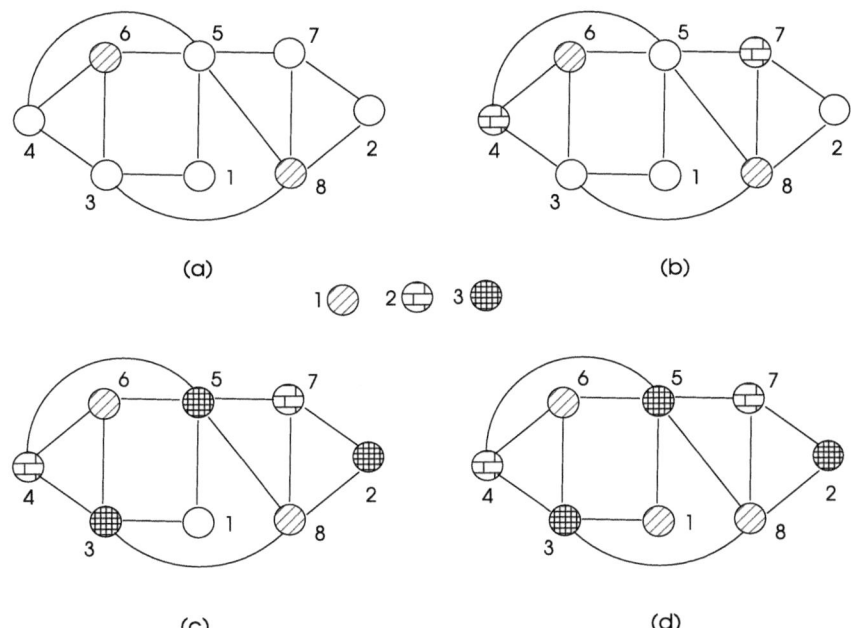

(a)   (b)

1 ◍   2 ⊞   3 ⊕

(c)   (d)

Fig. 11.6 Distributed coloring of a sample graph Algorithm 11.7. In four rounds, three colors assigned to eight nodes

1. **if** $i = root$ **then**
2. $c_i \leftarrow 0$
3.   **send** *color(1)* to children
4. **else receive** *color(c)*
5. $c_i \leftarrow 1 - c$
6.   **if** $i \neq leaf$ **then**
7.   **send** *color(c_i)* to children

Each node should be aware whether it is *root*, an *intermediate* or a *leaf* node in this algorithm. Figure 11.7 depicts a tree constructed using the above procedure. We can run this algorithm in synchronous rounds in the SSI model or asynchronously. The running time of the algorithms is the depth d of the tree $O(d)$ in both cases. The total number of messages exchanged in both cases is the number of edges of the tree which is n-1. The algorithm proposed by Cole and Vishkin can be used to color trees in a distributed way.

Fig. 11.7 Distributed coloring of a sample tree with two colors

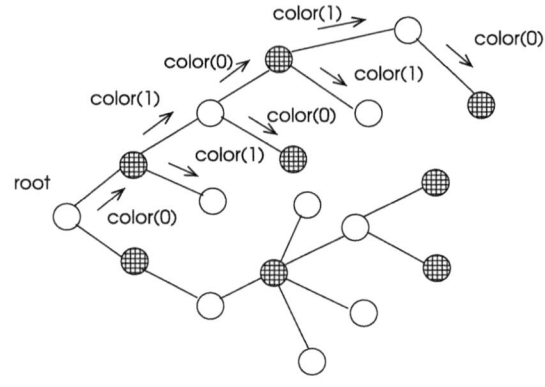

11.3   Edge Coloring

In the edge coloring of a graph, edges are assigned colors such that adjacent edges have different colors.

Definition 11.4 (*Edge coloring*) An *edge coloring* of a graph $G = (V, E)$ is an assignment $\phi' : E \rightarrow C$ such that any two edges e_i and e_j that are incident on the same vertex have $\phi'(e_i) \neq \phi'(e_j)$, where C is a set of colors elements of which are commonly the elements of $\mathbb{N}^+$.

In *proper edge coloring*, adjacent edges are assigned distinct colors. *Edge coloring problem* is finding the minimum number of colors to color edges of a graph. In the decision version of this problem, we try to find an answer to whether the edges of a graph can be colored with at most k different colors. A graph is said to be *k-edge colorable* if there is a coloring $\phi : E \rightarrow C$ such that $|C| = k$. The *edge chromatic number* or the *chromatic index* $\chi'(G)$ of a graph G is the minimum value of k such that G is k-edge colorable. In other words, G is *k-edge chromatic* if $\chi'(G) = k$.

Theorem 11.6 *For any simple graph G,*

$$\chi'(G) \geq \Delta(G). \tag{11.4}$$

Proof Let v be the vertex with the maximum degree in G. Every edge incident to v must be colored with a different color, therefore $\chi(G)$ must be at least equal to $\Delta(G)$.

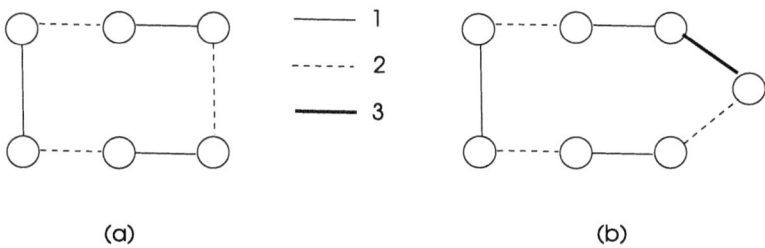

Fig. 11.8 Edge coloring of even-and odd-cycle graphs

Strong lower and upper bounds exist for edge coloring of graphs. Vizing has shown that for every nonempty simple graph G with no multiple edges and no loops [19],

$$\chi'(G) \le 1 + \Delta(G) \tag{11.5}$$

Based on Eqs. 11.4 and 11.5,

$$\Delta(G) \le \chi'(G) \le 1 + \Delta(G), \tag{11.6}$$

which means for every nonempty simple graph G, either $\chi'(G) = \Delta(G)$ or $\chi'(G) = 1 + \Delta(G)$. The simple graphs G where $\chi'(G) = \Delta(G)$ are called *Class 1* graphs and graphs which have $\chi'(G) = \Delta(G) + 1$ are called *Class 2* graphs.

Remark 11.6 A star graph S_n has $n - 1$ edges. Since all of these edges are adjacent to each other, $\chi(S_n) = n - 1$. Therefore, S_n is a Class 1 graph.

Remark 11.7 An even-cycle graph can be edge-colored with two colors. An odd-cycle graph on the other hand, needs three colors as there will be at least one edge that needs a third color as shown in Fig. 11.8. Hence, even-cycle graphs are *Class 1* graphs and odd-cycle graphs are *Class 2* graphs.

11.3.1 Relation to Graph Matching

In a proper edge-coloring of a graph G, the edges of the same color constitute a matching in G since these edges are not adjacent to each other. In other words, given any two edges (u, v) and (w, y) of a matching, u or v cannot be neighbors to either w or y. We can, therefore, define edge coloring EC of a graph G as the union of a set of disjoint matchings $M_1 \cup M_2 \cup \ldots, M_k$ of G. For a graph G with m edges and a maximum matching size $\alpha'(G)$, every edge coloring of G must use at least $m/\alpha'(G)$ colors.

Theorem 11.7 *For a graph* $G = (V, E)$ *with size m greater than 1,*

$$\chi'(G) \geq \frac{m}{\alpha'(G)}. \tag{11.7}$$

Proof Let us assume $\chi'(G) = k$ and $E_1, E_2, \ldots, E_k$ are the color classes of G. $\forall E_i \in E, 1 \leq i \leq k; |E_i| \leq \alpha'(G)$ since edges in each color constitute a matching of G and $\alpha'(G)$ is the size of the maximum matching of G. Therefore,

$$m = |E(G)| = \sum_{i=1}^{k} |E_i| \leq k\alpha'(G) \tag{11.8}$$

hence, $\chi'(G) \geq \frac{m}{\alpha'(G)}$. □

A possible method to find EC is then find a maximal matching M_i of graph G and color it with a new color, remove all edges of M_i from G and repeat this process until all edges are colored as shown in Algorithm 11.8. Note that we find an unused color simply by incrementing the index i.

Algorithm 11.8 *MM_Ecolor*

1: **Input** : $G = (V, E)$
2: **Output** : $\phi' : E \rightarrow C$ where $C = \{1, 2, ..., k\}$
3: $G'(V', E')' \leftarrow G(V, E)$
4: $i \leftarrow 1$
5: **while** $E' \neq \varnothing$ **do**
6:     $M_i \leftarrow Find\_MM(G')$
7:     **color** the edges in M_i with i
8:     $E' \leftarrow E' \setminus M_i$
9:     $i \leftarrow i + 1$
10: **end while**

The performance of this algorithm clearly depends on how finding the maximal matching is performed. The operation of this algorithm is depicted in Fig. 11.9, where the MM edges of a sample graph found at each step are colored with a new color.

11.3.2   A Greedy Sequential Algorithm

The greedy algorithm for edge coloring of a graph can be sketched similar to the greedy algorithms we have seen. An uncolored edge e is picked randomly and colored with the minimum legal color that does not conflict with the assigned colors of the adjacent edges of the edge e. This edge is then removed from the graph and the process is repeated until there are no more uncolored edges left. Algorithm 11.9 displays the pseudocode for this algorithm.

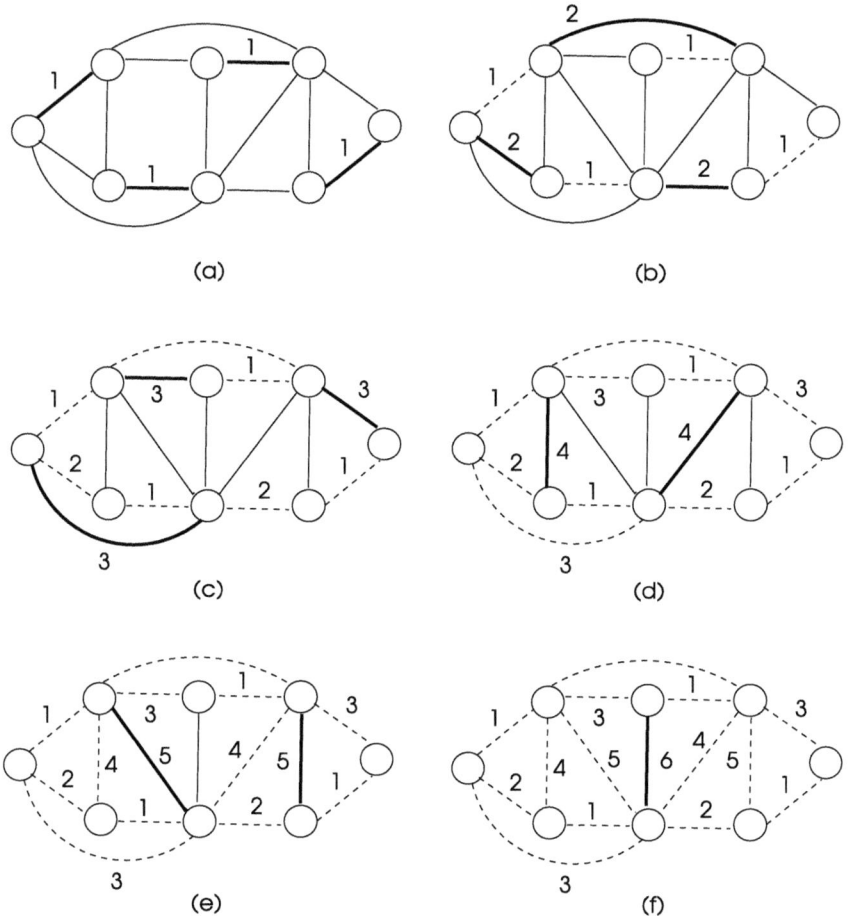

Fig. 11.9 Operation of Algorithm 11.9 in a small sample graph. Selected edges at each iteration are shown in bold with an assigned color number next to the edges. We have colored the edges of this graph with $\Delta + 1 = 6$ colors

---

Algorithm 11.9 *Greedy_Ecolor*

1: **Input** : $G = (V, E)$
2: **Output** : $\phi' : E \rightarrow C$ where $C = \{1, 2, ..., 2\Delta - 1\}$
3: $E' \leftarrow E$
4: **while** $E' \neq \varnothing$ **do**
5:    **select** an edge $(u, v) \in E'$
6:    $\phi'((u, v)) \leftarrow$ the smallest legal color from C
7:    $E' \leftarrow E' \setminus \{(u, v)\}$
8: **end while**

The colors required for this algorithm can be as high as $2\Delta - 1$ as shown in Fig. 11.10, where two vertices with Δ degrees are connected by an edge that we want to color. The only available color in this case, is $2\Delta - 1$.

Operation of this algorithm in a simple sample graph is depicted in Fig. 11.11. The *while* loop is executed m times to color all edges. There is also Δ time needed to find an available color for each vertex resulting in $O(\Delta m)$ time complexity.

The edge-coloring of a graph G is equivalent to the vertex coloring of the line graph $L(G)$ of G. Given Δ for the maximum degree of G, the maximum degree of $L(G)$ is $2\Delta - 2$ which means we can color the vertices of $L(G)$ using $2\Delta - 1$

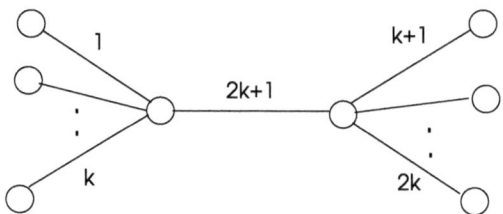

Fig. 11.10 Two vertices with Δ degrees are colored with all available colors and the only remaining color for the edge between them is $2\Delta - 1$

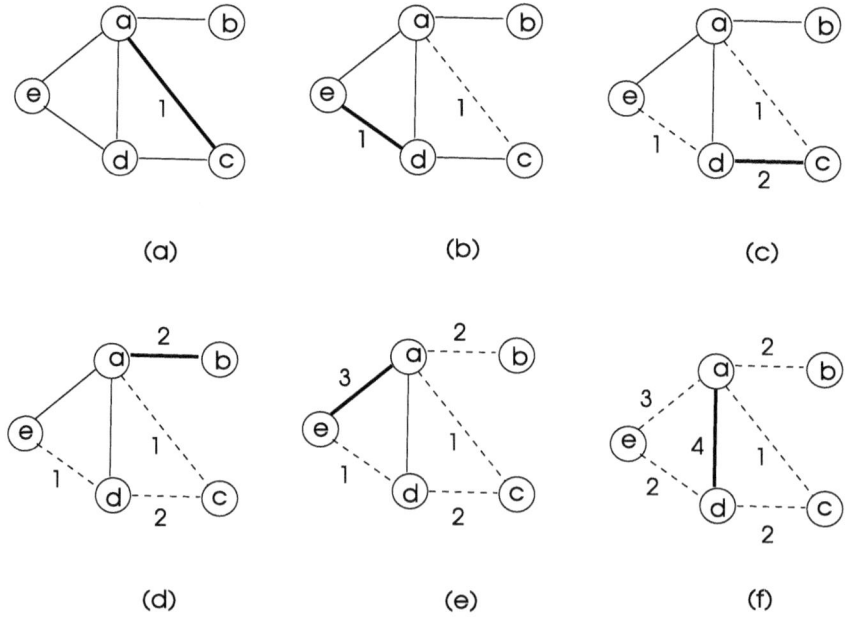

Fig. 11.11 Operation of Algorithm 11.9 in a small sample graph. Selected edges at each iteration are shown in bold with assigned color number next to the edges. We could color this graph with Δ colors

colors using Brook's Theorem. Hence, G can be edge-colored with $2\Delta - 1$ colors since each edge of G corresponds to a vertex in $L(G)$.

11.3.3 Bipartite Graph Coloring

The edges of bipartite graphs can be colored with Δ colors as proved by König [14] which means,

$$\chi'(G) = \Delta(G). \tag{11.9}$$

Every bipartite graph G is contained in a $\Delta(G)$-regular graph. We can add vertices and edges to a bipartite graph $G = (V_1, V_2, E)$ to make it a Δ-regular bipartite graph G' with $V_1 = V_2$. Such formed graph has always a perfect matching. Then, we can find a perfect matching M_1 (1-factor) such that every vertex of G' is incident to an edge of M_1 and color this perfect matching with the first available color. Removing M_1 from G and finding another perfect matching M_2 of the new graph and continuing in this manner, we can have Δ colors for the edges of G. This method motivates an algorithm to color the edges of a bipartite graph as shown in Algorithm 11.10.

Algorithm 11.10 *Bipartite_Ecolor*

1: **Input** : $G = (V_1, V_2, E)$
2: **Output** : $\phi' : E \rightarrow C$ where $C = \{1, 2, ..., \Delta\}$
3: $E' \leftarrow E$
4: $G' \leftarrow G \cup \{\text{added edges}\} \cup \{\text{added vertices}\}$
5: **for** $i = 1$ to $\Delta(G)$ **do**
6:     **find** a perfect matching M of G'
7:     $\forall (u, v) \in M : \phi'((u, v)) \leftarrow i$
8:     $G' \leftarrow G' - \{M\}$
9: **end for**
10: **remove** added edges from G' to get edge colored G

Finding a perfect matching in a k-regular bipartite graph takes $O(km)$ time (see Chap. 9), hence Algorithm 11.10 provides edge coloring of a bipartite graph with a maximum degree Δ in $O(\Delta m)$ time. The execution of this simple algorithm in a small bipartite graph is shown in Fig. 11.12.

11.3.4 Edge Coloring of Complete Graphs

A complete graph K_n can be colored with $n - 1$ colors if n is even, otherwise, n colors are needed when n is odd. We can find edge coloring of a complete graph K_n by iteratively selecting disjoint maximal matchings of it as depicted in Fig. 11.13. Edge coloring of even and odd complete graphs are shown in Fig. 11.14.

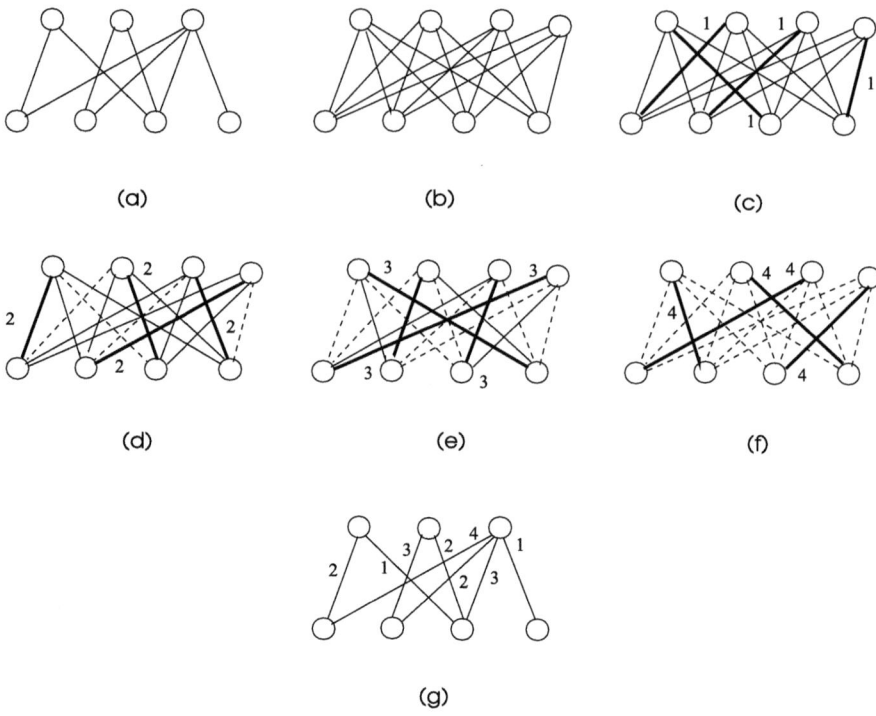

Fig. 11.12 Coloring edges of a bipartite graph. The graph in **a** is made full-bipartite by adding edges to obtain the graph in **b**. After three iterations of disjoint perfect matchings, the original graph is edge colored with $\Delta = 4$ colors as shown in **g**

11.3.5   A Parallel Algorithm

Edges of the same color in a properly colored graph G establish a maximal matching of G as we have seen. We have reviewed in Chap. 9 how to perform parallel matching of a graph and hence, we can use these algorithms to perform parallel edge coloring. Algorithm 11.11 displays the pseudocode of such an algorithm which performs finding MM and removing these edges from the graph in addition to assigning colors to the edges of the maximal matching.

Algorithm 11.11 *Par_Ecolor*

1: **Input** : $G = (V, E)$
2: **Output** : $\phi' : E \to C$ where $C = \{1, 2, ..., k\}$
3: $G'(V', E')' \leftarrow G(V, E)$
4: **while** $E' \neq \varnothing$ **in parallel do**
5:     $M_i \leftarrow Find\_MM(G')$
6:     **color** the edges in M_i with a new color
7:     $E' \leftarrow E' \setminus M_i$
8: **end while**

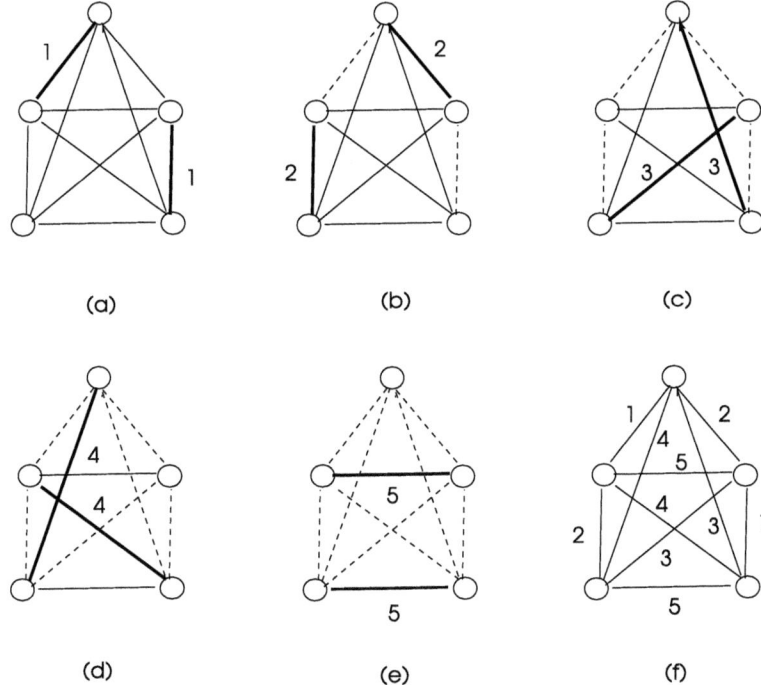

Fig. 11.13 Edge coloring of K_5. Disjoint maximal matchings are selected at each iteration with each matching assigned a new color. The final colored graph shown in **e** has 5 distinct colors

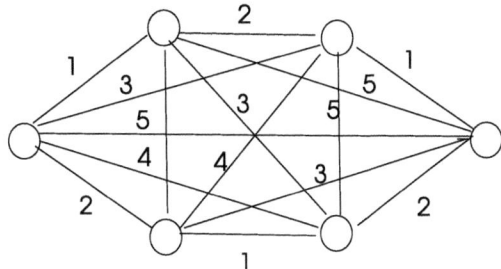

Fig. 11.14 Coloring of K_6. 5 colors suffice to color this graph

11.3.6 A Distributed Algorithm

In a network setting, each node should act independently to color the edges adjacent to it. We need a way to break symmetries in a distributed environment as we commonly implement. Let us define a heuristic to be the node with maximum degree decides what color to be assigned to its adjacent edge. Different than distributed vertex coloring problem, we need to consider the case when a node with a lower degree receives requests from two maximum degree neighbors. In such a case, we will assume it selects the one with the larger degree. The messages to be used in the algorithm we propose are as follows.

- *degree(i,deg(i))*: Sent by node i to inform neighbors of its degree *deg(i)*.
- *propose(i,col)*: Sent by node i to request coloring of the edge (i, j) adjacent to its neighbor j.
- *ack(i)*, *nack(i)*: Sent by node i to accept message and to reject the incoming *request* message
- *decide(i,col)*: Sent by node i to node j confirm the edge coloring of edge (i, j) with color *col*.

We assume each node is aware of its neighbors and hence, its degree initially. The variable list *curr_neighs* holds the identifiers of currently active neighbors of a node i such that $\forall j \in curr\_neighs$, the edge (i, j) is not yet colored. The algorithm starts by each node exchanging its degree with neighbors. Thereafter, as long as there are uncolored edges adjacent to a node i, it checks its degree with the current neighbors. If it finds it has the maximum degree among them, then node i broadcasts a *propose* message to all of its active neighbors. A neighbor that receives more than one request message responds to the sender with the largest degree by the *ack* message as shown in a coarse sketch of this algorithm in Algorithm 11.12.

Algorithm 11.12 *Dist_Ecol*

1: **Input**: neighbor list $N(i)$
2: **Output**: set of edge colors *edge_cols* of node i in the network
3: **message types**: $degree(x, deg(x)), propose(x, col), ack(x), nack(x), decide(x, col(x))$
4: $curr\_neighs \leftarrow N(i)$ ▷ active neighbors initialized
5: $neigh\_colors \leftarrow \varnothing$
6: **send** $degree(deg(i))$ to $N(i)$ ▷ exchange degree values with neighbors
7: **receive** $degree(deg(j))$ from all $j \in N(i)$
8: **while** $curr\_neighs \neq \varnothing$ **do**
9:     **if** $deg(i) > deg(j) \,\forall j \in curr\_neighs$ **then**
10:         **for all** $j \in curr\_neighs$ **do**
11:             **send** minimum available color c in $neigh\_cols$ to j in $propose(i,c)$
12:         **end for**
13:         **receive** $ack/nack$ messages from all nodes in $curr\_neighs$
14:         **update** $neigh\_cols$
15:         **update** $curr\_deg$
16:         **select** one of the senders (j) of ack messages arbitrarily
17:         **select** first available color $c \notin edge\_cols$
18:         **send** $decide(i, c)$ to j
19:     **else if** $propose$ received from neighbor j or more than one neighbor **then**
20:         **select** the highest degree requesting neighbor j
21:         **send** $ack(i)$ to j
22:         **send** $nack(i)$ to all others
23:         **receive** $decide(j, c)$
24:     **end if**
25:     $neigh\_colors \leftarrow neigh\_cols \cup \{c\}$
26:     $curr\_neighs \leftarrow curr\_neighs \setminus \{j\}$
27:     **broadcast** $info(curr\_deg, neigh\_cols)$ to $curr\_neighs$
28: **end while**

The operation of this algorithm in a small sample network is depicted in Fig. 11.15, where nodes with highest degrees in their neighborhoods decide to color edges incident to them by proposing to their neighbors. The maximum colors used will be $2\Delta(G) - 1$ since there will always be a color in this range proposed by the highest degree nodes. Grable et al. proposed a synchronous randomized distributed edge coloring algorithm in which each edge (u, v) picks a random color c from its palette of $(1 + \varepsilon)\max\{deg(u), deg(v)\}$ colors in each round [10]. If color c is not in conflict with any of the selected colors of neighbors, c is determined to be the color of the edge (u, v). It is shown this algorithm colors the edges of a graph with $(1 + \varepsilon)\Delta(G)$ colors for any $\varepsilon > 0$ in $O(\log \log n)$ rounds for graphs with sufficiently large degrees.

11.4   Chapter Notes

Vertex coloring is the process of coloring the vertices of a graph such that no two adjacent vertices have the same color. It can form the basis of more complicated graph algorithms and also has various practical implementations such as assigning

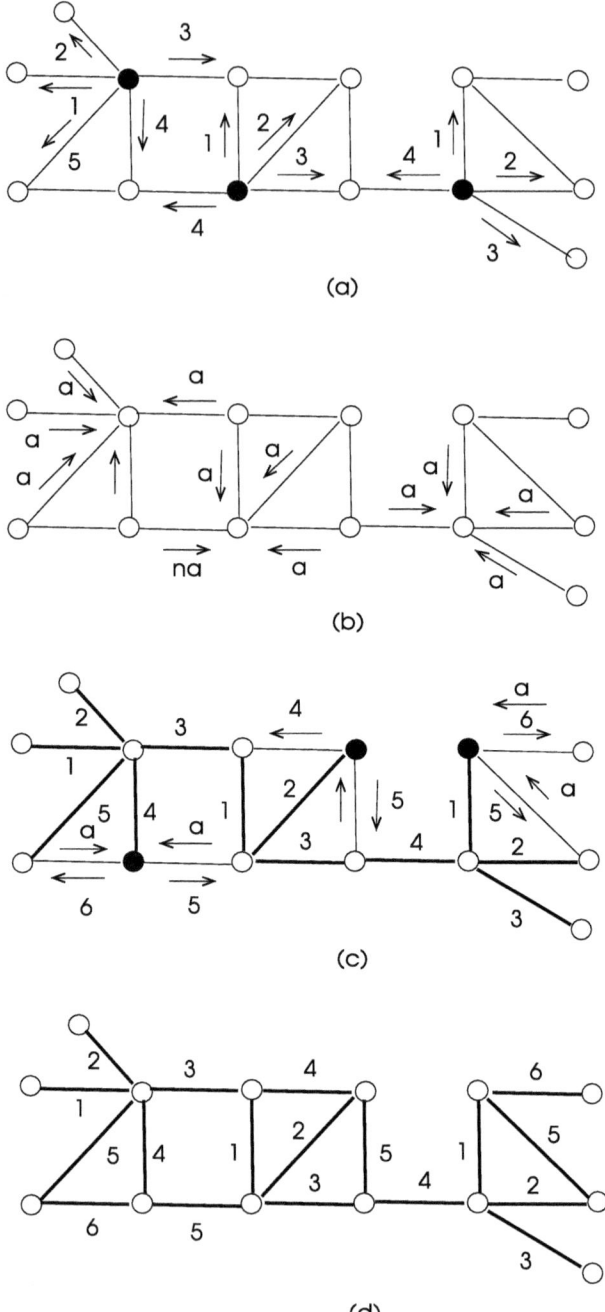

Fig. 11.15 Edge coloring of a small network *by Dist_Ecol*. The black nodes have the maximum current degree among their neighbors. They propose edge colors for their uncolored edges from their neighbors which respond by *ack* (a) or *nack* (na) messages. Colored edges are shown in bold. Broadcasting of *info* messages are not shown. The final coloring requires 6 colors

Fig. 11.16 Sample graph for
Exercise 1

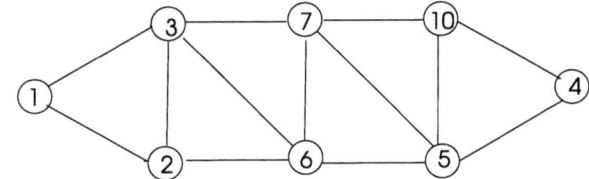

channel frequencies in wireless networks. We reviewed theoretical aspects of vertex coloring and then described sequential, parallel, and distributed algorithms for
this purpose. Coloring vertices of a graph with its chromatic number is NP-hard
but otherwise coloring with $\Delta + 1$ coloring is straightforward.

A simple parallel algorithm makes use of the independent set property where
no two vertices in such a set are adjacent. We can, therefore, assign the same color
to the vertices of an independent set. We can find an independent set in parallel
as was shown in Chap. 10, hence, this algorithm can be used to assign colors in
parallel. There are various distributed vertex coloring algorithms. We described
a basic rank-based algorithm which is slow as its execution time in rounds is
dependent on the number of nodes in the network and another simple algorithm to
color trees.

Edge coloring refers to assigning colors to the edges of a graph such that
no two edges with common endpoints receive the same color. We described a
simple sequential algorithm and parallel and distributed algorithms for edge coloring. Edge coloring can make use of matching algorithms since edges of the
same class in a graph, that is edges with the same color, constitute a matching in
that graph. This way, we can perform edge coloring in parallel using a parallel
maximal matching algorithm as we reviewed. Distributed algorithms for edge coloring require careful consideration as we need to check colors of edges incident
to neighbors of a node v when coloring an edge incident to v. Total graph coloring
requires both the vertices and edges of a graph to be colored.

Exercises

1. Color the vertices of the graph in Fig. 11.16 using the greedy algorithm that
   selects the first uncolored lowest index.
2. Partition the sample graph of Fig. 11.17 into all possible disjoint maximal independent sets and color each set with a new color to find the vertex coloring of
   this graph.
3. Find the vertex coloring of the graph shown in Fig. 11.18 using the greedy
   distributed method of Algorithm 11.6. Assign identifiers to nodes arbitrarily.
4. Color the edges of the bipartite graph shown in Fig. 11.19 by first forming a
   Δ-regular bipartite graph and then finding disjoint maximal matchings of this
   graph and coloring edges of each matching with a new color.
5. Use the disjoint maximal matching method to color the edges of the graph
   depicted in Fig. 11.20.

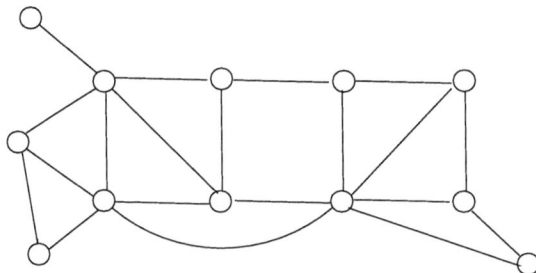

Fig. 11.17 Sample graph for Exercise 2

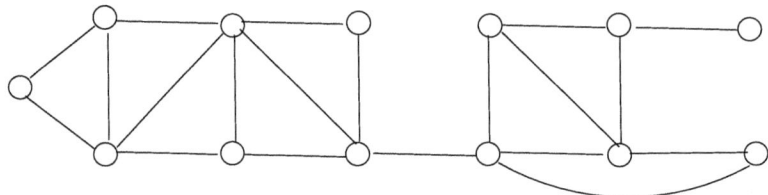

Fig. 11.18 Sample graph for Exercise 3

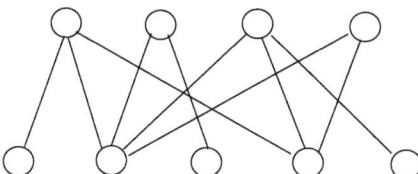

Fig. 11.19 Sample graph for Exercise 4

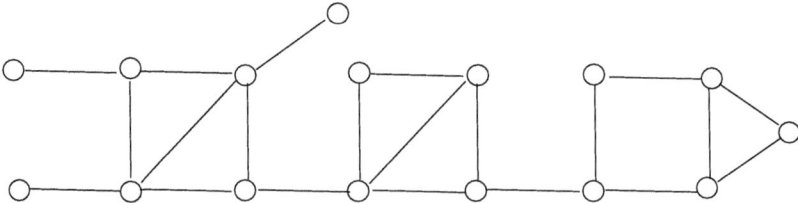

Fig. 11.20 Sample graph for Exercise 5

6. Work out the edge coloring of the graph shown in Fig. 11.21 using the greedy distributed method of Algorithm 11.12.

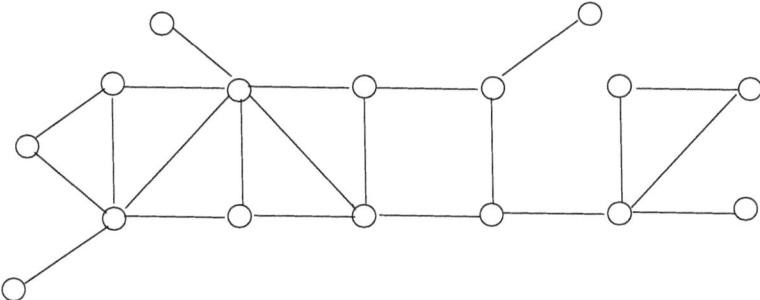

Fig. 11.21 Sample graph for Exercise 6

References

1. Allwright JR, Bordawekar R, Coddington PD, Dincer K, Martin CL (1995) A comparison of parallel graph coloring algorithms. Technical report SCCS-666, Northeast Parallel Architecture Center, Syracuse University
2. Barenboim L, Elkin M (2013) Distributed graph coloring. Monograph, Ben Gurion University of the Negev
3. Brelaz D (1979) New methods to color the vertices of a graph. Commun ACM 22(4):251–256
4. Brooks RL (1941) On colouring the nodes of a network. Proc Camb Philos Soc Math Phys Sci 37:194–197
5. Cole R, Vishkin U (1986) Deterministic coin tossing with applications to optimal parallel list ranking. Inf Control 70(1):32–53
6. Coleman TF, More JJ (1983) Estimation of sparse Jacobian matrices and graph coloring problems. SIAM J Numer Anal 20(1):187–209
7. Gandham S, Dawande M, Prakash R (2005) Link scheduling in sensor networks: distributed edge coloring revisited. In: Proceedings of the 24th INFOCOM, vol 4, pp 2492–2501
8. Garey MR, Johnson DS (1979) Computers and intractability. W.H. Freeman, New York
9. Gebremedhin AH (1999) Parallel graph coloring. MS thesis, Department of Informatics University of Bergen Norway
10. Grable D, Panconesi A (1997) Nearly optimal distributed edge-coloring in $O(\log \log n)$ rounds. Random Struct Algorithms 10(3):385–405
11. Grimmet GR, McDiarmid CJH (1975) On coloring random graphs. Math Proc Camb Philos Soc 77:313–324
12. Halldorsson MM (1991) Frugal methods for the independent set and graph coloring problems. Ph.D. thesis, The State University of New Jersey, New Brunswick, New Jersey, October 1991
13. Jones MT, Plassmann PE (1993) A parallel graph coloring heuristic. SIAM J Sci Comput 14(3):654–669
14. König D (1931) Graphen und Matrizen. Math Lapok 38:116–119
15. Lovasz L (1975) Three short proofs in graph theory. J Comb Theory Ser B 19:269–271
16. Luby M (1986) A simple parallel algorithm for the maximal independent set problem. SIAM J Comput 15(4):1036–1055
17. Matula DW, Marble G, Isaacson JD (1972) Graph coloring algorithms. Academic Press, New York
18. Nishizeki T, Terada O, Leven D (1983) Algorithms for edge-coloring of graphs. Tohoku University, Electrical Communications Department, Technical report, TRECIS 83001

19. Vizing VG (1964) On an estimate of the chromatic class of a p-graph. Diskret Anal 3:25–30 (in Russian)
20. Welsh DJA, Powell MB (1967) An upper bound for the chromatic number of a graph and its application to timetabling problems. Comput J 10:85–86

Part III

Advanced Topics

Algebraic and Dynamic Graph Algorithms

12

12.1 Introduction

Algebraic graph theory is the study of algebraic methods to solve graph problems. Linear algebra and group theory are the two of the mostly referred areas of algebra while dealing with graphs. Algebraic graph algorithms using linear algebra commonly make use of the matrices associated with a graph to solve various problems in graphs. By using this approach to form graph algorithms for many of the problems, we have seen has a number of benefits. First of all, we can use various existing matrix operations for this task which results in simpler algorithms which can be converted to executable codes with ease in general. As another advantage, parallel matrix operations and such software environments for them are readily available making parallel formation of these tasks simpler.

Our purpose in the first part of this chapter is to introduce this paradigm, and give examples of solving some of the graph problems we have investigated in Part II of this book. We start with a short review of matrices that are used to represent graphs. We then review algebraic graph algorithms for some graph problems, which include graph traversals, shortest paths from a single source, all-pairs shortest paths, connectivity, and matching. We also discuss parallel implementations of various algorithms presented. A thorough treatment of this topic, namely, algebraic graph algorithms can be found in [13].

In the second part of this chapter, we look at dynamic graph algorithms. Many real-life networks are represented by dynamic graphs in which new vertices/edges may be inserted and some vertices/edges may be deleted as time goes by. For example, new interactions are possible in protein interaction networks and two friends of a person in a social network may be acquainted to form an edge between them in such a network. Rather than running the algorithms, we have seen in Part II from scratch in such dynamic graphs it is beneficial to use new algorithms

© The Author(s), under exclusive license to Springer Nature Switzerland AG 2026
K. Erciyes, *Guide to Graph Algorithms*, Texts in Computer Science,
https://doi.org/10.1007/978-3-032-05294-0_12

with better performances as we will investigate. Lastly, we review dynamic algebraic graph algorithms which are used for dynamic graphs using linear algebraic techniques.

12.2   Graph Matrices

An $m \times n$ matrix contains mn real numbers organized as m rows and n columns. We show the element at Ith row and jth column of a matrix A as a_{ij} or sometimes $A[I, j]$ in algorithms. When the number of rows and the number of columns of a matrix A are equal, A is called a *square matrix*. Basic operations such as addition, multiplication by a scalar, and matrix multiplication are described in Appendix B.3. An $n \times 1$ matrix is called a *column vector* and $1 \times n$ matrix is called a *row vector*. The three main matrices associated with a graph are its incidence matrix, adjacency matrix, and its Laplacian matrix, as we saw in Chap. 2, which we will now briefly review.

Incidence Matrix
Let us assume an undirected or directed graph G with n vertices $v_1, \ldots, v_n$ and $E = \{e_1, \ldots, e_m\}$ edges. The incidence matrix $Q(G)$ of a graph G is defined as $q_{ij} = 0$ if e_j is not incident to vertex v_i; $q_{ij} = 1$ if e_j starts from v_i and $q_{ij} = -1$ if e_j ends at vertex v_i. The following properties of $Q(G)$ can be stated [1].

* The column sums of $Q(G)$ is zero, therefore the rows of $Q(G)$ are linearly independent.
* Rank of $Q(G)$ is -1 for a connected graph G.
* $Q(G) = n - k$ for a graph with k components.

Adjacency Matrix
We have used the adjacency matrix $A(G)$ of a graph G in various algorithms. Let us briefly review basic algebraic properties of $A(G)$ as regards to graphs. The entry a_{ij} of $A(G)$ is equal to 0, if vertices v_i and v_j are not adjacent and is 1 if they are neighbors. An entry a_{ii} is 0 in $A(G)$ and hence $A(G)$ has zeros in its diagonal. We observe the following properties of $A(G)$.

* $A[i, j]^k$ is the number of paths of length k from vertex I to vertex j.
* Let us form A^{-1} using Boolean multiplication and addition. Entry (I, j) in this matrix is 1 if vertex I is connected to vertex j. We have used this property to form the connectivity matrix of the graph G.
* If distance $d(I, j)$ between the two vertices I and j is k, then the matrices $I, A, \ldots, A^k$ are linearly independent [1].

Eigenvalues

Let us consider the equation $Ax = \lambda x$ where A is a non-singular square matrix, x is a vector, and λ is a constant. When a vector is multiplied by a matrix, its direction changes in general. Some vectors such as x are different since they do not change direction. These vectors are called *eigenvectors* of the matrix A and the number λ is called an *eigenvalue* of A. When A is the identity matrix, all vectors are the eigenvectors of A with all eigenvalues being 1. Let us rewrite the equation $Ax = \lambda x$.

$$Ax - \lambda x = 0 \qquad (12.1)$$

$$(A - \lambda I)x = 0,$$

therefore,

$$det(A - \lambda I)x = 0. \qquad (12.2)$$

The equation $det(A - \lambda I) = 0$ is called the *characteristic polynomial* of A. This provides us the method to find the eigenvalues and eigenvectors of a graph by implementing the following steps.

1. Compute the determinant of $A - \lambda I$. This gives a polynomial of degree n in λ.
2. Solve $\det(A - \lambda I) = 0$ to find n roots $\lambda_1, \ldots, \lambda_n$ which are the n eigenvalues of A.
3. Solve $(A - \lambda_i I)x = 0$ for $I = 1, \ldots, n$ to find the corresponding eigenvectors for each eigenvalue.

Laplacian Matrix

The Laplacian matrix $L(G)$ of an undirected unweighted graph G without multiple edges is defined as

$$L(G) = D(G) - A(G), \qquad (12.3)$$

where D is the degree matrix with an entry d_{ii} as the degree of vertex I with all other elements equal to 0 and $A(G)$ is the adjacency matrix of G. The other entries in L are -1 if vertex I is adjacent to vertex j and 0 otherwise. The Laplacian $L(G)$ of a graph G is related to its incidence matrix $Q(G)$ as follows:

$$L(G) = Q(G)Q^T(G), \qquad (12.4)$$

where $Q^T(G)$ is the transpose of the incidence matrix. The normalized Laplacian $\mathcal{L}$ of G is defined as

$$\mathcal{L} = D^{-1/2}LD^{-1/2} = D^{-1/2}(D - A)D^{-1/2} = I - D^{-1/2}AD^{-1/2}. \qquad (12.5)$$

The normalized Laplacian $\mathcal{L}$ then has the following entries:

$$l_{ij} = \begin{cases} d_i & \text{if } I = j \\ -1 & \text{if } I \text{ and } j \text{ are neighbors} \\ 0 & \text{otherwise .} \end{cases} \qquad (12.6)$$

The set of all eigenvalues of the Laplacian matrix of a graph G is called the *Laplacian spectrum* or just the *spectrum* of G. We will see shortly the eigenvalues of the Laplacian matrix will provide vital information about the connectivity of a graph.

12.3 Algebraic Graph Algorithms

Algebraic graph algorithms employ various operations using the three main matrices associated with a graph, its adjacency matrix A which is sparse in general, its incidence matrix I, and the graph Laplacian L. We provide algebraic algorithms for sample graph problems in this section.

12.3.1 Connectivity

The second smallest eigenvalue of the Laplacian matrix of a graph G, called the *Fiedler value*, provides information on how well G is connected. This value is greater than 0 if and only if G is connected. Moreover, the larger this value is, the more connected G is and the number of 0s in the Laplacian eigenvalues of a graph G is the number of connected components of G. The Fiedler value of a graph G, shown by $\alpha(G)$, is called the *algebraic connectivity* of G and has been used in numerous applications involving spectral graph theory and combinatorial optimization problems. Let $\kappa(G)$ denote the vertex connectivity of G; Fiedler showed that [9]

$$\kappa(G) \geq \alpha(G). \qquad (12.7)$$

The Laplacian matrix $L(G)$ of a graph G can also be used to enumerate the spanning trees of the graph G according to the theorem below.

Theorem 18 (Matrix-Tree Theorem): *Let u and v be the vertices of a graph $G = (V, E)$ and let $L(G)(u, v)$ be the submatrix resulting from the deletion of row u and column v from its Laplacian matrix $L(G)$. Then $|detL(G)(u, v)|$ is equal to the number of spanning trees of G.*

Strongly Connected Components

A strongly connected digraph has paths between any of its $\{u, v\}$ vertex pairs in both directions. A strongly connected component (SCC) of a directed graph G is a subgraph G' of G, such that each vertex in G' is strongly connected to all other vertices in G'. We reviewed two main algorithms to detect SCCs of a directed graph due to Tarjan and Kosaraju in Chap. 8. The first one used back edges in a DFS tree of a graph G to detect SCCs, and the latter performed a DFS in the graph by putting vertices visited in a stack according to their last visit times, then obtained transpose of G and formed DFS subtrees of G by popping vertices from G. Each subtree rooted at popped vertices is then an SCC of G.

We can structure a simple algorithm using the connectivity matrix C of a directed graph to find its SCCs. The connectivity matrix C has entries $c_{ij} = 1$ if there is a path from a vertex I to j and $c_{ij} = 0$ otherwise. The key observation here is that if we form the transpose of C, then C^T should have a path from its I to j showing the path from j to I in C for I and j to be in the same SCC. We can then form a new matrix C' which is formed by taking logically *and* of each element of C with C^T. This matrix C' has entries $c_{ij} = 1$ if and only if there is a path from I to j and from j to I meaning vertices I and j are in the same SCC. Let us illustrate this method by the graph in Fig. 12.1.

The connectivity matrix C and its transpose C^T can be structured as follows:

$$
C = \begin{array}{c} \\ 1 \\ 2 \\ 3 \\ 4 \\ 5 \\ 6 \\ 7 \\ 8 \end{array}
\begin{array}{c} 1\,2\,3\,4\,5\,6\,7\,8 \\ \left(\begin{array}{cccccccc}
1 & 1 & 1 & 0 & 0 & 0 & 0 & 0 \\
1 & 1 & 1 & 0 & 0 & 0 & 0 & 0 \\
1 & 1 & 1 & 0 & 0 & 0 & 0 & 0 \\
1 & 1 & 1 & 1 & 1 & 1 & 1 & 0 \\
1 & 1 & 1 & 1 & 1 & 1 & 1 & 0 \\
1 & 1 & 1 & 1 & 1 & 1 & 1 & 0 \\
1 & 1 & 1 & 1 & 1 & 1 & 1 & 0 \\
1 & 1 & 1 & 1 & 1 & 1 & 1 & 0
\end{array}\right) \end{array},\;
C^T = \begin{array}{c} \\ 1 \\ 2 \\ 3 \\ 4 \\ 5 \\ 6 \\ 7 \\ 8 \end{array}
\begin{array}{c} 1\,2\,3\,4\,5\,6\,7\,8 \\ \left(\begin{array}{cccccccc}
1 & 1 & 1 & 1 & 1 & 1 & 1 & 1 \\
1 & 1 & 1 & 1 & 1 & 1 & 1 & 1 \\
1 & 1 & 1 & 1 & 1 & 1 & 1 & 1 \\
0 & 0 & 0 & 1 & 1 & 1 & 1 & 1 \\
0 & 0 & 0 & 1 & 1 & 1 & 1 & 1 \\
0 & 0 & 0 & 1 & 1 & 1 & 1 & 1 \\
0 & 0 & 0 & 1 & 1 & 1 & 1 & 1 \\
0 & 0 & 0 & 0 & 0 & 0 & 0 & 1
\end{array}\right) \end{array}
$$

Performing logical *and* of these two matrices gives us the new matrix C'. Any element I in this matrix is in the same SCC with an entry j if and only if $C'[i, j] = 1$. For this example of directed graph, we can see that the SCCs are $\{1, 2, 3\}$, $\{4, 5, 6, 7\}$, and $\{8\}$.

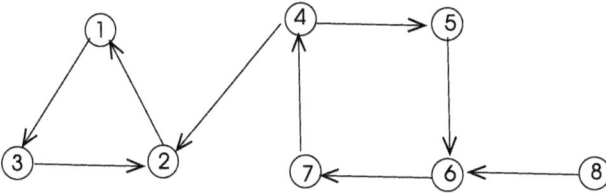

Fig. 12.1 Sample graph for SCC computation

$$
C' = C \wedge C^T = \begin{array}{c} \\ 1 \\ 2 \\ 3 \\ 4 \\ 5 \\ 6 \\ 7 \\ 8 \end{array}
\begin{array}{c} 1\,2\,3\,4\,5\,6\,7\,8 \\ \left(\begin{array}{cccccccc}
1 & 1 & 1 & 0 & 0 & 0 & 0 & 0 \\
1 & 1 & 1 & 0 & 0 & 0 & 0 & 0 \\
1 & 1 & 1 & 0 & 0 & 0 & 0 & 0 \\
0 & 0 & 0 & 1 & 1 & 1 & 1 & 0 \\
0 & 0 & 0 & 1 & 1 & 1 & 1 & 0 \\
0 & 0 & 0 & 1 & 1 & 1 & 1 & 0 \\
0 & 0 & 0 & 1 & 1 & 1 & 1 & 0 \\
0 & 0 & 0 & 0 & 0 & 0 & 0 & 1
\end{array}\right) \end{array}
$$

The connectivity matrix C can be formed by successively multiplying the adjacency matrix A of the directed graph to get A^{-1}, since the longest path in a graph may not be longer than -1. For example, for a graph with 12 vertices, we need to form $C = A^{11}$ which can be obtained by $A^2 = A \times A$; $A^4 = A^2 \times A^2$; $A^8 = A^4 \times A^4$ and $A^{11} = A^8 \times A^3$ for a total of 4 matrix multiplications using logical *and* and logical *or* operations instead of scalar multiplication and addition. We would need $\lceil \log n \rceil$ matrix multiplications and since an $n \times n$ matrix multiplication requires $O(n^\omega)$ operations, the complexity of this step is $O(n^\omega \log n)$. Taking the transpose of C and forming $C \wedge C^T$ both take $O(n^2)$ time, therefore total time complexity is $O(n^\omega \log n)$ with $\omega < 2.376$. Tarjan's or Kosaraju's SCC detection algorithms both use DFS and hence have better performances of $O(n + m)$. However, parallelizing DFS is difficult as discussed in Chap. 6, but matrix multiplication can be parallelized simply by distributing the rows or columns of matrices to a set of processes as we saw in Chap. 4.

12.3.2   Breadth-First Search

In breadth-first search (BFS), we explored vertices that are k hops away from a given source vertex before exploring the ones that are $k+1$ hops away as discussed in Chap. 6. This resulted in shortest distance paths in an undirected unweighted graph. Let us consider the sparse adjacency matrix A of an undirected unweighted graph $G = (V, E)$. We have a sparse vector X to show the source vertex position, for example, $X[3] = 1$ with all other elements 0 meaning vertex 3 is the source. Multiplying A^T by X gives us the vector that has 1s in all neighbors of vertex 3. Multiplying the product again by A^T provides neighbors that are two hops away and so on. We can now sketch an algebraic algorithm using this property as shown in Algorithm 12.4. We have the matrix A and vector X as input and we want to form the $n \times n$ matrix N, which shows the vertices that are I distance away in its Ith row. We need to provide a simple modification since the result of the multiplication shows all vertices that are at most I hops away.

Algorithm 12.1 BFS_Algeb

1: **Input** : $G(V, E)$, A, X ▷ connected, directed or undirected graph G, its adjacency matrix A
     and source vertex vector X

2: **Output** : $N[n, n]$ ▷ vertices that are 1 to n hops away

3:

4: **form** A^T

5: temporary matrix T

6: $N \leftarrow 0$

7: **for** $i = 1$ to $diam(G)$ **do**

8:      $Y^{(i)} \leftarrow A^T \times X$ ▷ find neighbors i hop away

9:      $T \leftarrow Y^{(i)} - Y^{(i-1)}$

10:      $Y^{(i-1)} \leftarrow Y^{(i)}$

11:      $N[i, *] \leftarrow T$ ▷ store neighbor identifiers

12:      $X \leftarrow Y$ ▷ update

13: **end for**

Python code that provides the connectivity matrix C and the SCC matrix C' of the graph depicted in Fig. 12.1 is shown in Listing 12.1 with the output in Listing 12.2. Note that we needed to logically *or* each power of the adjacency matrix with the current connectivity matrix C to get the contents of matrix C correctly since Ax displays connected nodes that are x hops away but not the nodes that are less hops apart. The output displays all of the SCCs for the example graph.

Listing 12.1 SCC computation

```python
1  import numpy as np
2  k = 8
3  A=np.array([[1,0,1,0,0,0,0,0],    # Adjacency Matrix
4              [1,1,0,0,0,0,0,0],
5              [0,1,1,0,0,0,0,0],
6              [0,1,0,1,1,0,0,0],
7              [0,0,0,0,1,1,0,0],
8              [0,0,0,0,0,1,1,0],
9              [0,0,0,1,0,0,1,0],
10             [0,0,0,0,0,1,0,1]],dtype=int)
11
12 def bool_mult(A,B,n):      # Boolean multiplication
13     H = np.zeros((n,n),dtype=int)
14     for i in range(0,n):
15         for j in range(0,n):
16             s = 0
17             for k in range(0,n):
18                 s = (A[i][k] and B[k][j]) or s
19             H[i][j] = s
20     return H
21
22 C = np.zeros((k,k),dtype=int)
23 X = A
24 for i in range(0,k-1): # Construct Connectivity matrix C
25     X = bool_mult(X,A,k)
26     C = C | X
27
28 print('Connectivity Matrix C:\n',C)
29 S = C & C.transpose()       # Compute SCC components
30 print('SCC Matrix S:\n',S)
```

Listing 12.2 SCCs in the graph of Fig. 12.1

```
Connectivity Matrix C:
 [[1 1 1 0 0 0 0 0]
 [1 1 1 0 0 0 0 0]
 [1 1 1 0 0 0 0 0]
 [1 1 1 1 1 1 1 0]
 [1 1 1 1 1 1 1 0]
 [1 1 1 1 1 1 1 0]
 [1 1 1 1 1 1 1 0]
 [1 1 1 1 1 1 1 1]]
SCC Matrix S:
 [[1 1 1 0 0 0 0 0]
 [1 1 1 0 0 0 0 0]
 [1 1 1 0 0 0 0 0]
 [0 0 0 1 1 1 1 0]
 [0 0 0 1 1 1 1 0]
 [0 0 0 1 1 1 1 0]
 [0 0 0 1 1 1 1 0]
 [0 0 0 0 0 0 0 1]]
```

Let us investigate how this algorithm works for the sample graph of Fig. 12.2. The transpose of the adjacency matrix A is itself since graph G is undirected.

Fig. 12.2 Sample graph to test algebraic BFS algorithm

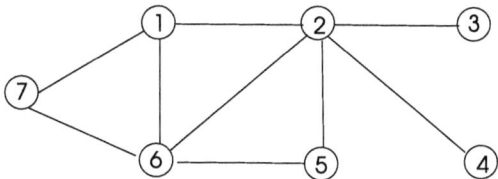

The matrices A and X formed for source vertex 7 in this graph and the resulting neighbor matrix $N[1, *]$ are as follows. We show the full matrix for comparison but only its Ith row shown in bold is modified in Ith iteration.

$$
A^T
\begin{matrix}
& 1\,2\,3\,4\,5\,6\,7 \\
1 & \begin{pmatrix} 0\,1\,0\,0\,0\,1\,1 \\ 1\,0\,1\,1\,1\,1\,0 \\ 0\,1\,0\,0\,0\,0\,0 \\ 0\,1\,0\,0\,0\,0\,0 \\ 0\,1\,0\,0\,0\,1\,0 \\ 1\,1\,0\,0\,1\,0\,1 \\ 1\,0\,0\,0\,0\,1\,0 \end{pmatrix} \\
2 \\ 3 \\ 4 \\ 5 \\ 6 \\ 7
\end{matrix}
\times X^{(1)}
\begin{pmatrix} 0 \\ 0 \\ 0 \\ 0 \\ 0 \\ 0 \\ 1 \end{pmatrix}
\to N^{(1)}
\begin{matrix}
& 1\,2\,3\,4\,5\,6\,7 \\
1 & \begin{pmatrix} \mathbf{1\,0\,0\,0\,0\,1\,0} \\ 0\,0\,0\,0\,0\,0\,0 \\ 0\,0\,0\,0\,0\,0\,0 \\ 0\,0\,0\,0\,0\,0\,0 \\ 0\,0\,0\,0\,0\,0\,0 \\ 0\,0\,0\,0\,0\,0\,0 \\ 0\,0\,0\,0\,0\,0\,0 \end{pmatrix} \\
2 \\ 3 \\ 4 \\ 5 \\ 6 \\ 7
\end{matrix}
$$

The second iteration of the *for* loop results in the following. Note that the result of the multiplication is the vector $(1\,1\,0\,0\,1\,1\,1)$ and we subtract the previous value $(1\,0\,0\,0\,0\,1\,0)$ from this product to obtain $(0\,1\,0\,0\,1\,0\,0)$, which becomes the second row of N.

$$
A^T
\begin{matrix}
& 1\,2\,3\,4\,5\,6\,7 \\
1 & \begin{pmatrix} 0\,1\,0\,0\,0\,1\,1 \\ 1\,0\,1\,1\,1\,1\,0 \\ 0\,1\,0\,0\,0\,0\,0 \\ 0\,1\,0\,0\,0\,0\,0 \\ 0\,1\,0\,0\,0\,1\,0 \\ 1\,1\,0\,0\,1\,0\,1 \\ 1\,0\,0\,0\,0\,1\,0 \end{pmatrix} \\
2 \\ 3 \\ 4 \\ 5 \\ 6 \\ 7
\end{matrix}
\times X^{(2)}
\begin{pmatrix} 1 \\ 0 \\ 0 \\ 0 \\ 0 \\ 1 \\ 0 \end{pmatrix}
\to N^{(2)}
\begin{matrix}
& 1\,2\,3\,4\,5\,6\,7 \\
1 & \begin{pmatrix} 1\,0\,0\,0\,0\,1\,0 \\ \mathbf{0\,1\,0\,0\,1\,0\,0} \\ 0\,0\,0\,0\,0\,0\,0 \\ 0\,0\,0\,0\,0\,0\,0 \\ 0\,0\,0\,0\,0\,0\,0 \\ 0\,0\,0\,0\,0\,0\,0 \\ 0\,0\,0\,0\,0\,0\,0 \end{pmatrix} \\
2 \\ 3 \\ 4 \\ 5 \\ 6 \\ 7
\end{matrix}
$$

The final value of N at third iteration is shown below. The first row has 1s at immediate neighbors of vertex 7, the second row has 1s at two-hop neighbors and the third row shows three-hop neighbors. Since the diameter of the graph is 3, we can stop at the third iteration.

$$
N^{(3)}
\begin{matrix}
& 1\,2\,3\,4\,5\,6\,7 \\
1 & \begin{pmatrix} 1\,0\,0\,0\,0\,1\,0 \\ 0\,1\,0\,0\,1\,0\,0 \\ \mathbf{0\,0\,1\,1\,0\,0\,0} \\ 0\,0\,0\,0\,0\,0\,0 \\ 0\,0\,0\,0\,0\,0\,0 \\ 0\,0\,0\,0\,0\,0\,0 \\ 0\,0\,0\,0\,0\,0\,0 \end{pmatrix} \\
2 \\ 3 \\ 4 \\ 5 \\ 6 \\ 7
\end{matrix}
$$

We need $diam(G)$ iterations of the *for* loop and also we need to perform $\Theta(n^2)$ multiplications at each iteration resulting in $\Theta(n^2 diam(G))$ time complexity for this algorithm. We can immediately see that this algebraic approach can be parallelized conveniently by 1D partitioning of A^T and X and distributing these to a number of processes.

12.3.3 Shortest Paths

We look at the algebraic versions of two main algorithms for shortest paths: Bellman–Ford SSSP algorithm and Floyd–Warshall APSP algorithm in this section.

12.3.3.1 Bellman–Ford Algorithm

Bellman–Ford algorithm is based on dynamic programming and builds shortest paths from a source vertex in an undirected or directed weighted graph progressively as we reviewed in Chap. 7. It can work in the presence of negative-weight edges of the graph, however, it will only report negative cycles if they exist. Given a weighted graph $G = (V, E, w)$ with $w : E \to \mathbb{R}$ and a source vertex s, it starts with an estimation of distance $d(v)$, $\forall v \in V$ using the adjacency matrix A and performs relaxation at each step until the shortest distances are found. Relaxation of an edge (u, v) is stated as providing $d(v) = min\{d(v), d(u) + w(u, v)\}$. An algebraic formulation of this algorithm will use the sparse adjacency matrix $A[n, n]$ and a vector $D[n]$ which shows shortest distance $d(I)$ from s, $\forall I \in V$ in the end as shown in Algorithm 12.2.

Algorithm 12.2 BF_Algeb

1: **Input** : $G(V, E)$, $A[n, n]$, $D[n]$ ▷ connected, directed or undirected weighted graph G, its
        adjacency matrix A and distance vector D
2: **Output**: $D[n]$ ▷ shortest distances
3:
4: $D \leftarrow \infty$
5: **for** $k = 1$ to n **do**
6:     $D \leftarrow D$ min.$+ A$ ▷ find neighbors i hop away
7:     **if** $D \neq d$ min. $+ A$ **then**
8:         **return** "negative cycle found"
9:     **end if**
10: **end for**

12.3.3.2 Floyd–Warshall Algorithm

This dynamic algorithm provided APSP among all vertices of a graph in $O(n^3)$ time as we have reviewed in Chap. 7. It used the relaxation method to update distances by working on the adjacency matrix of the graph. Hence, we can say it is an algebraic graph algorithm. The pseudocode of this algorithm using matrix notation is shown in Algorithm 12.5.

Algorithm 12.3 FW_Algeb

1: **Input** : $G(V, E)$, $A[n, n]$, $D[n, n]$ ▷ connected, directed or undirected weighted graph G, its
   adjacency matrix A and distance vector D
2: **Output:** $D[n]$ ▷ shortest distances
3:
4: $D \leftarrow A$
5: **for** $k = 1$ to n **do**
6:     $D \leftarrow D$ .min$[D(:, k)$ min.$+D(k, :)]$ ▷ find neighbors i hop away
7: **end for**

12.3.4   Minimum Spanning Trees

A minimum spanning tree (MST) of an undirected weighted connected graph $G = (V, E, w)$, $w : E \rightarrow \mathbb{R}$ is a spanning tree of MST with the minimum sum of weights of edges among all spanning trees of G. An MST of a graph is unique if all edge weights are distinct. A minimum spanning forest of an unconnected undirected weighted graph G consists of MSTs in each component of G. We will now reconsider Prim's algorithm we reviewed in Chap. 7, but this time using the algebraic version.

Prim's Algorithm
This algorithm starts with a set S consisting of any vertex of the graph $G = (V, E, w)$ initially and iteratively searches the minimum weight outgoing edge (MWOE) e from S. The edge e is included in the MST tree T and the endpoints of the edge e are added to S. The algorithm terminates when $S = V$ and has a time complexity of $O(m \log n)$ or $O(m + n \log n)$ when priority queue data structure is used.

In the algebraic version of this algorithm, we will assume vertices are numbered from 1 to n for convenience and describe an algorithm as in [13]. We have the sparse distance matrix D with entries $d_{ij} = 0$ when $i = j$, equals w_{ij} when $(i, j) \in E$ and is assigned ∞ when $(i, j) \notin E$. A $1 \times n$ vector M with entries m_i displays whether vertex $i \in S$ or not with $m_i = 0$ if $i \notin S$ and $m_i = \infty$ if $i \in S$. A $1 \times n$ vector Q with entries q_i shows the lightest edge weight connecting i to the set S and $q_i = 0$ if $i \in S$.

We start the algorithm by including the first vertex I in the set M making its value ∞ to mean it is an element of set S and assign Q to the first row of the distance matrix D. The variable wt stores the sum of the weights of the edges of the MST at any time. We then enter the *while* loop and continue until all elements of M are processed and have ∞ values. The *argmin* operation at line 8 is used to find the vertex u, which has the lightest edge to a vertex in S. This is actually searching the MWOE as in the classical algorithm. This vertex is included in S by setting its value to ∞ in M. We also need another vector π which holds the parent of each vertex to store the MST information. A tuple $< weight(u, v), u >$ can be defined to find the parent of the newly added vertex u with $v \in S$. The parent of the edge when it is included in the MST is kept in the variable x in line 10 and the parent of the vertex at the end of the new edge when its is entered in the MST is saved to assign to the parent value in π.

Algorithm 12.4 *Prim_Algeb*

1: **Input** : $G(V, E)$, $A[n, n]$, $D[n]$ ▷ connected, directed or undirected weighted graph G, its adjacency matrix A and distance vector D
2: **Output**: $D[n]$ ▷ shortest distances
3:
4: $M \leftarrow 0, wt \leftarrow 0$
5: $\Pi \leftarrow \emptyset, M[1] \leftarrow \infty$
6: $Q \leftarrow D[1, :]$
7: **while** $M \neq \infty$ **do**
8:     $u \leftarrow \text{argmin}\{M + Q\}$
9:     $M[u] \leftarrow \infty$
10:     $< w, x > \leftarrow Q[u]$
11:     $wt \leftarrow wt + Q[u]$
12:     $\pi[u] \leftarrow x$
13:     $Q \leftarrow Q .\min D[u, :]$
14: **end while**

We need $\Theta(n)$ for the *argmin* operation and the total time, therefore, is $O(n^2)$ which is worse than the original Prim algorithm for sparse graphs but comparable for very dense graphs. However, the algebraic approach is suitable for parallel processing as in other algebraic graph algorithms.

12.3.5 Algebraic Matching

A matching M of an undirected graph $G = (V, E)$ is a subset of its edges such that no edges in M share endpoints. In a perfect matching M, every $v \in V$ is incident to an edge $e \in M$. A maximum matching M' of G has the largest size among all matchings of G and a maximal matching M is not contained in any other matching of G.

Definition 12.1 (*Tutte Matrix*): The Tutte matrix $T(G)$ of an undirected simple graph $G = (V, E)$ is an $n \times n$ matrix with elements;

$$T_{ij} = \begin{cases} x_{ij} & \text{if } (i, j) \in E \text{ and } i < j \\ -x_{ij} & \text{if } (i, j) \in E \text{ and } i > j \\ 0 & \text{if } (i, j) \notin E, \end{cases}$$

where x_{ij} are formal variables. Tutte proved an important relation between the Tutte matrix and perfect matching of a graph [24] as follows.

Theorem 12.1 (Tutte) *Let* $G = (V, E)$ *be an undirected simple graph with a Tutte matrix* T. *Then,* G *has a perfect matching if and only if* $det(T) \neq 0$.

Tutte matrix consists of variables and its determinant is a polynomial of its variables. The determinant should be a polynomial with all zero parameters to have a perfect matching. We can, therefore, compute Tutte matrix T; compute its determinant, and check whether this is zero. However, computing T may take exponential time. Lovazs provided a randomized algorithm to test the perfect matching condition of a graph by substituting for each variable of Tutte matrix T from a polynomially large set of integers and then checking whether T is non-singular [14]. Lovazs also showed that the rank of Tutte matrix of a graph G provides the size of the maximum matching of G [15] shown by the following theorem.

Theorem 12.2 (Lovazs) *Let* $G = (V, E)$ *be an undirected simple graph with a Tutte matrix* T *and* k *be the size of the maximum matching of* G. *Then* $rank(T) = 2k$.

Rabin–Vazirani Algorithm
Once we know that the graph G has a perfect matching, we need an algorithm to find this matching. Let us assume an edge (u, v) that belongs to a perfect matching in G. The subgraph obtained by removing (u, v) and all of its adjacent edges from G has also a perfect matching. If we know how to find an edge e of a perfect matching, we can recursively build a perfect matching of G. Rabin and Vazirani found that the inverse T^{-1} of Tutte matrix provides this information as shown by the following theorem.

Theorem 12.3 (Rabin–Vazirani) *Let* $G = (V, E)$ *be an undirected simple graph with a Tutte matrix* T. *Then,* $(T(G')^{-1})_{i,j \neq 0}$ *if and only if* $G - \{i, j\}$ *has a perfect matching.*

Rabin and Vazirani developed a randomized algorithm based on this theorem, which computes Tutte matrix and its inverse and finds an edge (I, j) satisfying the property in Theorem 12.3. This edge belongs to the perfect matching and thus is added to the current matching. It is then removed from the graph with all its adjacent edges. The algorithm continues with the remaining graph until it becomes empty.

Algorithm 12.5 RV_Alg

1: **Input** : $G = (V, E)$ ▷ undirected simple graph G
2: **Output**: M ▷ perfect matching of G
3:
4: $M \leftarrow \emptyset$
5: $G' = (V', E') \leftarrow G = (V, E)$
6: **while** $G' \neq \emptyset$ **do**
7:     **compute** $T(G')$ and instantiate each variable with a random value from $\{1, \ldots, n^2\}$
8:     **compute** $T(G')^{-1}$
9:     **find** i and j such that $(v_i, v_j) \in E'$ and $(T(G')^{-1})_{i,j} \neq 0$
10:       $M \leftarrow M \cup \{(v_i, v_j)\}$
11:       $G' \leftarrow G' - \{v_i, v_j\}$
12: **end while**

Rabin and Vazirani showed that $n / 2$ matrix inversions are sufficient as each inversion provides one edge of matching. Matrix inversion which takes $O(n^\omega)$ time dominates the time taken for the algorithm and each trial to find the perfect matching takes $O(n^{\omega+1})$ time [19].

12.4   Dynamic Graph Algorithms

We have seen static graph algorithms that provide an output of some function on the graph data structure up to now. However, graphs that represent many real-life networks are not static going through modifications in time. A *dynamic graph* G may evolve with time due to changes in G such as insertion or removal of edges. Dynamic graphs represent many real-life networks, for example, the Internet, protein interaction networks, and social networks in which such changes occur frequently. A *dynamic graph algorithm* allows the following operations on dynamic graphs:

- *query*: We evaluate a certain property of the graph G. For example; "Is graph connected?"
- *insert*: An edge or an isolated vertex is inserted to the graph.
- *delete*: An edge or an isolated vertex is deleted from the graph.

The two latter operations are commonly called *update* procedures. We can perform these operations using the static graph algorithms we have seen up to now from scratch for the modified graph. However, the main goal of a dynamic graph algorithm is to provide more efficient solutions for these operations than the static algorithms. These algorithms are classified as follows:

- *Fully dynamic*: Insertions and deletion of edges and vertices are allowed.
- *Incremental*: Only insertions of edges and vertices are allowed.
- *Decremental*: Only deletion of edges and vertices are allowed.

The last two types of algorithms are named *partial* dynamic graph algorithms. Queries are allowed in all of the algorithms described. Intuitively, answering a query in a dynamic graph in general is simpler than performing an update operation. Another distinction is between undirected and directed graphs. A dynamic graph operation whether a query or an update is generally more difficult in a directed graph than an undirected graph. A dynamic graph algorithm, in general, provides a better performance than its static counterpart. We will see the design of advanced data structures is crucial when forming dynamic algorithms.

The *fully dynamic connectivity algorithm* in an undirected graph allows insertion and deletion of edges, and enables queries such as "is graph connected" or "are vertices u and v in the same component?". In the *fully dynamic minimum spanning tree* problem, we maintain a forest of minimum spanning trees when edges are inserted, deleted, and weights of edges change. The main problems in directed graphs are *dynamic transitive closure* and *dynamic shortest paths*. In the first problem, we keep information to evaluate whether a vertex v is reachable from a vertex u when edges are deleted and inserted. The shortest path problem involves providing and maintaining information about shortest paths when edges are inserted and deleted in a dynamic environment.

We start this section by first defining some methods to be used in designing efficient dynamic graph algorithms for undirected and directed graphs and then provide a brief survey of algorithms for two representative dynamic graph problems; connectivity and matching.

12.4.1 Methods

The methods for undirected graphs and directed graphs differ significantly. We will classify these methods as described in [7] for undirected and directed graphs.

12.4.1.1 Undirected Graphs
The main methods of designing algorithms for the dynamic undirected graph are clustering, sparsification, and randomization.

Clustering
The clustering method for dynamic graphs was proposed in [10], where the graph is partitioned to a number of clusters and update operation is performed in only the related cluster, for example, a spanning tree may be partitioned into a number of subtrees. It was shown in [10] that the minimum spanning forest of a graph can be maintained in $O(\sqrt{m})$ per update using this method.

Sparsification
Sparsification is a technique that is used as a black box in the design of dynamic graph algorithms as described in [8]. This is a divide and conquer method in which the graph G is partitioned into a set of $O(m / n)$ sparse subgraphs. Each subgraph information is kept in a certificate which is merged in pairs with other

subgraphs resulting in a balanced tree for the graph with each vertex having a sparse certificate. An update operation is then performed in $O(\log m/n)$ graphs with each having $O(n)$ edges [7, 8].

Randomization

Randomization is a powerful method for algorithm design; it also proved to be efficient in the design of dynamic graph algorithms for undirected graphs in [11]. They used random sampling to select a non-tree edge when a tree edge is removed from the graph while maintaining a spanning tree. It can also be combined with the graph decomposition.

12.4.1.2 Directed Graphs

Tools for directed graphs are outlined in [7] in terms of data structures used and methods employed. A special tree structure named *reachability tree* can be used to solve dynamic graph problems. The aim of using this data structure is to keep a BFS tree during edge deletions. Two types of operations supported are the *Delete(u,v)* which removes edge (u, v) from the graph and *Level(u)* which returns the level of a vertex u in the BFS tree. Matrix data structures can also be used to keep dynamic graph information with dynamic operations on rows and columns.

Demetrescu and Italiano proposed *locally shortest paths* defined as paths which have every proper subpaths as shortest paths [4]. Given a graph $G = (V, E)$, the *long paths property* means selecting $S \subset V$ at random results in sufficiently long paths of G having common vertices in S with high probability. Using this property, we can find a long path using short searches to design algorithms for transitive closure and shortest paths [5, 7].

12.4.2 Connectivity

In the dynamic connectivity problem, we need to test the connectivity of the graph when there are queries and updates. Typical queries would be testing whether graph G is connected (*connected(G)*) and whether vertices u and v are connected (*connected(u, v)*). For the update problem, we would need to perform these queries when an edge (u, v) is inserted or deleted from graph. We will consider two cases separately as their implementations are very different; the incremental and decremental dynamic connectivity.

12.4.2.1 Incremental Algorithms

We want to maintain information on the connected components of an undirected graph G dynamically with the use of following operations.

- *connected(u,v)*: Report whether u are v are in the same connected component of G
- *insert(u,v)*: Add edge (u, v) to G.

Let us recall the *union-find* data structure we have reviewed in Chap. 7. This structure maintains disjoint groups of data items with each group having a representative. It supports two operations; *find(x)* returns the representative of the set that *x* belongs and *union(x, y)* merges the groups of *x* and *y*. We can check whether two data items are in the same group by testing their representatives with *find* to see if they have the same representative. If they do, they are in the same group. We can also unite two groups by the *union* operation. The *union-find* data structure can be implemented in $O(\alpha(n))$ time where $\alpha(n)$ is the inverse of the very fast growing Ackermann function [22].

We can see this data structure is adequate to have a dynamic incremental connectivity algorithm. We can perform a DFS algorithm in the graph and store each tree of the forest in a group with the root of the tree being the representative of the tree. Each query can be realized by *find(x)* operations, which outputs the root of the tree that *x* is contained and an *insert(u,v)* operation is realized by the *union* operation which merges two trees if *u* and *v* are in different trees.

12.4.2.2 Decremental Algorithms

In the decremental connectivity problem, we want to remove an edge by *delete(u, v)* operation and check whether graph is still connected or not. The *union-find* structure does not work in this case and several studies aimed to find a solution to this problem. An update time of $O(\sqrt{m})$ using clusters was presented in [10] which was improved to $O(\sqrt{n})$ using the method of sparsification in [8]. A randomized algorithm with amortized $O(\log^2 n)$ expected time per operation was proposed in [11]. A deterministic algorithm with amortized $O(\log^2 n)$ per operation was presented in [12] and a randomized algorithm with expected $O(\log n(\log \log n)^3)$ amortized time per operation is described in [23]. We will take a closer look at a novel data structure called *Euler tour tree* that can be used for dynamic connectivity problem.

Euler Tour Trees

The Euler tour tree (ETT) was presented in [11] to store information about dynamic graphs. An Euler tour of a graph G is a path that traverses each edge of G exactly once. A tree does not have an Euler tour, in order to realize an Euler tour of a tree T, each edge is considered bidirectional and hence each edge is traversed twice and the tour starts and ends at the root vertex [11]. An Euler tour tree (ETT) associates a weighted or unweighted key for each vertex. An ETT is basically a balanced binary tree of an Euler tour of the tree T. We can think of an Euler tour of a tree T as a depth- first traversal of T. An Euler tour of a sample tree is shown in Fig. 12.3, where the BST stores the vertices in the order of their visit times and each vertex in the tree holds pointers to the vertices in the BST showing their first and last visited times.

The main idea of the connectivity algorithm based on ETTs is to store the Euler tour of a tree instead of storing the tree. Edge insertions or deletions can be performed by modifying Euler trees of the forest. Testing whether two vertices

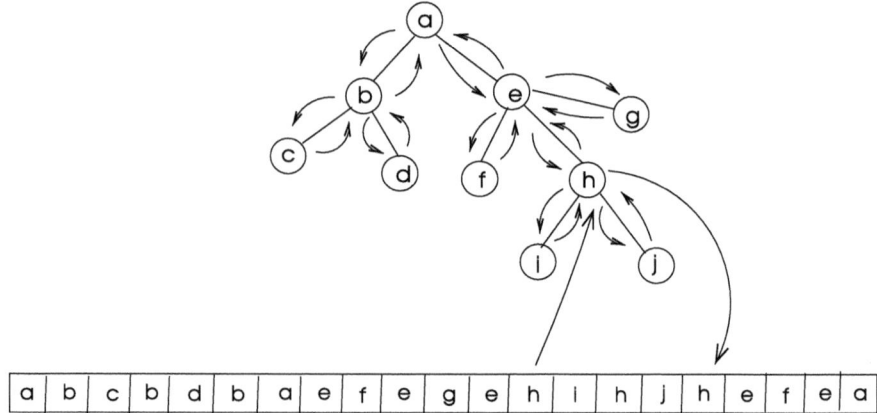

Fig. 12.3 Euler tour of a tree. Visit times for the vertex h is shown

are connected can be done by checking if these vertices are in the same ETT. The following main operations are provided in an ETT:

- *FindRoot(v)*: Finds the root of the tree that contains vertex v. Since the root is visited as the first element and the last element of the tree, the minimum or the maximum element of the tree is returned.
- *Cut(v)*: The subtree rooted at vertex v is cut from the tree it is contained. This can be implemented by dividing the BST into three segments; segment before the first visit to v, segment between the first and last visit to v, and the segment after the last visit to v. The second segment is shown as the bold box in Fig. 12.4. The first segment contains Euler tour of the tree before reaching vertex v, Euler tour of the subtree rooted at v and the Euler tour of the tree after v is visited last time. We can now merge the first and third segment to perform the cut operation. This procedure is illustrated in Fig. 12.4 for vertex h. Note that one occurrence of vertex e has to be deleted from the BST.
- *Link(u,v)*: The subtree rooted at vertex u is connected as a child of vertex v. We divide the BST into two segments; left segment S_l is from the beginning until before the last visit to v, and the second segment S_r is the rest of the BST. The ETT of the forest is then $ETT_F = \{S_1 \cup v \cup ETT_u \cup S_2\}$, where ETT_u is the ETT of the tree rooted at the vertex u.

There are various other procedures to modify the ETTs as described in [11]. In order to answer the *connected(u,v)* query, the roots of the ETTs that contain these vertices are found by the *FindRoot* procedure and checked whether they are the same. Insertion and deletion are performed by the reconstruction of Euler tours with changes in only $O(\log n)$ vertices of the balanced BST. Queries can be answered in $O(\log n / \log \log n)$ time and insertions take $O(\log^2 n / \log \log n)$ time using this method [11].

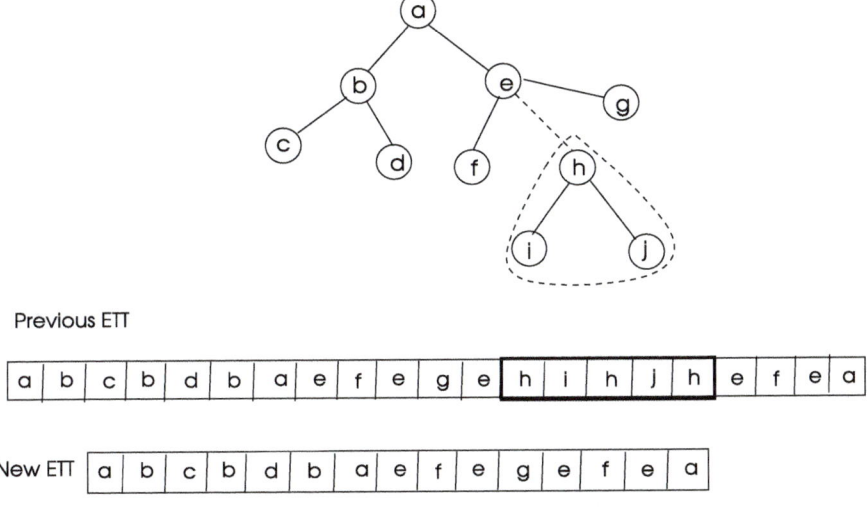

Previous ETT

a	b	c	b	d	b	a	e	f	e	g	e	h	i	h	j	h	e	f	e	a

New ETT

a	b	c	b	d	b	a	e	f	e	g	e	f	e	a

Fig. 12.4 Operation of the *Cut(h)* procedure on the sample graph of Fig. 12.3

12.4.3 Dynamic Matching

Dynamic matching of a graph G should maintain the maximal matching of G in the presence of edge insertions and removals. In order to do so, we need to ensure that there is no edge (u, v) of G with both free endpoints since such an edge should be included in the matching. We need to check the following while inserting and deleting edges to a graph with a maximal matching M.

- *insert(u,v)*: Whenever an edge (u, v) is to be inserted in G, we can check its endpoints and include this edge in the maximal matching M of G if both u and v are free. When an unmatched edge (u, v) is deleted, we do not need to do anything since this edge is not part of the maximal matching. For other cases, we need to check the neighbors of the vertices u and v and update the maximal matching as appropriate.
- *delete(u,v)*: We need to consider two cases when an edge (u, v) is deleted from the graph; when a matched edge (u, v) is deleted, we need to check $deg(u) + deg(v)$ neighbors of these vertices which are fine when degrees of vertices are small. For a graph with large degrees, randomization can be used in which a vertex is matched with a neighbor vertex selected at random in *random mating*. The simple approach described is combined with random mating that is presented in [2] to result in expected amortized $O(\log n)$ time per update.

We will take a closer look at a deterministic algorithm that works using the described logic to update the maximal matching of a graph.

Neiman and Solomon Algorithm

Neiman and Solomon presented a deterministic algorithm to find maximal matching of a graph that runs in $O(\sqrt{m})$ update time with a 3/2-approximation [17].
The main idea of this algorithm is to consider three cases when adding an edge
(u, v) to the graph G. If both endpoints of (u, v) are free, then (u, v) is added to
the existing matching. If both u and v are matched, then matching is not changed.
When one endpoint of edge is matched and other is not, neighbors of vertices u
and v are searched. When a matched edge (u, v) is deleted from the graph, the
neighbors of u and v are checked to see if an edge (u, w) or (u, y) or both can be
added to the matching. The algorithm works in rounds and the three invariants to
be maintained at the Ith round of the algorithm are as follows.

1. The degree $deg(v)$ of a free vertex v that can be matched at all times is $\leq$
   $\sqrt{2(m+n)}$.
2. For a free vertex v, $deg(v) \leq \sqrt{2m}$. When a high-degree vertex u becomes free
   and all of its neighbors are matched, a *surrogate* v' is searched in place of u.
   The vertex v' is matched to a neighbor v of u such that $deg(v') \leq \sqrt{2m}$. Then
   u and v can be matched and the low-degree vertex v' becomes free.
3. M is maximal.

Algorithm 12.6 displays the pseudocode of the *insert* procedure in this algorithm.
We have a procedure called *Surrogate* that is called when one endpoint of the edge
to be inserted is matched and the other is free. In this case, adding the edge (u, v)
may result in an augmenting path which means the maximal matching M can be
enlarged.

Algorithm 12.6 Inserting an edge

```
1: procedure INSERT((u, v))
2:     E ← E ∪ {(u, v)}
3:     m_i ← m_i + 1
4:     if both u and v are free then
5:         M ← M ∪ {(u, v)}
6:         if u is free and v is matched (or other way) then
7:             surrogate(u) (or v)
8:         end if
9:     end if
10: end procedure
```

The *Insert* procedure calls the procedure *surrogate* when one end of the added
edge (u, v) is free and the other is not. In the first case, if $(u, v) \notin M$, we simply
remove the edge (u, v) from the graph without changing the matching M. In the
second case, $(u, v) \in M$, and the edge (u, v) is deleted from the matching M.
This may result in forming new augmenting paths of length less than or equal to
3 which start either at u or v. In this case, we check whether there is a free vertex
w that is a neighbor of vertex u or v in which case edge (u, v) is added to the

matching M. Furthermore, the degree of vertex u under consideration is checked; two cases are when $deg(u) \leq \sqrt{2m}$ or $deg(u) > \sqrt{2m}$. In the first case, u may become free but a search for an augmenting path is carried. When $deg(u) > \sqrt{2m}$, u is not allowed to be free since its degree is high or it has no free neighbors. In this case, the procedure *Surrogate* is called to find a surrogate vertex that may become free instead of vertex u.

A recent work on dynamic deterministic approximate maximum matching with worst-case update time of $O(\log^3 n)$ time is presented in [3] and randomized 2-approximate matching algorithms are reported in [18, 21].

12.5   Dynamic Algebraic Graph Algorithms

A dynamic algebraic graph algorithm aims to solve a problem in a dynamic graph using algebraic methods. Our main focus here is again implementing linear algebra operations for the given graph problem. A dynamic matrix library to be used for this purpose can be formed and we will first list possible operations in such a library. We will then survey dynamic algebraic graph algorithms for two main problems we have been investigating in this chapter; connectivity and matching.

12.5.1   A Dynamic Matrix Library

A dynamic matrix operation can be defined as the procedure that performs a matrix function such as finding determinant or inverse of a matrix when a change such as the contents of a row or column occurs. This procedure should implement the required function without having to run the static counterpart from scratch, and therefore should provide a better performance. The following matrix operations can be performed dynamically as described in [20]:

- Determinant of a matrix
- Adjoint of a matrix
- Inverse of a matrix
- Matrix rank
- Characteristic polynomial of a matrix
- Linear system of equations.

It was shown in [20] that finding dynamic determinant of a matrix, computing matrix inverse, matrix adjoint, and solving linear system of equations with non-singular row and column updates, can be solved with the following costs:

- *Initialization*: $O(n^{\omega})$ arithmetic operations.
- *Update*: $O(n^2)$ arithmetic operations.
- *Query*: $O(1)$ arithmetic operations.

12.5.2 Connectivity

We have searched a solution to the connectivity problem using algebraic methods first, then reviewed the dynamic connectivity problem. We now want to investigate dynamic connectivity algorithms using linear algebra. Let us recall the two main algebraic methods for connectivity:

- *Laplacian Matrix*: The second smallest eigenvalue λ_2 of the Laplacian matrix called algebraic connectivity or the Fiedler value provides information on the graph connectivity. The Laplacian of an unweighted graph is a symmetric, positive semidefinite matrix that has a positive λ_2 if and only if the graph is connected. In order to find this value, we need to solve the equation $\det(A - \lambda I)x = 0$.
- *Adjacency Matrix*: The sum of the powers $C = \sum_{i=0}^{k} A^k$ of the adjacency matrix A for $k \leq -1$ provide information on the connectivity of a graph. The c_{ij} entry of C shows the number of paths of length k or less between the vertices I and j. If all entries of A^k are positive, then the graph is connected.

Finding the determinant of a matrix can be performed dynamically [20] and hence we have a dynamic method to find the connectivity using the Laplacian matrix. Similarly, matrix multiplication of dynamic matrices can be performed to result in a dynamic connectivity method using the adjacency matrix. Maintaining connectivity in a distributed system consisting of many autonomous computing elements has a number of applications. For example, providing connectivity in a mobile robot network is needed for the coordination of the robots and maintaining positive definiteness of positive entries of C is sufficient to provide connectivity in such a network [25].

12.5.3 Perfect Matching Using Gaussian Elimination

We described Rabin and Vazirani algorithm [19] for perfect matching in Sect. 12.3.5. The random adjacency matrix $A(G)$ of the graph G, called Tutte matrix T, is created first in this algorithm. Its inverse $A^{-1}(G)$ is then computed and an allowed edge e is found, this edge and its endpoints are removed from G and a new Tutte matrix T is created for the new graph. This loop continues until G becomes empty. The time to compute the matrix $T^{-1}(G)$ is $O(n^\omega)$ resulting in $O(n^{\omega+1})$ time in total.

Mucha and Sankowski found that computing the inverse of Tutte matrix in each iteration is not necessary since Tutte matrix at rth iteration, T_{r+1}, is T_r with two rows and columns corresponding to I and j deleted [16]. The following theorem was used to form a relation between T_{r+1}^{-1} and T_r^{-1}.

Theorem 12.4 (Elimination) *Let A be a non-singular n × n matrix. Then, A and its inverse can be written as*

$$A = \begin{pmatrix} a_{11} & v^T \\ u & B \end{pmatrix}, A^{-1} = \begin{pmatrix} \hat{a}_{11} & \hat{v}^T \\ \hat{u} & \hat{B} \end{pmatrix}$$

where $a_{1,1}$, $\hat{a}_{1,1}$ are numbers, u, v, $\hat{u}$, $\hat{v}$ are -1×1 vectors and $\hat{a}_{1,1} \neq 0$. Then, $B^{-1} = \hat{B} - \hat{u}\hat{v}^T / \hat{a}_{1,1}$.

Using this theorem, we can compute the inverse of a matrix dynamically after removing a row and a column without having to compute the inverse from scratch. We consider the columns as variables and the rows as the equations, and the described procedure eliminates the first variable using the first equation. Mucha and Sankowski provided a $O(n^3)$ algorithm shown in Algorithm 12.7 that finds the perfect matching of a simple undirected graph based on Rabin–Vazirani algorithm using the method we have described [16].

Algorithm 12.7 $MS\_AlgMatch$

1: **Input** : $G(V, E)$ ▷ undirected simple graph G
2: **Output**: M ▷ perfect matching of G
3:
4: $M \leftarrow \emptyset$
5: $A \leftarrow T_G^{-1}$
6: $G' = (V', E') \leftarrow G = (V, E)$
7: **while** $G' \neq \emptyset$ **do**
8:     **find** i such that $(v_i, v_j) \in E'$ and $a_{ij} \neq 0$
9:     $M \leftarrow M \cup \{(v_i, v_j)\}$
10:     $G' \leftarrow G' - \{v_i, v_j\}$
11:     eliminate i-th row and j-th column of A
12:     eliminate j-th row and i-th column of A
13: **end while**

12.6   Chapter Notes

We have reviewed first algebraic, then dynamic graph algorithms, and finally dynamic algebraic graph algorithms in this chapter. A class of algebraic graph algorithms relies heavily on matrices related to a graph and operations on them to solve a graph problem. We can see the already available matrix library functions can be used for such problems whether in sequential or parallel operations. The algebraic graph algorithms provided solutions which may have worse performances than the classical counterparts, but they are much easier to parallelize than the classical ones. Matrix multiplication is frequently used to solve various graph problems as we have seen. Since this basic matrix operation can be parallelized conveniently, we can deduce that these problems can be parallelized more easily than the traditional graph algorithms described until now. Dynamic graph algorithms provide more efficient solutions than the static ones when there is a change in the structure of the graph. Designing sophisticated data structures is crucial for dynamic graph algorithms since updates and queries depend largely on the data structures used. We reviewed basic methods for undirected graphs which include sparsification, randomization, and clustering; and directed graphs with reachability trees, matrix data structures, locally shortest paths, and long paths. We then reviewed algorithms for dynamic connectivity and dynamic matching. Our last topic of review was the dynamic algebraic graph algorithms which work on dynamic graphs using methods of linear algebra. We looked at two main problems again; connectivity and matching. We saw how a simple modification to Rabin–Vazirani algebraic matching algorithm using a basic method again from linear algebra led to a more efficient solution.

These topics are relatively more recent areas of study than the static graph algorithms we have seen until now and they have been the focus of many recent studies. Our main goal in the analysis of these topics is to provide a general survey with examples to give some idea on the related concepts rather than being comprehensive. A good review of algebraic graph algorithms is provided in [13]. A detailed survey of dynamic graphs and dynamic graph algorithms are provided in [6, 7] and algebraic theory related to graphs is presented in [1]. A thorough analysis of dynamic matrix operations for some graph problems is provided in [20].

Exercises

1. Find the Laplacian matrix for sample graph shown in Fig. 12.5 and work out the eigenvalues and eigenvectors of this graph.
2. Work out the algebraic BFS algorithm in the sample graph depicted in Fig. 12.6 for source vertex 2. Show the contents of neighborhood matrix N at each iteration. Describe how to run this algorithm in parallel using two processes p_0 and p_1 using this graph as an example.
3. Discover the SCCs of the directed graph depicted in Fig. 12.7 using its connectivity matrix and its transpose.
4. Form the *union-find* data structures as trees for the graph of Fig. 12.8. Show the operation of *find(c)*, *connected(d, f)*, and *insert(e, k)* using this data structure on this graph.
5. Sketch a parallel version of Rabin–Vazirani algorithm for perfect matching to run using k processes $p_0, \ldots, p_{k-1}$ on distributed memory computers. Write the

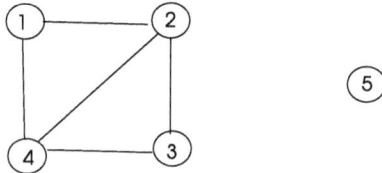

Fig. 12.5 Sample graph for Exercise 1

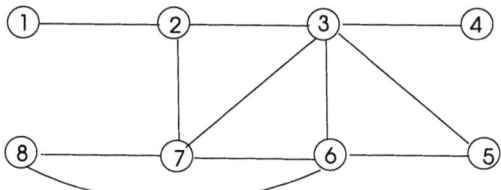

Fig. 12.6 Sample graph for Exercise 2

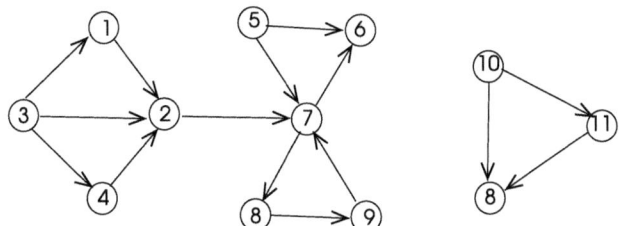

Fig. 12.7 Sample graph for Exercise 3

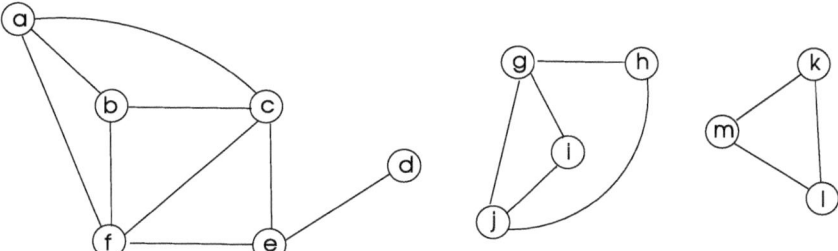

Fig. 12.8 Sample graph for Exercise 4

pseudocode for a process by showing the interprocess communication explicitly. You can assume a master/slave or a fully distributed model of a parallel processing system.

References

1. Bapat RB (2014) Graphs and matrices (Universitext), 2nd edn. Springer, Berlin (Chapters 3 and 4)
2. Baswana S, Gupta M, Sen S (2011) Fully dynamic maximal matching in O(log n) update time. In: 52nd annual IEEE symposium on foundations of computer science FOCS 2011, pp 383–392

3. Bhattacharya S, Henzinger M, Nanongkai D (2017) Fully dynamic approximate maximum matching and minimum vertex cover in $O(log^3 n)$ worst case update time. In: 28th ACM SIAM symposium on discrete algorithms (SODA17), pp 470–489
5. Demetrescu C, Italiano GF (2004) A new approach to dynamic all pairs shortest paths. J Assoc Comput Mach (JACM) 51(6):968–992
6. Demetrescu C, Italiano GF (2006) Fully dynamic all pairs shortest paths with real edge weights. J Comput Syst Sci 72(5):813–837
7. Demetrescu C, Finocchi I, Italiano GF (2004) Dynamic graphs. Handbook of data structures and applications. Computer and information science series. Chapman and Hall/CRC, Boca Raton (Sect. 36)
8. Demetrescu C, Finocchi I, Italiano GF (2013) Dynamic graph algorithms, 2nd edn. Handbook of graph theory. Chapman and Hall/CRC, Boca Raton (Sect. 10-2)
9. Eppstein D, Galil Z, Italiano GF, Nissenzweig A (1997) Sparsification a technique for speeding up dynamic graph algorithms. J Assoc Comput Mach 44:669–696
10. Fiedler M (1973) Algebraic connectivity of graphs. Czechoslovak Math J 23:298–305
11. Frederickson GN (1985) Data structures for on-line updating of minimum spanning trees, with applications. SIAM J Comput 14(4):781–798
12. Henzinger MR, King V (1999) Randomized fully dynamic graph algorithms with polylogarithmic time per operation. J ACM 46(4):502–516
13. Holm J, de Lichtenberg K, Thorup M (2001) Poly-logarithmic deterministic fully dynamic algorithms for connectivity, minimum spanning tree, 2-edge, and biconnectivity. J Assoc Comput Mach 48(4):723–760
14. Kepner J, Gilbert J (eds) (2011) Graph algorithms in the language of linear algebra. SIAM
15. Lovazs L (1979) On determinants, matchings, and random algorithms. In: Budach L (ed) Fundamentals of computing theory. Akademia-Verlag, Berlin
16. Lovazs L, Plummer M (1986) Matching theory. Academic Press, Budapest
18. Mucha M, Sankowski P (2006) Maximum matchings in planar graphs via Gaussian elimination. Algorithmica 45(1):3–20
19. Neiman O, Solomon S (2013) Simple deterministic algorithms for fully dynamic maximal matching. In: Proceedings of the ACM symposium on theory of computing (STOC'13), pp 745–754
20. Onak K, Rubinfeld R (2010) Maintaining a large matching and a small vertex cover. In: Proceedings of the ACM symposium on theory of computing (STOC'10), pp 457–464
21. Rabin MO, Vazirani VV (1989) Maximum matchings in general graphs through randomization. J Algorithms 10(4):557–567
22. Sankowski P (2005) Algebraic graph algorithms. Ph.D. thesis, Warsaw University, Faculty of Mathematics, Information and Mechanics
26. Solomon S (2016) Fully dynamic maximal matching in constant update time. In: Proceedings of FOCS, pp 325–334
27. Tarjan R (1975) Efficiency of a good but not linear set union algorithm. J ACM 22(2):215–225
28. Thorup M (2000) Near-optimal fully-dynamic graph connectivity. In: Proceedings of the thirty-second annual ACM symposium on Theory of computing. ACM Press, pp 343–350
29. Tutte WT (1947) The factorization of linear graphs. J Lond Math Soc s1-22(2):107–111
30. Zavlanos MM, Pappas GJ (2005) Controlling connectivity of dynamic graphs. In: Proceedings of the 44th IEEE conference on decision and control and European control conference, Seville, Spain, December 2005, pp 6388–6393

Analysis of Large Graphs

13

13.1   Introduction

Large graphs consist of thousands of vertices and tens of thousands of edges. Analysis of these graphs requires introduction of new parameters and methods conceptually different than the ones we have reviewed up to this point. Global description of large graphs is very difficult due to the large sizes involved. One way of tackling this problem is to select a sample and representative subgraph of a given graph, analyze it, and extend the results obtained to the whole graph. However, sample selection is a problem on its own and reliability of extrapolating the analysis results is another issue to be considered. Alternatively, and more commonly, we can analyze the local properties of all or majority of vertices and edges in these graphs and use the results obtained to have some idea on the overall structure of the graph.

We start this chapter by defining some new parameters for large graph analysis. Real large networks represented by large graphs have some interesting properties; these networks, commonly called *complex networks*, exhibit *small-world* and *scale-free* structures. The former means the distance between any two nodes in these networks is small compared to the number of nodes they have and the scale-free property is depicted by the existences of few very high-degree nodes and many low-degree nodes. These attributes are not found in random networks and we provide a brief review of these real-life network models. Main types of complex networks are the technological networks, biological networks, and social networks as we will analyze in the next chapter. We describe the centrality concept which provides assigning importance values to vertices and edges based on their usage in a network and we review the main algorithms to assess centralities in networks. Clustering provides grouping similar objects using some similarity measure. Graph clustering is the process of detecting dense regions of a given graph. We define parameters to assess the quality of the output of any clustering method and review

few basic algorithms to find clusters and cluster-like structures in graphs in the
last part of this chapter.

13.2   Main Parameters

Analysis of large graphs is difficult due to the size of data to represent them. We
can however analyze local properties in these graphs with the aim of deducing
their global properties. We need to define some new parameters for the assessment
of global properties of large graphs as described in the next sections.

13.2.1 Degree Distribution

Definition 13.1 (*Degree Distribution*): The degree distribution of a graph G displays
the ratio of the number of vertices with degree k to the total number of vertices.

This parameter basically shows the probability of a randomly selected vertex to
have a degree k. It can be formulated as follows:

$$P(k) = \frac{n_k}{n},\tag{13.1}$$

where n_k is the number of vertices with degree k and n is the number of vertices.
Plotting of $P(k)$ against the degrees provides visual analysis of the distribution
of the vertex degrees of the graph. For a random graph, we expect the degree
distribution to be binomial with peak around the average degree of the graph. For
graphs representing many real-life networks, we see rather interesting distributions
which are radically different than the binomial. A degree distribution of a simple
graph is depicted in Fig. 13.1.

In an *assortative network*, a node has a high probability of being neighbor of
a node with similar degree. For example, nodes of a social network are persons
and this property is exhibited in such networks since a person with many friends

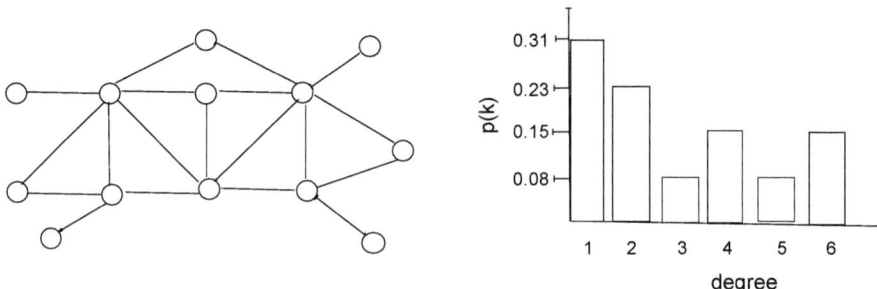

Fig. 13.1 Degree distribution of a simple graph

has a high chance of having another person with many friends as a friend, as in the case of celebrities who know each other. In *disassortative networks*, a high-degree vertex is commonly attached to vertex with a low degree as in the case of biological networks such as the protein interaction networks.

13.2.2 Graph Density

The density of a graph G, $\rho(G)$, is defined as the ratio of the size of its existing edges to the maximum possible of size of edges that can exist as follows:

$$\rho(G) = \frac{2m}{n(-1)}. \tag{13.2}$$

In an undirected graph, the sum of the degrees is equal to $2m$ by the handshake theorem (See Chap. 2) and the average degree of a graph, $deg(G)$, is the sum of all degrees divided by the number of its vertices is $2m / n$. Substitution in Eq. 13.2 yields

$$\rho(G) = \frac{deg(G)}{-1} \approx \frac{deg(G)}{n} \quad \text{when n is large}. \tag{13.3}$$

In a *dense graph*, $\rho(G)$ is stable when n is increased to very large values and $\rho(G)$ approaches 0 with large values of n in *sparse graphs*. The average degree of the graph in Fig. 13.1 is $38/13 = 2.9$.

13.2.3 Clustering Coefficient

Clustering coefficient is a local property defined for a vertex in a graph as follows.

Definition 13.2 (*Clustering Coefficient*): The clustering coefficient $CC(v)$ of a vertex v in a graph G is the ratio of the existing number of edges between the neighbors of v to the maximum possible number of connections between these neighbors.

Since the maximum possible number of connections between k vertices in a graph is $k(k-1)/2$, the clustering coefficient $CC(v)$ of a vertex v is defined as follows.

$$CC(v) = \frac{2n_v}{|N(v)|(|N(v)| - 1)}, \tag{13.4}$$

where n_v is the existing number of edges between the neighbors of vertex v. This parameter basically shows how well the neighbors of a vertex are connected and hence their tendency to forming a clique. In a social network for example, the clustering coefficient of a person provides evaluation of how much friends of that person are friends of each other. The average clustering coefficient of a graph G, $CC(G)$, is the mean value of all of the clustering coefficients of vertices as follows:

$$CC(G) = \frac{1}{n'} \sum_{v \in V} CC(v), \qquad (13.5)$$

where n' is the number of nodes that have a degree of two or more. For a vertex v with a degree less than two, $CC(v)$ is sometimes considered one or zero and in this case, the denominator in the above equation can be taken as n. If this parameter is high in a graph G, we can deduce that the vertices of G are well connected and therefore G is a dense graph. Clustering coefficients of vertices of a sample graph are depicted in Fig. 13.2.

Python code that computes the clustering coefficients of the nodes of this graph is shown in Listing 13.1 with the output in Listing 13.2. The function *clus_coe f* first stores the neighbors of the input node x in list *neighs* after which it finds the existing edges between them and finally calculates and returns the clustering coefficient for the input node. The output values are consistent with the manually calculated values on the graph.

Listing 13.1 Clustering coefficient computation

```
 1 import numpy as np
 2 k = 8
 3
 4 A=np.array([[0,1,0,0,0,0,0,1],   # Adjacency Matrix
 5             [1,0,1,1,0,0,1,1],
 6             [0,1,0,1,1,1,1,0],
 7             [0,1,1,0,0,0,0,0],
 8             [0,0,1,0,0,1,0,0],
 9             [0,0,1,0,1,0,1,0],
10             [0,1,1,0,0,1,0,1],
11             [1,1,0,0,0,0,1,0]],dtype=int)
12
13 def clus_coeff(x):      # Compute CC of node x
14     neighs = []
15     for j in range (0,k): # find neighbors
16         if A[x][j] == 1:
17             neighs.append(j)
18     l = len(neighs)           # l = max number of edges
19     n_edges = 0               # n_edges: existing edges
20     for i in range(0,l):
21         for j in range (i,l):
22             if A[neighs[i],neighs[j]] == 1:
23                 n_edges = n_edges + 1
24     cc = (2*n_edges)/(l*(l-1))    # calculate cc
25     return cc
26
27 for t in range (0,k):      # output ecah CC
28     print("CC({}): {} ".format(t,round(clus_coeff(t),2)),end=" ")
```

Fig. 13.2 Clustering coefficient values of the vertices of a sample graph

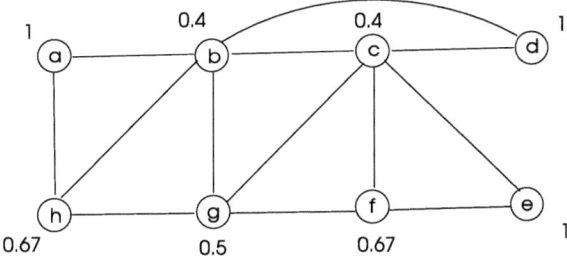

Listing 13.2 Cluster coefficents of nodes of the graph in Fig. 13.2

```
CC(0): 1.0   CC(1): 0.4   CC(2): 0.4   CC(3): 1.0   CC(4): 1.0   CC(5)
  : 0.67   CC(6): 0.5   CC(7): 0.67
```

The *transitivity* $T(G)$ of a graph $G = (V, E)$ as proposed in [12] assesses how well neighbors of the vertices of a graph are connected. Let a triangle subgraph of a graph G be $G_t = (V_t, E_t)$ with $V_t = \{v_1, v_2, v_3\}$ and $E_t = \{(v_1, v_2), (v_2, v_3), (v_1, v_3)\}$. A triplet is a three-vertex subgraph $G_r = (V_r, E_r)$ with $V_r = \{v_1, v_2, v_3\}$ and $E_r = \{(v_1, v_2), (v_2, v_3)\}$ with v_2 in the middle. Each triangle contains three triplets; the transitivity of a graph can now be defined as follows:

$$T(G) = \frac{3 \times \text{ number of triangles}}{\text{number of connected triplets}}. \tag{13.6}$$

A simple four-vertex graph is depicted in Fig. 13.3. The clustering coefficients are given next to vertices. Graph clustering coefficient is the average of these values, yielding $(2 + \frac{4}{3})/4 = 5/6$. There are two triangles in this graph and eight triplets, counting three triplets per triangle and including $\{d, a, b\}$ and $\{a, b, c\}$ triplets giving a total of eight triplets. Hence, the transitivity of this graph is $3 \times 2/8 = 0.75$.

Fig. 13.3 Comparison of transitivity and the clustering coefficient in a sample graph

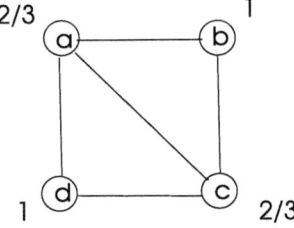

13.2.4 Matching Index

One way of assessing similarity between the two vertices of a graph is to find the number of their common neighbors. *Matching index* defined below is used to quantify this similarity.

Definition 13.3 *(Matching Index)*: The *matching index* of two vertices u and v in a graph G is defined as the ratio of the number of their common neighbors to the number of the union of their neighbors.

In Fig. 13.2, the matching index of vertices b and f, m_{bf}, is 0.33 since the union of their neighbor set is $N(b) \cup N(f) = \{a, c, d, e, g, h\}$ with a size of 6 and they have two common neighbors, c and g. The vertices in a graph may be far apart especially in the case of very large networks in which case common neighbors may not exist. In such a case, we propose the *k-hop matching index* parameter which is basically the ratio of common neighbors of two vertices in k-hop neighborhood to all of their neighbors in such a neighborhood. We can evaluate this parameter sequentially, and also in a distributed setting. The following distributed algorithm steps for a node I of a distributed system can be employed for this purpose. The algorithm can be implemented in k synchronous rounds under the control of a supervisor using SSI model.

1. $degs[1] \leftarrow \deg(I)$
2. **for** $I = 1$ to k
3.   **send** $degs[k]$ to $N(I)$
4.   **receive** $degs[k + 1]$ from $N(I)$
5. **end for**
6. $comm \leftarrow$ common neighbors in $degs[1..k]$
7. $all \leftarrow$ neighbors in $degs[1..k]$
8. $m_i \leftarrow \frac{|comm|}{|all|}$

13.3   Network Models

The network models can be classified as random networks, small-world networks, scale-free networks, and hierarchical networks.

13.3.1 Random Networks

A random network model assumes that the edges between nodes are inserted randomly. In the basic random graph model, $G(n, p)$, proposed by Erdos–Renyi, there are n vertices and each edge is successively added between two vertices with probability p independently [7, 8]. In order to generate a random network based on this $G(n, p)$ model, the following steps are applied:

1. There are n isolated nodes initially.
2. Select a node pair u and v from these and generate a random number $0 \le x \le 1$. If $x \ge p$ then connect u and v to form edge (u, v), otherwise leave them unconnected.
3. Repeat Step 2 for each of the $n(-1)/2$ vertex pairs.

We will have a different random network with the same values of n and p for each generation. The degree distribution in random networks is binomial centered around the average degree and these networks also exhibit small average clustering coefficient with low diameter with respect to the number of nodes.

13.3.2 Small-World Networks

A *small-world network* has a small diameter compared to its size which means reaching from any node to any other in such a network can be performed by few number of hops. This property is observed in many real large networks such as social networks and biological networks. This is a useful property in large networks as fast communication between any two nodes is possible. The small-world property is characterized by a low value of *average path length l* defined as the mean of distances between all pairs of n nodes in a graph $G = (V, E)$ as below:

$$l = \frac{1}{n(n-1)} \sum_{u,v \in V, u \ne v} d(u, v). \tag{13.7}$$

In a small-world network, the value of l should be bounded by $\log n$. This property in complex networks was commonly associated with a high average clustering coefficient indicating the presence of clusters as shown by Watts and Strogatz [4, 18].

13.3.3 Scale-Free Networks

An interesting property observed in large graphs representing real-life networks is the degree distribution displayed by $P(k) \approx Ak^{-\gamma}$ with $2 < \gamma < 3$ instead of a binomial distribution of a random network. In practical terms, this means there are few nodes called *hubs* with very high degrees with most of the nodes having small degrees in these networks. This power-law degree distribution of vertices can be generated following the rules of *growth* and *preferential attachment* as in the following algorithm:

1. *Growth*: A new vertex v is generated and connected to one of the existing vertices with the following rule.

2. *Preferential attachment*: Vertex v is attached to one of the existing vertices, say u, with a probability related to the degree of vertex u.

Starting with a small number of vertices, a scale-free graph is obtained when these two steps are repeated sufficient number of times as proven by Barabasi and Albert [1].

Let us define the clustering function $C(k)$ of a graph G as the average clustering coefficient of the nodes with degree k. In various biological networks, $C(k) \approx k^{-1}$ showing the clustering coefficient parameter is higher in lower degree nodes than the higher degree nodes. This basically means higher degree nodes have neighbors that are less connected than neighbors of lower degree nodes. This new model was named *hierarchical networks* in which low-degree nodes are densely clustered and these regions are connected by high-degree nodes [6, 15].

13.4   Centrality

Centrality of a vertex or an edge in a graph denotes a value showing its importance in the graph. Importance may have different meanings; for example, the centrality of an edge may reflect the percentage of shortest paths through it. We review the main centrality parameters for graphs in the next sections.

13.4.1 Degree Centrality

The degree of a vertex is denoted as its *degree centrality*. A vertex with a higher value than the average degree of a graph gives some insight to its importance in the network. In general, it is difficult to predict the overall structure of a graph from the value of its average degree. However, plot of its degree sequence does provide information about the general organization of the graph. Given the adjacency matrix A of a graph $G = (V, E)$, we can form the centrality vector DC which has the centrality value $DC(I)$ for vertex I as follows:

$$DC = A \times [1]. \tag{13.8}$$

In a distributed setting, we can have a root process p_i gathering all of the degree values of nodes over a previously constructed spanning tree T using the convergecast operation with the SSI model. This process can then compute the average graph degree $deg(G)$ and the centrality vector DC. Upon reception of a *start* message from the root over the tree T, the role of nodes is to simply convergecast their degrees over the edges of T to the root.

Fig. 13.4 Closeness
centrality values for vertices
a, b, c, d, and *e* in this graph
are 1/10, 1/9, 1/16, 1/13, and
1/13 respectively

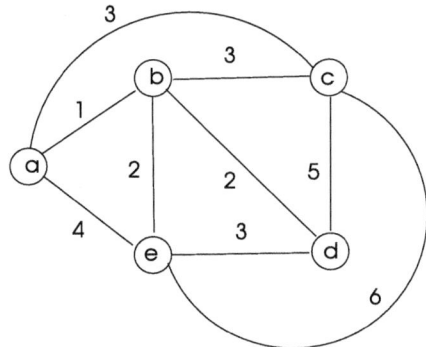

13.4.2   Closeness Centrality

In the *closeness centrality* attribute, we evaluate distance of a vertex v to all other
vertices in the graph, assuming a vertex that is closer to all other vertices in the
graph is more important than a vertex which is not so close to all others. The
closeness value of a vertex is defined as follows:

$$CC(v) = \frac{1}{\sum_{u \in V}} d(u, v). \tag{13.9}$$

We sum the distance of every vertex to vertex v and take the reciprocal of this
value. A possible way to evaluate the closeness centralities of all vertices in a
graph is then to compute APSP routes in a graph using a modified version of
Dijkstra's algorithm we saw in Chap. 7 for this purpose. We need to add the calculation of the sums of distances while assigning a vertex to the set of decided
routes (see Exercise 4). In a parallel or distributed setting, we can use the algorithms described in Chap. 7 with the modification described (see Exercise 5). If the
graph is unweighted, the BFS algorithm can be used with a similar modification.
Figure 13.4 depicts a sample undirected weighted graph from which the closeness
centrality values can be computed using the shortest paths.

13.4.3   Betweenness Centrality

In another attempt to assess the importance of a vertex or an edge in a graph, we
can evaluate the usage of a vertex or an edge when communications between vertices of a graph are needed. The vertex centrality value of a vertex is the percentage
of shortest paths between all pairs of vertices that run through that vertex. Similarly, the edge betweenness value of an edge refers to the percentage of shortest
paths that go through that edge. We will describe formal definitions and algorithms
to find these parameters in graphs in the next sections.

13.4.3.1 Vertex Betweenness Centrality

The *vertex betweenness* of a vertex v in a graph G is the ratio of the shortest paths that run through v to the total number of shortest paths in G as follows:

$$BC(v) = \sum_{s \neq t \neq v} \frac{\sigma_{st}(v)}{\sigma_{st}}, \qquad (13.10)$$

with $\sigma_{st}(v)$ as the total number of shortest paths between vertices s and t that run through vertex v, and σ_{st} as the total number of shortest paths between vertices s and t. For an unweighted graph, we can simply run the BFS algorithm for every vertex, count the number of shortest path through each vertex, and divide this number by the total number of shortest paths in the graph. For a weighted graph, we need to run a APSP algorithm. The vertex betweenness values of a sample graph are depicted in Fig. 13.5.

We can count the number of shortest paths through each vertex excluding the starting and ending vertices to find 3, 0, 7, 0, 0, and 9 for vertices a, b, c, d, and e,

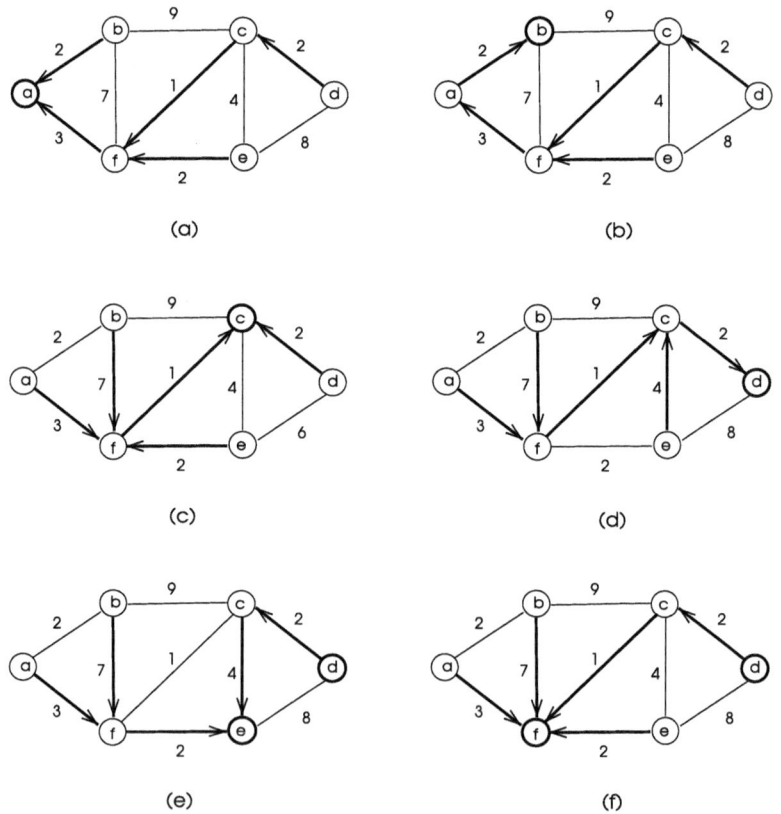

Fig. 13.5 Vertex betweenness value calculations of a sample graph

respectively. There are a total of 15 shortest paths between each vertex pairs and the vertex betweenness values for these vertices are then 0.2, 0, 0.47, 0, 0, and 0.6, respectively, in lexicographical order. We can conclude vertex f is the most influential vertex since the largest number of paths pass through it which can in fact be detected visually.

13.4.3.2 Edge Betweenness Centrality

The *edge betweenness* of an edge e in a graph G is the ratio of the shortest paths that run through e to the total number of shortest paths in G as follows:

$$BC(e) = \sum_{s \neq t \neq v} \frac{\sigma_{st(e)}}{\sigma_{st}}. \qquad (13.11)$$

We will now describe an algorithm due to Newman and Girvan [11] to compute edge betweenness centrality values of edges in a graph. This algorithm has two parts: a vertex weight assignment and edge weight assignment phases. Vertices are labeled with the number of shortest paths that go through them first and then, this information is used to find edge weights to yield edge centrality values later. The first part of the algorithm is depicted in Algorithm 13.1 which works for a source node s that has a distance of 0 and a weight of 1 initially [6]. The vertex weight assignment procedure then iteratively assigns distance values to the vertices as in a BFS algorithm with additional vertex weight values. However, if a vertex u is visited before and has a weight one more than the weight of its ancestor v, its weight is made equal to the sum of its previous weight and the weight of its ancestor. This is needed since an alternative shortest path to the source vertex s is discovered through the ancestor vertex v. We need to repeat this procedure for all vertices with each one as the source vertex in an iteration and the vertex weight of a vertex v is the sum of all of the values assigned to it at each run of the procedure.

Algorithm 13.1 *Label_Vertex*

1: **Input** : $G(V, E)$
2: **Output** : $\forall v \in V$: w_v weight of vertex v
3: $d_s \leftarrow 0; w_s \leftarrow 1$ ▷ initialize source vertex s
4: **for all** $v \in N(s)$ **do**   ▷ initialize neighbors of vertex s
5:     $d_v \leftarrow d_s + 1, w_v \leftarrow 1$
6: **end for**
7: **repeat**
8:     **for all** $u \in N(v)$ **do**
9:        **if** u is not assigned a distance **then**
10:           $d_u \leftarrow d_v + 1; w_u \leftarrow w_v$
11:        **else if** $d_u = d_v + 1$ **then**   ▷ test multiple shortest paths
12:           $w_u \leftarrow w_v + w_u$
13:        **end if**
14:     **end for**
15: **until** all vertices have assigned distances

Fig. 13.6 Edge betweenness values of the edges of a sample graph for source vertex h. Vertex distance, weight values obtained from the first procedure are shown next to vertices with edge betweenness values obtained by the second procedure displayed on the edges

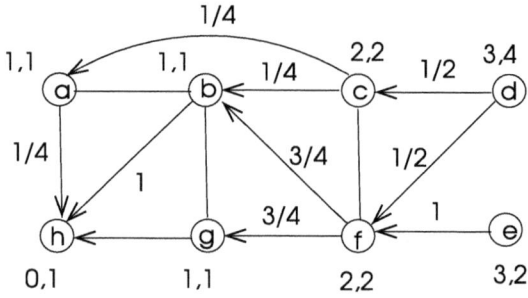

The second procedure of the algorithm uses the vertex weights assigned to denote edge weights. It starts by assigning weights to edges that end up in leaf vertices as the ratio of the vertex weight of its ancestor to the weight of itself. We are basically denoting weights that show the percentage of shortest paths to the leaf edges. Then, as we move upward in the tree toward the source vertex s, each edge (u, v) is assigned a weight that is the sum of all edge weights below (u, v) multiplied by the ratio of the weight of vertex u to the weight of vertex v. Again the sum of all shortest paths through edge (u, v) is scaled to give the percentage of shortest paths through that edge as shown in Algorithm 13.2. This process has to be repeated for all vertices considering each of them as the source and the edge betweenness value of an edge is determined as the sum of all values found for that edge. Total time needed is $O(nm)$ considering n vertices. The edge betweenness values of the edges of a sample graph for a single source vertex is depicted in Fig. 13.6.

Algorithm 13.2 *Edge_Betweenness*

1: **Input** : $G(V, E)$
2: **Output** : $\forall v \in V : w_v$ weight of vertex v
3: **for all** $v \in V$ which is a leaf vertex **do**
4:     **for all** $u \in N(v)$ **do** $w_{uv} \leftarrow w_u/w_v$ ▷ initialize leaf edges
5:     **end for**
6: **end for**
7: **repeat**
8:     move upwards to the source vertex such that u is closer to s:
9:     $w_{uv} \leftarrow (1 + \text{sum of the edge weights below } v) \times w_u/w_v$
10: **until** vertex s is reached

13.4.4 Eigenvalue Centrality

The general idea of *eigenvalue centrality* is to attribute importance to a node if it has important neighbors as frequently happens in a social network. The importance scores of the neighbors of a node I affect its score as follows [6]:

$$x_i = \frac{1}{\lambda} \sum_{j \in N(i)} x_j = \frac{1}{\lambda} \sum_{j \in V} a_{ij} x_j, \tag{13.12}$$

where $N(i)$ is the set of neighbors of node i and a_{ij} is the ij-th entry of the adjacency matrix A of the graph $G = (V, E)$, and λ is a constant. We can rewrite this equation in matrix notation as follows:

$$Ax = \lambda x, \tag{13.13}$$

which turns out to be the eigenvalue equation of the matrix A. There will be n eigenvalues and corresponding eigenvectors associated with the adjacency matrix A. The eigenvalue centrality values of vertices are determined by the eigenvector corresponding to the largest eigenvalue of A as shown by Perron–Probenius theorem [14]. We can now state the steps of an algorithm to find the eigenvalue centralities of the vertices in a graph $G = (V, E)$ as follows:

1. Construct the adjacency matrix A of G.
2. Compute eigenvalues $\lambda_1, \ldots, \lambda_n$ of A using the equation $\det(A - \lambda I) = 0$ and select the largest eigenvalue λ_m from these eigenvalues.
3. Compute the eigenvector X_m associated with λ_m.

Every vertex $v_i \in V$ has an eigenvalue centrality value $x_i \in X_m$. Eigenvalue centrality can be used for page ranking on the Web and also to investigate gene–disease associations [13].

13.5   Dense Subgraphs

We will now look at ways of finding dense subgraphs of a given graph. These subgraphs, often termed *clusters*, indicate a region of dense activity in the network represented by the graph. In the extreme case, a clique is a subgraph that is fully connected. However, clique-like or cliquish structures are more commonly encountered in practice than cliques. We will review methods to find cliques and these structures in this section.

13.5.1   Cliques

A subgraph in which every vertex is a neighbor to all others in this subgraph is called a *clique*. A clique of a graph G is a densely connected region in G which may indicate a special kind of activity in the network that is represented by G, for example, a clique of friends in a social network. Therefore, detecting cliques is a commonly required task in such graphs representing real-life phenomena. In

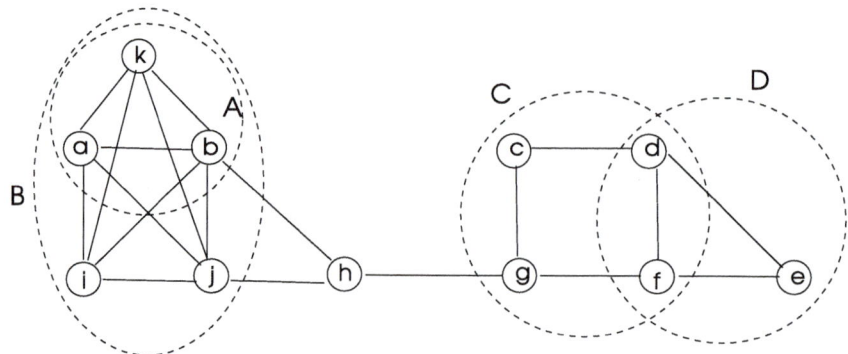

Fig. 13.7 Cliques of a sample graph; $A = \{a, b, k\}$ is a clique but not maximal as it included in the larger clique $B = \{a, b, k, i, j\}$ which is also the maximum clique of the graph. $C = \{c, d, f, g\}$ and $D = \{d, e, f\}$ are maximal cliques as they are not a subset of larger cliques

real-life networks, one often finds clique-like structures than full cliques in graphs representing real-life networks such as protein interaction networks of the cell and social networks and hence we present algorithms to discover such structures in graphs.

Definition 13.4 *(Clique)* A *clique* of a graph $G = (V, E)$ is a subset C of its vertices such that every vertex in C is adjacent to all other vertices in C. In other words, a clique is an induced complete subgraph of G.

A *maximal clique* of a graph G is the clique that is not a subset of a larger clique. The *maximum clique* of a graph G is the clique of G with the largest order among all cliques of G as shown in Fig. 13.7. The order of the maximum clique of G is denoted by $\omega(G)$ and is called the *clique number* of G. Finding the maximum clique in a graph is called the *maximum clique problem* and is NP-hard [9], and therefore various heuristics are commonly used to compute an approximation to this parameter. The decision version of this problem, *k-clique problem*, is determining whether a graph has a clique with at least k vertices. Finding all maximal cliques of a graph is also NP-hard as the number of such cliques grows exponentially with the order of the graphs.

Detecting cliques in a graph has various implications as these represent intense interaction among the entities represented by the nodes of the graph. A clique in a network would represent an intimate group detecting of which may help to understand the general behavior in such a network. However, we may be interested to find clique-like structures rather than cliques in graphs since cliques occur less frequently and also finding them is not a trivial task. Clique-like structures we may search are as follows [6]:

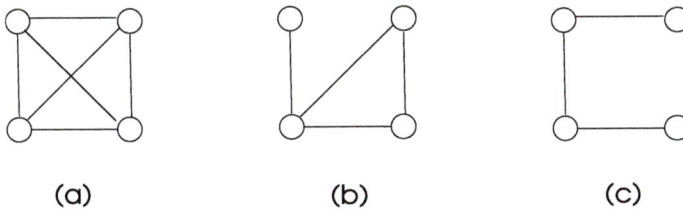

Fig. 13.8 *k*-clique examples, **a** 1-clique **b** 2-clique **c** 3-clique

Definition 13.5 *(k-clique)* A *k*-clique of a graph G is exhibited by a subgraph G_c, where the shortest path between any two vertices in G_c is at most k. Paths may consist of vertices and edges external to G_c.

Some *k*-clique examples are shown in Fig. 13.8.

Definition 13.6 *(k-club)* A *k*-club of a graph G is a subgraph G_c in which $diam(G_c) \leq k$.

Every *k*-club of a graph is also its *k*-clique. However, not every *k*-clique of a graph is its *k*-club since a *k*-clique may contain vertices outside of G_c. *The maximum k-club problem is to find the largest k-club of a graph and is NP-hard.*

Definition 13.7 *(k-plex)* In a *k*-plex subgraph $G_c \subseteq G$, each vertex is connected to at least $n - k$ other vertices in G.

Definition 13.8 *(k-core)* A *k*-core subgraph G_c of G consists of vertices that have at least a degree of k. A clique is a $(k\text{-}1)$ core.

We can find some clique-like structures such as *k*-cores in polynomial time. We first describe a recursive algorithm that has exponential time complexity which has been used as basis for various parallel algorithms as we will see. We then present an algorithm to find *k*-cores of a graph which works in polynomial time.

13.5.1.1 Bron and Kerbosch Algorithm
Bron and Kerbosch provided a simple backtracking procedure that finds all maximal cliques of an undirected graph [5]. The algorithm uses three sets of vertices, R is the set of vertices to be included in the clique, X is the set to be excluded from the clique, and P is the set to be decided. The set R is initially empty and is expanded using vertices in P and without using vertices in X. The following steps comprise the algorithm:

1. Select a candidate vertex $v \in P$.
2. Add v to R.

3. Create new sets P and X from the old sets by removing all vertices not connected to R.
4. Call the extension operator on the newly formed sets.
5. Upon return, remove v from R, add it to X.

The pseudocode for this algorithm is shown in Algorithm 13.3 [6]. We need set P to be empty so R cannot be extended and X to be empty to ensure that the clique is not included in another clique to have a maximal clique in R. This condition is checked at each recursive call to the procedure first. Otherwise, a recursive call is made that adds a vertex in P to R and its neighbors in P and X. Experimentally, the time complexity was found to be $O(4^n)$ and a second version using pivoting resulted in time complexity of $O(3.14^n)$ [5]. Various parallel implementations of this algorithm exist; using MPI in [10], using thread pools in Java in [3], and on a Cray XT supercomputer in [17].

Algorithm 13.3 *Bron Kerbosch Algorithm*

1: **procedure** *BronKerbosch(R, P, X)*
2:     $P \leftarrow V$ includes all of the vertices and $R, X \leftarrow \emptyset$
3:     **if** $P = \emptyset \wedge X = \emptyset$ **then**
4:         **return** R as a maximal clique
5:     **else**
6:         **for all** $v \in P$ **do**
7:             *BronKerbosch*$(R \cup \{v\}, P \cap N(v), X \cap N(v))$
8:             $P \leftarrow P \setminus \{v\}$
9:             $X \leftarrow P \cup \{v\}$
10:        **end for**
11:    **end if**
12: **end procedure**

13.5.2 *k*-cores

Let us consider a graph $G = (V, E)$; a subgraph $H = (V', E')$ of G induced by the vertex set V' is a k-core of G if and only if $\forall v \in V'$, $deg(v) \geq k$ and H is the maximum graph with this property [6]. *Main core* of a graph G is a core of G that has the maximum order and the *coreness value* or the *core value* of a vertex v is the highest order of a core that includes the vertex v [2]. Cores of a graph are naturally nested and a subgraph core with a larger core value is nested within cores with smaller values as shown in Fig. 13.9.

Finding cores of a graph is useful in detecting dense regions of a graph as each vertex in a k-core has at least a degree of k. Hence, rather than trying to find cliques or clique-like structures of a graph to discover dense regions in that graph, we can find cores of a graph which can be accomplished in polynomial time due to an algorithm described in the next section.

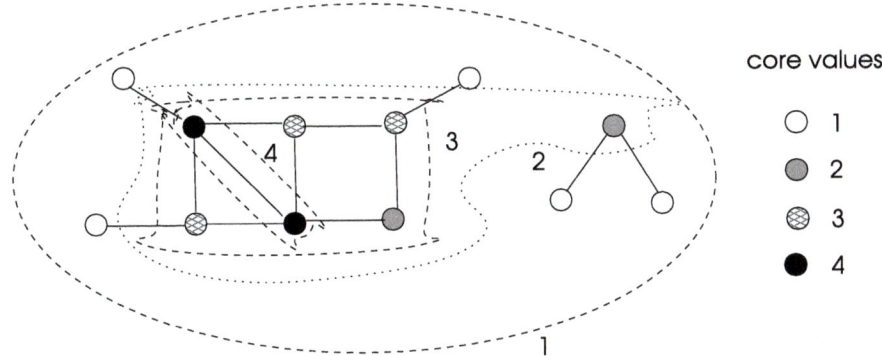

Fig. 13.9 Cores of sample graph with two components. 1, 2, 3, and 4 cores are encircled and the core values of vertices are shown with filled colors

13.5.2.1 Batagelj and Zaversnik Algorithm

In the algorithm proposed by Batagelj and Zaversnik [2], the core values for each vertex is determined in linear time. The general idea of this algorithm is to remove all vertices having degree less than k from the graph under consideration to obtain the k-core of the graph. The algorithm starts by sorting all vertices in the graph with respect to their degrees and the vertices are placed in a queue in increasing degrees. The vertex v from the front of the queue is then removed, is labeled with the core value that equals its degree, and the degrees all of the neighbors of v that have degree more than v are decremented by one. The degree of any neighbor u with the same degree of v is not decremented as this would result in u to be included in a class of vertices with a lower k value than itself. The pseudocode of this algorithm is depicted in Algorithm 13.4 [2, 6]. The authors showed that the time complexity of this algorithm in a general graph is $O(max(m, n))$, and this becomes $O(m)$ in a connected network as $m \geq -1$ in such a graph.

Algorithm 13.4 *Batagelj_Alg*

1: **Input** : $G = (V, E)$
2: **Output** : core values of vertices
3: $Q \leftarrow$ sorted vertices of G in increasing degrees
4: **while** $Q \neq \emptyset$ **do**
5:     $v \leftarrow$ front of Q
6:     $core(v) \leftarrow deg(v)$
7:     **for all** $u \in N(v)$ **do**
8:         **if** $deg(u) > deg(v)$ **then**
9:             $deg(u) \leftarrow deg(u) - 1$
10:        **end if**
11:    **end for**
12:    **update** Q
13: **end while**

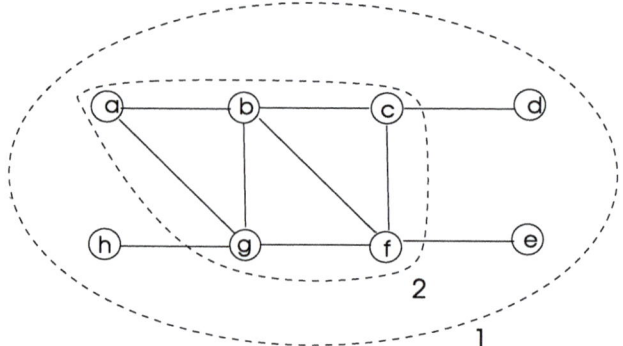

Fig. 13.10 Sample graph with two cores

Table 13.1 BZ algorithm execution

Iterations	Queue sorted in ascending degree				Coreness values	
	1	2	3	4	$k=1$	$k=2$
1	h, d, e	a	c	b, f, g	$\{h\}$	$\{\emptyset\}$
2	d, e	a	c, g	b, f	$\{h, d\}$	$\{\emptyset\}$
3	e	a, c	g	b, f	$\{h, d, e\}$	$\{\emptyset\}$
4	–	c, g	g, f	b	$\{h, d, e\}$	$\{a\}$
5	–	g, b, f	b, f	–	$\{h, d, e\}$	$\{a, c\}$
6	–	b, f	–	–	$\{h, d, e\}$	$\{a, c, g\}$
7	–	f	–	–	$\{h, d, e\}$	$\{b, a, c, g\}$
8	–	–	–	–	$\{h, d, e\}$	$\{b, f, a, c, g\}$

Let us see the step-by-step operation of this algorithm in the sample graph of Fig. 13.10. The contents of the sorted queue along with the removed and labeled vertices are shown in Table 13.1 for running the algorithm in this simple graph.

13.5.3 Clustering

Clustering is the process of grouping of similar objects based on some metric. When the nodes of a graph are used to represent objects and edges depict their relations, this process is equivalent to finding dense subgraphs of the graph representing the network. In this case, the aim of a clustering algorithm is to divide the graph into subgraphs such that density within a subgraph is maximized with minimum number of edges among clusters. When edges are weighted, we need to maximize the total number of edge weights within a cluster and minimize the total weight of edges among the clusters.

We can have a vertex belonging to more than one cluster in *graph clustering* and a vertex may belong to only one cluster in *graph partitioning*. We need to assess the quality of the clusters obtained after using a clustering method. A convenient way to achieve this is to compare the densities of clusters and the average density of the graph. The density $\rho(G)$ of an unweighted, undirected simple graph G is the ratio of the size of its edges to the size of maximum possible edges in G as $\rho(G) = \frac{2m}{n(-1)}$.

The edges inside a cluster are called *internal* edges of the cluster and the edges connecting the cluster vertices to the other vertices of the graph are called *external edges*. We can examine whether a vertex v is appropriately placed in a cluster by examining the ratio of the number of internal edges incident to v to the number of external edges on it. We can now define *intracluster density* of a cluster C_i as the ratio of all internal edges in C_i to all possible number of edges in C_i as follows [16]:

$$\deg_{int}(C_i) = \frac{2\sum_{v \in C_i} \deg_{int}(v)}{|C_i||C_i - 1|}. \tag{13.14}$$

The intracluster density of the whole graph as the average of all intracluster densities as follows [6]:

$$\deg_{int}(G) = \frac{1}{k} \sum_{i=1}^{k} \deg_{int}(C_i), \tag{13.15}$$

where k is the number of clusters obtained. Clearly, we would need $\deg_{int}(G)$ as high as possible when compared with graph density for proper clustering. A sample graph divided into three clusters is depicted in Fig. 13.11 with calculated intracluster densities.

We can now define the *intercluster density* $\deg_{ext}(G)$ as the ratio of the size of the intercluster edges to the maximum allowed number of edges between the clusters as shown below [16] and we need this parameter to be significantly lower than the graph density for a good quality clustering.

$$\deg_{ext}(G) = \frac{2 \times \text{sum of inter cluster edges}}{n(n-1) - \sum_{i=1}^{k}(|C_i||C_i - 1|)} \tag{13.16}$$

The intercluster density of the sample graph in Fig. 13.11 is 0.092 which is significantly lower than the graph density 0.28 and hence we can consider the clustering in this graph is favorable. Similar cluster parameters for edge-weighted graphs can be defined. Let us first consider the density of an edge-weighted graph which is the ratio of the sum of edge weights to the maximum possible number of edges as below:

$$\rho(G(V, E, w)) = \frac{2\sum_{(u,v) \in E} w_{(u,v)}}{n(-1)}. \tag{13.17}$$

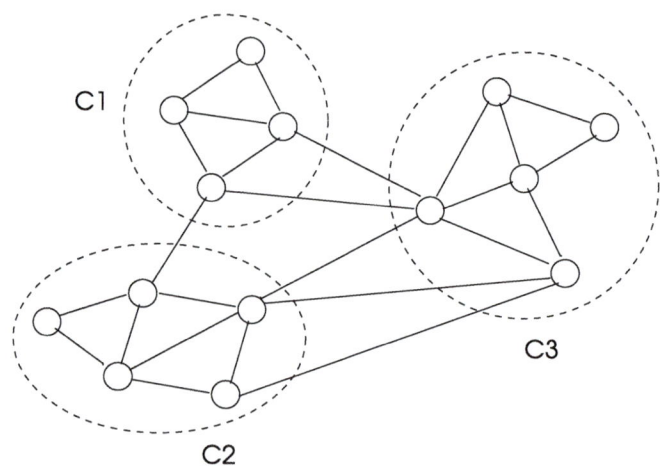

Fig. 13.11 Sample graph divided into three clusters C_1, C_2, and C_3. The intracluster densities are $\delta_{int}(C_1) = 0.83$, $\delta_{int}(C_2) = 0.7$, and $\delta_{int}(C_3) = 0.6$ with the average graph intracluster density of 0.71. Graph density is 0.28

The intracluster densities now are formed as the ratio of the sum of edge weights in a cluster to the maximum possible number of edges in that cluster and the graph intracluster density is the average value of all the clusters contained in the graph. In both unweighted and weighted graphs, a good clustering should result in average graph intracluster values which are significantly higher than the graph densities. The intercluster density in the case of a weighted graph is the ratio of the sum of weights of all edges between each pair of clusters to the maximum possible number of edges between clusters.

13.6   Chapter Notes

Analysis of large graphs as a whole is difficult due to their huge sizes. As one approach to overcome this problem to some extent, local properties around vertices can be assessed and global properties may then be approximated using these local properties. For example, the degree of a vertex and its clustering coefficient are local properties; degree distribution and the average clustering coefficient are global properties of a graph that give some insight on its overall structure. We reviewed these parameters along with the matching index of two vertices and the density of a graph in the first part of this chapter.

Real-world networks such as technological networks, biological networks, and social networks are termed complex networks and graphs are commonly used to represent them. We then described main models of these networks which are small-world and scale-free networks along with random networks. Distance between any two vertices is small compared to network size in small-world networks which also

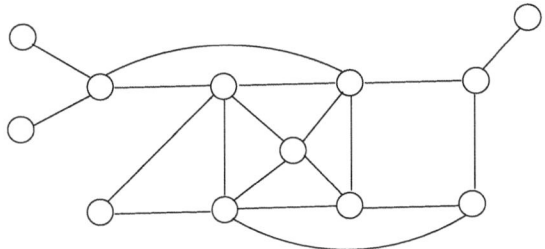

Fig. 13.12 Sample graph for Exercise 1

have high average clustering coefficients than expected. Scale-free networks have few high-degree nodes and the rest of the nodes have only few degrees with high clustering coefficients. Complex networks commonly exhibit these two properties but in the case of biological networks, other models such as hierarchical networks may be needed.

Centrality is a key concept in the analysis of large graphs and again, we can estimate global properties of a graph from the various forms of this local parameter. Degree centrality of a vertex is simply its degree and closeness centrality shows the importance of a vertex based on its distance to all other vertices in the graph. Betweenness centrality of a vertex or an edge manifests the percentages of shortest paths that run through them and eigenvalue centrality is based on the spectra of the graph. Edge betweenness and eigenvalue centralities are used in practice more to analyze and also to discover clusters in graphs than other centrality measures.

Detection of dense subgraphs in large graphs is needed as these regions may indicate heavy activity there. We reviewed algorithms discovery of cliques and clique-like structures and described few methods of clustering in large graphs.

Exercises

1. Plot the degree distribution of the graph depicted in Fig. 13.12.
2. Work out the graph density, intracluster densities for three clusters, and the intercluster density and the average cluster density for the graph in Fig. 13.13.
3. Find the clustering coefficient for each vertex in the graph shown in Fig. 13.14 and also work out the average clustering coefficient.
4. The closeness centrality values of the vertices of a graph are to be computed. Modify Dijkstra's APSP algorithm to find the closeness centrality values.
5. Modify distributed Bellman–Ford Algorithm of Chap. 7 to find closeness centralities in a computer network.
6. Compute the vertex betweenness values for each vertex in the example graph depicted in Fig. 13.15.

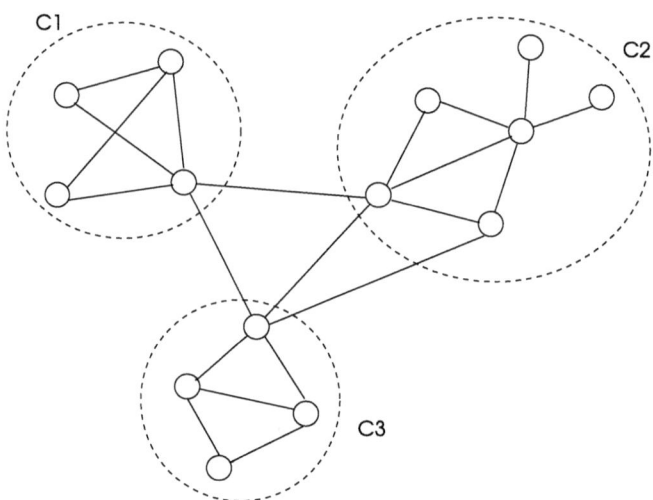

Fig. 13.13 Sample graph for Exercise 2

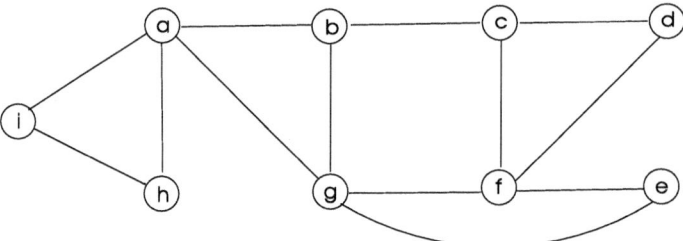

Fig. 13.14 Sample graph for Exercise 3

Fig. 13.15 Sample graph for
Exercise 6

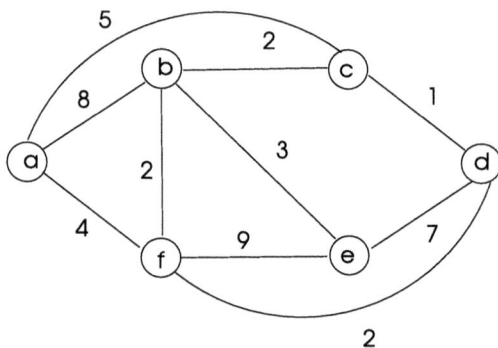

References

1. Barabasi A-L, Albert R (1999) Emergence of scaling in random networks. Science 286:509–512
2. Batagelj V, Zaversnik M (2003) An O(m) algorithm for cores decomposition of networks. CoRR (Computing research repository), arXiv:0310049
3. Blaar H, Karnstedt M, Lange T, Winter R (2005) Possibilities to solve the clique problem by thread parallelism using task pools. In: Proceedings of the 19th IEEE international parallel and distributed processing symposium (IPDPS05)Workshop 5 Volume 06 in Germany
4. Boccaletti S, Latorab V, Morenod Y, Chavez M, Hwang D-U (2006) Complex networks: structure and dynamics. Phys Rep 424:175–308
5. Bron C, Kerbosch J (1973) Algorithm 457: finding all cliques of an undirected graph. Commun ACM 16(9):575–577
6. Erciyes K (2015) Distributed and sequential algorithms for bioinformatics. Springer, Berlin (chapters 10–11)
7. Erdos P, Renyi A (1959) On random graphs. Publicationes Mathematicae 6:290–297
8. Erdos P, Renyi A (1960) On the evolution of random graphs. Publ Math Inst Hung Acad Sci 5:17–61
9. Garey MR, Johnson DS (1979) Computers and intractability: a guide to the theory of NP-completeness. W H Freeman and company, New York
10. Jaber K, Rashid NA, Abdullah R (2009) The parallel maximal cliques algorithm for protein sequence clustering. Am J Appl Sci 6:13681372
13. Newman MEJ, Girvan M (2004) Finding and evaluating community structure in networks. Phys Rev E 69:026113
14. Newman MEJ, Strogatz SH, Watts DJ (2002) Random graph models of social networks. Proc Natl Acad Sci USA 99:25662572
15. Özgr A, Vu T, Erkan G, Radev DR (2008) Identifying gene-disease associations using centrality on a literature mined geneinteraction network. Bioinformatics 24(13):277–285
16. Perron O (1907) Mathematische Annalen. Zur Theorie der Matrices 64(2):248–263
17. Ravasz E, Somera AL, Mongru DA, Oltvai ZN, Barabsi AL (2002) Hierarchical organization of modularity in metabolic networks. Science 297:15511555
18. Schaeffer SE (2007) Graph clustering. Comput Sci Rev 1:2764
19. Schmidt MC, Samatova NF, Thomas K, Park B-H (2009) A scalable, parallel algorithm for maximal clique enumeration. J Parallel Distrib Comput 69:417428
21. Watts DJ, Strogatz SH (1998) Collective dynamics of small-world networks. Nature 393:440442

Complex Networks

14

14.1 Introduction

Complex networks consist of tens of thousands of nodes and hundreds of thousands of edges connecting these nodes. The graphs used to model these networks are large and we will see special methods are commonly needed for the analysis of these networks. In theory, we can use the algorithms we have reviewed in Part II for some of the problems encountered in complex networks but even the linear time algorithms may require substantial computation time and hence the heuristic algorithms with lower complexities are usually preferred. Yet many problems in these networks are NP-hard making the use of heuristics as the only choice.

We will review the main real-life complex networks which are biological networks, social networks, technological networks, and information networks with brief description of the algorithms needed to solve some problems in these networks in this chapter. Clustering or community detection is a problem in all of these networks and we will describe fundamental sequential, parallel, and distributed algorithms for this purpose with emphasis on distributed clustering in computer networks. Efficient search algorithms are needed in the Web and we will review two such algorithms in the last part of this chapter.

14.2 Biological Networks

Biological networks are the networks of organisms with nodes representing biological entities and edges showing the interactions among them. The cell is the basic biological unit of all organisms. Biological networks can be classified as networks within the cell and networks outside the cell at a more macroscopic level. Cells are mainly of two types: *eukaryotic cells* which have nuclei carrying the genetic information in chromosomes and *prokaryotic cells* which do not have nuclei. A *gene* is the basic unit of hereditary and consists of a string of *nucleotides* which are small molecules that make deoxyribonucleic acid (DNA) in a double helix structure.

Genes are decoded to make *amino acids* which are chained to make *proteins* which are large molecules outside the nucleus of the cell carrying all vital functions related to life. This process is called the *central dogma of life*. Proteins are large molecules and their main functionality depends on their amino acid sequences, their 3D shape, and also the interaction with other proteins. These interactions can be represented by graph edges to obtain protein–protein-interaction (PPI) networks with proteins as the nodes of the network. Figure 14.1 displays the PPI network of *T. pallidum*. Other main networks in the cell are the *gene regulation networks* (GRNs) formed by interacting proteins and genes and *metabolic networks* represent biochemical reactions in the cell to generate metabolism [8]. Other biological networks outside the cell include brain networks, neural networks, phylogenetic networks, and the food web [8]. The main problems encountered in biological networks are clustering, network motif search and network alignment.

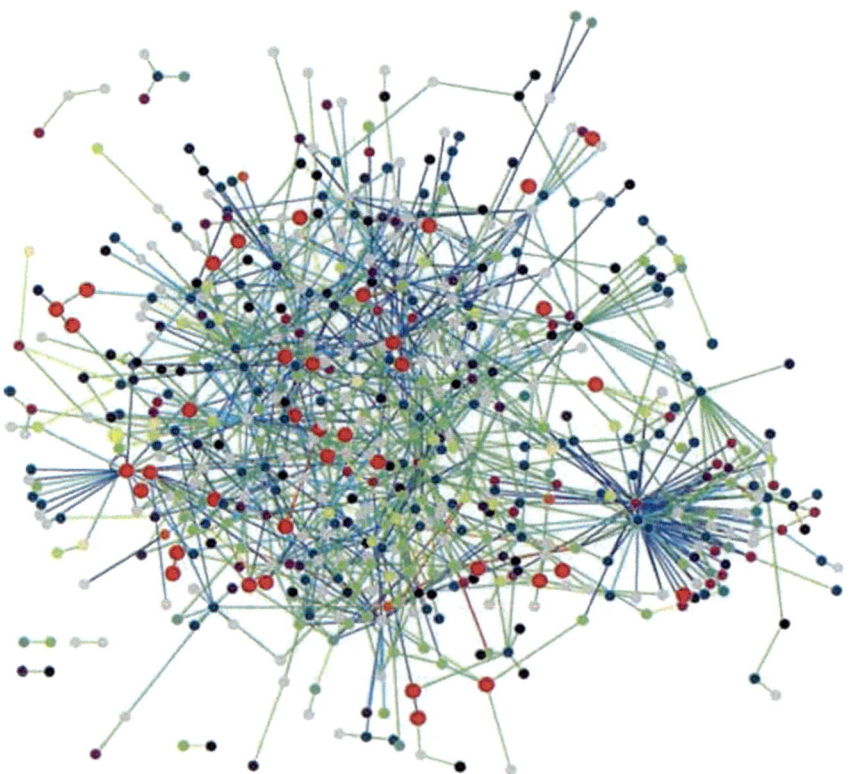

Fig. 14.1   The PPI network of T. pallidum taken from [22]

14.2.1 Clustering

Finding clusters in biological networks is needed as these dense regions of biological activity bears some significance, sometimes showing disease states of an organism. Clustering algorithms in biological networks can be broadly classified as *hierarchical clustering, density-based clustering, flow-based clustering*, and *spectral clustering* [8]. We have already reviewed density-based clustering as searching clique-like structures in Chap. 10, and we will review sample methods of clustering in the next sections.

14.2.1.1 MST-Based Clustering
Hierarchical clustering is a classical, simple and a popular method of clustering. An example of this method is the MST-based clustering algorithm which is based on the idea that two nodes that are farthest in the MST may be assigned to two different clusters. Since MST is acyclic, removing of an edge from MST divides it into two clusters. This process may be repeated until a certain cluster quality criterion is met. The algorithm is implemented using the following steps:

1. **compute** the MST T of the graph $G = (V, E)$
2. **repeat**
3. $e_{uv} \leftarrow max\{\forall (u, v) \in T, w_{uv}\}$
4. $G \leftarrow G - \{e_{uv}\}$
5.   **label** vertices of new clusters
6. **until** a quality criterion is met

Note that MST is computed only once and the heaviest edges are removed iteratively. Labeling of the newly formed clusters can be done simply by the BFS algorithm in linear time for the two clusters. We may remove a number of edges at each step from the MST that are more than a threshold distance apart. Computation of MST can be performed in parallel using any of the algorithms described in Chap. 7. Clustering through MST in parallel (CLUMP) is a proposed method of clustering for biological data [20] in which bipartite subgraphs are constructed in parallel.

14.2.1.2 Spectral Clustering
We saw the adjacency matrix representation of a graph G and its unnormalized Laplacian matrix is $L = D - A$ where A is the adjacency matrix and D is the degree matrix with degrees of vertices in the diagonal. The normalized Laplacian is $L = 1 - D^{1/2}AD^{-1/2}$. L in both forms is real and symmetric and its eigenvalues are therefore real. The second smallest eigenvalue of L bears information about its connectivity as we saw in Chap. 12. We can also use this value called the *Fiedler value* and its eigenvector called *Fiedler vector* [9] to divide the graph G into two clusters C_1 and C_2 as below [12]:

1. $G = (V, E)$ with $V = \{v_1, \ldots, v_n\}$
2. $C_1 \leftarrow \emptyset, C_2 \leftarrow \emptyset$
3. $L \leftarrow D - A$
4. **compute** Fiedler vector F of L
5. **for all** $F[i]$ of F **then**
6.   **if** $F[i] \leq \tau$
7. $C_1 \leftarrow C_1 \cup \{v_i\}$
8.   **else** $C_2 \leftarrow C_2 \cup \{v_i\}$
9.   **end if**
10. **end for**

The threshold value τ is commonly taken as 0, and a recursive implementation of this algorithm will yield k clusters. Other methods of clustering in biological networks include a flow-based approach called Markov Clustering Algorithm which uses the idea that random walks in a graph will end up in a cluster [6]. There are various other algorithms for clustering in biological networks most of which use heuristics.

14.2.2 Network Motif Search

A *network motif* is a recurring connected subgraph in a biological network. These are considered as building blocks of such networks and commonly assumed to have some basic function associated with them. Discovery of motifs aid to understand the structure of biological networks and also various networks representing organisms can be compared to seek whether they have common ancestors. Figure 14.2 displays some motif examples.

Motif discovery is one of the fundamental research areas in biological networks. It is a complicated process where heuristics are commonly used and consists of the following steps:

1. *Motif Search*: Either all instances of a motif are searched or *sampling* is used where only a sample subgraph is searched and the results are projected to the whole graph. For very large graphs, the latter is commonly employed.

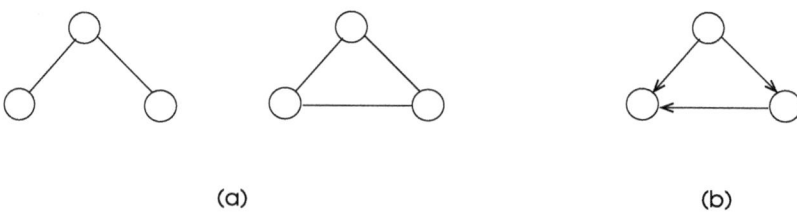

(a)   (b)

Fig. 14.2 Network motif examples of order 3

2. *Isomorphic Classes*: Different looking subgraphs may be isomorphic; therefore, they need to be grouped together for correct processing.
3. *Statistical Significance*: We need to determine the statistical significance of discovered subgraphs. This process is usually performed by generating a set of random graphs similar in structure to the target graph and performing the search in these graphs. The results obtained in two cases can then be compared to determine whether the found subgraphs are actual motifs.

14.2.3 Network Alignment

It is often required to compare two biological networks as a whole or partly to determine their relatedness which may help to understand the evolutionary process better. We can compare two biological networks in *pairwise alignment* or a number of biological networks in *multiple alignment* to search their similarities. In *global alignment*, whole networks are compared for similar species whereas similar subnetworks are searched in *local alignment* for diverse species.

In order to assess similarity of two networks, we can evaluate *node similarity* or *topological similarity* of the networks, in many cases however, both metrics are used with different weights. The former may reflect the internal structure of nodes such as amino acid sequences of proteins in PPI networks. Let us assume we have to graphs G_1 and G_2 representing two biological networks. A similarity matrix R can be formed that has entry r_{ij} with a weight depicting the similarity between node i from G_1 and node j from G_2. Based on the entries of R, a complete weighted bipartite graph can be constructed and the problem of network alignment is then reduced to maximal bipartite matching problem we have studied in Chap. 9.

14.3 Social Networks

Social networks consist of individuals or groups of people with relationships among them. A graph representing a social network has its nodes as persons or groups and edges depict the relationships. Social networks have small-world and scale -free properties like other complex networks.

14.3.1 Relationships and Equivalence

Some terminologies related to the relationships in a social network are as follows.

- *Triadic Closure*: Let us assume A and B are two friends in a social network. There is a good chance that a friend of A (or B) who does not know B (or A) will become friends with that person in future. This property is called *triadic closure* as there will be a triangle formed among the three persons in future by the composition of an edge between the two people who have not met before.

- *Homophily*: The homophily property observed in dynamic social networks is that the individuals or groups have tendency to arrange relationships with other individuals or groups like themselves. The similarity could be the age or the philosophy or something else in common.
- *Relations*: In a friendship social network, we can label edges between two persons as positive (+) meaning they like each other or negative (−) showing they dislike each other, assuming these relations are symmetric. In a small social network of triangle structure with three people A, B, and C, we can have four cases:

1. A, B, and C are mutual friends.
2. A, B, and C all dislike each other.
3. One of them is friends with two others but these two do not like each other.
4. Two of the three are friends with each other but neither of them is a friend with the third one.

The first case and the last case are balanced relations while the second and third are not. With three people who all dislike each other, there is a tendency for two to become friends against the third one and hence this case is unbalanced. Also, the third case implies two persons who do not want to be together but want to be together with a third person causing again an unbalanced situation. The last case is balanced as there is no conflict. In graph-theoretical terms, this means any triangle with one or three positive relations are balanced and triangles with zero or two positive edges are unbalanced. We can now find all triangles in a given social network with assigned relations, and if all of these are balanced, the whole social network is a *balanced* network; otherwise even if there exists one unbalanced triangle, the network is said to be *unbalanced*.
- *Structural balance*: A general method to determine whether a social network is balanced or not was proposed by Harary [13] in the Balance Theorem.

Theorem 14.1 *A complete social network is balanced when all pairs of its nodes have positive relations with each other or when its nodes can be partitioned into two sets V_1 and V_2 such that all nodes within these groups are friends with all other nodes in their groups and each node of one group has negative relations with all other nodes in the other group.*

A balanced social network according to this theorem is depicted in Fig. 14.3. It can be seen that any triangle which has two nodes in one group and one in the other has just one + label which means these triangles are balanced. The remaining triangles are all embedded in each group and have + labels on their edges, therefore they are also balanced resulting in a balanced network.

- *Equivalence*: Equivalence concept in a social network is related to the positions or roles of people or a person in a social network. If the positions of two

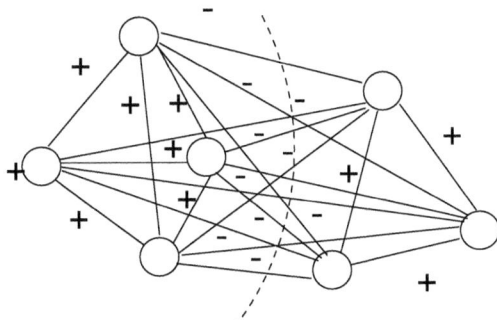

Fig. 14.3 A complete graph showing balanced social network separated into two sets of nodes by the dashed line. The three triangles in the two sets all have + labels on their edges and all other inter-group edges have − signs meaning this network is balanced according to structure balance theorem

equivalent persons in a social network are changed, the operation of the network should not change.

14.3.2   Community Detection

Detecting communities which are dense regions of activity in social networks have many implications; for example, we can analyze these clusters of persons or groups to understand their behavior. We will review two algorithms for this purpose which are implemented in social networks.

Edge Betweenness-Based Algorithm
The edge-betweenness value of an edge e in a graph G was the fraction of all, pairs-shortest-paths that pass through e. Intuitively, edges with high values have a greater probability of joining dense regions of the graph than edges with lower values. In the extreme case, a bridge removal of which disconnects a graph has a very high edge-betweenness value. Based on this observation, Girvan et al. proposed an algorithm to detect clusters in a complex network represented by a graph $G = (V, E)$ consisting of the following steps with a similar structure to MST-based clustering [11]:

1. **repeat**
2.   **compute** edge-betweenness value σ_{xy} for each edge (x, y) of graph G.
3. $e_{uv} \leftarrow$ the edge with maximum σ value
4. $G \leftarrow G - \{e_{uv}\}$
5. **until** a quality criterion is met

The most time consuming step for this algorithm is the calculation of the edge-betweenness values and hence it has low performance for graphs which have more than few thousand nodes. This method is also used for detecting clusters in biological networks.

14.3.2.1 Modularity-Based Algorithm

The quality of clusters formed by a clustering algorithm can be assessed by the *modularity* concept. Let us assume a graph G is divided into k clusters and e_{ii} is the fraction of edges in cluster i and a_i is the fraction of edges with at least one end in cluster i. The modularity Q of this graph can be determined as follows [19]:

$$Q = \sum_{i=1}^{k} (e_{ii} - a_i^2).$$

(14.1)

Using this equation, we are in fact evaluating the difference in probabilities of an edge being in module i and that a random edge would fall in module i and summing these values for all clusters. When the percentage of edges within clusters are much higher than the ones with one end in a cluster (inter-cluster edges), we expect a high value of Q, in fact the value of Q approaches unity when there are only few edges between clusters. A clustering algorithm based on the modularity concept can then be formed such that two clusters are combined to increase modularity as follows [19]:

1. Each node of the graph is considered a cluster initially
2. **repeat**
3.   **combine** the two clusters that will increase modularity most
4. **until** merging of clusters increase modularity.

This algorithm provides clusters as a dendogram which can be used to obtain the required number of clusters. The time complexity of this algorithm is $O((m+n)n)$ or $O(n^2)$ on sparse graphs [19].

14.4   Ad Hoc Wireless Networks

Computer networks consist of computational nodes such as personal computers, servers, routers and phones connected by an interconnection network. The communication medium between the nodes may be coaxial cable, fiber link, a wireless medium, or more frequently a combination of all of these. Data to be transferred between the nodes of a computer network is commonly divided into *packets*. These data units are delivered to destination using a *packet-switching network* where each packet is commonly routed independently to the destination and then the whole data consisting of a number of packets is reassembled in the destination before being delivered to the user or the application. This way, many users/applications may share the same communication medium. In contrast, a part of the communication medium is dedicated to the application during data transfer in *circuit switching*. Network packets may carry control information or data information. The control packets are used by the *communication protocols* to perform tasks such as routing and synchronization of communicating parties.

The wired medium may be twisted pair, coaxial cable, or fiber optic links in the order of performance and cost from low to high. Wireless communication links commonly employ microwave, satellite, and also optical communication technologies to transfer data. Network nodes consist of repeaters, bridges, switches, routers, and modulator/demodulators (MODEMs). Since an electrical signal is weakened and distorted through distance, a repeater is needed to remove noise from an electrical signal and transfer it by empowering it at regular distances. These devices operate on the physical layer of the ISO OSI 7-layer model [14]. Bridges and switches work at the data link layer of the OSI model to filter traffic and forward packets. Routers working at network layer are the key devices to determine the path the packet is to be transferred.

An ad hoc network is formed without any central administration with nodes serving as hosts and routers at the same time. Wired or wireless, a computer network may be conveniently modeled as a graph with the vertices of the graph representing the nodes and the edges showing the communication links between the nodes. In wireless networks, a link between two nodes exists if they are within communication range of each other. Note that we may need a digraph to represent such a network if the communication capabilities of two nodes are not the same. Two types of widely used wireless ad hoc networks are the mobile ad hoc networks (MANETs) where nodes move dynamically and wireless sensor networks (WSNs) with small sensing computational nodes. A WSN can also be mobile but mostly, these networks have fixed wireless topologies. The nodes of such ad hoc networks use multi-hop communication where data packets are relayed to other nodes toward destination.

A wireless ad hoc network does not need a fixed network infrastructure using devices such as routers and access points. Such a structure has many advantages as we do not have to install and maintain expensive infrastructure equipment like routers, access points and the wired communication medium. However, we are faced with new problems such as managing mobility in a MANET and the need for power-efficient algorithms in a WSN as battery power is scarce. A *vehicular network* is an example of a MANET where vehicles communicate and coordinate on the road to prevent accidents and exchange information. An example MANET is depicted in Fig. 14.4.

A WSN consists of small sensor nodes that detect a physical phenomenon such as heat, vibration, or motion and communicate with each other to convey this information to a special node with higher capabilities called the *sink*. WSNs are commonly employed for environmental monitoring, home automation, and e-health where health states of individuals can be monitored remotely. Figure 14.5 shows a WSN with a sink node. The fundamental graph-theoretic problems faced

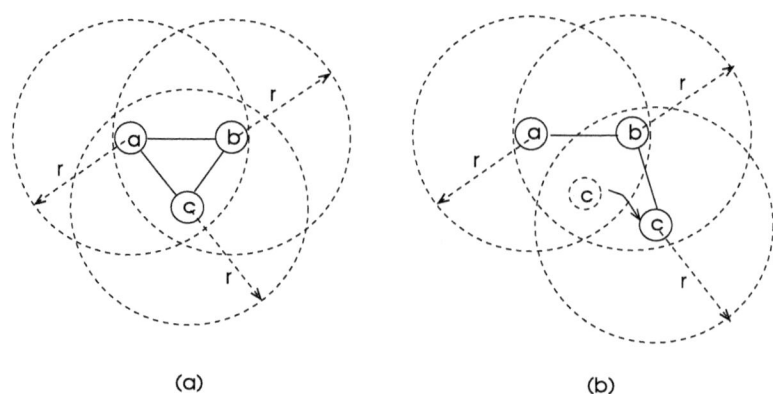

(a)   (b)

Fig. 14.4 A MANET with three nodes a, b, and c. They are all within ranges r of each other in **a** and node c moves to a new position which is out of range of node a in **b**; causing the edge (a, c) be deleted from the graph and the edge (b, c) to be modified

in MANETs and WSNs are the connectivity, backbone construction, clustering and routing as described in the next section.

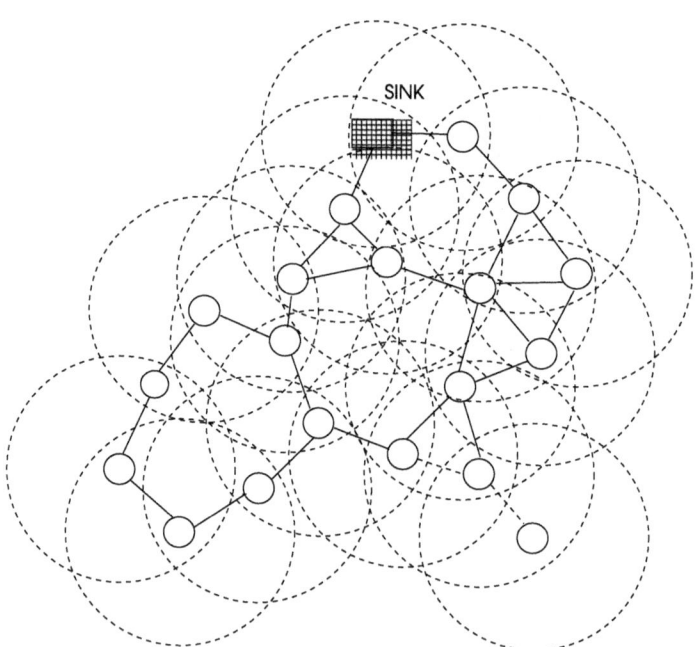

Fig. 14.5 A WSN with a sink node. Transmission ranges of all nodes are shown

We start this chapter with the graph-theoretic solutions to the problems encountered in ad hoc wireless networks which are: monitoring connectivity, clustering, backbone formation, and routing.

14.4.1 *k*-Connectivity

Connectivity in a graph $G = (V, E)$ meant there is a path between any two nodes u and v in G as we reviewed in Chap. 8. Connectivity is needed in any computer network for transfer of information between each pair of nodes but this problem is more eminent in a wireless network. For example, moving nodes in a MANET may disrupt connectivity easily and a sensor node that runs out of battery power may cause disconnection in such a network. The probabilities of these events are much higher than the failure of a router in a wired network.

Let us recall the *k*-connectivity problem: A network is *k*-connected if there is at least *k* disjoint paths between any two of its nodes. We can deduce that the failure of a minimum of *k*-1 nodes results in a disconnected and therefore non-functional network in a *k*-connected network. Clearly, the higher the value of *k*, the more strongly connected the network becomes. Hence, we can say a network with a higher *k* connectivity value is more reliable and fault tolerant against node failures than a network with a lower *k* value. In general, we have three main problems related to *k*-connectivity in an ad hoc wireless network [1]:

1. Placement the nodes of a WSN or a MANET at anytime so that *k*-connectivity is achieved.
2. Detection of the value of *k* in a given MANET or a WSN network.
3. Restoration of the value of *k*.

In search of a solution to the first problem, the neighbors of a node or the radio power of nodes are increased iteratively in various research studies. Even when this first step is accomplished, we need to monitor and determine the value of *k* to take remedial action when this falls below a desired value. When this happens, we can place new nodes in a WSN or move mobile nodes to new locations in a MANET to increase the value of *k*. We have already reviewed algorithms to determine the value of *k* in Chap. 8.

14.4.2 Clustering in Wireless Ad Hoc Networks

Clustering in a wireless ad hoc network is performed by grouping nodes that are within their radio transmission ranges. A specific node in a cluster is commonly assigned as the clusterhead (CH) and this node functions as the cluster manager. Clustering with CHs is advantageous for a number of reasons. First of all, a hierarchical structure is obtained in which various tasks can be assigned to CHs which can work in parallel to implement the required tasks in their clusters such as MAC layer functions including channel access, power measurements and maintain time division frame synchronization. The CHs also can be formed as directly or indirectly connected to each other and can be used to build a *backbone* for message transfers. This type of routing a message over the backbone until the closest CH to the destination node on the backbone is reached eliminates excessive message passing as shown in Fig. 14.6. There is no need to store and manage global data since each CH knows the identifiers of nodes in its cluster and every time a message is received on the backbone, a check is made with the destination field in the message and the cluster node identifiers. If there is a match, the message is relayed to the node in the cluster; otherwise, the message is transferred to the neighbor backbone node.

The nodes of a k-cluster are mutually reachable by a path of length at most k. A k-cluster with $k = 1$ is a clique. A k-hop cluster consists of nodes which are at most k-hop distance from their CH. The nodes in a clustered ad hoc network can be classified as CHs, *gateway nodes* that connect two clusters and *ordinary nodes* as shown in Fig. 14.6. Optimal clustering and the selection of optimal number of CHs are both NP-hard and various heuristics to perform these functions are commonly used. Once a cluster is formed, a CH can be selected by symmetry breaking among the nodes using their identifiers, their degrees, node mobility, node battery power, or various combinations of these parameters. After the construction of clusters and selecting CHs, clusters need to be maintained. Maintaining the cluster structure in a MANET is difficult due to the dynamicity of the nodes, and new CHs are needed

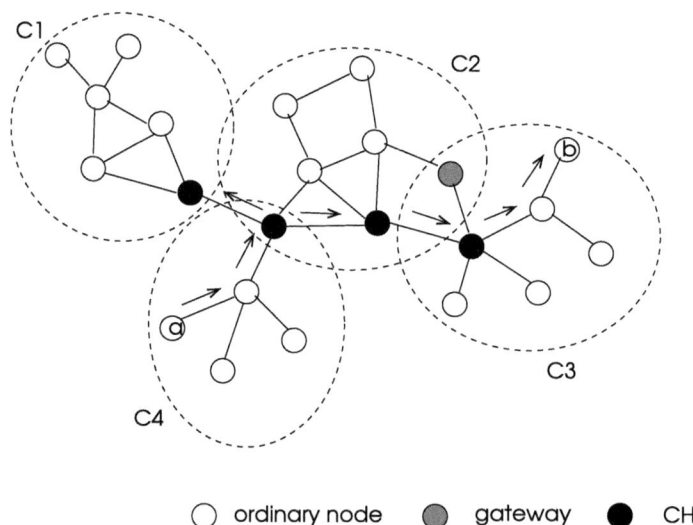

○ ordinary node   ● gateway   ● CH

Fig. 14.6 Example backbone in a wireless ad hoc network. Clusters C_1, C_2, C_3, and C_4 are shown inside dashed circles and CHs are shown in bold. Node a in C_4 sends a message to its CH which forwards the message along the backbone until the cluster of b which is C_3 is found

to be assigned by the nodes when the CH moves away. A moving node on the other hand, may join a new cluster it approaches.

The energy consumed by a CH is larger than an ordinary node due to its message relaying function in the network. For this reason, CH function may be rotated among ordinary nodes to provide load balancing. Mobility is the major concern in a MANET and the limited power of batteries should be considered when clustering in WSNs. We will review a simple algorithm to form clusters and CHs simultaneously in an wireless ad hoc network and review construction of a backbone using graph-theoretical approaches.

14.4.2.1 Algorithms

Gerla and Tsai proposed a clustering algorithm using the identifiers of nodes in wireless ad hoc networks [10]. A node in such a network broadcasts periodically the identifiers of the neighbors in its transmission range in its unit disk graph. After the broadcast, it listens to the medium for a while and does one of the following:

- A node that does not hear a node with a lower identifier than itself after a timeout decides to be a CH and broadcasts this condition.
- The lowest identifier neighbor that a node hears is assigned as its CH, unless that node voluntarily gives up its position as a CH.
- A node that hears the declaration of two or more CHs assigns itself as a *gateway* bridging two clusters.

As can be seen, the symmetry breaking condition is the choice of the node with the lowest identifier in the transmission range and hence the name of the algorithm. The node that hears all higher identifier nodes becomes the CH and broadcasts itself as the CH. Nodes that hear the CH declaration message become part of the cluster managed by that CH. A final correction in the algorithm involves selecting a node as a gateway when it hears two nodes as CHs. This simple algorithm creates clusters in linear time, however, a low identifier node joining a cluster results in reorganization of a cluster which may be costly.

The authors proposed another algorithm to form clusters called *highest connectivity cluster algorithm* which aims to select the node with the highest degree as the CH [10]. The following rules are applied in this algorithm after a node broadcasts the list of nodes it can hear including itself:

- A node that has elected a CH is *covered*; otherwise it is *uncovered*.
- The node with the highest connectivity (degree) among is uncovered neighbors is elected as the CH. Identifiers are used to break ties.
- A node that elects a CH gives up being a CH.

This time, nodes with higher degrees are selected assuming they can access the nodes in their clusters easily. In both algorithms, no two CHs will be adjacent to each other and the distance between any two nodes is at most two hops in a cluster. Figure 14.7 displays clusters formed by both algorithms. Total number of messages

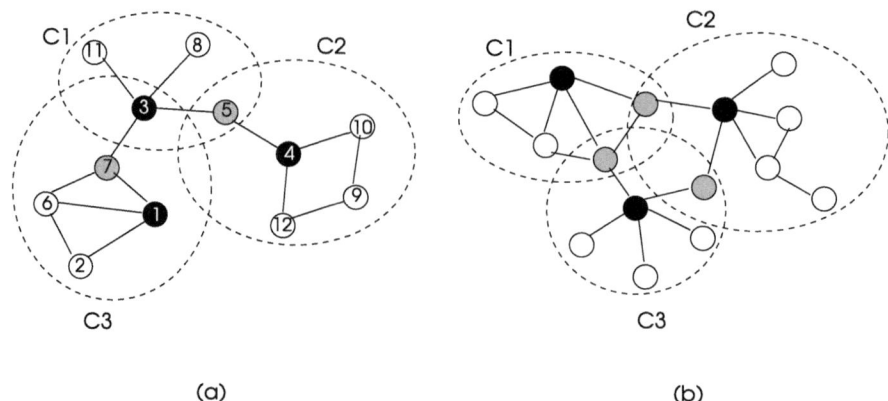

Fig. 14.7 **a** Lowest identifier algorithm, **b** highest connectivity algorithm implementations. CHs are shown in bold and gateway nodes are gray

communicated in the first algorithm is $2n$ as each node will broadcast one message (*update*) to its neighbors and another message (*i_am_chead* or *ordinary*) to inform whether it is a CH or an ordinary node.

14.4.3 Backbone Construction with Dominating Sets

Instead of grouping the wireless ad hoc network into clusters with CHs and then forming a backbone with these CHs, we can use connected dominating sets as the backbone. A dominating set D of a graph $G = (V, E)$ is a subset of its vertices such that $\forall v \in V$, either $v \in D$ or v is adjacent to a vertex $u \in D$. In a connected dominating set (CDS), there is a path between any two vertices in this set. Recall finding the minimum order connected or unconnected dominating set is NP-hard (see Chap. 3) we need to form a CDS to have correct operation of the backbone. Otherwise, we need to insert vertices between the elements of the dominating set. We also form clusters this way by denoting each element of D as CH and any neighbor connected to such a CH becomes the member of the cluster of this CH. For example, the CHs in Fig. 14.6 form a 3-hop dominating set with shown clusters around them. We can build a maximal independent set (MIS) of the graph and then connect the nodes in the MIS to obtain a CDS. We will describe a direct algorithm that finds the CDS in linear time in the next section. An evident requirement is that the backbone nodes should be connected and that every node should be in the backbone or a neighbor to a backbone node.

14.4.3.1 Pruning-Based Algorithm

The algorithm proposed by Wu and Lin [23] finds the minimal connected dominating set (MCDS) of a network with nodes having unique identifiers, using the neighbor information of nodes and hence is a local distributed algorithm consisting of two steps. The identifiers of neighbors of all nodes are exchanged in the first step and any node that finds it has at least two unconnected neighbors, nominates itself to be in MCDS by changing its color to *black*. Every node then sends its status to all of its neighbors in the second step after which the following pruning rules are applied to remove redundant nodes from the MCDS:

- **If** $\exists u \in N[v] | (color_u = black) \wedge (N[v] \subseteq N[u]) \wedge (id_v < id_u)$ **then** $color_u \leftarrow white$.
- **If** $\exists u, w \in N(v) | (color_u = color_v = black) \wedge (N(v) \subseteq (N(u) \bigcup N(w))) \wedge (id_v = min\{id_v, id_u, id_w\})$ **then** $color \leftarrow white$.

The first rule means a node that finds all of its closed neighbor set is covered by a neighbor with a lower identifier removes itself from the CDS. We have two nodes covering the same neighbors in this case and one of them can be removed; identifiers are used to break the symmetry. The second rule removes a node from CDS if its neighbors are covered by the union of the neighbors of two higher identity nodes in the CDS. In this case, we are looking at the union of nodes which may be partly covered by two neighbor nodes. As each node sends exactly two broadcast messages in a wireless ad hoc network in this implementation, total number of messages transmitted is $2n$ and it can be used conveniently in a MANET due to its low maintenance requirements. When a node moves away, only its neighbors need to update their states. Figure 14.8 shows an example network where a minimal CDS is formed in two steps.

Cokuslu and Erciyes modified this algorithm by incorporating the degrees of the nodes as well as their identifiers while pruning in the second step [4]. They compared their algorithm with Wu's algorithm and showed experimentally that it provides significantly smaller MCDSs. Das et al. [5] provided two algorithms that are the distributed versions of Guha–Khuller algorithms that we have seen in Chap. 10. In the first algorithm, nodes are assigned weights as their effective degrees which are the numbers of their non-CDS neighbors. Initially a small dominating set C is formed which may have several disconnected components. The forest consisting of the edges $\{v_1, v_2\}$ where $v_1 \in C$ and $v_2 \in N(v_1)$ is then connected in the second stage using a distributed minimum spanning tree (MST) algorithm. The CDS consists of the non-leaf nodes of the MST formed. This algorithm provides an approximation ratio of $2H\Delta + 1$ in $O((n + |C|)\Delta)$ time using $O(n|C| + m + n \log n)$ messages [5].

14.4.3.2 Greedy Distributed Algorithms

In the second algorithm, one or two-step paths emanating from the current CDS are investigated to find the node with the greatest number of white nodes in each round. A node or a pair of nodes with the highest number of span is added to the existing CDS as in [5]. This algorithm achieves an approximation ratio of $2H\Delta$ in $O(|C|(\Delta + |C|))$ time using $O(n|C|)$ messages [5].

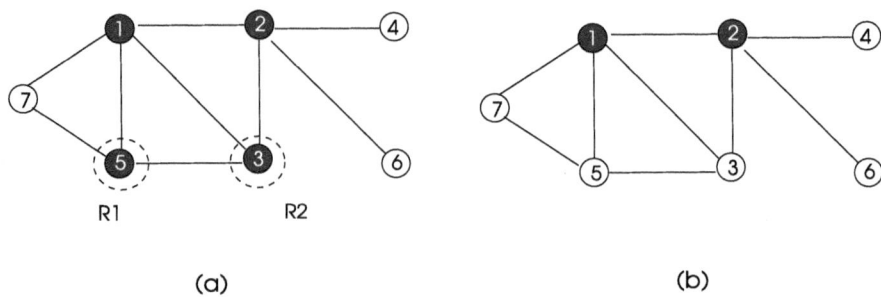

(a)   (b)

Fig. 14.8 Implementation of Pruning-based algorithm in a small graph. Nodes that mark them-selves *black* since they have two directly unconnected neighbors are shown in **a** and implementing Rule 1 and Rule 2 in nodes 5 and 3 respectively results in the smaller CDS shown in **b**

14.4.3.3 MIS-Based Distributed CDS Construction

Alzoubi et al. proposed a CDS construction algorithm based on UDGs that consists of three phases as follows [2]:

1. *Leader Election*: A spanning tree is constructed rooted at the leader and nodes notify leader that this phase is over.
2. *Level Calculation*: The leader starts this phase by sending a level 0 message and then each node increases the level received from parent and transfers level mes-sage to children if they exist. A convergecast operation by *complete* messages to the root concludes this phase.
3. *Color Marking*: The nodes in the MIS are colored *black* and all other nodes are colored *gray* at the end of this phase. The *dominator* message is sent by a node that marks itself *black*, and the *dominatee* is sent by a node that marks itself *gray*.

Initially, all nodes are *white*, and the algorithm is executed according to the following rules [2]:

1. A *white* node which receives a *dominator* message first time marks itself *gray* and broadcasts a *dominatee* to inform it has been dominated.
2. A *white* node that has received *dominatee* messages from all of the neighbors with lower ranks marks itself *black*, sends a *dominator* message to all of its neighbors, and assigns its parent in T as its dominator.
3. A *gray* node receiving a *dominator* message from a child node in T for the first time which has never sent a *dominatee* message, it changes its color to *black* and sends *dominator* message to all of its neighbors.
4. Whenever a *black* node finds that all of its neighbors are *black* and have lower ranks than itself, it changes its color to *gray* and sends *dominatee* message to all of its neighbors.

Rule 1 ensures that if the neighbor of a white node is included in the CDS, it colors itself gray to be a neighbor node of a CDS node. In Rule 2, if all the lower rank neighbors are gray, a node may be assigned as a CDS node. The second phase finishes when the leaves of the tree are marked. At the end of the first two phases, an MIS is formed and the nodes in this set are connected using *invite* and *join* messages. This algorithm has a time complexity of $O(n)$ and a message complexity of $O(n \log n)$, and the resulting CDS has a size of at most $8|\text{MinCDS}| + 1$ [2].

14.5 The Internet

The Internet is the largest computer network in the world consisting of billions of devices including personal computers, servers, mobile phones connected by various networks. The Internet is a complex network exhibiting small-world and scale-free properties as discovered experimentally [3]. The average distance between any two nodes is between 3 and 9 and a small fraction of nodes in the Internet have very high number of connections with most of the nodes having low degrees. In fact, the average degree of nodes in the Internet is between 2 and 8 [3].

Routing in Internet is needed to find efficient paths from a source to many destinations in the network. A routing protocol specifies a set of rules for efficient data transfer between sources and destinations. There are various choices for Internet routing protocols; the information can be stored *globally* or *decentralized*. We have all routers having complete topology information in the former and a router is aware of its neighbors and the link costs to these neighbors in the latter. The network may be *static* with routes changing slowly or *dynamic* with frequent route changes. Additionally, the routing may be sensitive to load in the network or not. Two representative routing algorithms in the Internet are the distance vector and link state algorithms. In both of these algorithms, it is assumed that the router is aware of the addresses of its neighbors and the cost of reaching to those neighbors. We will see the routing problem in the Internet can be solved efficiently with graph algorithms.

14.5.1 Distance Vector Protocol

The distance vector routing (DVR) protocol uses a local distributed and a dynamic algorithm that adapts to changes and link failures. It is based on the Bellman–Ford dynamic shortest path algorithm we reviewed in Chap. 3. The main idea of this algorithm is the diffusion of the shortest paths to neighbors. Each node i periodically sends its shortest distances to all other nodes in a message *update* including vector $length[1..n]$ with entry $length[j]$ showing its distance to node j. When a node receives these vectors from the neighbors, it updates its shortest paths to all other nodes in the network based on the values in $length$. When a node i receives an *update* message, it checks the entries in length vector and if there is a shorter route to a destination j in length than its own, it updates its local routing table with that of the one in *length*. The *count-to-infinity problem* is encountered in this protocol when a node becomes isolated due to a link failure or breaks down and all nodes start to increase their distances to this node.

14.5.2 Link State Algorithm

The link state protocol uses a global distributed algorithm in which each router is aware of the entire network topology and computes the shortest paths to all other nodes individually using the Dijkstra's single source shortest path (SSSP) algorithm we saw in Chap. 7. The network information is transferred by periodic link state packets (LSPs) that includes the costs of reaching neighbors, a sequence number and *time-to-live* field which is decremented at each hop. Nodes gather the information flooded through the network and use it to compute routes using the SSSP algorithm. The local routing tables need to be large as whole network information is to be stored.

14.5.3 Hierarchical Routing

Up to this point, we have assumed all routers are in a flat network structure which is not realistic. Hierarchical routing in the Internet is based on hierarchical placement of routers which is more reasonable than storing routing information for millions of destinations at a single node. Routers are clustered into autonomous systems (ASs) and routers in the same AS use the same routing protocol whereas routers in different ASs can run different routing protocols. There is an inter-AS routing protocol for data transfer among ASs. If a packet received by a router is destined for a destination in the same AS, the shortest route computed by the inter-AS routing algorithm is used. Otherwise, the packet is transferred to one of the gateway routers to be delivered to the destination AS. The required destination AS is computed by the inter-AS routing protocol. The standard inter-AS protocol of the Internet is the Border Gateway Protocol (BGP) [21].

14.6   The Web as an Information Network

The world wide web (WWW), or Web, is an information network formed by the references in Web documents. We can think of the Web as a higher-level structure over the Internet which consists of *Web pages* with links to each other. Such an organization of Web pages can be conveniently modeled by a digraph, commonly denoted as the *Web graph*. We can then search solution to the problems such as finding the most relevant page to a query encountered there using this digraph. The *hypertext transfer protocol* (http) is used for communication between the Web clients and Web servers and the links between Web pages are called *hyperlinks*.

The Web graph is very dynamic with numerous nodes (pages) being added and deleted at any time. It is a complex network consisting of millions of nodes bearing the commonly found complex network properties such as small-world and scale-free networks. In other words, there are only few hops between any two documents on the Web and only a small percentage of the nodes have very high degrees with most of the nodes having small degrees. The Web graph was found experimentally to have a very large strongly connected component called the *giant component* (GC) with other nodes grouped as follows:

- *IN*: This the set of nodes that have directed links to the GC.
- *OUT*: Nodes that have directed links from the GC form this component.
- *Tendrils*: A tendril has Web pages connected to either IN or OUT but are not part of IN, OUT or the GC.
- *Disconnected nodes*: Pages that cannot be accessed from any component.

These components in *IN-GC-OUT* sequence form a bow-tie structure. Our main goal from graph-theoretical point of view in the Web is to design efficient algorithms for convenient access to pages on the Web graph. There are two fundamental algorithms for this purpose: the HITS and PageRank algorithms described in the following sections.

14.6.1   HITS Algorithm

Kleinberg presented an algorithm called *hypertext-induced topic selection* (HITS) to assign values to pages on the Web for efficient search during a query [16]. The main idea of this algorithm is to give importance to some nodes on the graph based on their votes in pointing to a document during a Web query. The Web pages related to a query are divided to the *hubs* which cast votes and the pages pointed by hubs are called the *authorities*. Both of these pages have scores associated with them to reflect their importance. These scores can be assigned based on the following rules [16]:

- *Authority Update Rule*: The authority score of a page is the sum of the hub scores of all pages pointing to it.
- *Hub Update Rule*: The hub score of a page is the sum of the authority scores of all pages that it points to.

With these rules, we give more importance to authorities that are pointed by more hubs than others. Moreover, if a hub has pointed to authorities which have been pointed by many hubs, its importance is also raised. This feedback structure may be repeated in a loop to determine the importance of authorities which can then be presented to the user with respect to their priorities. A one-step implementation of this algorithm is shown in Fig. 14.9.

Fig. 14.9 Implementation of
HITS algorithm in a small
Web graph for the first
iteration. The scores next to
the authorities show the
number of hubs that point to
them. The hubs have a score
reflecting the total score of
authorities they point

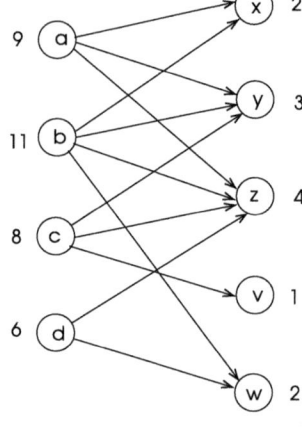

A possible pseudocode of this algorithm is depicted in Algorithm 14.1 in which
the hub and authority scores of the pages are initialized to 1 and the above rules are
applied iteratively [7]. The final scores at each iteration are calculated by dividing
the score value with the sum of the scores. It was shown in [16] the scores for hub
and authority pages converge as the number of iterations goes to infinity.

Algorithm 14.1 *HITS*

1: **Input** : $P = \{p_1, ..., p_n\}$ ▷ p is set of n pages
2:      k steps
3: **Output** : authority and hub values for all pages
4: **for all** $p \in P$ **do**   ▷ initialize authority and hub values
5:     $hub_p \leftarrow 1; auth_p \leftarrow 1$
6: **end for**
7: **for** $j \leftarrow 1$ to k **do**   ▷ apply rules for k steps
8:      **for all** $p \in P$ **do**
9:          **apply** *Authority Update Rule* to p to assign $auth_p$ values
10:      **end for**
11:      **for all** $p \in P$ **do**
12:          **apply** *Hub Update Rule* to p to assign hub_p values
13:      **end for**
14:      $auth\_total \leftarrow \sum_{p \in P} auth_p$ ▷ calculate total values
15:      $hub\_total \leftarrow \sum_{p \in P} hub_p$
16:      $auth_p \leftarrow auth_p / auth\_total$ ▷ calculate normalized values
17:      $hub_p \leftarrow hub_p / hub\_total$
18: **end for**

14.6.2 PageRank Algorithm

We have analyzed a dynamically formed Web graph in response to a query. This bipartite graph contained two different type of pages as hubs and authorities. The Web graph in general is not bipartite and can be viewed as a directed graph where the bipartite structure may exist only locally. *Page rank* is an attribute of importance of a page in the Web graph based on the number of pages that reference it. It is basically a score for a page based on the votes it receives from other pages. This is a sensible metric for the importance of a page since the relative importance and popularity of a page increases by the number of pages referencing it. Page rank can be considered as a fluid that runs through the network accumulating at important nodes. The page rank algorithm to find importance of pages in a Web graph assigns ranks of the pages in the Web graph such that the total page rank value in the network remains constant. It initially assigns rank values of $1/n$ to each page in an n node network as shown in Algorithm 14.1. The current rank value of a page is evenly distributed to its outgoing links and then, the new page rank values are calculated as the sum of the weights of the ingoing links of pages. Execution of this algorithm for k steps results in more refined results for page rank values as in the Authority and Hubs algorithm and the page rank values converge as $k \to \infty$.

Algorithm 14.2 *Page Rank Algorithm*

1: **Input** : $P = \{p_1, ..., p_n\}$ ▷ set of n pages
2:         k steps
3: **Output** : page rank values $rank_p, \forall p \in P$
4: $E_p(in) \leftarrow$ ingoing edges to page p
5: $E_p(out) \leftarrow$ outgoing edges from page p
6: **for all** $p \in P$ **do**   ▷ initialize page rank values
7:      $rank_p \leftarrow 1/n$
8: **end for**
9: **for** $r \leftarrow 1$ to k **do**   ▷ apply for k steps
10:      **for all** $p \in P$ **do**
11:          **for all** $e \in E_p(out)$ **do**
12:              $w_e \leftarrow rank_p/|E_p(out)|$
13:          **end for**
14:          $rank_p \leftarrow \sum_{e \in E_p(in)} w_e$ ▷ find sum of the weights of all links pointing to p_i
15:      **end for**
16: **end for**

Finding the initial edge weights using this algorithm in a small Web graph with five nodes is depicted in Fig. 14.10.

Running of the Page Rank algorithm for the graph of Fig. 14.10 for the first three iterations is shown in Table 14.1. We can see page 3 has the least rank as it is pointed by only one page and page 4 has slightly higher rank than others as it is the only page pointed by 3 pages. A page that does not point to many pages as other pages but has many input edges may have high scores after a significant

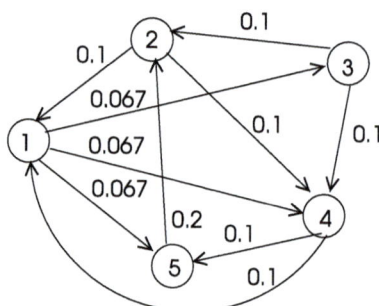

Fig. 14.10 Implementation of PageRank algorithm in a small Web graph to find weight values per edge

Table 14.1 Page Rank values of the Web nodes of Fig. 14.10

Vertices	1	2	3	4	5
n_{out}	3	2	2	2	1
$k = 1$: weight / edge	0.067	0.1	0.1	0.1	0.2
rank	0.2	0.3	0.067	0.267	0.167
$k = 2$: weight / edge	0.067	0.15	0.034	0.134	0.167
rank	0.284	0.201	0.034	0.251	0.201
$k = 3$: weight / edge	0.095	0.101	0.017	0.067	0.167
rank	0.168	0.184	0.095	0.213	0.162

number of iterations. This situation is corrected by the introduction of damping factor d which is used to scale down page rank values by $(1 - d)/n$ [16].

14.7   Chapter Notes

We have described and reviewed fundamental complex networks which are biological networks, social networks, technological networks and information networks. All of these networks exhibit small-world and scale-free properties of complex networks. We then took a closer look at some of the representative examples for these networks.

PPI networks are biological networks that exist outside the nuclei in the cell and three main problems encountered in these structures are detecting clusters, finding network motifs and aligning two networks. Clusters or dense subgraphs in these networks may indicate some dense activity in these regions and we reviewed two algorithms for the purpose of discovering clusters. Network motifs are repeating subgraphs and finding them is another important task and research area in PPI networks and other biological networks. These structures are assumed to have some basic functionality and are considered to be the basic building blocks of organisms. Moreover, finding similar network motifs in two or more organisms may indicate common ancestry. Alignment of two or more networks shows their similarities and is frequently used to compare various networks.

Social networks are formed by individuals or groups of individuals, and finding communities which are closely related groups provides insight to the structure of a social network. We reviewed two main algorithm for this purpose and also described relations, and balanced and unbalanced social networks. Wireless ad hoc networks are technological networks like the Internet. Two main types of these networks are the mobile ad hoc networks and wireless sensor networks. We described various clustering algorithms in these networks in detail which are all distributed algorithms executed by individual nodes of the network. We also reviewed main routing algorithms in the Internet which are extensions of the routing algorithms described in Chap. 7. We lastly reviewed the Web which is an information network and analyzed two algorithms to attribute importance to Web pages for efficient Web queries. The first algorithm called HITS divides the nodes to hubs and authorities during a query and assigns scores to these nodes based on the scores of nodes they point and are pointed by. The PageRank algorithm is more general and finds wide usage in the Web.

We can say clustering is a fundamental research area in all of these networks and heuristics are widely used to find the dense regions in the large graphs that represent these networks. There is not a single algorithm that fits all of the needs of the application and the experimental results obtained along with its complexity are commonly accepted as the goodness of an algorithm. Parallel and distributed algorithms are needed in all areas of research in these complex networks. A survey of parallel clustering algorithms with newly proposed ones are given in [8]. Distributed clustering algorithms in wireless ad hoc networks have been studied extensively but the same is not valid for clustering in other complex networks. There are only few studies for parallel network motif search and parallel network alignment which are potential research areas.

Exercises

1. Work out the MST of the graph shown in Fig. 14.11 using any algorithm. Then, divide this graph into three clusters by removing the heaviest two edges from the MST. Find the total cost of inter-cluster edges and the cost of edges within the clusters and compute cluster quality.
2. Find whether the social network depicted in the labeled graph of Fig. 14.12 is balanced or not by checking every triangle. Suggest what to do to make this network balanced.
3. Work out the scores of hub and authority pages of the Web query graph of Fig. 14.13 for three iterations of the HITS algorithm. Determine whether there is a convergence of score values or not.
4. A Web graph is given in Fig. 14.14. Implement the PageRank algorithm in this graph for four iterations by showing the page rank scores for each page.

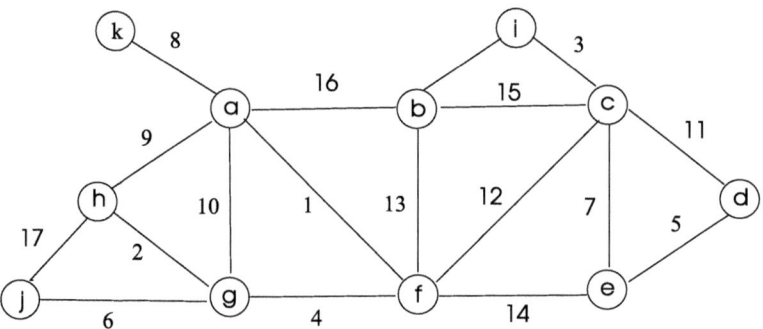

Fig. 14.11 Sample graph for Exercise 1

Fig. 14.12 Sample graph for
Exercise 2

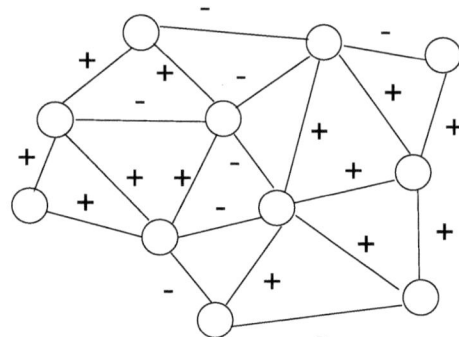

Fig. 14.13 Sample graph for
Exercise 3

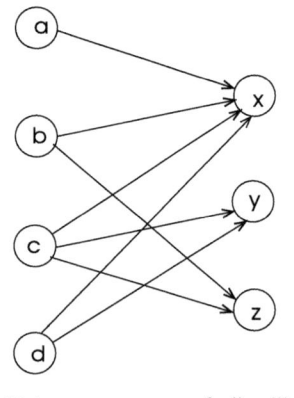

Fig. 14.14 Sample graph for Exercise 4

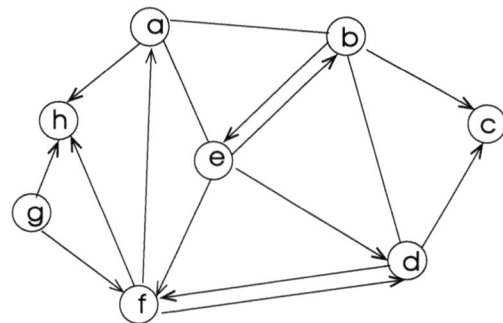

References

1. Akram VK, Orhan Dagdeviren O (2015) On k-connectivity problems in distributed systems. Advanced methods for complex network analysis. IGI Global
2. Alzoubi KM, Wan P-J, Frieder O (2002) New distributed algorithm for connected dominating set in wireless ad hoc networks. In: Proceedings of 35th Hawaii international conference on system sciences, Big Island, Hawaii
3. Caldarelli G, Vespignani A (2007) Large scale structure and dynamics of complex networks: from information technology to finance and natural science. Complex Systems and Interdisciplinary Science. World Scientific Publishing Company. Chapter 8, ISBN-13: 978-9812706645
4. Cokuslu D, Erciyes K, Dagdeviren O (2006) A dominating set based clustering algorithm for mobile ad hoc networks. Int Conf Comput Sci 1:571–578
5. Das B, Bharghavan V (1997) Routing in ad-hoc networks using minimum connected dominating sets. In: IEEE international conference on communications (ICC97), vol 1, pp 376380
6. Dongen SV (2000) Graph clustering by flow simulation. Ph.D. Thesis, University of Utrecht, The Netherlands
7. Erciyes K (2014) The Internet and the Web. In: Complex networks: an algorithmic perspective. CRC Press. ISBN-10: 1466571667, ISBN-13: 978-1466571662
8. Erciyes K (2015) Distributed and sequential algorithms for Bioinformatics, Springer, Berlin (Chaps. 10 and 11)
9. Fiedler M (1973) Algebraic connectivity of graphs. Czechoslov Math J 23:298–305
10. Gerla M, Tsai JTC (1995) Multicluster, mobile, multimedia radio network. Wirel Netw 1:255–265
11. Girvan M, Newman MEJ (2002) Community structure in social and biological networks. PNAS 99:7821–7826
12. Hagen L, Kahng AB (1992) New spectral methods for ratio cut partitioning and clustering. IEEE Trans Comput Aided Des Integr Circuits Syst 11(9):1074–1085
13. Harary F (1953) On the notion of balance of a signed graph. Mich Math J 2(2):143–146
14. International Organization for Standardization (1989-11-15) ISO/IEC 7498-4:1989 – Information technology – open systems interconnection – basic reference model: naming and addressing. ISO Standards Maintenance Portal. ISO Central Secretariat. Retrieved 17 Aug 2015
15. Jorgic M, Goel N, Kalaichevan K, Nayak A, Stojmenovic I (2007) Localized detection of k-connectivity in wireless ad hoc, actuator and sensor networks. In: Proceedings of 16th international conference on computer communications and networks (ICCCN 2007), pp 33–38
16. Kleinberg J (1999) Authoritative sources in a hyperlinked environment. J ACM 46(5):604–632
17. Lin CR, Gerla M (1997) Adaptive clustering for mobile wireless networks. IEEE J Sel Areas Commun 15(1):1265–1275

18. Mount DM (2004) Bioinformatics: sequence and genome analysis, 2nd edn. Cold Spring Harbor Laboratory Press, NY. ISBN 0-87969-608-7
19. Newman M (2003) Fast algorithm for detecting community structure in networks. Phys Rev E 69:066133
20. Olman V, Mao F, Wu H, Xu Y (2009) Parallel clustering algorithm for large data sets with applications in bioinformatics. IEEE/ACM Trans Comput Biol Bioinform 6:344–352
21. RFC 4271 - A Border Gateway Protocol 4 (BGP-4). www.ietf.org
22. Titz B, Rajagopala SV, Goll J, Hauser R, McKevitt MT, Palzkill T, Uetz P (2008) The binary protein interactome of Treponema pallidum, the syphilis spirochete. PLOS ONE 3(5):e2292
23. Wu J, Li H (1999) On calculating connected dominating set for ef- ficient routing in ad hoc wireless networks. In: Proceedings of the third international workshop on discrete algorithms and methods for mobile computing and communications, pp 7–14

Advanced Graph Structures and Algorithms

15

15.1 Introduction

A graph $G = (V, E)$, even in the simplest form of nodes and edges, provides a representation of numerous real-life networks with nodes representing entities and the edges showing the relationships between the nodes. Modern graph theory, however, provides graph representations with more information embedded in a graph. One way of achieving enclosing of some information is the labeling of the edges or vertices with some data. A knowledge graph is a digraph which provides this information in the form of source, relation, target with the label showing the relationship between the source and the target. For example, "Sean is a doctor" statement would be represented as Sean a node, doctor as another node, and a directed edge labeled with "is."

In many real-life networks, the existence of an edge between two nodes can not be determined precisely. Uncertain graphs are proposed to overcome this difficulty by labeling an edge between the two nodes in such a graph with a probability that approximates its existence. Quantum graphs … Graph mining is the process of discovering structures and relationships in graphs. We review these three relatively recent graph representations and algorithms associated with them together with main graph mining methods in this chapter.

15.2 Knowledge Graphs

A knowledge graph (KG) is a directed labeled graph with nodes representing entities such as persons, objects, or places and the edges between the nodes displaying the relation between them with the arrow of an edge pointing to the node that the

© The Author(s), under exclusive license to Springer Nature Switzerland AG 2026
K. Erciyes, *Guide to Graph Algorithms*, Texts in Computer Science,
https://doi.org/10.1007/978-3-032-05294-0_15

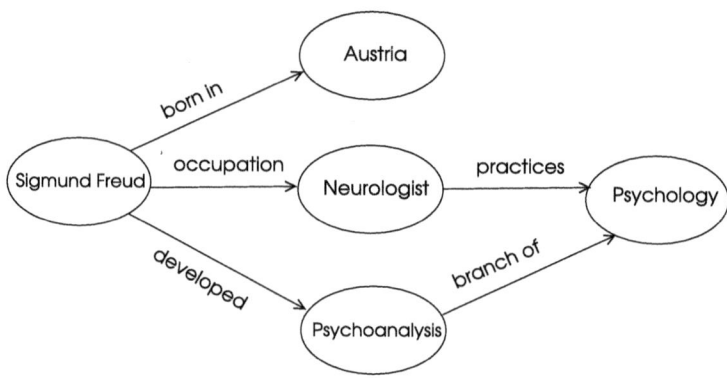

Fig. 15.1 Knowledge graph example

association is made. A KG can be expressed as a triple (head, relationship, tail), and an entity in a KG can be a place, a firm, a person, or an abstract concept. The KG in Fig. 15.1 displays the neurologist Sigmund Freud with his associations.

Construction of a KG first involves data collection from various resources such as web pages, databases, and documents. Relevant information such as relationships and attributes are then extracted from data using methods such as natural language processing and machine learning. This information is now to be represented in a suitable format for processing after which a KG can be used to answer queries and find new relationships which is called *data reasoning*.

A standard KG does not display time information of the event it represents. A *temporal knowledge graph* is more descriptive than KG as it shows the timestamp of the event represented. A KG may be input to a machine learning algorithm by first converting it to numerical vector called *embeddings* as domain knowledge.

15.2.1 Analysis

We will consider some cities in the world and their relationships to their countries, etc., as an example analysis of a knowledge graph. The *pandas* library of Python is used to build a data frame from the start points, relationships, and ending points of the directed edges in this graph. For example, Istanbul is a city of Türkiye and is a type of "city" class as shown in Listing 15.1. A KG is then constructed using this data frame and displayed.

Listing 15.1 *Large cities of the world knowledge graph example*

```
1  import pandas as pd
2  import networkx as nx
3  import matplotlib.pyplot as plt
4
5  # starting node, relationship and ending node
6  start = ['US', 'Turkiye', 'Italy', 'New York', '
       Istanbul', 'Rome','New York', 'Istanbul', 'Rome'
       ]
7  relation = ['belongs to', 'belongs to','belongs to'
       , 'city of', 'city of', 'city of','type of','
       type of','type of']
8  end = ['country','country','country','US','Turkiye'
       ,'Italy','city','city','city']
9
10 # Create a dataframe
11 datafr = pd.DataFrame({'start': start, 'relation':
       relation, 'end': end})
12
13 # Add edges of the knowledge graph from data frame
14 G = nx.Graph()
15 for _, row in datafr.iterrows():
16     G.add_edge(row['start'], row['end'], label=row['
       relation'])
17
18 # Display the knowledge graph
19 posit = nx.spring_layout(G, seed=32, k=0.9)
20 labels = nx.get_edge_attributes(G, 'label')
21 plt.figure(figsize=(8, 6))
22 nx.draw(G, posit, with_labels=True, font_size=12,
       node_size=3100, node_color='lightblue',
       edge_color='black', alpha=0.6)
23 nx.draw_networkx_edge_labels(G, posit, edge_labels=
       labels, font_size=12, label_pos=0.3,
       verticalalignment='baseline',connectionstyle='
       arc3,rad=0.2')
24 plt.title('Knowledge Graph of Cities and Countries'
       )
25 plt.show()
```

The graph output of this algorithm is depicted in Fig. 15.2 with nodes as cities, countries, and city and country classes. Once a KG is built, it can be used to answer queries like "Is Rome a city?".

We will now analyze this KG using graph analysis methods described in Chap. 13 which are degree centrality, closeness centrality, and (vertex) betweenness centrality. The code in Listing 15.2 is added to the main code to obtain the results of these analysis.

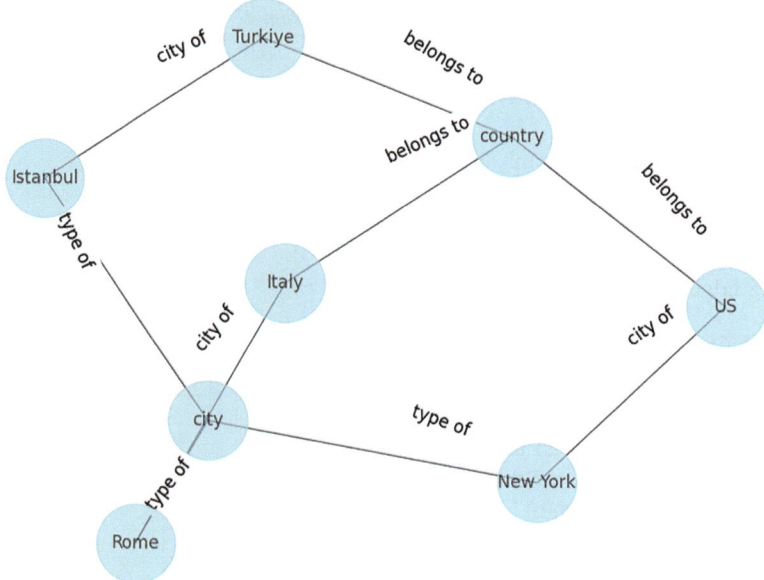

Fig. 15.2 Output of knowledge graph algorithm

Listing 15.2 *Large cities of the world knowledge graph example*

```
num_nodes = G.number_of_nodes()
num_edges = G.number_of_edges()
print(f'Number of nodes: {G.number_of_nodes()}')
print(f'Number of edges: {G.number_of_edges()}')
dc = nx.degree_centrality(G)
bc = nx.betweenness_centrality(G)
cc = nx.closeness_centrality(G)
print('Degree Centralities:')
for node, centrality in dc.items():
    print(f'{node}: {centrality:.2f},',end=" ")
print('\nCloseness Centralities:')
for node, centrality in cc.items():
    print(f'{node}: {centrality:.2f},',end=" ")
print('\nBetweenness Centralities:')
for node, centrality in bc.items():
    print(f'{node}: {centrality:.2f},',end=" ")
```

The output from the code in Listing 15.2 shown in Listing 15.3 shows that the *city* and the *country* nodes have the highest degrees as expected as they are central to almost all of the nodes. The closeness centrality parameter shows how central a node is, that is, in closer distance to all nodes. In this case, we can see that *city* and *country* nodes are the closest nodes to all nodes; however, the values for most of the nodes are close to each other. This is observed as the example graph is small, and all nodes are in vicinity of each other.

Listing 15.3 *Output of KG analysis*

```
Number of nodes: 8
Number of edges: 9
Degree Centralities:
US: 0.29, country: 0.43, Turkiye: 0.29, Italy:
    0.29, New York: 0.29, Istanbul: 0.29, Rome:
    0.29, city: 0.43,
Closeness Centralities:
US: 0.50, country: 0.58, Turkiye: 0.50, Italy:
    0.50, New York: 0.50, Istanbul: 0.50, Rome:
    0.50, city: 0.58,
Betweenness Centralities:
US: 0.11, country: 0.29, Turkiye: 0.11, Italy:
    0.11, New York: 0.11, Istanbul: 0.11, Rome:
    0.11, city: 0.29,
```

The vertex betweenness centrality values show that *city* and *country* nodes have the highest values again, as they are the nodes that have the highest number of shortest paths that run through them.

15.2.2 Applications

KGs are used in a number of diverse applications some of which are as follows:

- *Search Engines*: Search results are improved by providing users context close to their queries. Google Knowledge Graph employs KGs for this purpose.
- *Healthcare*: KGs may be used to integrate patient data with medical information to help identify diseases and provide treatment protocols. Google Health and Mayo Clinic systems employ KGs for healthcare.
- *Natural Language Processing (NLP)*: NLP applications are improved by the KGs which provide information about words and language structures.
- *Link Prediction*: Link prediction aims to predict missing or future connections between nodes in a graph. KGs may be used for this purpose to forecast missing links in a network.
- *Cybersecurity*: A copy of the network environment can be constructed using KGs which can be analyzed by data scientists for vulnerability, weaknesses, and malicious attacks.

KGs can also be used for processing of large volumes of data; some examples of KG databases that can process large KGs are IBM Watson [6], Amazon Neptune [9], and DBPedia [4].

15.3 Uncertain Graphs

Many graphs representing real-life networks are not deterministic. For example, a
protein-protein-interaction (PPI) network is formed by experimental evaluations of
protein interactions in a cell which may be erroneous due to noisy environments; the
relationship of two individuals shown by an edge between them in a social network
may be predicted wrongly. A wireless network connection may not function correctly
due to noise in the environment and assuming the validity of this connection may
provide wrong results.

In order to cater for this type of networks, uncertain graphs are proposed [8]. An
uncertain graph, informally, is a graph with uncertain edges that are labeled with a
probability to depict the chances of their existences.

Definition 15.1 (uncertain graph) An uncertain graph (UG) is denoted as $\mathcal{G} =
(V, E, P)$ where V is the set of vertices, E is the set of edges of the graph in the
usual sense, and P is the edge weight function that assigns each edge a probability,
that is, $P : E \rightarrow (0, 1]$.

An UG has an uncertainty adjacency matrix $\mathcal{A}$ defined as follows:

$$\mathcal{A} = \begin{pmatrix} \alpha_{00} & \alpha_{01} & \cdots & \alpha_{0n} \\ \alpha_{10} & \alpha_{11} & \cdots & \alpha_{1n} \\ \vdots & \vdots & \ddots & \vdots \\ \alpha_{n-10} & \alpha_{n-11} & \cdots & \alpha_{nn} \end{pmatrix}$$

where α_{ij} displays the probability of the existence of the edge (i, j). A *possible world*
of an uncertain graph denotes all possible patterns that may exist for that uncertain
graph.

Definition 15.2 (possible world) A possible world $G = (V, E')$ of an uncertain
graph $\mathcal{G} = (V, E, P)$ is a possible deterministic graph of $\mathcal{G}$ where $E' \subseteq E$.

The probability of a possible world can be calculated by Eq. (15.1), with the
assumption of independent edge existence probabilities.

$$\Pr(G) = \prod_{e \in E'} p(e) \prod_{e \in E \setminus E'} (1 - p(e)) \tag{15.1}$$

where $p(e)$ denotes the probability of the existence of an edge for that particular
possible world of $\mathcal{G}$. An UG is depicted in Fig. 15.3 with its adjacency matrix in
Fig. 15.4.

Two possible worlds of the graph of Fig. 15.3 are shown in Fig. 15.5 with probabilities $\Pr(G_1) = 0.0036$ and $\Pr(G_1) = 0.012$.

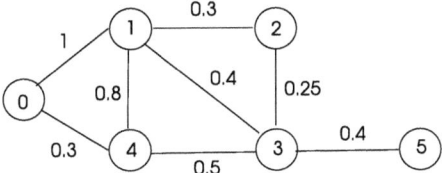

Fig. 15.3 Example uncertain graph

$$A = \begin{pmatrix} 0 & 1 & 0 & 0 & 0.3 & 0 \\ 1 & 0 & 0.3 & 0.4 & 0.8 & 0 \\ 0 & 0.3 & 0 & 0.25 & 0 & 0 \\ 0 & 0.4 & 0.25 & 0 & 0.5 & 0.4 \\ 0.3 & 0.8 & 0 & 0.4 & 0 & 0.3 \\ 0 & 0 & 0 & 0.4 & 0 & 0 \end{pmatrix}$$

Fig. 15.4 Its adjacency matrix

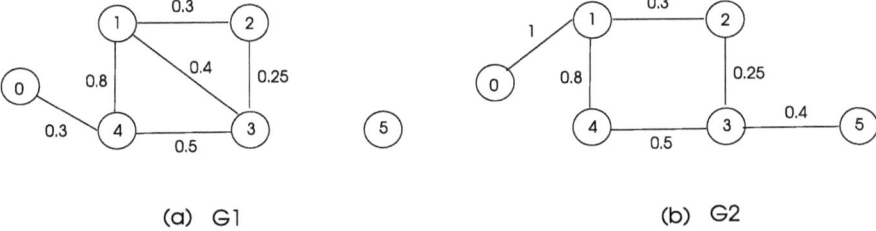

(a) G1   (b) G2

Fig. 15.5 Two possible worlds of the graph of Fig. 15.3

An UG may be weighted or unweighted; a weighted graph is attributed by a weight function that assigns weights to edges of the graph, that is, $W : E \rightarrow \mathbb{R}$. Also, an uncertain graph may be directed or undirected. An algorithm for a given graph problem has its counterpart in the uncertain graph domain. We will describe two clustering algorithms and a minimum spanning tree algorithm for UGs in the following section.

15.3.1 Clustering Uncertain Graphs

Clustering is the process of grouping objects that have some common attribute such as close distance to each other or some similarity feature. This process is one of the most studied topics in Computer Science and also in Artificial Intelligence as grouping similar entities provides discovery of meaningful structures.

Clustering UGs may be performed using several approaches two of which we will describe. In the simplest case, we can select an arbitrary vertex u of the UG $\mathcal{G}$, form a cluster around it by inserting any of its incident edges to its cluster that have a probability more than a specified threshold as shown in Algorithm 15.1 [10]. This

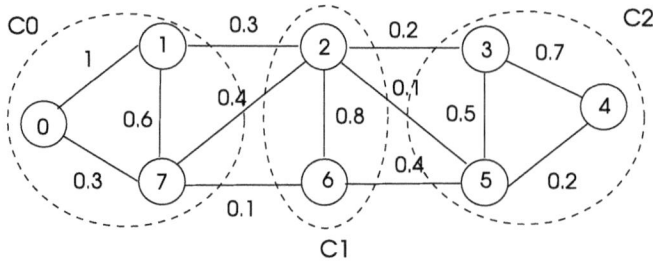

Fig. 15.6 Clustering using Algorithm 1

cluster vertices may then be removed from the vertex set, and the process is repeated until all vertices are assigned to clusters. Note that a selected neighbor w of vertex u may already be assigned to another cluster $C(w)$ in which case we need to merge these two clusters as depicted in lines 9 and 10 of the algorithm.

Algorithm 15.1 *Cluster 1*

1: **while** $V \neq \emptyset$ **do**
2:     **Select** $u \in V$ randomly
3:     $C(u) \leftarrow u$ ▷ insert u in its cluster
4:     **for all** $(u, v) \in E$ **do**   ▷ test adjacent edges
5:         **if** $p(u, v) \geq 0.5$ **then**
6:             $C(u) \leftarrow C(u) \cup \{v\}$
7:         **end if**
8:     **end for**
9:     **if** $\exists w \in C(u)$ such that $w \in C(w)$ **then**   ▷ merge clusters
10:         $C(w) \leftarrow C(w) \cup C(u)$
11:     **end if**
12:     $V \leftarrow V - C(u)$ ▷ remove u from V
13: **end while**

The sample UG of Fig. 15.6 is used to test Algorithm 15.1 using the Python code in Listing 15.6. The output of this code displayed in Listing 15.7 shows that the clusters are consistent with the manually found ones in the figure.

Listing 15.4 *UG clustering Algorithm 1 code*

```
1  import numpy as np
2  import random
3
4  A = np.array([[0,1,0,0,0,0,0,0.3],
5               [1,0,0.3,0,0,0,0,0.6],
6               [0,0.3,0,0.2,0,0.1,0.8,0.4],
7               [0,0,0.2,0,0.7,0.5,0,0],
8               [0,0,0,0.7,0,0.2,0,0],
9               [0,0,0.1,0.5,0.2, 0,0.4,0],
```

```
10                        [0,0,0.8,0,0,0.4,0,0.1],
11                        [0.3,0.6,0.4,0,0,0,0.1,0]],dtype=float)
12 n = len(A)
13 V = list(range(n))
14 C = []
15
16 while len(V) > 0 :
17     flag = False
18     Cu = []
19     u = random.choice(V)  # select a random vertex
20     Cu.append(u)             # start its cluster
21     for j in range(0,n): # include in cluster any
       neighbor with weight >= 0.5
22             if A[u][j] >= 0.5:
23                 Cu.append(j)
24     for w in Cu:            # check if a neighbor
       belongs to another cluster
25             for x in range(0,len(C)):
26                 if w in C[x]:
27                     C[x] = list(set(C[x]).union(Cu))
28                     flag = True
29     if not flag:
30         C.append(Cu)
31     V = list(set(V) - set(Cu))
32 print("Final C: ",C)
```

Listing 15.5 *Output of UG clustering Algorithm 1*

```
Final C:   [[2, 6], [1, 0, 7], [3, 4, 5]]
```

For the edge weight values displayed in Fig. 15.6, the clusters obtained are consistent as there is no merging of clusters. However, changing the weight of the edge (2, 7) to 0.7 may result in the merging of Cluster 0 and Cluster 1 although this mode of operation is not deterministic. It may be possible to form Cluster 2 by selecting node 6 first in which case the clusters will be the same as in the figure, otherwise selecting node 2 results in nodes 0, 1, 2, 6, 7 to be in one cluster (see Ex. 3).

Another clustering algorithm may be formed with the following logic [10]; each node of the UG $\mathcal{G}$ is a cluster initially, the main loop of the algorithm then finds the maximum average probability between the clusters, and if this value is larger than a threshold value, two cluster are merged as shown in Algorithm 15.2. This process continues until the maximum probability between the clusters is less than the threshold value.

Algorithm 15.2 *Cluster 2*

1: $thr \leftarrow 0.5$ ▷ threshold value
2: **for all** $u \in V$ **do**
3:     $C(u) \leftarrow \{u\}$ ▷ each node is a cluster initially
4: **end for**
5: **while** $p_m \geq thr$ **do**
6:     **for all do**
7:         Let $C(a)$ and $C(b)$ be two clusters with maximum p_{av} between them
8:         **if** $p_{av} \geq thr$ **then**
9:             **Merge** $C(a)$ and $C(b)$
10:        **end if**
11:    **end for**
12: **end while**

Implementation of Algorithm 15.2 in the sample UG of Fig. 15.6 with a threshold value of 0.5 is depicted in Fig. 15.7. Edges (0, 1), (2, 6), and (3, 4) shown in bold are selected to merge their endpoints to form the clusters enclosed in dashed regions in (a) in the first three iterations of the algorithm. The fourth iteration of the algorithm finds the maximum probability to be between nodes 1 and 7, and the clusters $C(7)$ and $C(0, 1)$ are merged. The final iteration selects edge (3, 5) to form the cluster $C(3, 4, 5)$.

15.3.2 Uncertain MST Algorithm

Uncertain MST algorithm is implemented on a weighted UG with edges having two labels; a probability of existence and the cost of traversing the edge between its endpoints. We propose a simple approach to compute the MST of such a weighted UG with the assumption that we need to select edges to be in the MST of UG that have higher probability and smaller costs than others. A simple way to address this problem is to assign new single weights to the edges by a combination these two values as below.

$$new\_weight_e = weight_e * (1 - \Pr(e))$$

Thus, edges with high probabilities and small weights have a higher priority to be in the MST. The new distance matrix of the UG can be calculated using these values after which the computation of the MST can be performed as in the classical Prim's algorithm as shown in Algorithm 15.3. A root vertex s of the MST is selected arbitrarily, and the lightest edge incident on this vertex is included in the MWOE list $mwoes$. The main loop of the algorithm iterates by selecting the smallest weight edge from this list, including it in the MST until all vertices are visited.

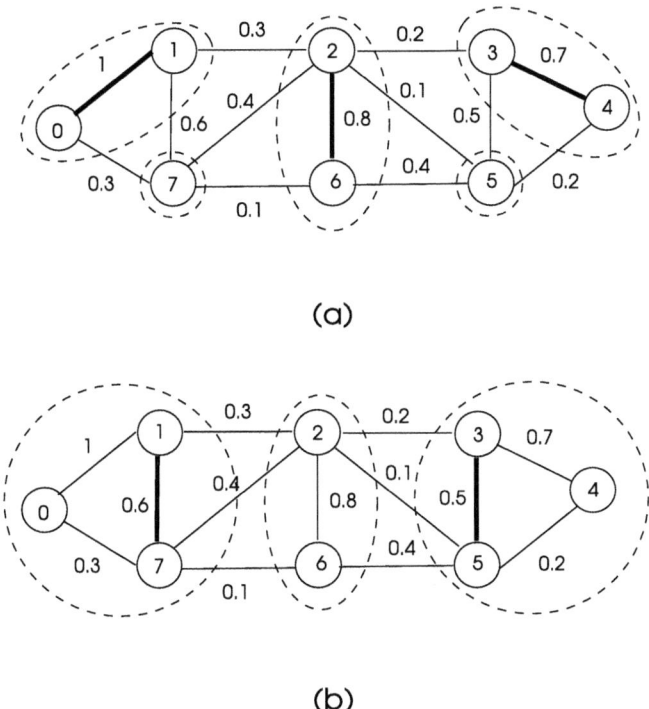

(a)

(b)

Fig. 15.7 Clustering using Algorithm 2

Algorithm 15.3 *Uncertain MST algorithm*

1: **Input**: $\mathcal{G} = (V, E, P)$ an UG
2: **Output**: MST T of $\mathcal{G}$
3: $G \leftarrow \mathcal{G}$ with new weights
4: $T \leftarrow \varnothing$
5: $S \leftarrow \{s\}$ ▷ s: root of MST
6: $mwoes \leftarrow$ lightest edge on s
7: **while** $S \neq V$ **do**
8:     **select** minimum edge (u, v) from $mwoes$ ▷ $u \in S, v \notin S$
9:     $S \leftarrow S \cup \{v\}$ ▷ mark v as visited
10:     $T \leftarrow T \cup \{(u, v)\}$ ▷ include (u, v) in MST
11:     $mwoes \leftarrow mwoes \cup$ lightest edge on v
12: **end while**

A sample weighted UG is depicted in Fig. 15.8a with edges labeled x, y; x showing the probability of existence of the edge and y denoting the cost of traversing the edge. The new weights of the graph are calculated in Fig. 15.8b after which running Prim's algorithm results in the MST shown in bold. The steps of executing Prim's algorithm are shown in parenthesis next to the edge weights.

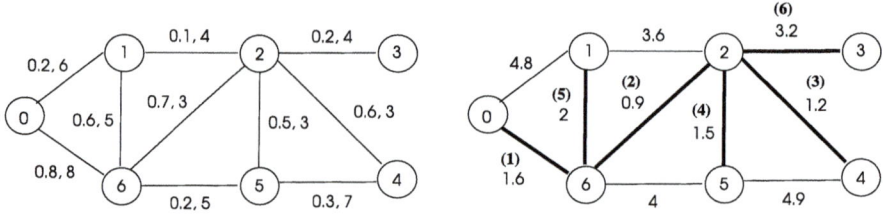

Fig. 15.8 **a** Sample UG graph, **b** its classical graph equivalence

Shortest path algorithms on UGs can be structured similarly, assigning weights to edges based on their probabilities and traversal weights. Then, the classical Dijkstra's single source shortest path algorithm or the Belman–Ford algorithm that can input negative weight edges may be applied to the newly labeled graph. Floyd–Warshal algorithm implemented in the new graph will find all pairs shortest paths (exercises).

The matching of a graph $G = (V, E)$ is the set $E' \subseteq E$ that are independent, that is, they do not have any common endpoints. Computing the minimum spanning tree (MST) of a weighted graph is a classical problem in graph theory. Searching an MST of a UG involves considering edges that have weights, and they exist with some probability [14]. In such a case, an edge of an UG will have two weights: first for its existence probability and the second for the cost of traversing that edge as in a weighted deterministic graph. The k-nearest neighbor search in an UG answers a query which requires to find the closest neighbors around a graph node which may represent a data point. A clique in a deterministic graph $G = (V, E)$ is defined as a subset V' of its vertices such that every vertex in V' is adjacent to all other vertices in V'. A k-clique is a clique with k vertices. A survey of computational problems for UGs can be found in [2].

15.4 Graph Mining

Data mining is the process of extracting useful hidden information from a large database. Common data mining methods include clustering, classification, and discovering frequent data patterns. Graph mining on the other hand is the process of extracting insightful knowledge from graph data. Hidden relationships between data is sought in this method through analysis. Fundamental methods for graph mining are listed in Fig. 15.9. Motif discovery which is the search of frequently occurring subgraphs of a graph we reviewed in Chap. 14 and finding clusters' detection in graphs are also considered as graph mining methods; however, graph mining is more general than these two methods with specific methods and algorithms for this purpose.

Graph embedding is the process of representing graphs in a low-dimensional vector space which can be used for link prediction and graph classification. Graph adjacency matrix or Laplacian matrix can be factorized to achieve this goal. Random

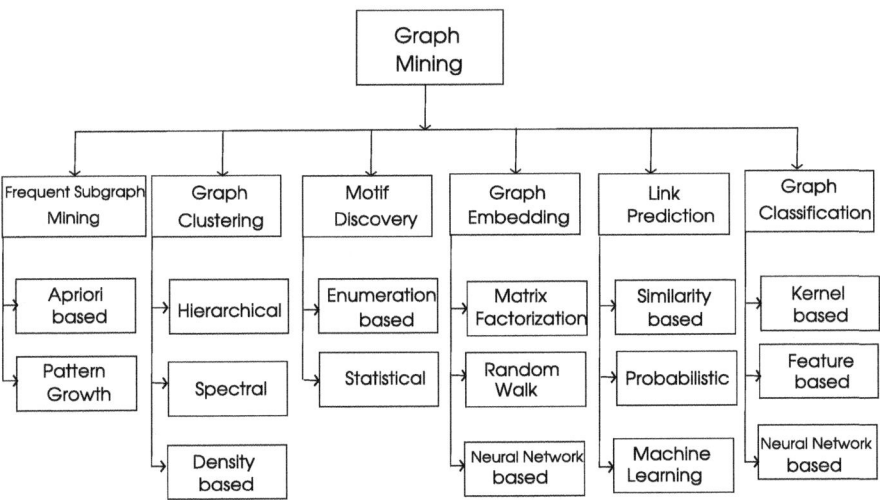

Fig. 15.9 Graph mining methods

walk-based methods use these walks to generate node embeddings and graph neural network embeddings (GNNs) we will review in Chap. 16. *Link prediction* method aims at discovering missing links (edges) or predicting future connections in a graph. Link prediction algorithms may evaluate node similarity scores using probabilistic methods and machine learning techniques. For example, two close nodes with high similarity scores in a social network may indicate the forming of a friendship link in future. Graphs are classified into predefined categories in *graph classification.* Kernel-based methods use graph kernels which are functions to compare and classify graphs. Feature-based methods extract graph features for graph classification and GNNs learn graph representations to group them. Some important terminology of graph mining are as follows. We will review the frequent subgraph mining methods for graph mining in the next section.

- *Support*: The support of a subgraph is the number of graphs in a graph dataset that contain the subgraph.
- *Frequent Subgraphs*: A subgraph is considered frequent if its support value is larger than a specified support threshold value.
- *Subgraph Isomorphism*: This is the process of determining whether a subgraph is contained in a given graph.

15.4.1 Frequent Subgraph Mining

The main goal of frequent subgraph mining (FSM) is to extract all of the frequent subgraphs in a given set of graphs that occur more than a specified threshold [11].

This method employs either BFS or DFS algorithm to grow candidate graphs in the first step and then tests whether these candidate graphs occur significantly in the dataset in the second step. Generation of duplicate graphs should be avoided in the first step; testing the occurrence of the candidate in the target graph is called *isomorphism checking*. The FSM method employs the following steps in general, to discover frequent subgraphs:

- Generate subgraphs, commonly by starting from small graphs and adding an edge or vertex at each iteration.
- Evaluate the support of the generated subgraph using subgraph isomorphism.
- Include a subgraph in the discovered frequent subgraph set if its support is greater than a threshold support value.

Two main methods of FSM are the *apriori-based approaches* and *pattern-growing approaches*.

15.4.1.1   Apriori-Based Approach

In this method of graph mining, which is also called pattern-merging method, a small graph is formed and an extra vertex, edge, or a path is added at each step to generate candidate graphs. A check for isomorphic subgraphs to the candidate graphs is then made, and if the number of discovered isomorphic graphs is greater than a threshold value, the candidate subgraph is considered to be frequent. This process continues in a bottom-up manner until a threshold size is reached. Frequent vertices or edges of order/size 1 are first detected in the graphs of the graph dataset $\mathcal{G}$ and are included in the frequent subgraph set $\mathcal{F}$ as shown in Algorithm 15.4. The size parameter k is incremented, and candidate subgraphs of size $k - 1$ are generated in the $while$ loop of the algorithm at line 8. For each generated candidate c, subgraph isomorphism test is performed for each graph G_i in the graph dataset $\mathcal{G}$ at line 11, and if this test is positive, the occurrence of the candidate is incremented. If the number of occurrences of the candidate c is greater or equal to the threshold value min_sup, the candidate is included in the frequent subgraph set F_k which is further included in the final output set $\mathcal{F}$ at the end of the kth iteration.

Algorithm 15.4 *Apriori FSM algorithm*

1: **Input**: A graph data set $\mathcal{G} = \{G_1, G_2, \ldots, G_n\}$, minimum support min_sup
2: **Output**: $\mathcal{F} = F_1, F_2, \ldots, F_k$: set of frequent subgraphs of order 1 to k
3: $\mathcal{F} \leftarrow \emptyset$, candidate set $C \leftarrow \emptyset$
4: $F_1 \leftarrow$ all frequent subgraphs of size 1 in $\mathcal{G}$
5: $\mathcal{F} \leftarrow \{F_1\}$
6: $k \leftarrow 2$
7: **while** $F_{k-1} \neq \emptyset$ **do**
8:    $F_k \leftarrow \emptyset$
9:    $C_k \leftarrow Candidate\_Gen(F_{k-1})$
10:   **for all** $c \in C_k$ **do**
11:      **for all** $G_i \in \mathcal{G}$ **do**
12:         **if** $subgraph\_isomorphism(c, G_i)$ **then**
13:            $support(c) \leftarrow support(c) + 1$
14:         **end if**
15:      **end for**
16:      **if** $support(c) \geq min\_sup \wedge c \notin F_k$ **then**
17:         $F_k \leftarrow F_k \cup \{c\}$
18:      **end if**
19:   **end for**
20:   $\mathcal{F} \leftarrow \mathcal{F} \cup F_k$
21:   $k \leftarrow k + 1$
22: **end while**

Key to the operation of this algorithm is the candidate generation and subgraph isomorphism algorithms. The former is performed by joining two $k - 1$ order subgraphs to generate a subgraph of size k, commonly by performing a BFS procedure.

15.4.1.2 Pattern-Growth Approach

Pattern-growth approaches for graph mining differ from apriori approaches that use *generate-and-test* method resulting in high computation costs due to candidate generation and subgraph isomorphism procedures. Pattern-growth algorithms eliminate candidate generation phase by generating subgraphs from the existing frequent subgraphs. These algorithms start by discovering frequent single-edge subgraphs initially and enlarge any frequent subgraph recursively, commonly by a DFS procedure although BFS growth is also possible. Any generated subgraph is checked for its support at each step, and whenever a grown subgraph does not meet the minimum support, another subgraph is processed using backtracking as displayed in Algorithm 15.5.

Algorithm 15.5 *Pattern-growth FSM algorithm*

1: **Input**: A graph data set $\mathcal{G} = \{G_1, G_2, \ldots, G_n\}$, minimum support m_s
2: **Output**: $\mathcal{F} = F_1, F_2, \ldots, F_k$: set of frequent subgraphs of order 1 to k
3: $\mathcal{F} \leftarrow \emptyset$, candidate set $C \leftarrow \emptyset$
4: **for all** $G_i \in \mathcal{G}$ **do**   ▷ generate single-edge frequent subgraphs
5:    **for all** each edge $e \in G_i$ **do**
6:       **if** $support(e) \geq min\_sup$ **then**
7:          $S \leftarrow S \cup \{e\}$
8:       **end if**
9:    **end for**
10: **end for**
11: **for all** $s \in S$ **do**
12:    **if** $support(e) \geq min\_sup$ **then**
13:       $F \leftarrow F \cup \{s\}$
14:    **end if**
15: **end for**
16: **for all** $g \in F$ **do**
17:    Perform DFS-based pattern growth from g
18: **end for**

Key to the operation of this algorithm is the extension of the frequent subgraphs obtained which can be done by adding a new edge in every possible direction. In order to prevent the discovery of the same graph, the gSpan [13] algorithm uses a rightmost extension method, where extensions to the discovered subgraphs occur only on the rightmost path. A rightmost path is found by going as deep as possible by performing a DFS from a vertex in the subgraph.

Python Implementation

A simple Python implementation of the Pattern-Growth FSM algorithm is displayed in Listing 15.6. This procedure simply calculates the support of a given subgraph in the graph database.

Listing 15.6 *UG clustering Algorithm 1 code*

```
1  # graph database
2  G1 = {(0,1), (1,2), (0,2)}
3  G2 = {(0,1), (1,2), (0,2), (2,3), (0,3)}
4  G3 = {(0,1), (0,2),(2,3)}
5
6  Graph_DB = [G1, G2, G3]
7  min_support = 2
8
9  # Find all unique initial edges and sort them
10 init_edges = set()
11 for graph in Graph_DB:
12     for edge in graph:
13         init_edges.add(edge)
14 init_edges = sorted(init_edges)
15
```

```
16  # Initialize candidates with single-edge subgraphs
17  candidates = [(edge,) for edge in init_edges]
18  freq_subgraphs = []
19
20  while len(candidates) > 0:
21      subgraph = candidates.pop(0)   # get first
        candidate
22      support = 0   # calculate its support
23      for g in Graph_DB:
24          if all(edge in g for edge in subgraph):
25              support += 1
26      if support >= min_support:
27          # Add to frequent subgraphs
28          freq_subgraphs.append((list(subgraph),
        support))
29          # Get last edge of candidate and extend
        over it
30          last = subgraph[-1]
31          indx = init_edges.index(last)
32          for edge in init_edges[indx+1:]:
33              new = list(graph) + [edge]
34              new = tuple(new)
35              candidates.append(new)
36
37  print("Frequent Subgraphs:")
38  for subgraph, support in freq_subgraphs:
39      print(f"Subgraph: {subgraph}, Support: {support
        }")
```

The graph database consists of three graphs: G_1, G_2, and G_3 displayed in Fig. 15.10. The output of the program in Listing 15.6 is listed in Listing 15.7 showing all possible subgraphs in the graph database that have a support more than or equal to the minimum support value of 2 (Fig. 15.11).

Listing 15.7 *Output of pattern-growth algorithm*

```
Frequent Subgraphs:
Subgraph: [(0,1)], Support: 3
Subgraph: [(0,2)], Support: 3
Subgraph: [(1,2)], Support: 2
Subgraph: [(2,3)], Support: 2
Subgraph: [(0,1), (0,2), (2,3), (0,2)], Support: 2
Subgraph: [(0,1), (0,2), (2,3), (2,3)], Support: 2
Subgraph: [(0,1), (0,2), (2,3), (2,3)], Support: 2
Subgraph: [(0,1), (0,2), (2,3), (2,3)], Support: 2
Subgraph: [(0,1), (0,2), (2,3), (2,3)], Support: 2
```

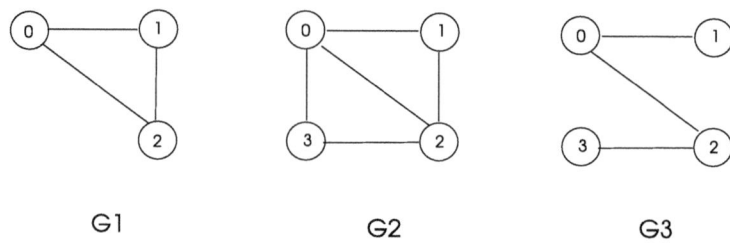

G1   G2   G3

Fig. 15.10 Graph database of three sample graphs

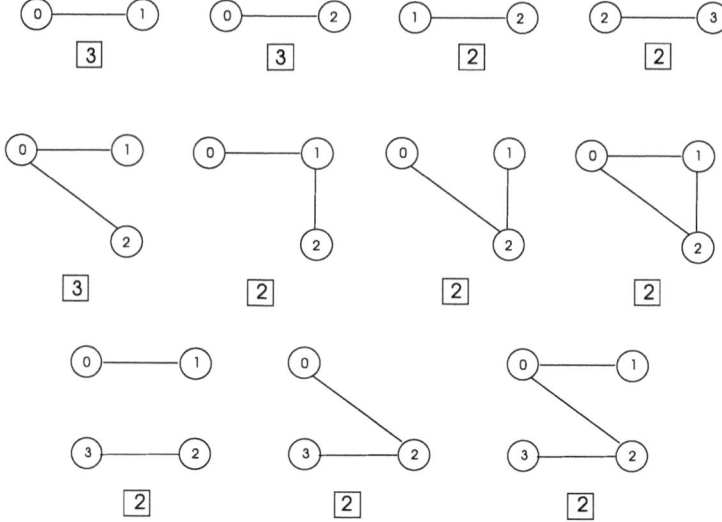

Fig. 15.11 Output of the Python program in Listing 15.6

15.4.2 Uncertain Graph Mining

Discovering clusters and forming MSTs in uncertain graphs (UGs) may be considered as methods of mining in these graphs. Querying UGs to detect patterns of subgraphs can be performed by the usual mining procedures but considering the probability threshold value of edge existences while calculating the support values.

Another well-studied topic in the area of mining UGs is the link prediction where snapshots $G_1, G_2, \ldots, G_T$ of the UG that represents a network are taken at times $t = 1, 2, \ldots, T$, for example, and the snapshot of G_{T+1} at time $T + 1$ is predicted [14]. Formally, this problem can be defined as follows:

Link prediction in UGs

Input: Snapshots of UG $G_t(V, E_t, P_t)$ at time instances $t = 1, 2, \ldots, T$
Output: Similarity matrix S where $S[i, j]$ shows the probability of existence of the edge $(i, j) \in E(G_{T+1})$

The link prediction in UGs can be applied to social networks to predict future friendship relationships and computer network analysis to anticipate link status. Other mining problems in UGs may be stated as *flow maximization*, where expected flow to a subset of vertices in a UG is maximized, and *node classification* in which labels of a subset of vertices are predicted given the labels of an initial subset of vertices in UG [2].

15.4.3 Knowledge Graph Mining

Knowledge graph (KG) mining imposes additional challenges than the well-investigated methods such as frequent subgraph mining of regular graphs. Dealing with the heterogeneity of representations, various labeling schemes of the vertices and edges in KGs are the fundamental difficulties encountered when mining these graphs. Moreover, unlike the regular graphs targeted in graph mining, KGs are commonly directed and have a single-connected component. Some studies involve converting the KG to a set of transactions and then implementing itemset-mining algorithms on these data to reduce the complexity [5,12]. Evaluating centrality values in KGs provides useful information about the structures of these graphs and these methods may be considered as an important class of mining in KGs. A number of tasks in mining KGs may be classified as below [1]:

- *Link Prediction*: As in UGs, link prediction in KGs involves evaluating the probability of existence of links in KGs. A commonly pursued method in this process involves evaluating the similarity between the nodes of a KG, assuming that similar nodes have a high probability of being connected in future.
- *Clustering*: This method aims at grouping nodes of a KG based on their structural and semantic similarities. Due to the complexity and heterogeneity of the node and edge data, this process is challenging. Moreover, user-comprehensible labels for the clusters need to be generated. For example, a KG may be formed for authors and their books and a cluster in such a KG may include authors who write on similar book topics.
- *Node Classification*: A node in a KG may have attributes shown with labels. Node classification involves attributing labels to the unlabeled nodes of a KG, commonly using deep learning methods of artificial intelligence.

Minimum Description Length principle of information theory may be used for graph mining of KGs [7]. This method is implemented by selecting the minimum description length data that best describes the KG and then implementing mining over this data [3].

15.4.4   Graph Mining Applications

Common applications of graph mining can be summarized as follows:

- *Social Network Analysis*: Graph mining in these networks aims commonly at discovering communities that are closely related. Examples of social networks are e-mail networks and collaborative networks. Partitioning of a social network is another area of investigation in social networks in which the graph representing a social network is divided into groups that have minimum interactions among them.
- *Web Mining*: A Web graph consists of a directed graph with Web pages as nodes and hyperlinks as edges between these pages. Web graphs are huge and dynamic making mining these graphs a difficult task. Web mining method aims at discovering patterns on the Web graph. It involves discovering relationships between Web pages and also the layout relationships between the pages in a Web location. Common mining tasks include finding clusters in Web graphs which is the discovery of Web pages related to a specific topic of interest: link prediction and anomaly detection such as finding spam pages. Link structures may be combined with content data such as text and images may improve mining of Web graphs. Mining of Web graphs requires scalable, distributed algorithms to manage huge and dynamic graphs representing the Web.
- *Cybersecurity Graphs*: Nodes of a cybersecurity graphs are computers and users, and edges are the interactions between the nodes. Mining of these graphs may provide information to detect threats, vulnerabilities, or anomaly behavior. This task commonly involves finding clusters for anomaly detection and evaluating centrality parameters to classify nodes. An important node found using centrality measures in a cybersecurity network may be critical for the operation of the network and thus needs to be protected. Anomaly detected can be a sign of a user trying to login to many servers in a short duration which may indicate a brute-force attack.

15.5   Chapter Notes

We reviewed two advanced graph structures in this chapter which are the knowledge graphs, uncertain graphs with algorithms implemented on these graphs for common problems of deterministic graphs. Knowledge graphs provide relationship information between their endpoints of their edges which make them suitable data structures to be used in artificial intelligence and natural language processing. Many real-life networks such as biological networks and wireless communication networks are constructed using unreliable data due to instability of the environments they exist. Uncertain graphs are used to model these networks using edges that have weights showing their existence probabilities. We described uncertain graphs and two clustering algorithms to build close groups in these graphs. Contemporary graph theory and graph algorithms research is mainly centered around these structures and their usages to model real-life networks and find solutions to classical graph problems in these relatively newly introduced graph models. Moreover, these structures find many implementations in machine learning as we review in Chap. 16.

Programming Exercises

1. Provide the pseudocode for a graph traversal algorithm for the KG of Fig. 15.8.
2. Provide pseudocode for Dijkstra's single source shortest path algorithm for UGs and then write the Python code for this algorithm and implement it on the UG graph example of Fig. 15.8a.
3. Implement the Python code in Listing 15.6 for edge weights $w(2, 7) = 0.7$ and $w(2, 5) = 0.9$ of Fig. 15.6 and display the clusters obtained.
4. Write the Python code for the Clustering Algorithm 2 for UGs and implement it on the graph of Fig. 15.7.
5. Provide Python code for Prim's MST algorithm for the UG graph example of Fig. 15.8a and show that the results are the same as in (b) of this figure.

References

1. Akef I. Knowledge graph mining: a survey of methods, approaches, and applications
2. Banerjee S (2022) A survey on mining and analysis of uncertain graphs. Knowl Inf Syst 64(7):1653–1689
3. Bariatti F, Cellier P, Ferre S (2023) KG-MDL: mining graph patterns in knowledge graphs with the MDL principle. Semant Web
4. Bizer C, Lehmann J, Kobilarov G, Auer S, Becker C, Cyganiak R, Hellmann S (2009) DBpedia—a crystallization point for the web of data. J Web Semant 7(3):154–165
5. Bobed C, Maillot P, Cellierand P, Ferré S (2020) Data-driven assessment of structural evolution of RDF graphs. Semant Web 11(5):831–853
6. Ferrucci D, Levas A, Bagchi S, Gondek D, Mueller ET (2013) Watson: beyond jeopardy! Artif Intell 199:93–105
7. Galbrun E (2022) The minimum description length principle for pattern mining: a survey. Data Min Knowl Discov
8. Gao X, Gao Y (2013) Connectedness index of uncertain graphs. Int J Uncertain Fuzziness Knowl Based Syst 21(1):127–137
9. https://docs.aws.amazon.com/neptune/
10. Kassiano V et al (2016) Mining uncertain graphs: an overview. In: International workshop of algorithmic aspects of cloud computing. LNCS. Springer, pp 87–116

11. Kuramochi M, Karypis G (2001) Frequent subgraph discovery. In: First IEEE international conference on data mining (ICDM'01), p 313
12. Ramezani R, Saraee M, Nematbakhsh M et al (2014) SWApriori: a new approach to mining association rules from semantic web data. J Comput Secur
13. Yan X, Han J (2002) gSpan: graph-based substructure pattern mining. In: Proceedings of 2nd IEEE international conference on data mining (ICDM'02), pp 721–724
14. Zhang A, Zou Z, Li J, Gao H (2016) Minimum spanning tree on uncertain graphs. In: International conference on web information systems engineering. Springer, pp 259–274

Graph Machine Learning

16

16.1 Introduction

Artificial intelligence (AI) is a multi-disciplinary field of science that aims to mimic human intelligence by performing various human functions. AI employs algorithms and models using high-performance computers to enable machines to learn and solve various tasks. AI systems input large amounts of data, analyze this data for patterns and correlations, and use this analysis to predict future states. The main human-like tasks performed by AI systems are outlined as follows:

- *Learning*: AI systems learn by training and investigating data to achieve the required solution. Learning by trial and error is the simplest method of learning in AI; when a solution after arbitrary trials is found, it can be stored to be used when such an input is encountered in future. Applying the past solutions to similar problems is called *generalization*. *Deep learning* employing neural networks is used to analyze large complex data.
- *Reasoning and Problem Solving*: Reasoning is the process of drawing inferences from situations, and it can be deductive or inductive. Problem solving involves a systematic search of possible actions to reach the required solution. AI can form hypothesis based on its input and make decisions through reasoning and problem solving; it uses algorithms to find the optimal solution to a problem through trial-and-error procedures.
- *Natural Language processing* (NLP): AI systems understand, translate, and generate human language using various methods such as text analysis and machine translation.
- *Perception*: This process is sensing, interpreting, and deriving information from environment data. In the case of AI, data are input from sensors, cameras, signals and analyzed using image and signal processing techniques. Speech recognition and computer vision are some common AI-based perception methods.

© The Author(s), under exclusive license to Springer Nature Switzerland AG 2026
K. Erciyes, *Guide to Graph Algorithms*, Texts in Computer Science,
https://doi.org/10.1007/978-3-032-05294-0_16

AI implementations are based on two main approaches: *symbolic* and *connectionist* approaches. Symbolic methods follow a top-down approach, whereas connectionist methods use a bottom-up approach using artificial neural networks to imitate neural networks of the brain.

In this chapter, we first review machine learning which is a fundamental component of any artificial intelligence system. We then describe *knowledge graphs* and *graph neural networks* which are two graph-based methods used in contemporary AI research and applications.

16.2 Machine Learning

Machine learning (ML) is a branch of artificial intelligence (AI) where a machine learns from experimenting with data and makes future predictions. Data input to a ML system may be in various forms such as numerical or categorical. A numerical data may be expressed by a number such as the weight of an entity and categorical data may simple be the gender of a person. From a different angle, data input to a ML system can be in the form of a *training*, a *validation* or a *testing set*. The training set is used to train the model used in ML, and test set provides the evaluation of the performance of the model. Data are transferred into *information* which provides meaningful interpretation of data to the application. *Knowledge* is higher-level interpretation of data by combining information, learning, and experience. The main methods in ML are supervised learning, unsupervised learning, and reinforcement learning. Deep learning is a fundamental component of machine learning, commonly implemented using neural networks. Deep learning is a subset of machine learning which is a subset of artificial intelligence as depicted in Fig. 16.1.

ML is used in a number of diverse applications ranging from health care to entertainment. Doctors can use ML to detect and diagnose diseases; self-driving cars employ ML systems to drive and navigate the vehicle reliably. Marketing decisions

Fig. 16.1 Relationships between artificial intelligence, machine learning, and deep learning

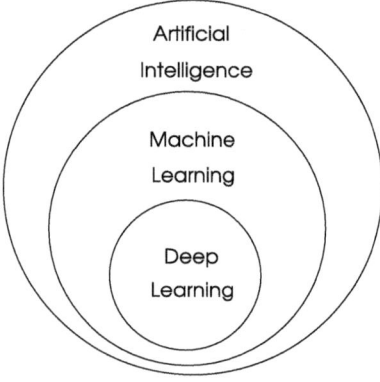

may be made by analyzing customer behavior using ML methods, and preferred movies and personalized music may be presented to users based on analyzing their customary behaviors. An important area of ML application is natural language processing (NLP) where ML techniques are used to for translation between languages and understanding human talk.

16.2.1 Supervised Learning

Supervised learning (SL) method of ML uses labeled datasets for training to classify data and predict the outputs. The training dataset to SL includes inputs and correct outputs to provide learning to the model. SL can be of two basic types:

- *Classification*: This process involves assigning input data to classes that are defined beforehand. Data are assigned one of two distinct classes in *binary classification*, for example, determining an incoming e-mail is spam or not may be decided in this method. The algorithms used for this method of supervised learning include logistic regression and support vector machines. In *multi-class classification*, input data are assigned to one of multiple classes. Decision trees, random forests, and neural networks are commonly employed for multi-class classification.
- *Regression*: The predicted output values obtained in this method of supervised learning are continuous. Let us consider the base area and the price of houses in a particular area as the given training set to a ML system that employs supervised learning through regression. Given any input house with base area, the ML system should predict a possible price for that house. This method uses various algorithms including linear regression, logistic regression, decision trees, support vector machines, and k-nearest neighbors.

16.2.2 Unsupervised Learning

Unsupervised ML algorithms are given unlabeled data and attempt to find relationships in this data without any prior knowledge and, thus, are called unsupervised and sometimes *self-learning* algorithms. The main tasks implemented by an unsupervised learning algorithm are *clustering*, *association rule mining*, and *dimensionality reduction*.

Clustering

The clustering method aims at grouping objects based on their similarities and may be implemented by various algorithms outlined as follows:

- *Exclusive Clustering*: A data point belongs to only one cluster in this method of clustering. A commonly implemented algorithm is k-means clustering algorithm in which k as the number of clusters needed is specified by the user.
- *Overlapping Clustering*: Clusters may overlap such that a data point may belong to more than one cluster.
- *Hierarchical Clustering*: A cluster may be embedded in a larger cluster. In *agglomerative clustering*, each data point is a cluster initially; points with similar attributes are merged to form clusters. Alternatively, the whole data is a single cluster initially in *divisive clustering*; this cluster is divided into smaller and smaller clusters until certain requirements are met.
- *Probabilistic Clustering*: Clusters are created by using probability distribution in this method. A data point with a low probability in being a cluster may be corrupted.

Association Rule Learning

Association rules can be used to find associations between different subjects in a dataset. For example, based on historical data, one can find out what items are usually bought together. Associations between different items in a dataset may be used to form association rules. For example, a customer buying cornflakes is likely to buy milk. This type of rules may be represented by the conditional *if X then Y* where $X =$ buying cornflakes and $Y =$ buying milk. Once these rules are established, predictions may be made such as placing milk bottles next to cornflakes. An *apriori algorithm* is used to find the most frequent items in a dataset to form an association rule.

Dimensionality Reduction

Large data can make the visualization difficult and also affect the performance of ML algorithms, and thus, there is a need to compress data. *Dimensionality reduction* method reduces the number of data inputs to smaller sized while preserving data integrity. A commonly used techniques for this purpose are principal component analysis (PCA) which reduces redundancies in data and compresses datasets. In another approach, singular value decomposition (SVD) method factorizes a matrix into three low-rank matrices to reduce dimension. Unsupervised learning can be used in various applications such as image and video processing, customer profile creation in business, and anomaly detection in data.

16.2.3   Reinforcement Learning

Reinforcement learning (RL) is a type of unsupervised learning method by interacting with the environment and receiving feedback in the form of rewards or penalties. The goal in this method is to maximize the cumulative reward over time. The main application areas of this unsupervised learning procedure are robotics, autonomous systems, and game systems. Some of the commonly used terms in RL are as follows:

- *Agent*: Agent is the learning entity that makes decisions on the environment.
- *Environment*: This is any entity that the agent interacts with.
- *Action*: An action is performed by the agent on the environment.
- *State*: A state is determined by the environment after each action.
- *Reward*: Feedback from the environment after an agent performs an action.
- *Policy*: Based on the current state, an agent uses policy to determine the next action which will cause a change in the state.

In *value-based* RL, the agent attempts to find the optimal value function in a state under any policy. The optimal policy that yields maximum future rewards is searched by actions in *policy-based* approaches of RL. These methods may be further divided into *deterministic* and *stochastic* types where the same action is formed at any state in the former and the action is provided with a probability in the latter. The *model* is the representation of the environment to predict future states and rewards.

An agent in RL learns by trial and error and by getting rewards for the correct actions. As an example, let us consider a robot through a maze which has a reward in one of the locations and various hazards such as fires in few locations. The robot makes a move and learns its way through trial and error to reach the reward.

16.3   Neural Networks

Artificial neural networks (ANNs or simply NNs) are inspired by biological neural networks of the brain. A neuron in a NN receives inputs and processes them typically by calculating the weighted sum of the inputs after which an activation function is applied to this sum to produce the output. A neuron in this model either fires or not depending on the final value of the processed inputs which may be below a threshold or not. The first artificial neuron model that can adjust weights applied to the inputs called the *perceptron* was developed by Rosenblatt [9]. The perceptron takes n inputs $\{x_1, ..., x_n\}$ and provides a single binary output. A single neuron in this model consists of the following components as shown in Fig. 16.2.

- The synapses or connecting links that have weights, w_i, to the input values, x_i for $i = 1, ...n$
- An adder that sums the weighted input values and a bias value.
- A function σ that processes the sum and produces the output.

The initial perceptron model allowed only binary input values $x_i \in \{0, 1\}$ and a step function as the activation function.

Formally, the operation of a single neuron can be represented in Eq. 16.1 where w_i are the weights associated with x inputs, b is the input bias value, and $\sigma : \mathbb{R} \to \mathbb{R}$

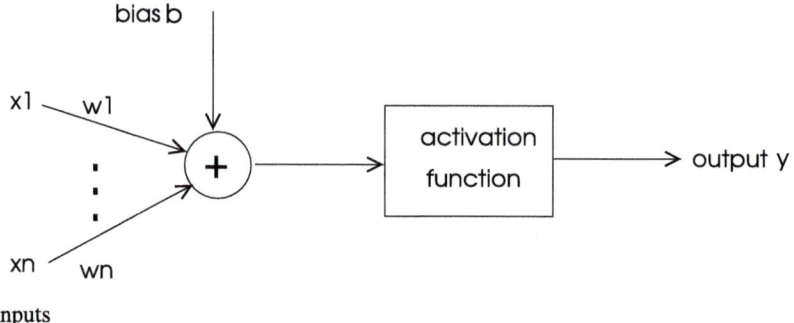

Fig. 16.2 Single neuron model

Fig. 16.3 Sigmoid function

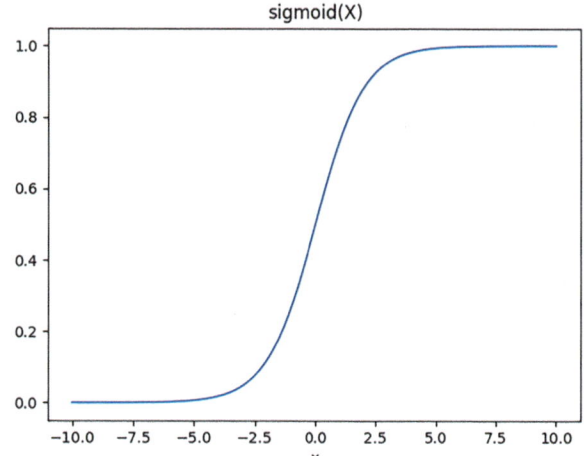

is the *activation function*. This function is used to decide whether the neuron should be activated or not.

$$y = \sigma \left(\sum_{i=1}^{n} w_i x_i + b \right) = \sigma (W^T x + b). \tag{16.1}$$

Commonly used activation functions are the sigmoid and the hyperbolic tangent functions shown in Eq. 16.2 with their plots in Figs. 16.3 and 16.4.

$$\sigma(y) = \frac{1}{1 + \exp(-y)}, \quad \sigma(y) = \tanh(y) = \frac{\exp(y) - \exp(-y)}{\exp(y) + \exp(-y)}. \tag{16.2}$$

A NN consists of layers of neurons connected in sequence: the *input layer*, *hidden layers*, and the *output layer*. The input layer receives the raw data, hidden layers are the intermediate layers that process data, and the output layer which delivers the output of the NN. Neurons in this networks are connected with edges that are

Fig. 16.4 Tanh function

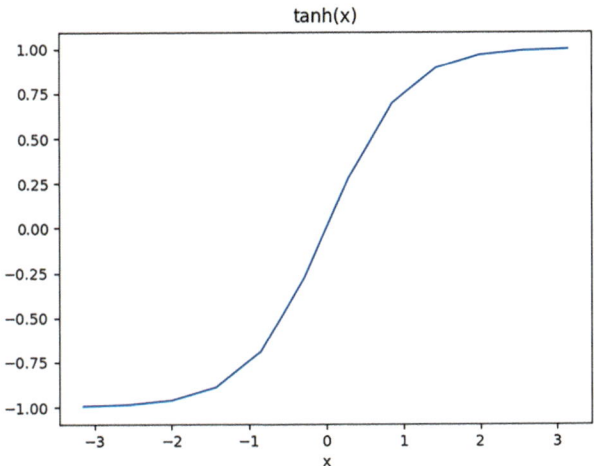

labeled with weights. Each neuron in this structure operates as a single neuron, delivering its output based on Eq. 16.1 to a neuron in the next layer. The weights in these connections are adjusted to provide the required output during the *training* or *learning* process.

16.3.1 Feedforward Neural Networks

A feedforward neural network (FNN) consists of layers of neurons where information flows in one direction from the input layer to the output layer through hidden layers, without any cycles or loops. A FNN with one hidden layer is shown in Fig. 16.5a, and another FNN with two hidden layers is displayed in (b) of the figure.

Data in a FNN are processed in sequence, from one layer to another until the output layer which is called *forward propagation*. Each neuron computes the weighted sum of its inputs and a bias value, applies an activation function to this sum, and transfers its output to the neurons of the next layer. In supervised learning, a FNN may start from a random choice of weight values and the obtained output is compared to the correct output supplied, using a *loss function* such as mean squared error for regression and cross-entropy loss for classification. The gradients of the loss function with respect to the weights and biases are evaluated, and the values of weights and biases are updated accordingly which is called the *training phase* or *backpropagation* of the FNN. These three steps, forward propagation, computation of loss, and backpropagation, are repeated for a number of times until the output values converge.

The Python code in Listing 16.1 shows how to create a single neuron using *numpy* library functions where the *sigmoid* function is used as the activation procedure to the output. The training function *train*, inputs the sample input, the supervised output for this input and the number of iterations for the training. It computes the output for the

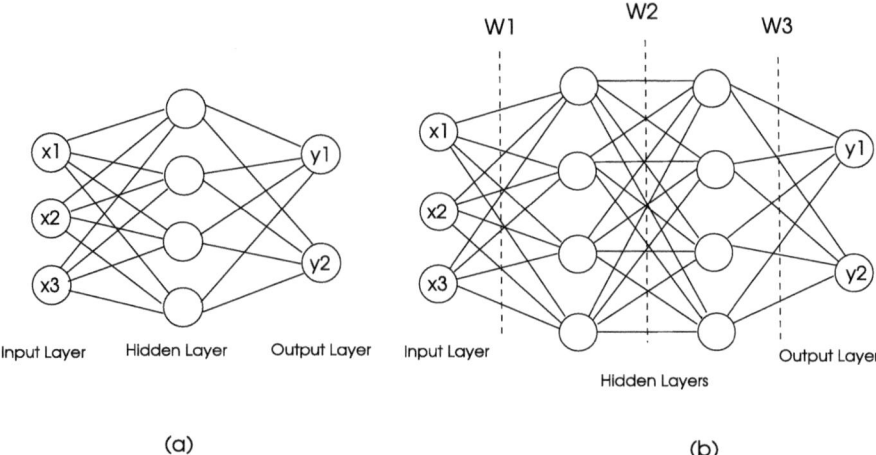

Fig. 16.5 a Neuron model

current weight values, calculates the error and the adjustment which is the product of the input matrix and the derivative of the sigmoid function, and updates weight values based on the adjustment. Performing these operations for a large number of times provides convergence after which weight matrix values become stable, concluding the learning phase. We then test the stable neuron circuit for three arbitrary inputs and observe that the outputs are stable approximating unity. The training input is the matrix I and the output is the array O as shown in the following:

$$I = \begin{pmatrix} 0 & 0 & 1 \\ 1 & 1 & 1 \\ 1 & 0 & 1 \\ 0 & 1 & 1 \end{pmatrix}, \qquad O = \begin{pmatrix} 0 \\ 1 \\ 1 \\ 0 \end{pmatrix}$$

Listing 16.1 Construction of a single neuron network

```
import numpy as np
import numpy as np

def sigmoid(x): # the sigmoid function
    return 1 / (1 + np.exp(-x))

def sigmoid_derivative(x): # derivative of the
    sigmoid function
    return x * (1 - x)

def fwd_propagate(inputs): # forward propagate
    inputs.astype(float)
    output = sigmoid(np.dot(inputs, weight_matrix))
    return output
```

```
14
15 def train(train_inputs, train_outputs,num_train):  #
16     global weight_matrix
17     for i in range(num_train):
18         output = fwd_propagate(train_inputs)
19         # Calculate the error in the output.
20         error = train_outputs - output
21         adjustment = np.dot(train_inputs.T, error *
    sigmoid_derivative(output))
22         # Adjust the weight matrix
23         weight_matrix += adjustment
24
25 tr_inputs = np.array([[0, 0, 1], [1, 1, 1], [1, 0,
    1], [0, 1, 1]])
26 tr_outputs = np.array([[0, 1, 1, 0]]).T
27
28 np.random.seed(1)
29 weight_matrix = 2 * np.random.random((3, 1)) - 1
30 print("Initial Weight Matrix:\n",weight_matrix)
31 train(tr_inputs,tr_outputs,15000)
32 print("Trained Weight Matrix:\n",weight_matrix)
33 # Test the neural network for new input
34 print ("Input: [1,0,1]  output: {}".format(
    fwd_propagate(np.array([1,0,1]))))
35 print ("Input: [1,1,0]  output: {}".format(
    fwd_propagate(np.array([1,1,0]))))
36 print ("Input: [0,0,1]  output: {}".format(
    fwd_propagate(np.array([1,1,1]))))
```

The outputs for three inputs from the single neuron are shown in Listing 16.2

Listing 16.2 Output of the code in Listing 16.1

```
1 Initial Weight Matrix:
2  [[-0.16595599]
3  [ 0.44064899]
4  [-0.99977125]]
5 Trained Weight Matrix:
6  [[10.08740896]
7  [-0.20695366]
8  [-4.83757835]]
9 Input: [1,0,1]  output: [0.99477899]
10 Input: [1,1,0]  output: [0.99994884]
11 Input: [0,0,1]  output: [0.99358625]
```

Example Python code for the forward propagation of the three-layer neural network of Fig. 16.5b is listed in Listing 16.3 in Python. The Python *numpy* library functions are used to provide the inputs to the layers by summing the input weights and the bias. The three input values, three bias values, and the values of weight matrices W_1, W_2, and W_3 are assigned randomly. The first weight matrix W_1 between the input layer and the first hidden layer has a dimension of (4×3) since the hidden layer has four neurons each of which has three inputs. Similarly weight matrix W_2

is a (4 × 4) matrix as the second hidden layer has four neurons with four inputs each from the first hidden layer; and the last weight matrix W_3 is (2 × 4) as the output layer has two neurons with four inputs each from the second hidden layer. A single forward propagation from this neural network resulted in the output displayed in Listing 16.4 with $h1$ and $h2$ as the input values to the first and second hidden layers.

Listing 16.3 Forward propagation example

```
 1  import numpy as np
 2
 3  f = lambda x:1.0/(1.0+np.exp(-x))    # sigmoid
        function
 4  np.random.seed(1)
 5  x = 2*np.random.random((3,1))-1    # generate random
        values
 6  b = 2*np.random.random((3,1))-1
 7  W1 = 2*np.random.random((4,3))-1
 8  W2 = 2*np.random.random((4,4))-1
 9  W3 = 2*np.random.random((2,4))-1
10
11  h1 = f(np.dot(W1,x) + b[0])
12  print("h1 \n",h1)
13  h2 = f(np.dot(W2,h1) + b[1])
14  print("h2 \n",h2)
15  out = np.dot(W3,h2) + b[2]    # output values for a
        single run
16  print("out",out)
```

Listing 16.4 Output of the code in Listing 16.3

```
 1  h1:
 2   [[0.44500093]
 3    [0.29947763]
 4    [0.72730733]
 5    [0.34476137]]
 6  h2:
 7   [[0.39006863]
 8    [0.51555574]
 9    [0.1720145 ]
10    [0.39596837]]
11  out:
12   [[-0.52669219]
13    [-0.56848351]]
```

FNNs are efficient and simple to implement and thus are used in many applications such as classification of objects as in image classification, speech recognition and regression. FNNs are the basic building blocks of more complex neural networks structures like convolutional neural networks and recurrent neural networks. *Deep learning* is a subset of ML that focuses on using neural networks to process complex data.

16.3.2 Learning Process in Neural Networks

In summary, learning or training process in a NN involves adjusting the weights of the neuron connections based on the input data and the target output. Specifically, the learning consist of the following steps:

- *Forward Propagation*: The input data are passed through the network layers to compute and produce the output.
- *Calculation of Loss*: The loss is calculated as the difference between the predicted output and the obtained output.
- *Backward Propagation*: The loss or error is propagated back through the neural network, and the weights are adjusted to minimize the error.
- *Use of Activation Functions*: Activation functions introduce nonlinearity into the network and are used to decide on the value of the output. Some commonly used activation functions are the *sigmoid, tanh,* and *rectified linear unit* (ReLU).

16.3.3 Convolutional Neural Networks

Convolutional neural networks are extensively used in image processing tasks such as object detection and image classification, where the input data are typically large in size. These networks use convolution defined in Eq. 16.3 for two functions f and g, to reduce the size of image data.

$$f(t) * g(t) = \int_{\infty}^{\infty} f(t)g(x - t)dt \qquad (16.3)$$

This operation is performed by a sliding function g over another function f, multiplying them at each point of overlap, and summing up the results. Convolution is widely used in signal processing to filter some frequencies while suppressing some others. In image classification, convolution is performed by applying filters to the input image by way of convolution with the selected matrix kernels. The filter matrix is slid over the input image matrix, elements are multiplied, the computed products are summed, and the results are assembled to form the output image.

16.3.4 Recurrent Neural Networks

Recurrent neural networks (RNNs) are designed specifically to process sequential and ordered data where the current output may depend on previous time steps; they can have directed cycles to enable storage of the previous inputs. The main difference between a RNN and FNN or CNN is that the RNNs have memory as they are fed with information from prior inputs to compute the current input and the output. RNNs use forward propagation and *backward propagation through time* (BPTT) algorithms to compute the gradients. The BPTT algorithm works similar to classical backpropagation training by calculating errors and adjusting weights; however, the errors are summed at each time step in BPTT.

In the basic RNN model, the output at each time step depends on both the input and the hidden state from the previous time step. They are commonly implemented for tasks that process data sequentially in language modeling, speech recognition, and real-time such as in sensor systems where output predictions are to be made immediately. Long short-term memory (LSTM) RNN has cells consisting of an input gate, an output gate, and a forget gate in the hidden layers of the neural network. These gates control the flow of information depending on the input, such as estimating the next word of a sentence. Encoder–decoder RNNs encode an input sequence to a vector, and the decoder uses this vector to generate output sequence as used in machine translations.

Neural networks are used in a wide range of applications including image and video processing for facial recognition and object detection; in natural language processing enabling language translation, understanding and generating human language; in healthcare for disease diagnosis, medical data analysis; and in finance for fraud detection and stock market prediction.

16.4 Graph Neural Networks

Graph neural networks (GNNs) are a type of neural networks that can input, process, and output graph data. The need for this type of networks has risen since many real-life systems such as biological networks and social networks to be processed by machine learning systems are in the form of graphs. A GNN node collects information from its neighbor nodes through a process called *message passing* after which it can update its information to obtain an *embedding* or a *feature vector*. Each node of a GNN is initialized with a feature vector. The main steps of operation of a GNN may be summarized as below:

- *Message Passing*: Each node exchanges feature messages with its neighbors.
- *Aggregation*: Each node aggregates neighbor messages using functions such as calculating mean or sum of received values.
- *Updating*: Each node combines its features with the aggregated features from the neighbors using the neural network.
- *Propagation*: The steps of this process are repeated through a number of GNN layers.

The update rule for a node of a GNN at layer k is given in Eq. 16.4

$$h_v^{k+1} = Update(h_v^k, Aggregate(\{h_v^k : u \in N(v)\}))$$
(16.4)

where $N(v)$ is the set of neighbors of node v, h_v^k is the representation of node v at layer k, *Aggregate* function aggregates features of the neighbor nodes of node v, and *Update* function is used to update the representation of node v.

Graph data input to a GNN consists of the adjacency matrix of the graph and the node feature matrix. The adjacency matrix $A^{n \times n}$ has elements $A_{ij}=1$ if nodes i and j are adjacent in the usual sense. The node feature matrix $X^{n \times m}$ has n rows for n nodes and m columns for m features. For example, if the input graph represents a social network of n individuals, each row of X may represent {age, gender, country}. GraphSAGE is a framework for inductive representation learning on large graphs [1] using GNNs. GraphSAGE learns representation of a node in the graph based on some combination of its neighboring nodes, parametrized by h. Each node v has a feature vector x_v, h_v^0 is its initial node embedding representation, h_v^k is its embedding representation at iteration k as shown in the pseudocode of this algorithm in Alg. 16.1, and the final output for a node v is z_v. The main idea of GraphSAGE is to represent a node based on information from its neighbors.

Algorithm 16.1 GraphSAGE

Algorithm

1: **Input**: Adjacency matrix A of graph $G = (V, E)$ and node feature matrix x
2: **Output**: Vector representations $z_v, \forall v \in V$
3: $h_v^0 \leftarrow x_v, \forall v \in V$ ▷ Set initial node embeddings to feature vector
4: **for** $k = 1$ to N **do**
5:     **for all** $v \in V$ **do**
6:         $h_{N(v)}^{k+1} = Aggregate(h_u^k, \forall u \in N(v)))$ ▷ Aggregate and Update iteration
7:         $h_v^{k+1} = \sigma(W \cdot Concat(h_u^k, h_{N(v)}^{k+1})))$
8:     **end for**
9:     $h_v^k \leftarrow h_v^k / \left\| h_v^k \right\|_2, \forall v \in V$ ▷ Normalise node embeddings
10: **end for**
11: $z_v \leftarrow h_v^N, \forall v \in V$

GNNs are commonly used for graph-related machine intelligence tasks such as link prediction between the two nodes of a network, community detection in a network, graph embedding, and graph generation. GNNs are influenced by convolutional neural networks (CNNs) that are commonly used for image processing as noted. Recurrence neural networks (RNNs) are typically employed for text processing as described. GNNs can be classified as a combination of these networks with graphs as follows:

- *Graph Convolutional Networks* (GCNs): A GCN is a type of GNN that inputs a graph and explores information about the nodes using its neighbors and updates the representation of nodes based on the features of their neighbors. The formula in Eq. 16.5 is modified to reflect neighbor nodes features to provide the linear transformation h_i for node i, note that the effect of nodes with high degrees is

reduced by dividing the sum of neighbor features with the degree of the node
$\deg(i)$.

$$h_i = \frac{1}{\deg(i)} \sum_{j \in N(i)} \mathbf{W} x \tag{16.5}$$

- *Recurrent Graph Neural Networks* (R-GNNs): These networks combine the strong
  features of GNNs and RNNs to handle sequential or dynamic graph data. The
  Update function in R-GNNs may be expressed as follows:

$$h_v^k = \text{RNN}(h_v^k, \text{Aggregate}(\{h_u^k : u \in N(v)\})) \tag{16.6}$$

where h_v^k is the hidden state of node v at time k. As stated in Eq. 16.6, The
Update function is in fact implementing RNN operation over the aggregated features. R-GNNs are used for dynamic social network analysis, traffic prediction
and recommendation systems.
- *Gated Graph Neural Networks* (GGNNs) : These networks use gating mechanisms
  to allow the network to update and forget information. Neighbor node features
  are aggregated and updated as in a GNN; however, gating function is used to
  determine how much of the new information is to be used to form the new state
  of a node. GGNNs are implemented for node classification, link prediction, and
  graph classification tasks.
- *Graph Auto-encoder Networks* (GAENs): These networks are a type of GNNs
  with an encoder and a decoder component; the encoder maps the input graph into
  a set of node embeddings or graph-level embeddings. The decoder reconstructs
  the graph from the representations formed by the encoder. The training of a GAEN
  is accomplished by minimizing a loss function evaluating the difference between
  the original graph and the constructed graph. GAENs are used for tasks such as
  link prediction and anomaly detection in networks.

Graph Neural Networks Using Knowledge Graphs

GNNs may be effectively used with knowledge graphs as they can learn meaningful
representations of nodes and relationships on the edges. Some common implementations of GNNs over knowledge graphs are as the following:

- *Node Classification*: GNNs may classify nodes based on their attributes; for example, nodes may be grouped into cities or countries in a knowledge graph showing
  what city belongs to which country.
- *Graph Classification*: GNNs may be used to detect similar subgraphs within a
  knowledge graph which is in fact another and efficient method of graph mining in
  knowledge graphs.
- *Link Prediction*: GNNs can learn embeddings of nodes and edges in a knowledge
  graph, thus they can predict the likelihood of a relationship between the nodes
  which may provide detection of missing or potential links in such a graph.

- *Knowledge Graph Reasoning*: GNNs may be implemented in a knowledge graph for tasks such as answering to a query or inferring knowledge based on the existing information.

In order to implement a GNN over a knowledge graph, the graph is first constructed, and node and edge features are initialized. GNN layers are then implemented on this graph through message passing, and node and edge features are updated. Note that we have additional updates of edge features in the case of using a knowledge graph as input to a GNN. The node and edge features can now be used for specific tasks such as link prediction, query answering, and classification.

16.5 Quantum Computing

Computers today are based on classical physic principles with the assumption that a system can be in one state only at any given time. A quantum system however can be in a *superposition* of many states at the same time as defined in quantum physics. *Quantum computing* is based on computers that work using quantum mechanical principles and thus merges two principles; quantum physics and computer science. Data in a classical computer are stored in bits which can be 0 or 1. Data unit in a quantum computer is *qubit* which is a superposition of 0 and 1. Comparison of classical computing and quantum computing is summarized in Table 16.1.

Time complexities of some time consuming tasks using matrix computations in classical computing and quantum computing are displayed in Table 16.2 where significant speedups are obtained using quantum computations as can be seen.

Table 16.1 Comparison of classical computing and quantum computing

	Classical computing	Quantum computing
Input	Binary bits	Quantum bits (qubits)
Hardware	Digital circuits	Quantum circuits
Algorithm	Classical algorithms	Quantum algorithms
Output	Program output	Quantum measurements

Table 16.2 Comparison of classical computing and quantum computing

	Classical computing	Quantum computing
Matrix inversion	$O(n^2)$	$O(\log n)^2$
Eigenvectors	$O(n^2)$	$O(\log n)^2$
Fast Fourier transform	$O(n \log n)$	$O(\log n)^2$

16.5.1 Quantum Circuits

A *quantum circuit* works on qubits using *quantum gates*. In other words, a quantum circuit is like a logical circuit of *and*, *or*, and *not* gates, but the gates used in this circuit are quantum gates.

16.5.1.1 Basic Quantum Gates

These Identity (I), Not (X), and Hadamard (H) gates comprise the fundamental building blocks for quantum circuits [15]. The I matrix does not change the state of a qubit; thus, it can be used as a buffer between two circuits. The X gate inverts the state of the qubit, that is, $|0\rangle$ changes to $|1\rangle$ and $|1\rangle$ changes to $|0\rangle$. The H gate transforms a qubit into superposition of two states, and it can be used to define the inputs of a quantum circuit for all possible qubit states. These gates in matrix form are shown in Eq. 16.7. The Y and Z gates perform qubit amplification and qubit phase shifting.

$$I|0\rangle = \begin{pmatrix} 1 & 0 \\ 0 & 1 \end{pmatrix}, \quad X|0\rangle = \begin{pmatrix} 0 & 1 \\ 1 & 0 \end{pmatrix}, \quad H|0\rangle = \frac{1}{\sqrt{2}} \begin{pmatrix} 1 & 1 \\ 1 & -1 \end{pmatrix} \quad (16.7)$$

The Controlled-NOT gate (CNOT gate) operates on two qubits: a control qubit and a target qubit. The CNOT gate flips the state of the target qubit if and only if the control qubit is in the state $|1\rangle$; otherwise, the target qubit remains unchanged if the control qubit is in the state $|0\rangle$. The possible transform of a qubit using a CNOT gate is the following: $|00\rangle \rightarrow |00\rangle$, $|01\rangle \rightarrow |01\rangle$, $|10\rangle \rightarrow |11\rangle$, $|11\rangle \rightarrow |10\rangle$, thus inverting the target qubit if the control qubit is $|1\rangle$.

IBM's qiskit tool [2] can be used to build quantum circuits. The Python code using this tool shown in Listing 16.5 builds a simple quantum circuit with one qubit, one control bit, and one classical bit. The *QuantumCircuit* function creates a quantum circuit with these properties. The qubit is initialized to $|0 >$ after which it is inverted and a measurement storage is attached. Then, a H gate is applied, and another measurement storage is attached to the output of the H gate. Lastly, the CNOT gate initialized to $|1 >$ is attached to the output.

Listing 16.5 Quantum circuit construction

```
1  from qiskit import QuantumCircuit
2
3  # Create a quantum circuit with two qubits and one
      classical bit
4  qc = QuantumCircuit(2, 1)
5
6
7  qc.reset(0)      # Initialize the qubit to the state
      |0>
8  qc.x(0)          # Apply the X gate to the qubit
9  qc.measure(0, 0) # Measure the qubit
10 qc.h(0)          # Apply H gate to the qubit
```

```
11  qc.measure(0, 0)  # Measure the qubit
12  qc.cx(1,0)   # set control qubit q1
13
14  # Draw the circuit
15  print(qc.draw())
```

The output of this algorithm is depicted in Fig. 16.6.

16.5.2 Quantum Algorithms for Graphs

One way to solve graph problems employing quantum computation is to employ a quantum procedure in the classical solution of the problem. A favorable candidate for a quantum procedure that can be used in a number og-f graph algorithms is the search algorithm by Grover. The following is a list of procedures commonly used in quantum algorithms to solve graph problems:

- **Grover's Search Algorithm**: Provides a quadratic speedup for unstructured search problems, useful for finding specific elements or patterns in graphs.
- **Quantum Walks**: Generalization of classical random walks, used for graph traversal and search.
- **Quantum Annealing**: Used for optimization problems like graph partitioning and maximum cut.
- **Amplitude Amplification**: Enhances the probability of finding correct solutions in quantum algorithms.
- **Quantum Linear Algebra**: Used for solving linear systems and performing matrix operations, which are often involved in graph algorithms.

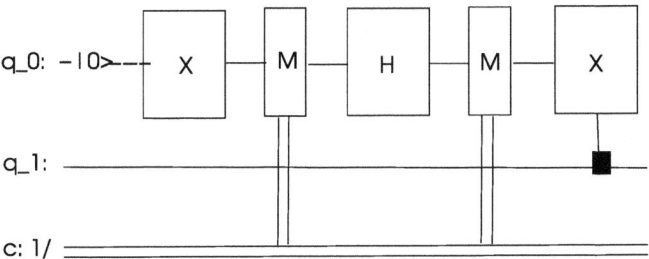

Fig. 16.6 A quantum circuit

16.5.2.1 Grover's Algorithm

Grover's (Search) algorithm is a fundamental quantum algorithm to solve the unstructured search problem in which we search for an element in an unstructured list of N elements. This problem can be solved in $O(N)$ steps using a classical algorithm; however, Grover's algorithm provides the solution in $O(\sqrt{N})$ steps time with high probability [3]. The main idea of this algorithm is to increase the chances of obtaining the right solution by amplitude amplification using an oracle. A high-level description of Grover's algorithm is as follows:

1. *Prepare the Initial State*: A quantum register as a superposition of all possible states is prepared, each representing a potential solution. Hadamard gates are used in this stage to provide an equal superposition of all states.
2. *Implement the Oracle*: Grover Oracle is a quantum operation that marks the possible correct solution.
3. *Apply the Grover Diffusion Operator*: This step amplifies the probability amplitude of the correct solution while reducing the probability of other solutions, thereby increasing the chance of finding the required item.
4. *Repeat Grover Iteration*: The iteartion consisting of steps 2 and 3 is repeated approximately ($\frac{\pi}{4}\sqrt{N}$) times to reach a target state that is amplified significantly.
5. *Measure the Output*: Measure the qubits that correspond to the correct solution.

This algorithm may be used as a basis for various other problems such as solving optimization problems, in cryptography and discovering patterns in data in quantum machine learning.

16.5.2.2 Quantum Minimum Spanning Tree Algorithms

Computing MST of a weighted graph has numerous applications such as clustering and building road maps. We reviewed three classical algorithms for this problem as Kruskal's algorithm, Prim's algorithm, and Boruvka's algorithm in Chap. 7. The quantum versions of these algorithms make use of the quantum procedures described. The classical Kruskal's algorithm starts by sorting the edges of the weighted graph with respect to their weights and includes edges starting from the lightest one in the MST as long as they do not form a cycle with the current partial MST. The quantum version of this algorithm has a similar structure but uses Grover's search algorithm to find the minimum-weight edge that does not form a cycle as shown below.

Quantum Kruskal's Algorithm

Input: An weighted graph $G = (V, E, w)$ with $w : E \rightarrow \mathbb{R}$
Output: An MST of G
Steps:

- **Repeat**

  1. Sort the edges with respect to their weights. (Classical or using Quantum Sort)
  2. Use Grover's search to find the minimum-weight edge e that does not form a cycle with the edges already in the MST.
  3. Add edge to MST.

- **Until** all vertices are processed.

Complexity: reduced from $O(2^n)$ to $O(2^{\sqrt{n}})$ using Grover's algorithm.

Moreover, the union-find data structure is commonly used in Kruskal's MST algorithm, and quantum version of this data structure may be used to improve Kruskal's algorithm. Prim's algorithm to construct and MST of a weighted graph starts from any vertex and adds the minimum-weight outgoing edge to the partial MST at each iteration. The quantum version of this algorithms works similarly but search for MWOE is performed using a quantum algorithm due to Durr and Hoger, reducing the time complexity for this operation from $\log E$ to $\sqrt{E}$ as shown in the pseudocode below. Total time complexity of this algorithm may be reduced to $O(n\sqrt{m})$ from $O(m \log n)$ using the quantum approach.

Quantum Prim's Algorithm

Input: An weighted graph $G = (V, E, w)$ with $w : E \rightarrow \mathbb{R}$
Output: An MST of G
Steps:

- **Repeat**

  1. Start with any initial vertex $v \in V$.
  2. Use Durr and Hoyer's quantum minimum finding algorithm to find the minimum outgoing edge (MWOE) from the current MST.
  3. Add edge to MST.

- **Until** all vertices are processed.

Complexity: reduced from $O(\log E)$ to $O(\sqrt{E})$ at each step to find MWOE due to quantum minimum finding algorithm.

16.5.2.3   Sample Algorithms

We review sample classical graph problems which are matching, the minimum spanning tree (MST), and shortest path computation which can be solved by quantum algorithms in this section.

Quantum Maximum Matching Algorithm

A matching of an unweighted graph $G = (V, E)$ is a subset E' of its edges that do not share any endpoints. A maximal matching MM can not be enlarged by the addition of a new age, and the maximum matching MaxM has the largest cardinality among all matchings of the graph G. We reviewed matching algorithms for graphs in <Chap. The quantum matching algorithms provide significant speedups compared to classical matching algorithms.

Quantum Maximum Matching Algorithm

Input: An unweighted graph $G = (V, E)$
Output: Maximum matching MM of G
Steps:

1. Encode all possible matchings into a superposition of quantum states.
2. Use Grover's search to amplify the states corresponding to valid matchings.
3. Iteratively apply amplitude amplification to find the maximum matching MM.

Complexity: reduced from $O(2^n)$ to $O(2^{\sqrt{n}})$ using Grover's algorithm.

Quantum Breadth-First-Search (BFS) Algorithm

The quantum BFS algorithm provides improvement over the classical BFS algorithm which has a time complexity of $O(n + m)$. The BFS algorithm is a graph traversal algorithm that explores the vertices at each level iteratively from a given source vertex. The quantum BFS algorithm can perform the traversals in parallel using superposition. Moreover, Grover's algorithm may be used to find all of the neighboring vertices in parallel. We describe a high-level description of a quantum walk-based BFS construction in the following pseudocode.

Quantum Breadth-First-Search Algorithm

Input: An unweighted graph $G = (V, E)$ and a source vertex s
Output: BFS tree of G rooted at vertex s
Steps:

1. **Initialization**: Represent vertices as quantum states and edges as transitions.
2. Initiate a *quantum walk* through multiple levels of the graph simultaneously.
3. Perform *amplitude amplification* so that the probability of measuring states corresponding to vertices at the next level of the BFS tree is improved

Complexity: Visiting all unvisited neighbors of a vertex set is reduced from $O(n)$ to $O(\sqrt{n}$.

Quantum Depth-First-Search (DFS) Algorithm

The classical DFS algorithm traverses a graph starting from a source vertex s and goes as deep as possible before backtracking. This algorithm has a time complexity of $O(n+m)$ which can be improved using the quantum parallelism and superposition. It can be performed using quantum walks or Grover's algorithm. The quantum walks-based algorithm has a similar structure to quantum walk-based BFS algorithm where the graph is encoded to a quantum system by representing the vertices as quantum states and edges as transitions after which a quantum walker is initialized at the source vertex. A quantum walker is then performed in parallel to explore the graph. Amplitude amplification provides improved probability of unvisited neighbor state values.

16.5.3 Quantum Programming Tools

Various tools and simulators to program quantum computers exist some of which are described as follows:

- **Qiskit** [2]: Qiskit is a widely used open-source tool developed by IBM Quantum. It provides facilities to develop quantum algorithms, develop quantum circuits, and access to quantum processors.
- **Cirq** [4]: Cirq provided by Google enables construction and simulation of quantum circuits. This tool can be employed by users who want to control quantum hardware.
- **PyTorch** Integration [5]: The PyTorch machine learning tool is integrated with Qiskit to use both the machine learning facilities of PyTorch and the quantum circuit construction of Qiskit.
- **TensorFlow Quantum** [6]: This tool is implemented by providing an extension to the machine leaning framework TensorFlow supplying improved machine learning algorithms using quantum computing.

Public-key cryptography algorithms such as RSA system relies on the difficulty of factoring very large integers; the Diffie–Hellman algorithm is based on discrete logarithm problem. Both of these systems can be broken by Schor's quantum algorithm which means that new crypto algorithms which will be difficult to be broken by quantum computers are to be designed in future.

16.6 Quantum Machine Learning

Machine learning (ML) has been applied to numerous fields including natural language processing, finance, image processing, speech recognition. In spite of these successful applications, computational power and scalability are the main barriers to overcome by the ML algorithms. The huge data size in a classical ML algorithm requires increased computational resources such as processor power and memory size, resulting in decreased performance when these are not available. Also, classical neural networks have limitations when handling training on a small dataset and performing optimizations.

Quantum machine learning (QML) combines basic principles of quantum computing and ML to enable solving typical machine learning problems using quantum computations to achieve improved performance. QML uses quantum computers to achieve fast processing of large data sets. This recent research area also involves employment of ML methods to improve quantum theory and computation in the reverse direction. A common approach pursued by researchers in this emerging field is to investigate implementation of core supervised or unsupervised ML algorithms such as clustering and pattern matching algorithms using quantum computations. QML can be divided into three main branches: classical ML with quantum data, quantum speed-us for classical ML, and quantum algorithms used on quantum data [7]. The first QML class involves feeding quantum data into a classical ML system and using classical ML algorithms on this data. Classical data are encoded as quantum states and subroutine of a classical algorithm is converted to a quantum algorithm in the second class. The last class of QML systems work with quantum data and quantum algorithms; a common implementation of these systems is the quantum neural networks.

Common quantum ML algorithms include quantum support vector machines (QSVMs) as the quantum version of the classical support vector machine algorithm, quantum k-Means clustering which is the quantum version of the k-means clustering algorithm, and quantum neural networks. Some of the common QML applications are drug discovery and modeling where QML may be used for the analysis of chemical systems and to predict molecular properties; clustering and pattern recognition for grouping of similar entities and detecting anomalies, and natural language processing.

16.6.1   Quantum Neural Networks

Quantum neural networks (QNNs) employ quantum neurons which are represented by qubits as the basic unit to perform computations [8]. These neurons exist in superposition of states enabling then to process multiple inputs in parallel which results in faster training than the classical neural networks. Input layer to a QNN is a set of qubits converted into quantum states from classical data using methods such as amplitude encoding and basis encoding. The hidden layers comprise quantum circuits constructed using quantum gates and the output layer commonly provides the measurement of qubits which can then be transferred to classical information to produce the final output.

A variational quantum circuit (VQC) uses rotation operator gates with free parameters to perform various numerical tasks, such as optimization, approximation, and classification. A VCQ consists of three components as a standard quantum circuit: an initial state, a quantum circuit, and a measurement. A VCQ functions as a hybrid quantum and classical processing, and the algorithms are executed on a quantum computer; however, the optimization process is performed using classical algorithms. A neural network's activation function can be replaced with quantum entanglement layers as quantum gate operations are reversible linear operations. This modification provides a multi-layer architecture to quantum circuits making them suitable for neu-ral network like functioning. A variational quantum algorithm (VQA) uses VQCs to solve a variety of problems including optimization and eigenvalue problems. VQAs are the basic building blocks of quantum neural networks.

A VQC-based QNN first encodes the input data to a corresponding qubit state using amplitude encoding or base encoding as stated; this qubit state is then transformed using rotation gates and entangling gates, and the transformed state is measured by calculating the expected value of the system Hamiltonian operator such as Pauli gates. The output is then decoded to an appropriate output data format.

Applications of Quantum Neural Networks

QNNs provide significant speedups for various computational tasks than the tradi-tional NNs. Some common applications of QNNs are as follows:

- *Pattern recognition*: QNNs can recognize patterns in high-dimensional data and thus can be used for pattern search and image processing.
- *Quantum Chemistry*: QNNs are used to simulate molecular and atomic interac-tions, and thus provide means of drug discovery.
- *Optimization Problems*: QNNs can be used to solve complex optimization prob-lems in finance and materials science.
- *Cryptography and Security*: QNNs can detect anomalies in cybersecurity data.
- *Natural Language Processing* (NLP): QNNs may be used for various NLP tasks including text classification and translation.
- *Biomedical Applications*: QNNs may be used to analyze complex biological data to infer phylogenetic relationships.

In conclusion, QNNs provide significant advantages in terms of performance over classical NNs for a variety of tasks. However, development of reliable quantum hard-ware and efficient quantum algorithms have challenges such as noise and implemen-tation which makes realization of hybrid classical-quantum computers more practical than building a full quantum computer.

And there is still need for a significant research to overcome these problems. Ongo-ing research in this area is commonly pursued in building novel QNN architectures and designing quantum algorithms

16.6.2 Quantum Graph Neural Networks

GNNs can process data represented as graphs present in networks such as social networks and sensor networks, as noted. GNNs process information by iteratively using aggregation of neighbor nodes. These networks have high computational and space complexity and

Quantum graph neural networks (QGNNs) are tailored to process graph data using QNNs. In other words, a QGNN is a combination of a GNN and quantum comput-ing. A QGNN employs quantum encoding of graphs into quantum states, quantum message passing, and quantum measurements and evaluations. More specifically, the following key tasks are performed in these networks:

- *Quantum Graph Encoding*: Graphs are encoded into quantum states using methods such as qubit representation of nodes or edges and amplitude encoding.
- *Quantum Message Passing*: Quantum operations among neighbors are performed to update and aggregate data.
- *Post Processing*: Commonly, quantum outputs are measured and fed into classical neural networks to finalize the outputs.

QGNNs can be implemented conveniently in *quantum chemistry* for drug discovery; complex network analysis tasks such as community detection and routing in large networks.

16.7   Chapter Notes

A core component of artificial intelligence is machine learning which implements deep learning, commonly using neural networks. In this chapter, we briefly reviewed basic artificial intelligence concepts and terminology starting with supervised learning, unsupervised learning, and reinforcement learning. The expected output for a data set is provided to a machine learning system in supervised learning with the expectation that this system learns and implements the rules on new data. The unsupervised learning systems do not have the expected output and can be used in a variety of applications that use clustering, associative rule mining, and dimensionality reduction.

We described basic neural networks which form the basis of deep learning systems with Python examples. A neural network consists of an input layer, a number of hidden layers, and an output layer. It performs forward propagation, computation of loss function, and backpropagation operations to achieve convergence to the given outputs through a number of iterations. The weight values of the neuron inputs are adjusted accordingly to achieve the approximation to the outputs. We reviewed very briefly convolutional neural networks that are mostly used for image classification and Recurrent Neural Networks with main implementations in text and natural language processing.

A graph neural network (GNN) is a type of neural network that inputs graph data and uses basic principles of neural networks to process this data to obtain graph output. These networks are commonly used in graph-related tasks such as link prediction and cluster detection. A GNN which inputs knowledge and uncertain graphs is another recent area of research that GNNs are implemented. Lastly, we described quantum computation basics with quantum gates and fundamental quantum graph algorithms. Quantum machine learning combines machine learning with quantum computation and is a promising research area providing significant improvements over classical machine learning processes. We investigated quantum neural networks which are also a recent research topic in machine learning. These networks implement machine learning tasks using quantum circuits with better performances than the classical neural networks. Quantum graph neural networks use the principles of both graph neural networks and quantum computation, thus providing high performance to solutions of tasks represented and modeled by graphs. They act as the convergence of these networks assuming graph data as input but convert this data to qubits, provide solutions using quantum circuits and transform the quantum output to graph data outputs. Most of the topics we have outlined briefly in this chapter are contemporary research areas at the intersection of graph theory, algorithms, machine intelligence, and quantum computations.

Exercises

1. Extend the Python code of a single forward propagation of Listing 16.3 to a full learning neural network that learns with the following input I and output O values for 10000 iterations. Run the algorithm for arbitrary inputs and observe that the output has stable values.

$$I = \begin{pmatrix} 1 & 0 & 0 \\ 1 & 1 & 0 \\ 1 & 0 & 1 \\ 0 & 1 & 0 \end{pmatrix}, \quad O = \begin{pmatrix} 0 & 1 \\ 1 & 1 \\ 1 & 0 \\ 1 & 0 \end{pmatrix}$$

2. Compare graph neural networks with convolutional neural networks by outlining their differences and similarities.
3. Describe the working principles of QNNs. What are the main differences between a QNN and a NN?
4. Describe an example of VQC and VQA.
5. Compare GNNs with QGNNs and state the main application areas of each.
6. Build and display a quantum circuit with one qubit that implements a X gate to the input.

References

1. William LH, Ying R, Leskovec J (2017) Inductive representation learning on large graphs. In: 31st Conference on neural information processing systems (NIPS 2017), Long Beach, CA, USA
2. https://www.ibm.com/quantum/qiskit
3. Grover, Lov K (1996-07-01) A fast quantum mechanical algorithm for database search. In: Proceedings of the twenty-eighth annual ACM symposium on Theory of computing - STOC '96. Philadelphia, Pennsylvania. Association for Computing Machinery, USA, pp 212–219
4. https://quantumai.google/cirq
5. https://qiskit.org/ecosystem/machinelearning/tutorials/05 torch connector.html
6. https://www.tensorflow.org/quantum

7. K. Beer (2022) Quantum neural networks. Ph.D. thesis, University of Hannover
8. Kak SC (1995) Quantum neural computing. In: Hawkes PW (ed) Advances in imaging and electron physic: Elseviers, pp 259–313
9. Rosenblatt F (1958) The perceptron: a probabilistic model for information storage and organization in the brain. Psychol Rev 65(6):386

Epilogue

17

17.1   Introduction

We have described the fundamentals of sequential, parallel, and distributed graph algorithms in Part I. We then reviewed basic graph algorithms which may be used as building blocks to solve more complicated problems related to graphs in Part II which formed the core of the book. We looked at sequential, parallel, and distributed algorithmic solutions to the problems studied in this part. In the final part, we first briefly reviewed algebraic and dynamic graph algorithms. Then, our emphasis was on large graphs that represent real-life large networks commonly called complex networks. Graph algorithms for such large networks required new analysis approaches and efficient algorithms as we noted.

Our aim in this final chapter is to briefly provide a guide when dealing with a new problem which can be modeled by a graph. We attempt to sketch a road map when a graph algorithm is needed for a task we have and we may not have an existing graph algorithm to solve the problem. Once we have a workable method for the problem at hand, we need to investigate the options of using sequential, parallel, or distributed algorithms for this purpose. Moreover, relatively new methods of algebraic and dynamic graph algorithms may be conveniently used. Algebraic graph algorithms allow the use of existing matrix library functions in sequential or parallel form and dynamic graph algorithms are needed when edge deletions and insertions occur as in many real-life networks. We describe where to use these methods and conclude the chapter with a case study.

17.2   Road Map for Difficult Problems

We know by now most of the problems encountered in graph world are NP-Hard except few ones such as the matching problem where we searched for the disjoint edges with the maximum size. Any new problem we face will probably not have a solution in polynomial time. We will attempt to specify the steps to follow in such a case as follows.

- In-depth understanding of the basic graph algorithms is very helpful. These algorithms such as the DFS and BFS algorithms may be used as the building blocks to solve a more complicated problem. In many cases, a modified form of the basic algorithm can be used. We saw how simple DFS algorithm with some modifications can be used for various problems such as finding articulation points, bridges, strongly connected components, and the blocks of a graph.
- When dealing with an NP-hard graph problem, we can search an approximation algorithm if it exists. In many cases however, approximation algorithms are rare and attempting to design a new one is not a trivial task. After all, if one can come up with a new approximation algorithm that has a better proven approximation ratio than existing ones, this can be published in an article. In some cases, we may opt to use an approximation algorithm that has a slightly worse approximation ratio than the best available one, due to the complexity of implementing a better algorithm. For example, finding the minimal vertex cover of a graph using matching is a simple algorithm with an approximation ratio of 2.
- In the more common case, use of heuristics is unavoidable. Choice of a heuristic largely depends on the nature of the problem at hand. When we are to design a new graph algorithm or modify an existing one, we have only few properties to begin with. Especially in the case of altering an existing algorithm for our purpose, degree of nodes can be incorporated to break symmetries or to directly select a node to work on. Degree of neighbors or degree of k-hop neighbors can also be used. We can define new parameters which use degrees and neighbors and their relationship. The clustering coefficient of a vertex for example shows how well-connected neighbors of that vertex are.
- For problems related to large graphs, we may opt for an approximation algorithm with better performance than a polynomial algorithm to solve the problem due to the high execution times involved. Moreover, parallelization is always helpful to improve performance.
- A computer network consists of autonomous nodes that function independently. For network problems, we need efficient distributed algorithms which are executed by the network nodes. These algorithms commonly use neighbor information to find local solutions which are then used to find a global solution to a problem.

17.3   Are Large Graph Algorithms Different?

We can use any of the algorithms we have developed in Part II for large graphs. However, even a linear time algorithm may be problematic due to the size of these large graphs. Employment of the following techniques is frequently needed in such large graphs.

- *Use of Heuristics*: Heuristics are commonly used in solving NP-hard graph problems as we noted. In some cases, one may opt for a heuristic algorithm that has, for example, $O(n)$ complexity using a heuristic than a deterministic algorithm that has $O(n^2)$ complexity. This improvement in performance may be significant for a large graph to decide to use the heuristic solution.
- *Scalable Parallel Algorithms*: Parallel algorithms are needed in the analysis of large graphs representing complex networks due to the magnitude of the graph. This method is a necessity rather than a choice in such implementations.
- Distributed Algorithms: In the case of a large computer networks such as the Internet or a wireless sensor network, distributed algorithms are needed.

17.4 Conversions: When Are They Useful?

We have three modes for graph algorithms: sequential, parallel, and distributed as emphasized throughout the book. We may then have the following possible conversions:

- *Sequential to Parallel*: We can either design a parallel algorithm from scratch or convert an existing sequential algorithm to a parallel one. Two commonly used methods in the latter are distributing data (data parallelism) and distributing code (functional parallelism) when we have a distributed memory parallel processing system as we have reviewed in Chap. 4. While using the shared memory model of parallel processing, we need to provide control of shared address space using locks, semaphores, or other mechanisms. Message passing is widely used for parallel processing in distributed memory computers.
- *Sequential to Distributed*: Distributed algorithms run at the nodes of a computer network to solve a problem related to the network. Each node cooperates by communicating with its neighbors in finding the result and a global solution is commonly reached by a special node which may then transfer the result to individual nodes. We can design a distributed algorithm from scratch or convert an existing sequential algorithm to a distributed one as in the parallel case. Converting a sequential algorithm to a distributed one requires detailed analysis of what to communicate and when. A well-known problem in a computer network is routing where we search for shortest paths between each pair of nodes. We described how to convert the sequential Bellman–Ford algorithm that uses dynamic programming to find shortest paths from a source node. The main difference between the sequential and the distributed algorithms is having each node compute its shortest distance to the root based on its current information in each round. We need synchronization in each round which can be performed by a special node. In summary, we need to make sure synchronization is flawless while using a synchronous distributed algorithm. Asynchronous distributed algorithms are more flexible but in general are more difficult to design than the synchronous ones.

- *Distributed to Parallel*: Let us assume we have a network and a distributed algorithm A to solve a problem B related to the network and each vertex of the graph represents a computing node of the network in the general sense. Our goal is to provide a solution to P in a parallel computing environment and then transfer the results to each node. A possible conversion of algorithm A to a parallel algorithm B may involve the following steps. We coarsen the graph iteratively using a suitable method such as edge contraction or star contraction to obtain supernodes each of which represents a number of nodes with incident edges of the original graph. We can then assign each supernode to a parallel processing system where each process p solves the problem for the supernode it is assigned. It may communicate with the neighbors of the supernode in the coarsened graph using the distributed algorithm A. The results may be collected at a root node which computes the global result by merging the results from each process p and transfers the results to each node. We need to consider the border vertices between each partition carefully as symmetry breaking such as using unique identifiers may be needed to assign a border vertex to a neighbor supernode for convenience. Let us consider the matching problem in a computer network and we want to find the solution in a parallel computer with k processes. We should have each network node to know whether any one of its incident edges is included in the final matching M. We can coarsen the graph to k subgraphs of supernodes $G_1, \ldots, G_k$ and assign each G_i to a process p_i to have each p_i work out the matching M_i in its supernode G_i. The processes can now communicate with neighbor supernodes as if they are the nodes of a network to find the global solution either using a supervisor or in fully distributed manner.
- *Parallel to Distributed*: Given a parallel algorithm B to solve a graph problem P, we want to know whether converting B to a distributed algorithm A is more convenient than designing a distributed algorithm from scratch or convert an existing sequential algorithm to a distributed one for problem P. Our strategy in this conversion can be to work in the reverse direction of the method used in distributed algorithm to parallel algorithm conversion this time. Instead of coarsening, we attempt to refine the parallel algorithm to the level where each parallel process p_i is responsible for a single vertex of the graph and thus we can partition each row of the adjacency or the distance matrix to a distinct process. The restriction we have is that the process can only communicate with its neighbors. If this is possible, we have a distributed algorithm that can run on autonomous nodes of the network.

17.5   Implementation

Once we have some way of solving the problem, we need to consider the implementation choices. We saw three ways of implementations that consider data flow in an algorithm for a given problem: sequential, parallel, or distributed. From another perspective, the data structures used and the environment the algorithm is practiced play an important role to classify the algorithm as classical, algebraic, dynamic, or sometimes all.

17.5.1 Sequential, Parallel, or Distributed

We have devoted most of the book to sequential graph algorithms with sample parallel and distributed algorithms to specific graph problems. The size of the problem and the environment it is implemented is crucial in deciding whether we should look for a parallel or a distributed algorithm. For a large graph representing a complex network, parallel algorithms are commonly required to provide efficiency. On the other hand, if we are to solve a network problem in which network nodes participate in finding the solution, we need to search a distributed algorithm for the task at hand. In many cases, these boundaries are not so clear. For example, we may have a wireless sensor network with a cluster of computing nodes used as the sink and hundreds of sensing nodes. Solving a problem such as routing or more complicated problems can be handled by nodes performing some local operation and sending their data to the sink using a distributed algorithm; the sink finding the solution efficiently using parallel computing and then sending the result to the individual nodes using the distributed algorithm.

As a general rule, we can say that parallel computing is required in solving problems related to complex networks represented by large graphs. These problems such as network motif search or network alignment are NP-hard in many cases which require use of heuristics, and even such implementations take considerable time due to the huge size of the graph. The speedup obtained by such a parallel algorithm is the ratio of the sequential time to the parallel computing time and the efficiency is defined as the speedup divided by the number of processing elements used.

When we are dealing with a problem in which network nodes represent graph vertices, we should look for efficient distributed algorithms. We saw single initiator synchronous distributed algorithms are frequently used in such cases due to their relatively ease of implementation. The number of rounds and the total number of messages exchanged to terminate the algorithm provide us a good indication of its performance.

17.5.2 Classical, Algebraic, Dynamic, or All?

The main part of this book including parts I and II and most of Part III is dedicated to graph algorithms that can be considered as classical algorithms in a sense that traditional algorithmic techniques such as greedy, dynamic, and divide and conquer methods are employed. We reviewed alternative and relatively more recent method of algebraic graph algorithm design in Chap. 12. This method makes use of main matrices associated with a graph: adjacency matrix, incidence matrix, and the Laplacian matrix. The review of solutions to few graph problems showed the performances of such algorithms are commonly inferior to their classical counterparts. However, the algebraic method provides simpler algorithms and more importantly, a vast library of matrix operations in sequential and parallel form are readily available for use in such algorithms. For example, if the algebraic algorithm involves matrix multiplication, we already have a method to perform this operation in parallel by row, column, or block partitioning. Therefore, we do not need to spend a lot of time to find a parallel algorithm for the problem studied if we can form the algebraic solution using basic matrix operations.

Dynamic graph algorithms are a necessity rather than a choice since real networks are almost always dynamic with frequent addition and deletion of edges. Examples of these networks are the Internet, the Web, social networks, and biological networks of the cell. Instead of running a known static (classical or algebraic) algorithm for the modified network from scratch, it is sensible to design algorithms that make use of the existing network information and solve the problem faster. We noted the main challenge in the design of dynamic graph algorithms lies in the design of clever data structures so that modifications can be handled quickly.

Lastly, we investigated dynamic algebraic graph algorithms which are even a less investigated area than algebraic and dynamic graph algorithms. When we have an algebraic method to solve a graph problem, there is the possibility to have a dynamic version of this technique by using some known result from linear algebra and also by using a dynamic matrix library operation. We have seen such a procedure in dynamic algebraic matching when an algebraic solution to this problem was combined with dynamic matrix operations and a theorem from linear algebra.

In conclusion, graph algorithm design using the traditional approaches will be used for small to moderate size problems. Moreover, they commonly provide the basic design methods to be used in algebraic or dynamic algorithms for graphs. However, when we are searching for a solution in a large graph, parallel processing is needed and such operation can be handled more conveniently by algebraic graph algorithms. Dynamic graph algorithms are needed for dynamic networks for better performance. In many cases, dynamic networks such as protein interaction networks, social networks, and the Internet are large. We can therefore conclude dynamic algebraic graph algorithms will continue to be an effective research area in the future.

17.6   A Case Study: Backbone Construction in WSNs

Let us elaborate on a case study to illustrate the guidelines we have expressed until now. We need to cluster nodes of a WSN for the general benefits obtained from such process. Electing a cluster head (CH) eases various tasks such as routing since the CH may perform these tasks on behalf of the nodes in its cluster. This hierarchical structure is clearly useful in managing any establishment including countries. However, we need a spanning tree in the WSN to broadcast various commands from the root and also aggregate the data of the sensors to the root. We may use two distinct algorithms for our purpose but a closer look reveals that two tasks can be performed by one algorithm in a more efficient way. We will call this process *spanning tree-based clustering*. The general idea of the new algorithm is to build clusters and the spanning tree simultaneously. Each cluster will be a subtree in the spanning tree.

We now search for a spanning tree algorithm and see if we can modify this algorithm for our purpose. We saw how to build a spanning tree using flooding in Chap. 5. Erciyes et al. presented an algorithm that builds a spanning tree and forms clusters simultaneously using flooding [2]. The main idea of this algorithm is to keep a record of the depth of the spanning tree obtained during the iterations and assign the nodes of the tree within every *d* hops to a cluster.

The root node starts the algorithm by sending the first *probe* message, and the nodes receiving this message first time mark the sender as their parent and send back an *ack* message; otherwise, the sender is replied with a *nack* message as in the original flooding algorithm. Additionally, the depth of subtree cluster is determined prior to execution in the variable *max_depth* and at every message reception by the nodes, the variable *count* is incremented and checked against *max_depth*. The end of the cluster is marked when this is reached and another cluster is started as shown in Algorithm 17.1 [1]. An unvisited node that receives a message *probe* with a 0 in count field becomes a CH of its subtree. All nodes other than the root are classified as CH, *ordinary* or *leaf* at the end of the algorithm. Clusters formed with this algorithm in a sample WSN are shown in Fig. 17.1.

Algorithm 17.1 *ST_Clust*

1: **int** *parent* ← ⊥
2: **set of int** *childs* ← {∅} , *others* ← {∅}
3: **message types** *probe, ack, reject*
4: **states** *CH, ordinary, leaf*
5: **if** *i* = *root* **then**   ▷ root initiates tree construction
6:     **send** *probe*(0, 0) to *N*(*i*)
7:     *parent* ← *i*
8: **end if**
9:
10: **while** *childs* ∪ *others*) ≠ (*N*(*i*)\{*parent*} **do**
11:     **receive** *msg*(*j*)
12:     **case** *msg*(*j*).*type* **of**
13:         *probe*(*cid, n_hops*) : **if** *parent* = ⊥ **then**   ▷ *probe* received first time
14:                                     *parent* ← *j*
15:                                     **send** *ack* to *j*
16:                                     **if** *count* = 0 **then**   ▷ i am the clusterhead
17:   *state* ← *CH*
18:   *cid* ← *i*
19:                                     **else if** *count* = *ds* **then**
20:   *state* ← *leaf*
21:                                     **else** *state* ← *ordinary*
22:                                     *count* ← (*count* + 1) MOD *max_depth*
23:                                     **send** *probe*(*cid, count*) to *N*(*i*)\ {*j*}
24:                                 **else send** *reject* to *j*   ▷ *probe* received before
25:         *ack*   : *childs* ← *childs* ∪ {*j*}   ▷ include *j* in children
26:         *reject*   : *others* ← *others* ∪ {*j*}   ▷ include *j* in unrelated
27: **end while**

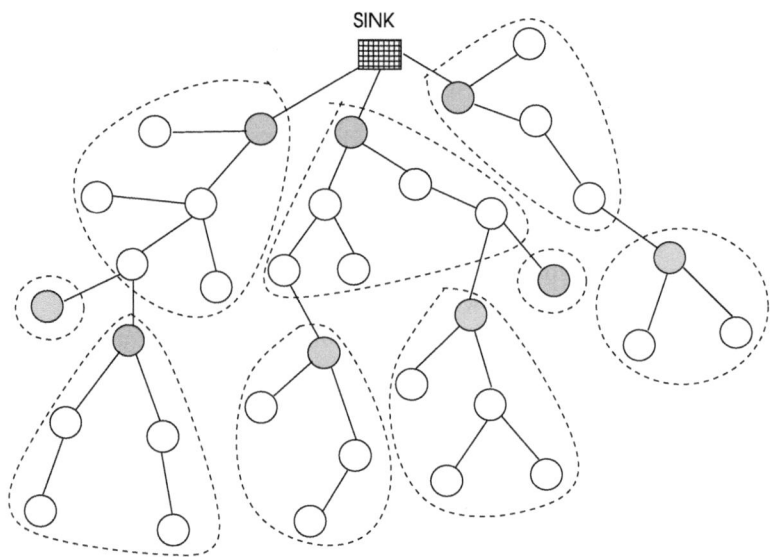

Fig. 17.1 Cluster formation in a sample WSN with a sink node. The maximum hop count is 2, and CHs are shown in gray with clusters in dashed regions

Theorem 17.1 *Time complexity of ST_Clust is O (d) , where d is the diameter of the network and its message complexity is O (m).*

Proof The diameter of the graph is the upper bound on the time required for the algorithm as it is the farthest distance between any two vertices. Since each edge will be traversed twice by either *probe/ack* or *probe/reject* message pairs, the message complexity of the algorithm is $O(m)$.

Banerjee and Khuller [1] also proposed a protocol based on a spanning tree by grouping branches of a spanning tree into clusters of an approximate target size.

17.7 Advanced Graph Structures and Graph Machine Intelligence

There has been two major advancements in computational methods to solve difficult problems including graph related ones in the last decade or so. One is the widespread use of artificial intelligence approaches to solve these problems, commonly implemented by deep learning which is typically realized by multi-layered neural networks. These networks existed and their operation was kwon and experienced many decades ago, but the availability of big data and achieved computational power in recent years laid the pavement for efficient applications of deep learning methods to a wide range of problems. However, problems to be solved still needs huge computation power and one way to tackle this problem is to partition and distribute the task to a number of neural networks known as *distributed deep learning.*

On the other hand, quantum computation which uses radically different principles than classical computing started to emerge. This approach relies on quantum mechanics to achieve very fast processing, although building a quantum computer is still at its infancy. However, quantum algorithms already have been shown to work a multitude of times faster than their equivalent classical algorithms. The convergence of these two relatively recent approaches to computation was inevitable and we are witnessing this development in various patterns; one method with promising results is quantum neural networks. These structures work using principles of neural networks but use quantum circuits consisting of quantum gates to achieve performance far superior than classical neural networks.

Graphs were not excluded from these recent developments as expected since many real-life data such as of social networks, biological networks and computer networks can be conveniently represented by graphs. The classical graph structure $G = (V, E)$ is augmented by more information in two commonly used relatively new graph structures: knowledge graphs and uncertain graphs. The former conveys relationships as edge labels and the latter provide the probability of existing of an edge as its label. Knowledge graphs are found to be convenient for a wide range of machine learning applications such as natural language processing and are conveniently incorporated into neural network structures. A graph neural network inputs a graph structure, works on this structure to solve a graph related problem using the principles of deep learning. A very recent approach is to integrate quantum computing with graph neural networks which is basically inputting graph data to a quantum circuit that basically works as in deep learning.

We are witnessing all these recently new approaches of deep learning, quantum computation, and advanced graph structures being integrated into robust, reliable, and efficient machine learning systems that can be used to solve a wide range of difficult problems. There is still a long path of research to be done in these areas and graph algorithms for complex networks is one area that needs to be addressed by these systems to achieve favorable performance in the processing of this big data.

17.8 Conclusions

We can briefly summarize the steps to follow when we need to decide on an algorithm for the graph problem we have:

1. Investigate carefully whether the problem can be solved by a combination of basic graph algorithms such as direct or modified BFS and DFS. We saw BFS is used effectively to find the edge betweenness values in a graph and DFS for various connectivity problems. If this is not possible, we can search for an approximation algorithm that works in linear time. If we cannot find an

appropriate algorithm, our best choice will be to use some heuristics that give good results most of the time.

2. If the graph is large, it is always worthwhile to attempt to parallelize the algorithm. In this case, we saw partitioning of the adjacency matrix of the graph yields feasible solutions in many cases. We can also partition the graph by first contracting its vertices to obtain a simpler graph and then partition the simple graph. For such graphs, using algebraic graph algorithms provides easiness in parallelization as parallel matrix operations are already available.

3. If we need to design a distributed algorithm for a computer network such as a WSN, we may attempt to convert the sequential algorithm to a distributed one as a general approach. This may not be a trivial task especially if the sequential algorithm relies heavily on global data since the distributed algorithms commonly work using local data around the nodes.

References

1. Banerjee S, Khuller S (2000) A clustering scheme for hierarchical routing in wireless networks. Technical report CS-TR-4103, University of Maryland, College Park
2. Erciyes K (2013) Distributed graph algorithms for computer networks. Computer communications and networks series. Springer, Berlin, pp 247–248. ISBN 978-1-4471-5172-2
3. Erciyes K, Ozsoyeller D, Dagdeviren O (2008) Distributed algorithms to form cluster based spanning trees in wireless sensor networks. ICCS 2008. LNCS. Springer, Berlin, pp 519–528

Pseudocode Conventions

A.1   Introduction

We show the conventions of pseudocode writing used throughout the book here. We follow the mainstream adaptations such as in [1, 2]. Main points to be emphasized are as follows:

- Every algorithm starts with the declaration of its input and the output produced by it.
- The lines of the algorithm are numbered for reference.
- We use indentations to show blocks which are executed within control structures.
- A procedure that is used from the main body of the algorithm is shown explicitly.

The data structures, control structures, and distributed algorithm structures are described in the next sections.

A.2   Data Structures

Each line of an algorithm is a *statement* which commonly evaluates an *expression* or performs a specific function such as calling a procedure. An expression consists of constants, variables, and operators. Declaration of a variable is performed as shown in the following examples:

$$\textbf{booolean} \quad visited \leftarrow false$$

Here, a Boolean variable *visited* is declare and initialized to *false* value.

$$\textbf{setofint} \quad vertices \leftarrow \emptyset$$

K. Erciyes, *Guide to Graph Algorithms*, Texts in Computer Science, https://doi.org/10.1007/978-3-032-05294-0

Table A.1 General
algorithm conventions

Notation	Meaning
$a \leftarrow b$	Assignment
$=$	Comparison of equality
$\neq$	Comparison of inequality
$true, false$	Logical true and false
$null$	Nonexistence
$\triangleright$	Comment

A set of *vertices* which will contain integer values is declared and initialized as empty.

The assignment in the above examples is performed by using the $\leftarrow$ operator. The value on the right of this operator is evaluated and assigned to the variable on the left in the usual sense. Sometimes, we have two or more short expressions which are placed in the same line of the algorithm separated by semicolons as follows. Note that a statement line does not end with a semicolon.

$$i \leftarrow 5; \quad j \leftarrow 8$$

Pseudocode conventions used in the book are shown in Table A.1.

Arithmetic and logical operators used are shown in Table A.2.

We use sets instead of arrays to show a collection of variables. An element x is contained in a set A is performed by the *union* ($\cup$) operator as follows:

$$A \leftarrow A \cup \{x\}.$$

And removing an element y from the set A is performed by using the *setminus* ($\backslash$) operator as follows:

$$A \leftarrow A \backslash \{v\}.$$

Table A.3 shows the set operations used in the text with their meanings.

Table A.2 Arithmetic and
logical operators

Notation	Meaning
$\neg$	Logical negation
$\wedge$	Logical and
$\vee$	Logical or
$\oplus$	Logical exclusive-or
a/b	a divided by b
$a \cdot b$ or ab	Multiplication

Table A.3 Set operations

Notation	Meaning
$S|$	Cardinality of S
$\emptyset$	Empty set
$u \in S$	u is a member of S
$S \cup R$	Union of S and R
$S \cap R$	Intersection of S and R
$S \setminus R$	Set subtraction
$S \subseteq R$	S is a subset of R
$S \subset R$	S is a proper subset of R
$\max/\min\{\ldots\}\, S$	Maximum/minimum value of a set of values

A.3   Control Structures

Control structures are used to alter the flow of execution. Selection and repetition are two main modes of control as described below.

Selection
Selecting one of few alternative flows is commonly performed by the *if-then-else* construction. The Boolean expression after the *if* statement is evaluated and the branch after *then* is taken if this expression yields a *true* value. We can specify an *else* block to specify the alternative flow when the expression yields a *false* value. An example is depicted in Algorithm A.7 where we want to test which of the given two integers a and b is greater than the other or whether they are equal to each other. We see line 7 is executed in this example.

Algorithm A.1 *if-then-else structure*

1: $a \leftarrow 3; b \leftarrow 5$
2: **if** $a > b$ **then**   ▷ test condition
3:     **output** "a is greater"
4: **else if** $a = b$ **then**   ▷ *else if* of *if* statement
5:     **output** "$a = b$"
6: **else**
7:     **output** "b is greater"
8: **end if**   ▷ end of *if* statement

We can select a specific flow of execution from a number of alternative flows again by the evaluation of a number of expressions using the *case-of* structure. A simple calculator that can perform addition, subtraction, multiplication, and division is shown in Algorithm A.3.

Algorithm A.2 *A Simple Calculator*

1: **input** *operator*
2: **input** *a* and *b*
3: **case** *operator* of
4:       "+" : $c \leftarrow a + b$
5:       "-" : $c \leftarrow a - b$
6:       "*" : $c \leftarrow a * b$
7:    **default** $c \leftarrow a/b$
8: **output** c

Repetition

We use the loop constructs *for*, *while*, and *repeat until* to implement a statement for a number of times. The *for-do* loop is commonly used when the number of iterations is known beforehand. The example shown in Algorithm A.3 finds the sum of the elements of a matrix with integer elements.

Algorithm A.3 *Sum of a Matrix*

1: **Input**: int $A[n] = \{...\}$
2: **Output**: *sum* of the elements of A
3: **for** $i = 1$ to n **do**
4:    $sum \leftarrow sum + A[i]$
5: **end for**

When we are dealing with sets and do not know the size of the set, the *for all* loop can be conveniently used. Commonly, we arbitrarily select an element of the set and perform an operation on this element as shown in Algorithm A.4 where we simply output each element of set S which consists of integers.

Algorithm A.4 *Output Elements of a Set*

1: **Input**: set of int $S = \{...\}$
2: **Output**: elements of S
3: **for all** $u \in S$ **do**
4:    **output** u
5: **end for**

There are cases when we want to enter a loop based on a condition. The *while* loop can be used for such implementations, and this type of loop may be entered 0 or more times based on the evaluation of a Boolean expression as shown in Algorithm A.5 where the sum of numbers entered is calculated until 99 is entered. Note that 99 may be entered as the first input causing no execution of the block inside the loop.

Algorithm A.5 *Sum of Integers*

1: **Input**: **set of int** S
2: **Output**: elements of S
3: **input** a
4: **while** $a \neq 99$ **do**
5:    $sum \leftarrow sum + a$
6:    **input** a
7: **end while**

The last loop structure we use in algorithms is the *Repeat ... Until* loop where the decision to execute the loop is made after the loop is run. This type of loop is used when we know the loop is to run at least once as shown in Algorithm A.6, where we implement the above example of adding numbers entered. Note that we do not need the input statement before the loop this time since we know the loop will execute at least once.

Algorithm A.6 *Sum of Integers*

1: **Input**: **set of int** S
2: **Output**: elements of S
3: **input** a
4: **repeat**
5:    **input** a
6:    $sum \leftarrow sum + a$
7: **until** $a \neq 99$

A.4   Distributed Algorithm Structure

A distributed algorithm is executed by a node of a network, and the action to be done is commonly decided by the type and contents of the message received. We have a *case* structure after the message is received and then each necessary action is decided by the type and then the contents of the received message. The code is executed by node i, and we frequently omit writing the identity of the node i in various operations performed for simplicity. For example, *send msg_x* to j means sending of message msg_x from node i to node j. We need to check reception of message condition, for example, when messages are received from all neighbors or only one message received from the parent in a tree, etc. In synchronous distributed algorithms, we need a Boolean variable to show us that the round is over and commonly, we have a *while* loop that tests this condition. When this condition becomes true, we do not wait for any more messages as shown in Algorithm A.6.

Algorithm A.7 *Distributed Algorithm Structure 1*

```
 1: int i, j                          ▷ i is this node; j is the sender of a message to i
 2: while ¬flag do                    ▷ all nodes execute the same code
 3:    receive msg(j)
 4:    case msg(j).type of
 5:           a₁  :  action₁
 6:           ...  :  ...
 7:           aₙ  :  actionₙ
 8:    if messages received from all neighbors then
 9:       flag ← true
10:    end if
11: end while
```

References

1. Cormen TH, Leiserson CE, Rivest RL, Stein C (2001) Introduction to algorithms. MIT Press, Cambridge
2. Erciyes K (2013) Distributed graph algorithms for computer networks. Springer, Berlin

Linear Algebra Review

B

B.1 Introduction

A graph can be represented by its adjacency matrix or incidence matrix. The Laplacian matrix of a graph provides information about the spectral properties of a graph. The algebraic graph theory is based on applying algebraic methods to graph problems and commonly, the matrices associated with a graph are used for this purpose. Linear algebra is a branch of mathematics that deals with matrices. We provide a very brief and partial review of linear algebra sufficient to be a background for the spectral graph properties and algebraic graph algorithms described in the book.

B.2 Basic Matrix Types

A matrix is a set of elements organized into rows and columns.

- A general matrix with m rows and n columns can be written as

$$A_{m,n} = \begin{pmatrix} a_{1,1} & a_{1,2} & \cdots & a_{1,n} \\ a_{2,1} & a_{2,2} & \cdots & a_{2,n} \\ \vdots & \vdots & \ddots & \vdots \\ a_{m,1} & a_{m,2} & \cdots & a_{m,n} \end{pmatrix}.$$

- *Transpose of a Matrix*: This matrix is obtained by writing rows as columns and columns as rows. The transpose of the above matrix is

$$A_{n,m}^T = \begin{pmatrix} a_{1,1} & a_{2,1} & \cdots & a_{n,1} \\ a_{1,2} & a_{2,2} & \cdots & a_{n,2} \\ \vdots & \vdots & \ddots & \vdots \\ a_{1,m} & a_{2,m} & \cdots & a_{n,m} \end{pmatrix}.$$

© The Editor(s) (if applicable) and The Author(s), under exclusive license to Springer Nature Switzerland AG 2026
K. Erciyes, *Guide to Graph Algorithms*, Texts in Computer Science,
https://doi.org/10.1007/978-3-032-05294-0

- *Diagonal Matrix*: All entries except the diagonal values of this matrix are 0. For a 4×4 matrix

$$A = \begin{pmatrix} a_{1,1} & 0 & 0 & 0 \\ 0 & a_{2,2} & 0 & 0 \\ 0 & 0 & a_{3,3} & 0 \\ 0 & 0 & 0 & a_{4,4} \end{pmatrix}.$$

- *Identity Matrix*: The identity matrix I a diagonal matrix with all diagonal values of unity. I_4 is shown below:

$$A = \begin{pmatrix} 1 & 0 & 0 & 0 \\ 0 & 1 & 0 & 0 \\ 0 & 0 & 1 & 0 \\ 0 & 0 & 0 & 1 \end{pmatrix}.$$

- *Symmetric Matrix*: The values symmetric to the diagonal are equal in this matrix, which means this matrix is equal to its transpose.

$$A = \begin{pmatrix} a_{1,1} & a_{1,2} & a_{1,3} & a_{1,4} \\ a_{1,2} & a_{2,2} & a_{2,3} & a_{2,4} \\ a_{1,3} & a_{2,3} & a_{3,3} & a_{3,4} \\ a_{1,4} & a_{2,4} & a_{3,4} & a_{4,4} \end{pmatrix}.$$

- *Upper Triangular Matrix*: Matrix A is upper triangular if $A_{ij} = 0$ when $i > j$ as shown below:

$$A = \begin{pmatrix} 3 & 1 & 5 \\ 0 & 7 & 1 \\ 0 & 0 & 4 \\ 0 & 0 & 0 \end{pmatrix}.$$

- *Lower Triangular Matrix*: Matrix A is lower triangular if $A_{ij} = 0$ when $i < j$.

$$A = \begin{pmatrix} 2 & 0 & 0 & 0 \\ 4 & 1 & 0 & 0 \\ 3 & 6 & 4 & 0 \end{pmatrix}.$$

- *Submatrix*: A *submatrix* of a matrix is formed by deleting a set of rows and/or a set of columns. Deleting row 1 and column 2 of a matrix results in the shown submatrix.

$$A = \begin{pmatrix} 2 & 1 & 0 & 4 \\ 5 & 3 & 2 & 1 \\ 0 & 2 & 4 & 3 \end{pmatrix} \rightarrow \begin{pmatrix} 5 & 2 & 1 \\ 0 & 4 & 3 \end{pmatrix}.$$

- Vector: A vector is a $n \times 1$ matrix.

B.3   Matrix Operations

- *Addition*: Two matrices can be added if they have the same dimension. The corresponding items of the matrices can then be added to form the sum matrix. The addition of two 3×3 matrices A and B to get matrix C is as below:

$$C = A + B$$

$$
\begin{pmatrix} c_{1,1} & c_{1,2} & c_{1,3} \\ c_{2,1} & c_{2,2} & c_{2,3} \\ c_{3,1} & c_{3,2} & c_{3,3} \end{pmatrix} = \begin{pmatrix} a_{1,1} & a_{1,2} & a_{1,3} \\ a_{2,1} & a_{2,2} & a_{2,3} \\ a_{3,1} & a_{3,2} & a_{3,3} \end{pmatrix} + \begin{pmatrix} b_{1,1} & b_{1,2} & b_{1,3} \\ b_{2,1} & b_{2,2} & b_{2,3} \\ b_{3,1} & b_{3,2} & b_{3,3} \end{pmatrix}
$$

$$
= \begin{pmatrix} a_{1,1} + b_{1,1} & a_{1,2} + b_{1,2} & a_{1,3} + b_{1,3} \\ a_{2,1} + b_{2,1} & a_{2,2} + b_{2,2} & a_{2,3} + b_{2,3} \\ a_{3,1} + b_{3,1} & a_{3,2} + b_{3,2} & a_{3,3} + b_{3,3} \end{pmatrix}.
$$

- *Multiplication with a Scalar*: Two matrices can be added if they have the same dimension. The corresponding items of the matrices can then be added to form the sum matrix. The addition of two 3×3 matrices A and B to get matrix C is as below:

$$C = k \cdot A$$

$$
C = k \cdot \begin{pmatrix} a_{1,1} & a_{1,2} & a_{1,3} \\ a_{2,1} & a_{2,2} & a_{2,3} \\ a_{3,1} & a_{3,2} & a_{3,3} \end{pmatrix} = \begin{pmatrix} k \cdot a_{1,1} & k \cdot a_{1,2} & k \cdot a_{1,3} \\ k \cdot a_{2,1} & k \cdot a_{2,2} & k \cdot a_{2,3} \\ k \cdot a_{3,1} & k \cdot a_{3,2} & k \cdot a_{3,3} \end{pmatrix}.
$$

- *Matrix Multiplication*: Multiplication of a matrix by the identity matrix does not change it $AI = A$

$$C = A \times B$$

$$
= \begin{pmatrix} a_{1,1} & a_{1,2} \\ a_{2,1} & a_{2,2} \end{pmatrix} \times \begin{pmatrix} b_{1,1} & b_{1,2} \\ b_{2,1} & b_{2,2} \end{pmatrix}
$$

$$
= \begin{pmatrix} a_{1,1}b_{1,1} + a_{1,2}b_{2,1} + a_{1,1}b_{1,2} + a_{1,2}b_{2,2} \\ a_{2,1}b_{1,1} + a_{2,2}b_{2,1} + a_{2,1}b_{1,2} + a_{2,2}b_{2,2} \end{pmatrix}.
$$

For example,

$$
\begin{pmatrix} 5 & 4 \\ -3 & -2 \end{pmatrix} = \begin{pmatrix} 1 & 2 \\ -1 & 0 \end{pmatrix} \times \begin{pmatrix} 3 & 2 \\ 1 & 1 \end{pmatrix}.
$$

Multiplying an $m \times n$ matrix by the $n \times n$ identity matrix I_n does not change it.

$$AI_n = A = I_m A$$

For an $n \times n$ matrix A, if there exists a $n \times n$ matrix B such that

$$AB = A = BA = I$$

Then, the matrix B is called the *inverse* of A written as A^{-1}, and A is called a *non-singular* matrix. When such an inverse matrix can not be determined, the matrix A is called *singular*. A square matrix is singular if and only if its determinant is 0.

Determinant of a Matrix

The determinant of a square matrix A, shown by $\det(A)$ or $|A|$, is used for various operations including in finding the inverse of the matrix A. Determinant of a 2×2 matrix A is calculated as follows:

$$|A| = \begin{vmatrix} a & b \\ c & d \end{vmatrix} = ad - bc.$$

Determinant of a 3×3 matrix A can be found selecting a row or a column and multiplying each element of the selected row/column by the determinant of the subgraph obtained by deleting the row/column and column of that element from the matrix as below:

$$|A| = \begin{vmatrix} a & b & c \\ d & e & f \\ g & h & i \end{vmatrix} = a \begin{vmatrix} e & f \\ h & i \end{vmatrix} - b \begin{vmatrix} d & f \\ g & i \end{vmatrix} + c \begin{vmatrix} d & e \\ g & h \end{vmatrix}$$

$$= aei - bfg + cdh - ceg - bdi - afh.$$

The *minor* M_{ij} of an element a_{ij} of a square matrix A is the determinant of the submatrix obtained by deleting ith row and jth column from A. The *cofactor* of a_{ij}, C_{ij}, is obtained by multiplying its minor by $(-1)^{i+j}$. In the above example, $\det(A)$ is calculated as the sum of the cofactors multiplied by row elements of the first row of A. The inverse of a matrix can be computed using various methods. When the matrix A is not large, its inverse can be determined as follows:

$$A^{-1} = \frac{1}{\det(A)} C^T$$

where C is the cofactor matrix of A. For a 2×2 matrix A, its inverse is calculated as below:

$$A^{-1} = \begin{pmatrix} a & b \\ c & d \end{pmatrix} = \frac{1}{\det(A)} \begin{pmatrix} d & -b \\ -c & a \end{pmatrix} = \frac{1}{ad - bc} \begin{pmatrix} d & -b \\ -c & a \end{pmatrix}.$$

B.4   Properties of Matrix Operations

The following properties of matrix operations are valid assuming the matrices are of appropriate sizes:

- $A + B = B + A$.
- $A(B + C) = AB + AC$.
- $(A^T)^T = A$.
- $(A + B)^T = A^T + B^T$.
- $(AB)^T = B^T A^T$.
- $(AB)^{-1} = B^{-1} A^{-1}$.
- $k(A + B) = kA + kB$ for a scalar k.

B.5   Linear Equations

A system of linear equations of the form shown below can be solved using matrix operations.

$$a_{11}x + a_{12}x_2 + \cdots + a_{1n} = b_1$$

$$\cdots$$

$$a_{m1}x + a_{m2}x_2 + \cdots + a_{mn} = b_m.$$

Let A be an $m \times n$ matrix of coefficients, and x is an $n \times 1$ vector representing m variables. We can write these equations in matrix form as follows:

$$\begin{pmatrix} a_{1,1} & a_{1,2} & \cdots & a_{1,n} \\ a_{2,1} & a_{2,2} & \cdots & a_{2,n} \\ \vdots & \vdots & \ddots & \vdots \\ a_{m,1} & a_{m,2} & \cdots & a_{m,n} \end{pmatrix} \begin{pmatrix} x_1 \\ x_2 \\ \cdots \\ x_m \end{pmatrix} = \begin{pmatrix} b_1 \\ b_2 \\ \cdots \\ b_m \end{pmatrix}.$$

The solution vector x is then solution to the equation,

$$x = A^{-1}b.$$

Let us consider the following linear equation with two variables x_1 and x_2:

$$2x_1 + 3x_2 = 4$$

$$x_1 + 2x_2 = 3.$$

We can write this equation using matrix notation as $Ax = b$ as follows:

$$\begin{pmatrix} 2 & 3 \\ 1 & 2 \end{pmatrix} \begin{pmatrix} x_1 \\ x_2 \end{pmatrix} = \begin{pmatrix} 4 \\ 3 \end{pmatrix}.$$

We can now compute A^{-1} and then $x = A^{-1}b$ as below:

$$\begin{pmatrix} x_1 \\ x_2 \end{pmatrix} = \begin{pmatrix} 2 & -3 \\ -1 & 2 \end{pmatrix} \begin{pmatrix} 4 \\ 3 \end{pmatrix} = \begin{pmatrix} -1 \\ 2 \end{pmatrix}$$

to yield values $x_1 = -1$ and $x_2 = 2$.

Gaussian Elimination

Another method to solve a system of linear equations is to first form the matrix equation $Ax = b$ as before. We then form the augmented matrix equation as below:

$$A_{m,n}^G = \begin{pmatrix} a_{1,1} & a_{1,2} & \cdots & a_{1,n} & |b_1 \\ a_{2,1} & a_{2,2} & \cdots & a_{2,n} & |b_2 \\ \vdots & \vdots & \ddots & \vdots & |\cdots \\ a_{m,1} & a_{m,2} & \cdots & a_{m,n} & |b_m \end{pmatrix}.$$

Next, the augmented matrix A^G is transformed into an upper triangular matrix A^U using elementary row operations. We then solve for x_m and then use the value of x_m to obtain the value for x_{m-1}, etc., using backward substitution. Let us consider the following system of equations with three variables x_1, x_2, and x_3:

$$x_1 + x_2 - x_3 = -2$$

$$3x_1 - 2x_2 + x_3 = 7$$

$$2x_1 - x_2 - 3x_3 = 9.$$

The augmented matrix is then as follows:

$$A_{3,3}^G = \begin{pmatrix} 1 & 1 & -1 & | & -2 \\ 3 & -2 & 1 & | & 7 \\ 2 & -1 & 3 & | & 9 \end{pmatrix}.$$

Multiplying the first row by -2 and adding it to the third row; then multiplying the first row by -3 and adding it to the second row yields the first matrix below.

The final upper triangular matrix obtained by multiplying the second row by $-3/5$ and adding it to the third row is as follows:

$$
\begin{pmatrix}
1 & 1 & -1 & | & -2 \\
3 & -2 & 1 & | & 7 \\
0 & -3 & 5 & | & 13
\end{pmatrix}
\rightarrow
\begin{pmatrix}
1 & 1 & -1 & | & -2 \\
0 & -5 & 4 & | & 13 \\
0 & -3 & 5 & | & 13
\end{pmatrix}
$$

$$
\rightarrow
\begin{pmatrix}
1 & 1 & -1 & | & -2 \\
3 & -2 & 1 & | & 7 \\
0 & 0 & 13/5 & | & 26/5
\end{pmatrix}.
$$

We can now determine $x_3 = 2$ by evaluating the last row. Backward substitution of x_2 in the second row provides $x_2 = -1$, and then substituting the values of x_1 and x_2 in the first row yields $x_3 = 2$.

Index

A

Algebraic graph algorithm, 378
  BFS, 380
  connectivity, 378
  matching, 386
    Rabin–Vazirani algorithm, 387
  matrices, 376
  minimum spanning tree, 385
  shortest path, 384
Algorithm
  algebraic, 376
  approximation, 62
  asymptotic analysis, 40
  complexity class, 53
    NP, 54
    NP-complete, 55
    NP-hard, 55
    P, 54
  divide and conquer, 69
  dynamic programming, 70
  graph
    minimum spanning tree, 67
  HITS, 443
  minimal dominating set, 323
  minimum spanning tree, 67
  NP-completeness, 53
  proof, 46
    contradiction, 47
    contrapositive, 47
    direct, 46
    induction, 48
    loop invariant, 49
    strong induction, 48
  randomized, 58
    Karger's algorithm, 58
  recursive, 44
  reductions, 50
  structures, 40

Alternating path, 268
Approximation algorithm, 62
Articulation point, 224, 229
Auction algorithm, 297
Auction-based algorithm
  parallel, 300
Augmenting path, 268

B

Backtracking, 63
Batagelj and Zaversnik algorithm, 417
Bellman–Ford algorithm, 71, 207, 210, 384
Berge's theorem, 270
Biconnected graph, 224
Biological network, 425
Bipartite vertex cover, 329
Bipartite weighted matching, 289
Block, 226
  decomposition, 233
    Hopcroft–Tarjan algorithm, 233
Boruvka's algorithm, 192, 198
Branch and bound, 64
Breadth-First Search (BFS), 380
Bridge, 237
  Tarjan's bridge algorithm, 237
Bron and Kerbosch algorithm, 415

C

Centrality, 408
  closeness centrality, 409
  degree centrality, 408, 409
  edge betweenness centrality, 411
  eigenvalue centrality, 412
  vertex betweenness centrality, 410
Clique, 310, 413
  Bron and Kerbosch algorithm, 415

GPSR Compliance

*The European Union's (EU) General Product Safety Regulation (GPSR)
is a set of rules that requires consumer products to be safe and our
obligations to ensure this.*

*If you have any concerns about our products, you can contact us on
ProductSafety@springernature.com*

In case Publisher is established outside the EU, the EU authorized
representative is:

Springer Nature Customer Service Center GmbH
Europaplatz 3
69115 Heidelberg, Germany

Batch number: 10202867

Printed by Printforce, the Netherlands